D1014648

DEVELOPING NEW FOOD PRODUCTS FOR A CHANGING MARKETPLACE

SECOND EDITION

DEVELOPING NEW FOOD PRODUCTS FOR A CHANGING MARKETPLACE

SECOND EDITION

EDITED BY:

AARON L. BRODY & JOHN B. LORD

CRC Press is an imprint of the
Taylor & Francis Group, an **informa** business

CRC Press
Taylor & Francis Group
6000 Broken Sound Parkway NW, Suite 300
Boca Raton, FL 33487-2742

Printed in the United States of America on acid-free paper
10 9 8 7 6 5 4 3 2 1

International Standard Book Number-13: 978-0-8493-2833-6 (Hardcover)

Library of Congress Cataloging-in-Publication Data

Developing new food products for a changing marketplace / edited by Aaron L. Brody and John B. Lord. -- 2nd ed.
p. cm.
Includes bibliographical references and index.
ISBN-13: 978-0-8493-2833-6 (alk. paper)
ISBN-10: 0-8493-2833-0 (alk. paper)
1. Food industry and trade. 2. New products. I. Brody, Aaron L. II. Lord, John B. III. Title.

HD9000.5.D47 2007
664.0068'5--dc22 2007014912

Visit the Taylor & Francis Web site at
http://www.taylorandfrancis.com

and the CRC Press Web site at
http://www.crcpress.com

Table of Contents

Forewords

Georgia is one of the fastest growing states in the rapidly growing Southeastern United States with a 2006 population of 9.4 million (ninth in the nation), an increase of 44% since 1990. The food products industry is the largest segment in the state's manufacturing sector with an output in 2006 of over $30 billion, nearly 19% of all manufacturing output. A number of multinational food companies are based or have regional headquarters in Georgia. At the other extreme, there are many more small-to-medium companies facing the challenges of developing new products or solving problems from production to marketing, often with inadequate or nonexistent research and development personnel and facilities. One responsibility of public research universities such as the University of Georgia is to sustain important industries with training of qualified personnel, conduct research to address problems and identify new opportunities, and outreach to communicate new knowledge to those industries.

The University of Georgia, part of the University System of Georgia, is home to a well-regarded Department of Food Science and Technology with 23 state-supported and 7 adjunct faculty members on 2 campuses. A group of Food Science and Technology faculty members, along with colleagues from other departments from other universities, and from the private sector have initiated a program, the purpose of which is to provide a broad range of services to the food industry in Georgia and beyond. A number of such university-based programs exist in other regions, but there is no other all-inclusive program in the southeast.

The University of Georgia Food Product Innovation and Commercialization Program (FoodPIC) is perhaps a unique partnership among academics, private consultants, and the local community's economic (Griffin City, Spalding County) development authority. The authority has pledged money toward construction of a multimillion food technology center on the University of Georgia's Griffin Campus in the greater Atlanta metropolitan area, and is leading the drive to raise the remainder from private and public sources. University of Georgia's FoodPIC is fortunate to have as founding members several of the contributors to this book and has especially benefited from the encouragement and vision of coeditor Aaron Brody. We are confident that the principles articulated in the following chapters will not only serve the food industry throughout the nation, but will also form the guiding framework by which University of Georgia FoodPIC will fulfill its mission and our vision for it.

R. Dixon Phillips
Food Product Innovation and Commercialization
Jenru Chen
Manjeet Chinnan
Yen Con Hung
Anna Resurreccion
Department of Food Science and Technology
University of Georgia
Griffin, Georgia

For well over 40 years, Saint Joseph's University has offered a program in food marketing, and our undergraduate major is one of the largest in the university and the largest of its kind in the United States. Since launching the BS in food marketing in the early 1960s, we have added an executive master's program and created the Center for Food Marketing, funded by a grant from USDA, which serves the industry via research, outreach, and continuing education. In addition to educating hundreds of undergraduates and hundreds of food industry professionals, Saint Joseph's University sponsors programs aimed at sustaining the agriculture community as well as helping to feed the poor, consistent with the Jesuit tradition of educating the whole person.

Aaron Brody and John Lord have assembled a "who's who" of academic and industry experts to create a book that combines both the art and the science of new food product development. The cross-disciplinary focus of the book reflects the nature of new product development, which requires cross-disciplinary activity in the firm. This book should serve the needs of students studying the principles that underlie and guide successful creation and marketing of new food products as well as the professional who needs to understand the latest tools and techniques of the trade.

In addition, a book that combines basic theory, scientific principles, and the business aspects of food product development reflects the commitment to a liberal education that is essential to the mission of Saint Joseph's University. As a university president and educator, I believe that the second edition of *Developing New Food Products for a Changing Marketplace* will play a valuable role in that the authors bring the science of new food product development to the food marketing classroom and the business of new food product development to the food science classroom, and will be a valuable reference tool for new product development professionals throughout the food industry.

Reverend Timothy R. Lannon, S.J.
President, Saint Joseph's University
Philadelphia, Pennsylvania

Successful food product development requires a multidisciplinary approach, and this book features the input of experts in each area involved. The plentiful chapters take the reader from beginning to end of food product development, commencing with understanding the reasons for developing new products through formulation, sensory and consumer testing, package design, and commercial production, to product launch and marketing.

New additions to this second edition are chapters on identifying and measuring consumer desires, generating product concepts, engineering scale-up, package design, and the increasingly important role of the research chef.

It is gratifying to see that many of the authors are members of the Institute of Food Technologists (IFT), a professional scientific organization that embraces all aspects of food science and technology. Through involvement in IFT activities, our members gain an enhanced understanding of the changing marketplace and cutting-edge technology. IFT's centers of core expertise are focused into 27 divisions almost all of which are pertinent to the product development dimensions covered in *Developing New Food Products for a Changing Marketplace*. All the IFT divisions give members significant learning opportunities that are applicable to new product development.

It is worth noting that seven of the authors have been elected an IFT Fellow, a unique professional distinction conferred for outstanding and extraordinary contributions in the field of food science and technology. Thus, the breadth of knowledge provided by these fellows and the other authors makes this book an effective tool-and a valuable addition-to the library of every food science and technology professional involved in food product development.

Margaret A. Lawson
D.D. Williamson & Co., Inc.
Past President, IFT,
Louisville, Kentucky

Preface to First Edition

DEVELOPING NEW FOOD PRODUCTS FOR A CHANGING MARKETPLACE

Hard to believe, but once upon a time, not so very long ago, new food products were the exclusive province of the white coat clad bench technical people in the recesses of their laboratories. They brewed their liquids, mixed their ingredients, blended their flavors, heated and cooled, conducted their private taste panels, and proclaimed, "This is our company's new food product! Go forth and sell it!"

Difficult to conceive, but ignoring both consumers and their marketing representatives within the corporate hierarchy was common practice and, in some places, technically driven new food products continue to be the preferred sourcing.

And then, there is the opposite pole: companies that employ hired MBA guns that are really New York City-based idea generators to "develop" new food products. To those pesto lovers, and panini lovers (this season), anything west of the Hudson River does not exist. And to them, the transformation of their pet focus group-tested concept to an edible reality is a simple overnight exercise by a lab "technician." And that type of thinking still prevails, whether the "product development" fountain is east or left coast.

We cannot possibly forget the emergence of packaging in this frequent scenario: after the product has been fully developed and the manufacturing folks have started engineering the line, someone remembers that this precious winning product must be thrown into a package.

"... and it has to be done by a week from Tuesday to meet the sales promotion schedule which coincides with the new TV ad campaign on the first pitch of the opening game of the Worlds' Series ... ."

"Oh, and don't forget to call in the ad agency this afternoon to design the package ... ."

"Who did the shelf life testing?"

"Where are the market launch print insert materials?"

"Has anyone checked with the legal department to find out if this new plastic is FDA approved and if the package copy claims are O.K.?"

"... I don't care, the price has to be $x. You are going to have to find a way to trim costs ... ."

"Never happened ...can't possibly happen now ...not in our company."

"Don't you believe it!"

When John Lord and Aaron Brody sat down at Saint Joseph's University Department of Food Marketing back in 1994, Professor Lord was teaching a senior course in product policy and Aaron Brody was consulting in food product and packaging technology.

Product policy was a marketing course embracing what were then perceived to be all those indispensable food product development elements such as identifying consumer needs, concept generation, concept evaluation, screening, targeting, product protocol, launch strategies, marketing research, advertising and promotion, post launch, and so forth. No technical development, no packaging; nothing that smacked of nutrition or chemistry or shelf life.

And universities such as University of Georgia were teaching food science and technology students about product formulation, pilot processing, flavor panel testing, instrumental analyses, and quality assurances. Hardly a nod toward consumers, marketing research and analysis, or product positioning, or anything that resembled marketing.

At Michigan State University School of Packaging, some students were learning package structural design and testing, materials, permeation, shelf life, migration, and protection, and all those other good subjects that together summed to the physical package.

And, of course, Harvard Graduate School of Business Administration and its disciples were expounding on the theories to accelerate the new product development process. Schools of art were mentioning package design as an afterthought as if it were a wayward son, . . . but we imagine that you can imagine the situation back then . . . .

That day in Philadelphia, John and Aaron conceived a revolutionary paradigm to teach product policy to food marketing majors as the food product development integral of food marketing, food technology, and food packaging. Everyone in class might not become an expert in every facet, but each would know that many components together constitute food product development-and would be at least cognizant of the various inputs required. Since the food marketing department encouraged course innovation and individual faculty "owned" their courses, the concept and syllabus (which we did not have in the beginning) did not have to go through the academic gauntlet.

As one of our next steps, we searched for a text book for students to study on days between those early morning classes. We read Merle Crawford's classical work, but it was not food and hardly touched on what happened on the technical side. Bob Cooper's book certainly hit on reasons for success and failure, but where was the organization or the plan? And the Gruenwald book with its accompanying video tape was fascinating reading, if you were an advertising person. From the realm of food science and technology came several "product development" books, one on future technology, another with lots of information on formulation, but you get the idea.

Even the old book on *Principles of Package Development* with Aaron listed as one author turned out to be archaic and sophomoric.

There we were: a course concept that definitely fitted the needs of our food marketing soon-to-be-employee-and-manager students. We had also identified that the concept to which we had given birth was equally important to the food industry product development enterprises that were only just then recognizing that teams from all disciplines were the most effective and efficient implementation route to new product success. But, not a single coherent hard copy document capturing the universe of coherent new food product development as we envisioned it.

Our solution was simple: for each class session, we prepared notes from our experience (and we certainly had lots of them, good and bad) and copied selected references from the literature (probably violating all sorts of copyright laws and rendering ourselves criminals) and recited from the gospel according to the writers-and invited a host of experts from industry to demonstrate and lecture on their specialties. Professionals conceived and formulated products before the students' eyes, showed slides and videos of their new food product development projects, and then asked the students to taste their outrageous foods, ranted over distribution, went into rapture over the launch tactics and sampling, pontificated on nutritional values, and on and on it went.

A few students were bored; some were overwhelmed; some were underwhelmed; but a major portion (most of whom are now practicing what they learned with food companies and consultancies) grasped the concept that new food product development was a systematic integration of many diverse disciplines. That perpetuated fact of a too high percentage of new food product failure may, in large measure, be attributed to a paucity of participatory elements, and not just a failure to identify consumer needs or to meet consumer expectations and so on, as the conventional wisdom asserts. And this product policy course highlighted *all* the elements, and enhanced and enriched them with anecdotes and sea stories from the salts who had lived them. And capstoned with a term project in which student teams actually developed a new food product with its marketing plan, including a budget, packaging, formulae, consumer test results, and all the rest of the appurtenances that would make for a successful new food product if anyone had financing.

Not to be outdone (and with some encouragement from the dynamic duo team of teachers), news media heard of the experiment and the course was given 5 minutes or so on channel 6 at 11 p.m. and articles in *Food Processing* and in *Packaging Technology and Engineering* with pictures. The course in new food product development encompassing so many disparate but necessarily related elements was a success and continues to be taught, not only to seniors, but also to weekend executive education and a short course with University of Georgia and now regularly in conjunction with Pennsylvania State University.

But until now, without a text. We have invested several years of our lives in formalizing our notes and persuading our nonacademic and nonwriting colleagues to contribute to this volume and so this text for students and industry folks exists. With much effort, we have some of our original course lecturers such as Stan Segall from Drexel University on laboratory procedures; and Roy Parcels, a Packaging Hall-of-Famer on package design, in print and pictures. And we have some of the real pioneers such as Bob Smith, the force behind Nabisco SnackWells new food product organization; Al Pyne from CPC International on laboratory testing; and Howard Moskowitz on sensory testing; and some world class pros such as University of Georgia's Romeo Toledo on food science and Eric Greenberg on legal issues and Tetra Pak's Gordon Robertson on shelf life testing; and folks like Jean Storlie who condensed her classic work on labeling to fit this work; Liz Robinson of New Product Sightings on the need; Purdue's John Connor to establish the framework of our food industry; and University of Missouri's Steve Raper and Arthur D. Little's Marv Rudolph on quality function deployment; and the thoughts and ideas of so many others who taught us and shared with us. No more knowledgeable and seasoned group anywhere or anytime could have been assembled to define new food product development.

Sorry, but this is not the definitive book on the topic of new food product development. Rather, it is the best book of the end of this millennium and the beginning of the next-because it is by far the most complete and most comprehensive.

But, it will be supplanted-by the next edition, if our publishers permit, and if we are still ambitious and possess the strength for another undertaking of this nature. It will be supplanted because our colleagues and students and former students will tell us how to do it, phrase it, express it, organize it, and just plain make it better. Might we paraphrase Winston Churchill, this book is the worst of all on food product development, except for all the others. It is the most nearly perfect recital of how to best to develop new food products-until the next one.

We are proud of this accomplishment! Our contributors are proud of their participation! We all hope that we have interpreted our market accurately and have targeted this market with a product that meets every need and fulfills every promise-and that our publisher launches and markets the book effectively to our audience. We envision that this tome will become the standard food product development reference for all food industry marketing and new product and brand and product development executives and practitioners-and the standard textbook in university business and food science and technology departments around the world.

Aaron L. Brody
Duluth, Georgia

John Lord
Philadelphia, PA

Preface to Second Edition

DEVELOPING NEW FOOD PRODUCTS FOR A CHANGING MARKETPLACE

Prophecy might have been a core value of our first edition since it stressed change, and, as is so transparent, evolution and revolution in food products, marketing, packaging, and development protocols that have been our mainstay. In reviewing our first edition, which was nearly the last book of the twentieth century, we could not find mention of nanotechnologies, sustainability, active packaging, the Wal-Mart phenomenon, radio frequency identification (RFID), food security, dashboard dining or the explosive growth of ethnic foods through immigration and its fallout. We forecasted the concepts of innovation and change because this dynamic underlies our every notion in new food product development.

When first we ventured into codifying the dreams, drama, imagination, and rigor that constitute new food product development, we intended to set forth principles from all disciplines and skill sets that play upon this dimensionally volatile landscape. Food product development is a visceral drill in daring to touch the unknown plus empowerment plus kinetics plus basic science plus bold application of knowledge blended into a mix that forges new alliances with that most important member of the value chain, the consumer who eats food. We debuted a concept that food product development embraced a multiplicity of diverse inputs, each contributing a chunk or two or three, to the final whole. Applying the cliché, everyone has something of value to contribute, and without that component, the entire structure is vulnerable to those too-frequent maladies of new food product introduction: disappointing sales, failure, and withdrawal.

Even with each team member participating fully, the probability of reaching and delighting the target consumers is statistically not great. As in the baseball metaphor, not every at bat is a home run or even a single or a foul, but if you do not swing the lumber, you are never going to get to first base or round third and head for home. That we achieved our goal in part is reflected in the repeated use of our first edition in undergraduate and graduate food science and technology, marketing, and other classrooms, and in short courses and food industry laboratories, conference spaces and offices, and by consultants who adopted and marketed our uncanny hard-edged interdisciplinary philosophies.

Just as telling in our success was the formation around North America of food product development centers and facilities in independent organizations, at universit-ies, and other sites. To cite all of these innovation tanks would probably slight some we might overlook, and so we shall focus on our own University of Georgia (UGA) Food

Product Innovation and Commercialization (FoodPIC™) group in Griffin's Department of Food Science and Technology. Aspiring to become a full-fledged center in the vast university, FoodPIC is populated by a host of the greats in world food product development, some of whom are authors of this second edition: editors and authors John Lord from St. Joseph's University and Aaron Brody of University of Gerogia; Anna Resurreccion and Manjeet Chinnan of University of Georgia; Traci Morgan of Morgan Consultants; Howard Moskowitz of Moskowitz-Jacobs; and Mark Thomas of MDT Associates. Each is a member of the university-industry team that constitutes FoodPIC at University of Georgia. An all-star lineup of food technology, food packaging, food marketing, food psychology, food engineering, food Culinology™, food sensory, food management, food shelf life, food regulation and food economics, lots of other professionals have been assembled to hard wire the vision of the first edition: holistic food product development. Founded in the middle of Georgia's expansive and expanding food arena by business and technical professionals, community and government leaders, FoodPIC has captured the physical and intellectual resources demanded to conquer the precarious path of food product development.

This second edition of *Developing New Food Products for a Changing Marketplace* is a reflection of the traction derived from both University of Georgia's Food PIC and St. Joseph's University Department of Food Marketing striving together to better enable professionals and would-be entrepreneurs to reach out and entrance consumers with superb new colors, forms, flavors, and mouth feels while simultaneously providing nutritional attributes. And all safely and affordably for the buyer, and profitably for the offerer. And, of course, to share our experience with students, practitioners, and aspiring food product developers.

Our second edition attempts to overcome many of the shortcomings of the first edition and enhances the principles set forth earlier-newer ideas on staffing for food product development; connecting into the consumer's mind and heart; the role of the culinologist in providing the basic recipe and "gold standard," scaling up from the kitchen, laboratory or even pilot plant; optimization; sensory analysis; package design for far beyond protection; applying insights from real life experience; and further probing into the retail environment. This is not 360° in a flat universe but rather 360° spherical-all dimensions are triangulated into a seamless whole.

We recognize that publication of this tome does not relieve us of the responsibility of incorporating even newer information and ideas into future editions should they ever be published. Food product development is an evolving discipline that will not be fully captured by any group of authors, however gifted and inured to the rigors of the practice. It remains a work-in-progress in which others, younger and wiser, will follow our lead in the future to further optimize the process, forever reaching for an elusive perfection for a constantly changing consumer.

John wishes to acknowledge his long-suffering wife of 33+ wonderful (for him) years, Joan, for putting up with the hours spent on the computer and away from the family, and, of course, his children, Megan, Sean, and Ryan, all sources of pride for mom and dad. In addition, John wants to acknowledge Pat Weaver and Sonia Jeremiah-Bennet of the Campbell Library at Saint Joseph's University, who have provided much valuable support and information over the years in his teaching and writing. Finally, he also wants to note that he hopes that before he passes from God's green earth, the Cubs finally win a championship.

Because he is older and more scarred by failure and occasional successes, Aaron has been laboring in this field longer than almost anyone. His wife, Carolyn, has endured more than five decades of turmoil, impact and the not infrequent brilliant bursts of world-changing action, just as we envisioned when we started our journey. She has loved and nurtured and been so patient, and continues to be loved plus qu'hier et moins que demain. Our eldest son Stephen and his supercharged wife, Susan and their precocious children and our grandchildren, Michelle Jennifer who has grown up much too fast, and her brother, Derek Jason, have graced the world with an aura of going beyond while observing grandpa's works from afar. Middle son Glen, the dynamo who is still a kid at heart, and his sobering wife Sharon are parents of Camryn Alexander and Skyler Alexis, two of the more talented and sparkling persons competing in this world. Sharon deserves special accolades because, while serving as Aaron's assistant, she has had the responsibility of assembling and tracking and integrating every word, graph, table, and picture in this book. Every reader owes her a debt of gratitude. Youngest son, Robyn, meticulous, demanding and loving and his marvelous wife Kellie—the six sigma black belt consummate professional—and both dote over Natalia Siena and Pierce Aaron—who are successfully trying to outperform their superb parents. To all of my family, I forever love you all for being you.

Aaron L. Brody
Duluth, Georgia

John Lord
Philadelphia, Pennsylvania

Editors

Aaron L. Brody is president and CEO, Packaging/Brody, Inc. and a consultant in food and food technology, packaging and marketing. Dr. Brody has been employed in his profession for more than 40 years. He teaches undergraduate and graduate food packaging and food product development and marketing courses at the University of Georgia and MBA strategic marketing and product development courses at St. Joseph's University.

His previous affiliations include General Foods; Raytheon Manufacturing Co., where he helped develop the first microwave oven; Whirlpool Corporation, where he led the development of modified atmosphere; Mars, Inc.; Mead Packaging, where he managed the Crosscheck aseptic packaging system; and marketing development manager with Container Corporation of America. Dr. Brody is author of hundreds of articles and eight textbooks in food and food technology, marketing and packaging including *Encyclopedia of Packaging Technology, Developing Food Products for a Changing Marketplace, Active Packaging for Food Applications*. He is a contributing editor for *Food Technology* and technical and marketing information editor for *BrandPackaging*.

Among his awards and honors are election to the Packaging Hall of Fame; the Nicholas Appert award, Institute of Food Technologists' (IFT) highest award; IFT's Industrial Scientist award and Riester-Davis award for lifetime achievement in food packaging; Institute of Packaging Professionals' (IoPP) member of the year and honorary life member. He is Fellow of IFT and IoPP.

He is a graduate of Massachusetts Institute of Technology with bachelor's and doctorate degrees. He also holds an MBA from Northeastern University, which honored him with its Outstanding Alumni Award.

In 2005, Michigan State University honored him by establishing the Aaron and Carolyn Brody Fund for Food Packaging Education and Research in its School of Packaging and in 2007, the Aaron Brody distinguished Lectureship in Food Packaging.

John B. Lord is a professor and chairperson of food marketing at Saint Joseph's University, where he has taught since 1975. Dr. Lord's primary teaching interests are in new food product development, strategic marketing management, and the economics and history of baseball. Dr. Lord has co-authored a book entitled *Developing New Food Products for a Changing Marketplace* with Aaron Brody, and served as contributing author for *Strategic Management: A Cross-Functional Approach*. Dr. Lord's professional interests focus on the changing eating patterns of the American consumer and how food companies are responding to the consumer challenge. His research currently focuses on the public policy and strategic implications of food advertising to children. He served as distinguished lecturer for the Institute of Food Technologists from 1999 to 2002, has written periodically for DeliBusiness magazine, is on the editorial board of *BrandPackaging* magazine, and is a frequent speaker at food industry meetings. His research has been published in the *Journal of Young Consumers, NFPA Journal, Journal of Advertising Research and Journal of Nutrition Education*.

Contributors

Christopher K. Bailey
The Bailey Group, Inc
Plymouth Meeting, Pennsylvania

Jacqueline Beckley
The U & I Group
Denville, New Jersey

Aaron L. Brody
Packaging/Brody Inc.
Department of Food Science and Technology
University of Georgia
Athens, Georgia

Manjeet Chinnan
Department of Food Science and Technology
University of Georgia
Griffin, Georgia

Michelle M. Depp
Morgan Consultants, Inc.
Atlanta, Georgia

Jeffrey Ewald
The Optimization Group
Bloomfield Hills, Michigan

John W. Finley
Kraft Foods
Glenview, Illinois

Eric F. Greenberg
Chicago, Illinois

John B. Lord
Department of Food Marketing
Saint Joseph's University
Philadelphia, Pennsylvania

Traci L. Morgan
Morgan Consultants, Inc.
Atlanta, Georgia

Howard R. Moskowitz
Moskowitz Jacobs, Inc.
White Plains, New York

Geralyn Christ O'Neill
The Bailey Group, Inc.
Plymouth Meeting, Pennsylvania

Alvan W. Pyne
Retired
International Corporation
Bradenton, Florida

Steven A. Raper
Department of Engineering Management
University of Missouri-Rolla
Rolla, Missouri

Anna V.A. Resurreccion
University of Georgia
Department of Food Science
Griffin, Georgia

Gordon L. Robertson
Retired
Brisbane, Australia

Marvin J. Rudolph
Arthur D. Little, Inc.
Cambridge, Massachusetts

Stanley Segall
Retired
Department of Nutrition and Food Science
Drexel University
Philadelphia, Pennsylvania

Matthias Silcher
Moskowitz Jacobs, Inc.
White Plains
New York

Robert E. Smith
Retired
Newport, Vermont

Mark Thomas
M.D.T., Ltd
Atlanta, Georgia

Romeo T. Toledo
Retired
Department of Food Science and Technology
University of Georgia
Athens, Georgia

1 The Food Industry in the United States

John B. Lord

CONTENTS

INTRODUCTION

The food processing industry in the United States often appears to be taken for granted—a feature of the economic landscape so unremarkable as to be nearly invisible. Our food industries may be viewed as commonplace, because the methods of production are in some cases quite ancient, or because food factories appear to be organized on a small scale. It may be that processed food products seem so familiar as to be humdrum, or that processing is believed to be merely a modest extension of agriculture or the consumer's own kitchen. Or perhaps our food product development may be chaotic or even archaic.

Some operations employed by food processors, as, for example, flour milling, are indeed prehistoric in their origins, but the methods and equipment used today bear little resemblance to colonial grist mills. Most food processing technologies in place today are the result of modern scientific discoveries and decades, if not centuries, of technological refinement. Moreover, even if our food processes do not match the technological elegance of chemical or pharmaceutical processing, they are still far ahead of the remainder of the world—and thus represent a major planetary resource to feed its population.

The majority of our country's approximately 15,000 food-processing plants are quite small, with fewer than 20 employees. Yet at the same time, some food factories rival the nation's largest. Small scale at the plant level belies a mode of business organization at the corporate level that is as modern as any in the manufacturing sector, if not more so. Scores of raw and semiprocessed foodstuffs must be assembled to finish a given processed food, and each input is, most likely, a unique procurement system. Once finished, the typical food or beverage is distributed through multiple channels, each of which has special business practices. All the while, new products are being fashioned and introduced into the network to satisfy the almost-insatiable demands of an ever-widening consumer population.

Food processing has been traditionally closely linked to agriculture or domestic household activities. Many processing industries were originally part of farm operations (such as butter or cheese making) or were skills found in domestic kitchens (pickling or baking). Today, contemporary food processing is similar to all other manufacturing, if not as sophisticated. The rupture between farming and processing is not yet complete, but it has gone far. It is not impossible to find otherwise-educated consumers who believe that chicken breasts originate in the backrooms of supermarkets or that mushrooms are grown in cans—although sometimes, one wonders.

As we shall demonstrate, the demand for the food processing industry's output is derived from the changing needs and wants of consumers. Hence, the nature of household demand for food shapes the strategies, activities, and market channels used by food processors. As consumers have become more numerous and diverse, multiple pathways or channels to market have evolved to cater to emerging segments. Attention has been devoted to these changes, especially those in food retailing and hotel/restaurant/institutional (HRI) or food service. In addition to researching consumers through the appropriate marketing channels, it is of paramount importance to have the right products available and to communicate the availability and attributes of these products to consumers.

Designing strategies to meet consumer needs does not occur in a vacuum. Product development and manufacturing are of necessity, dependent on the availability of inputs and the means by which both the cost and quality of food processing inputs can be managed. Because food is such a basic human need, its development, production, marketing, and now safety have long been considered areas of vital national interest.

Consequently, a myriad of regulations have evolved that covers many aspects of food processing and marketing and especially safety. New regulations instituted by the Food and Drug Administration (FDA) following the terrorist bombings on September 11, 2001 require those who process and distribute food to provide substantially more documentation for their activities.

WHAT IS THE FOOD PROCESSING INDUSTRY?

Commercial food processing is the branch of manufacturing that starts with raw animal, vegetable, or marine materials and transforms them into intermediate foodstuffs or edible products through the application of labor, machinery, energy, and scientific knowledge. Various processes are used to convert relatively bulky, perishable, and typically inedible agricultural and some other materials into ultimately more useful, convenient, waste-free, relatively shelf stable, and palatable foods or potable beverages. Heat, cold, water-removal, chemical and biological reactions, and other preservation techniques are applied to enhance distribution. Packaging confers portability as well as extending the shelf life by protecting against the environment. Changes in product forms, as well as package design, often reduce preparation time for consumers. Increasing sensory attributes, storability, portability, and convenience are all aspects of "adding value." In other words, food processors utilize factory systems to add economic value by transforming raw materials grown on farms or fished from the sea into useful products. Steers become meat; wheat becomes flour; corn becomes fructose; and freshly caught tuna becomes canned tuna.

FUNCTIONS OF FOOD PROCESSING

The principal and unique economic function of food processing is to convert various food materials into finished, consumer ready products, but the economic contributions do not end there. Food processors perform many value-added economic functions that are shared with other food marketing organizations: farm product assemblers, grocery wholesalers, food transporters, food retailers, and food services (HRI) operators. For example, food processors add value by transforming products through space and time, providing what economists term "place utility" and "time utility." That is, most processors are willing to deliver or arrange for delivery of their finished products to grocery wholesalers or retailer warehouses or even direct to retail stores. Moreover, the procurement operations of food manufacturers arrange for the orderly flow of the many material inputs and supplies required to manufacture foods. Often, although much less than in the past, food processors will maintain a significant inventory of material inputs and finished goods; in some food processing industries such as canned vegetables, processors typically have had a year's supply of canned goods on hand at the end of the canning season. This is changing with the advent of global raw material sourcing, financial pressures, and just-in-time operations. Many food processors act as wholesalers by purchasing finished products from other processors for resale. In short, processors perform several distribution functions essential for ensuring a steady food supply throughout the year and in all parts of the country.

Perhaps more important than physical storage and movement is the information function of food processing firms. Because of their central position in the U.S. food system, food processors have abundant, and at times unique, access to sources of information on the quality, quantities, and prices of processed foods. Food processors, especially those specializing in making semiprocessed food ingredients, spend significant corporate resources collecting, studying, and forecasting information on agriculture supplies in their region or possibly worldwide. The collective judgment of purchasing agents in a given processing industry can quickly drive farm prices up or down. Other food processors, particularly those manufacturing consumer products, employ teams of analysts that are experts on business practices in the food distribution industries and who follow consumer expenditure trends in minute detail. Information that is not generated internally within food processing companies can be purchased from companies, such as A.C. Nielsen and Information Resources Inc. (IRI) that specialize in collecting and analyzing food industry data.

The information collected about consumer demand and agricultural supply trends and conditions come together at the processing nexus. Tight supplies for a given food product at the retail level eventually translate into higher processor prices, a greater willingness to pay for key inputs, and a price signal to farmers to expand production or sell off their stored crops. It works in the other direction as well, with an unexpectedly short crop impelling manufacturing to encourage consumers to reduce their purchases by raising prices on those food products that incorporate the crop. In short, processors are in a strategic position to use price increases to signal lower than expected supplies to farmers, consumers, and all the other middleman in the food system; unexpected abundance is indicated by holding prices steady, price cuts, or more generous deals to retailers.

In addition to price-quantity information, processors are repositories of knowledge about food and agricultural product quality, such as the fine gradations in the flavor of coffee beans, the moisture content of a shipment of corn, the shelf life of products under various handling conditions, and nutritional characteristics, to name just a few. Thus, food processors are typically in the sole position to formulate and design foods, taking into account consumer preferences, distributor demands, ingredient availability, scientific knowledge of biological properties, technological feasibility, and profitability—a complex but critical task. Processors share some of their food quality information through ingredient labeling, nutrition labeling, and other programs that assist consumer choices. Through quality control and product testing, food processors assume most of the responsibility of protecting the safety of the nation's food supply.

OVERVIEW OF FOOD EXPENDITURES

According to data from United States Department of Agriculture (USDA), total expenditures for food climbed to nearly 1 trillion dollars in 2006, an average annual increase since 2002 of around 4–5%. Sales for food at home exceeded the half trillion

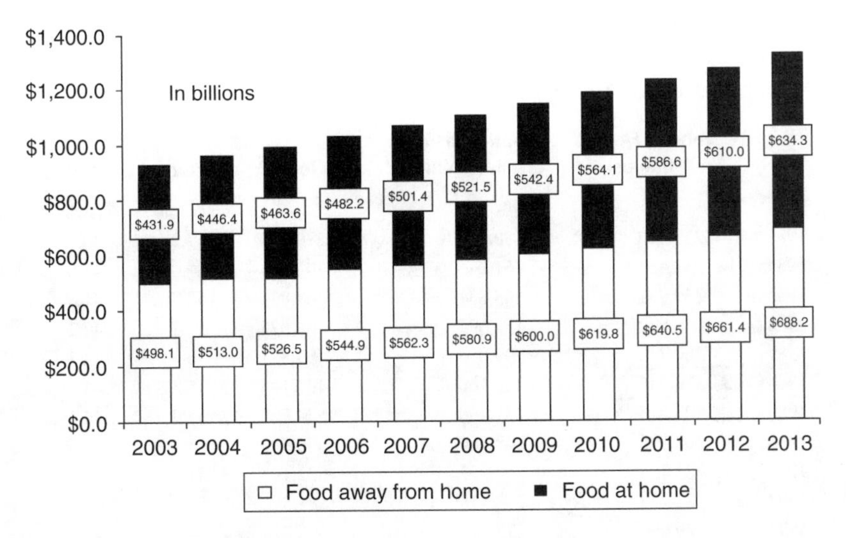

FIGURE 1.1 Food expenditures, 2000–2013. (From USDA, Economic Research Service.)

dollar mark while sales for food away-from-home were close to $500 billion. By comparison, total food sales were $825.6 billion in 2000, $655.1 billion in 1995, and $556.6 billion in 1990. The USDA projects expenditures for food to surpass $1.3 trillion by 2013 (see Figure 1.1). The share of food-away-from home reached 46.9% in 2004, a share that has remained relatively stable since 1998, but is up considerably relative to 20 years ago, when the share of food away-from-home was 41.0% (see Table 1.1).

The USDA figures include expenditures for all retail outlets (food stores, restaurants, and others) as well as service establishments (meals at lodging places and snacks at entertainment facilities, for example) plus allowances, on a retail dollar equivalent basis, for food served in institutions (schools, hospitals, and the like), in the travel industry (meals aboard airlines, e.g., although these are much less available than in the past), and in the military. It covers expenditures by individuals, families, and business, and government. It does not include expenditures for alcoholic beverages sold at stores or restaurants.

"REAL" CHANGES IN FOOD SPENDING

Annual changes in food expenditures and food prices for selected years since 1970 are displayed in Tables 1.2 and 1.3. These data show that generally, changes in total spending are more volatile in the away-from-home sector. Changes in spending for food-at-home more closely track population changes. Total spending on food advanced at a nominal (not adjusted for inflation) rate of nearly 5% from 2002 to 2003. Food at home expenditures increased by 3.7% while spending on food away-from-home rose at a 6.5% clip. For the most part, inflation-adjusted increases in food

TABLE 1.1
Expenditures for Food: 1983–2003

Year	Food at Home (Millions)	Food Away from Home (Millions)	Total	Percent of Total	
				Home	Away
1983	$219,525	$150,883	$370,408	59.3	40.7
1984	231,457	161,046	392,503	59.0	41.0
1985	239,991	168,831	408,822	58.7	41.3
1986	248,827	181,695	430,522	57.8	42.2
1987	260,776	199,055	459,831	56.7	43.3
1988	273,946	217,160	491,106	55.8	44.2
1989	293,213	231,490	524,703	55.9	44.1
1990	308,095	248,464	556,559	55.4	44.6
1991	321,797	260,397	582,194	55.3	44.7
1992	321,177	263,418	584,596	54.9	45.1
1993	329,690	278,684	608,374	54.2	45.8
1994	343,936	291,209	635,145	54.2	45.8
1995	352,722	302,419	655,141	53.8	46.2
1996	367,976	312,616	680,592	54.1	45.9
1997	375,707	328,725	704,432	53.3	46.7
1998	391,062	346,364	737,426	53.0	47.0
1999	417,938	362,419	780,356	53.6	46.4
2000	439,796	385,788	825,584	53.3	46.7
2001	469,055	400,911	869,966	53.9	46.1
2002	485,698	417,991	903,689	53.7	46.3
2003	503,460	445,108	948,567	53.1	46.9

Source: USDA.
http://www.ers.usda.gov/Briefing/CPIFoodAndExpenditures/Data/table1.htm

spending are predominantly dependent on population growth, which has been at rates just over 1% annually, and a cumulative 4.3% from April 1, 2000 to July 1, 2004, according to U.S. Census Bureau figures.

Food price increases have been relatively small in recent years. According to the Bureau of Labor Statistics, food prices for the at-home and away-from-home segments taken together increased by 3.1% in 2001, 1.8% in 2002, and 2.2% in 2003. Commodity prices tend to be more volatile than processed food prices. While food-at-home price increases have tended historically to be smaller than away-from-home price increases, the opposite was true in 2001 and 2003. Food prices were projected to rise at faster rates in 2004 and 2005 than in the previous three years. Food-at-home prices rose at a 3.8% rate during 2004, as reported by the Food Institute (2005), the fastest rise in 14 years, and well above the overall 2.7% increase in the All Items Consumer Price Index.

TABLE 1.2
Food-at-Home Expenditures, Food-at-Home Price Changes, 1970–1971 to 2002–2003

Annual Percent Change	Total	Expenditures (%)	Prices (%)
1970	78,907	5.1+	2.5+
1971	82,909		
1975	122,113	6.4+	2.1+
1976	129,949		
1980	188,067	6.6+	7.2+
1981	200,469		
1985	239,991	3.7+	2.9+
1986	248,827		
1990	308,095	4.4+	2.6+
1991	321,797		
1995	352,722	4.3+	3.7+
1996	367,976		
2000	439,796	4.4	3.3+
2001	469,055		
2002	485,698	3.7+	2.2+
2003	503,460		

Source: USDA.
http://www.ers.usda.gov/Briefing/CPIFoodAndExpenditures/Data/table2.htm

FOOD—A SMALL SHARE OF DISPOSABLE INCOME

Expenditures for food continue to account for a shrinking portion of disposable personal income as shown in Table 1.4. Since 1975, this figure has dropped from 13.8% to 10.1%, with 6.1% of disposable personal income spent on food at home and 4.0% on food away from home. Historical rates show just how efficient the food system has become as well as and how many other classes of goods are available in our economy of abundance. In 1929, families spent 20.3% of disposable personal income on food, in 1950, 17.0% and in 1960, 14.1%. This statistic is indicative of both the increasing efficiency of the food system and the changing mix of items purchased by households in a changing America.

TABLE 1.3
Away-from-Home Expenditures, Away-from-Home Food Price Changes, 1970/71–2002/03

Annual Percent Change	Total	Expenditures (%)	Prices (%)
1970	39,583	6.7+	5.1+
1971	42,251		
1975	68,109	12.8+	6.8+
1976	76,833		
1980	120,296	8.8+	9.0+
1981	130,914		
1985	168,831	7.6+	3.9+
1986	181,695		
1990	248,464	4.8+	3.4+
1991	260,397		
1995	302,419	3.4+	2.5+
1996	312,616		
2000	385,788	3.9+	2.1+
2001	400,911		
2002	417,991	6.5+	2.1 +
2003	445,108		

Source: USDA.
http://www.ers.usda.gov/Briefing/CPIFoodAndExpenditures/Data/table3.htm

CHANGING PATTERNS OF FOOD SPENDING

According to a report from the McKinsey and Company consultancy, the food service segment was expected to capture between 80% and 100% of food industry sales increment in the decade from 1995 to 2005, at the expense of traditional supermarkets. While actual results have not met these expectations, the food service sector has been increasing its share of the food business for decades and now accounts nearly half of total dollars spent for food in the United States. The battle for "share of stomach" among all types of food retailers has intensified as new formats, such as limited assortment stores, and competition from nontraditional formats, such as dollar stores and drug stores, continues to chip away at the supermarket share of the total retail food dollar.

TABLE 1.4
Food Expenditure as a Share of Disposable Income: 1975–1995

Year	Disposable Personal Income (Billions)	At Home	Expenditures for Food Away from Home	Total
1975	$1,159.2	9.9%	4.0%	12.9%
1976	1,273.0	9.7	4.1	13.8
1977	1,401.4	9.4	4.2	13.6
1978	1,580.1	9.2	4.3	13.5
1979	1,769.5	9.2	4.3	13.5
1980	1,973.3	9.1	4.3	13.4
1981	2,200.2	8.7	4.4	13.0
1982	2,347.3	8.5	4.5	12.9
1983	2,622.4	8.3	4.5	12.8
1984	2,810.0	7.9	4.3	12.2
1985	3,002.0	7.7	4.3	12.0
1986	9,187.6	7.5	4.3	11.8
1987	3,363.1	7.4	4.4	11.8
1988	3,640.8	7.2	4.3	11.5
1989	3,894.5	7.2	4.3	11.6
1990	4,166.8	7.3	4.3	11.6
1991	4,343.7	7.4	4.2	11.8
1992	4,613.7	7.0	4.2	11.1
1993	4,789.3	6.8	4.3	11.2
1994	5,018.8	6.9	4.3	11.2
1995	5,306.4	6.7	4.3	11.0

Source: USDA.

CHANNELS OF DISTRIBUTION

The basic channels of distribution for the food industry can be summarized as follows:

- Producers: agriculture, cattle/poultry/pork, and fishing
- Intermediate processors: ingredients and supplies
- Final processors: finished food product manufacturers
- Wholesalers and distributors
- Grocery retailers and HRI outlets
- Home or HRI consumers

There are many variations and complexities in the food channels and many types of operators at each level in the channel. There is also a wide range of different types of facilitating organizations. These organizations perform one or more vital activities to help food products move down the channel from the raw material production to availability of final products to consumers. They include common and contract carriers that provide transportation services, warehouses that provide storage and cross-docking services, plus advertising agencies, marketing research companies, information technology and communication providers, and financial institutions, such as lenders and insurers.

FOOD PROCESSING

Producers include all entities that provide products in an original or unprocessed state. Raw, unprocessed fruits and vegetables; freshly caught fish; beef "on the hoof"; and unprocessed grains are examples. The food-processing sector of the economy is highly complex and differentiated. Companies such as David Michael and Archer Daniels Midand (ADM) produce ingredients and goods used by other food processors. Companies such as Cultor Food Science and FMC Biopolynics are also primarily in the ingredients business, while McCormick, which specializes in spices, produces both intermediate and finished goods.

Many food processors operate at both the intermediate and final levels. For instance, Knouse Foods (Peach Glen, Pennsylvania) manufactures processed fruits and juices under several brand names (e.g., Lucky Leaf Applesauce) and provides fruit filling for bakers such as Tasty Baking in Philadelphia. Ocean Spray, like Knouse Foods, a growers' cooperative, manufactures cranberry products as ingredients for other processors and as final products for both the retail and food service sectors. Even Procter and Gamble, whose varied product lines include both nonfood and food items, manufactures products such as fats, oil, and Olean (the brand name for Olestra, an artificial fat replacer) for use by food processing companies such as Frito-Lay.

Some processors produce only for the retail sector, and some produce only for the food service sector, but many produce for both. Some manufactures produce exclusively branded food products, while others package exclusively private label—also called "store brand"—products for wholesalers, retailers, and even HRI outlets. Some companies offer both their own branded products and also pack on a contract basis for wholesalers and retailers who wish to market their "own brand" products, a phenomenon that is becoming increasingly important for retailers who want to market something beyond low prices.

Other companies that fit under the broad definition of processors include manufacturers (typically called contract or copackers) whose primary business includes contract manufacturing and packing to whom food companies outsource some portion or even all of their production. Most food processing companies purchase package materials from package-material suppliers. Other processors take a partially finished or intermediate product, such as filling, and then manufacture and produce the final product. A very few manufacturers outsource all packing to a packaging specialist. The April edition of *Stagnito's New Products Magazine* contains an exhaustive list of suppliers who serve the food processing industry.

WHOLESALING

The wholesaling sector includes different types of full-service wholesalers and distributors, specialty or limited assortment (such as those who feature frozen foods or produce), and rack jobbers who distribute products directly to the retail store shelves. As a traditional full-line merchant wholesaler, Supervalu supplies nearly 4000 grocery stores in 48 states with branded and private label goods. Supervalu sponsors a voluntary group of retailers, conducting store operations under three principal formats: superstores (Cub Foods and Shop 'n Save), limited assortment stores (Save-A-Lot), and food and drug combination stores (Farm Fresh and others) and also distributes to independent grocers. Supervalu achieved total sales of over $20 billion in 2004, an increase of over 5% from the previous year. Like Supervalu, both C&S and Nash Finch operate supermarkets and distributes to independent retailers, but on a more regional scale.

Some wholesalers, such as AWI, are jointly owned by a sponsoring group of retailers in the form of a cooperative. AWI services supermarkets and convenience stores with a full line of food and general merchandise. Retail ownership guarantees a strong retail focus. The boards of directors of food wholesale cooperatives usually consist of independent store operators elected by the membership.

Wholesalers such as Sysco and U.S. Foodservice primarily serve the food service trade. Club stores, such as Sam's Club, BJ's, and Costco, which we discuss in the following section on food retailing, serve small businesses and other organizations such as nonprofit church and community groups, in addition to being a shopping destination for household consumers.

Most of the large supermarket chains have vertically integrated supply operations and act as their own wholesalers. For example, most food manufacturers selling to Acme Markets, a division of Albertson's, the supermarket share leader in Philadelphia, Pennsylvania, sell direct or through brokers to Acme which purchases in wholesale quantities and stocks warehoused items in Acme warehouses for distribution to its stores in that area.

Companies whose primary business involves processing perishable commodities such as dairy and baked goods, plus beverages, often operate in what the industry refers to as "direct store delivery" mode. Local bottlers of Coca Cola and Pepsi Cola, snack food companies such as Frito Lay, and dairies such as Lehigh Valley (a regional dairy in eastern Pennsylvania), deliver product directly to retail stores and provide a full range of stocking and merchandising services. In some cases, the driver-salespeople are independent businesspersons, that is, owner-operators who purchase rights to service a defined geographic territory on an exclusive basis. Tasty Baking, based in Philadelphia, is one food processor which operates in this manner.

Brokers are also wholesale level organizations. Unlike wholesalers, which are merchant middlemen (i.e., they take title to the products they handle), brokers serve a facilitation function in that they bring buyers and sellers together, mainly serving in place of or to augment a manufacturer's sales organization. The broker industry has undergone some significant changes in recent years. First, there has been a significant amount of consolidation in the sector, as smaller regional brokers have been acquired by national organizations. Second, in order to insure their long-term viability as key players in the industry, large broker organizations such as Acosta Sales and Marketing (note the moniker) have become increasingly sophisticated and data-driven, expanding the range of services offered to retail selling execution, creative trade and consumer marketing programs, category management, logistics, customer service, and order processing. Brokers or sales and marketing organizations serve all classes of trade including food service.

FOOD RETAILING

The two primary divisions of food retailing are the traditional and nontraditional grocery channel. The traditional grocery channel includes conventional supermarkets, superstores (30,000 sq. ft and larger), food and drug combination stores (at least 25% of the merchandise mix in nonfoods and drugs), warehouse stores, limited assortment stores, convenience stores, and others. The nontraditional grocery channel includes hypermarkets, wholesale club stores, and deep discounters. Operators of supermarkets and superstores can be either chains (e.g., Kroger, Ralph's and Harris Teeter) or independents. Independent supermarkets can be members of wholesaler-sponsored voluntary groups, such as IGA, or cooperative buying groups, such as Associated Wholesalers.

The market share of different store formats continues to change. Supermarkets' share of the total retail food dollar continues to decline, with alternative formats such as supercenters (Wal-Mart, Fred Meyer, and Super Target), club stores, limited assortment stores (Aldi's and Save-A-Lot), and even drug and dollar stores gaining share. According to the Food Institute, traditional supermarkets accounted for 86.9% of grocery and consumable sales in 1998 and only 56.3% in 2003, with the decline expected to continue so that projected share in 2008 is just 49.0% (Food Institute, 2005). Wal-Mart has become the nation's leading food retailer, as well as the world's largest company. Wal-Mart operates approximately 2000 supercenters, a number which continues to grow. Willard Bishop Consulting estimates that nontraditional grocery outlets will grow from just over 40,000 in 2003 to over 48,000 by 2008, and their share of the grocery market will jump from 31.3% to 39.7% during the same period, while traditional grocery stores will lose almost 8% of their business, with stores dropping from 41,730 to 39,715 and share dropping from 56.3% to 48.3% of the grocery sector. At the same time, Food Marketing Institute (FMI) studies show that consumers continue to increase the extent to which they shop for groceries at discount stores such as Wal-Mart and BJ's. In FMI's Trends in the United States, 2004, 26% of shoppers reported they used discount stores for groceries in 2000, a number that grew to 31% in 2004; similarly the percentage of shoppers who shop for groceries at warehouse clubs grew from 14% to 16% over the same period.

Convenience stores such as 7-11, Wawa's, QuikChek, and others also play a very significant role in the food industry. Data from Willard Bishop Consulting show that in 2003, 129,000 "C-stores" in the United States accounted for almost $94 billion in sales of food and allied products. As Americans increasingly look for food "on-the-go" and solutions to last-minute meal problems, convenience stores will increase their share of the U.S. food retailing business.

The latest significant players in the grocery wars are drugstores (with over $33 billion in grocery and consumables sales in 2003) and dollar stores (almost $11 billion in 2003.) Most industry observers believe that shares of these segments will continue to grow as people increasingly look for food "anytime, anyplace" based primarily on convenience. Other interesting industry players include Trader Joe's, a limited assortment format featuring a large number of all-natural and unique products, mainly under Trader Joe's store brands, in a funky, friendly atmosphere offering some unique products at very reasonable prices and Whole Foods which emphasizes natural, organic, and in-store prepared foods.

HOME SHOPPING

During the Internet craze of the late 1990s, Web-based home shopping services such as Peapod, Webvan, and Shoppers Express became hot properties for investors and big news in the food retailing business. By early 2000, as the Internet bubble burst, none of these operations could sustain their business models. Webvan became a colossal failure, having raised significant amounts of capital but never getting close to profitable operation and the company had to dissolve. Peapod was acquired by Ahold USA, the U.S. division of the Royal Ahold Corporation (Netherlands), which operates supermarkets under banners such as Giant and Stop & Shop. However, the concept of Internet shopping coupled with home delivery is viable and appeals to shoppers with specific needs. Peapod is operating as a division of Ahold and provides home delivery service in selected market areas such as Baltimore-Washington, southern Connecticut, and Boston and recently announced plans to expand to other geographic locales.

Supermarket operators such as Safeway and Albertson's have created home delivery services that, instead of relying on separate food distribution centers and commissaries, create orders by picking products from their store shelves. Shoppers in selected cities served by Albertson's can click on Albertson's.com, order and have their customized supermarket orders delivered direct to their home for a modest delivery charge; the same service is available in specific geographic areas from Safeway.com. According to FMI statistics, 16% of shoppers reported their primary store offers Internet grocery shopping in 2004, a number which has leveled off after increasing from 9% in 2000 to 16% in 2002.

Fresh Direct operates in New York city and appeals to city residents who do not like the fresh food in the city and prefer not to drive an hour or more to North Jersey or Southern Connecticut to visit a supermarket that they like. Fresh Direct provides home delivery for about $5 (minimum order size: $40) within a 2 h window, the day after the order is placed. Much of the fresh food is prepared to order in a plant in Queens, and the time between production and delivery is much shorter than for conventional supermarkets and thus results in fresher product. This has led to a high degree of customer satisfaction, repeat business and plans to expand operations to new geographical areas, first in metro New York and then to other cities on the East Coast.

Online grocery shopping has been slow to catch on. According to FMI, only about 5% of shoppers indicate they have purchased groceries online in the past 12 months. According to the Food Institute, online grocery sales in the United States were projected to reach $8.8 billion in 2004, up from $200 million in 1999, but still only about 1.5% of total U.S. at-home-food sales.

HOTEL, RESTAURANT, INSTITUTIONAL, FOOD SERVICE

The HRI or food service industry is very broad and diverse and encompasses all meals and snacks prepared away from home, including all takeout food beverages. The food service sector is divided into three major components: commercial restaurants, non-commercial facilities, and the military. The commercial sector includes eating and drinking places (restaurants, ice cream stands, bars and taverns, delis and pizzerias, bagel shops, and social caterers), managed services (commercial food service management companies, such as ARAMARK, which have operations in plants, offices, schools and colleges, health care facilities, transportation facilities, and sports and recreation centers), and lodging places.

Restaurants come in several varieties, including full service restaurants (both "fine dining" and "family" restaurants) and quick service or fast-food restaurants, with the determining factors being breadth of menu and average check size per person. Noncommercial restaurant services include business, educational, institutional, or governmental organizations that operate their own food service operations. Military includes officers' and noncommissioned officers (NCOs) clubs plus military exchanges. An interesting development during the 1990s was the take-home restaurant, best characterized by Boston Market, which is in the business of providing home-cooked meals to people who lack the time, energy, ability, and inclination to cook, and provides food for both take-home and on-premise consumption, with a focus on "home-cooked" instead of typical fast-food fare such as burgers and tacos.

There are several significant drivers of food product development for food service establishments. First is a focus on more convenient or "speed scratch" items that require less preparation and therefore less labor in the back of the house. Products that require little or no preparation are more important today because of labor shortages and increasing labor expense, even in fine dining establishments. Second, with an increasing ethnically diverse population, we have a greater interest in foods that are "authentic ethnic" or fuse together different profiles, plus we have a greater interest in products with a higher flavor profile. Third, the increasing interest in health and wellness is reflected in healthier menu choices. It is important to note that people dining away from home typically are not willing to sacrifice flavor and that even healthy items must taste and look good, but the increasing concern over obesity has created more pressure for information about nutrient and caloric content of menu items, smaller portion sizes and menu items that allow individuals to follow a low carbohydrate or organic or natural diet. Until recently, you could go into a casual dining establishment, such as TGI Friday's, and find a full page of Atkins-friendly menu items. Today, menus include "natural" and organic items.

CONSOLIDATION

Consolidation in the food industry is driven by opportunities for increased buying power, substantial cost savings, smoother transition and blending of merged companies due to information technology, and bigger store brand programs. The late 1990s ushered in an era of consolidation in the food industry. Increasing pressures created by mature, slow growth markets and cost pressures led to an unprecedented number and scale of merger and acquisition activity. Consolidation was forced on the retail sector trying to respond to the increasing scale and buying power of Wal-Mart. In order to compete both in terms of cost structure and buying power, food retailers had to grow bigger. As a result, companies such as Safeway and Albertson's have made numerous acquisitions of smaller regional chains. By 2000, the percentage of the total grocery market controlled by the top ten chains had roughly doubled from the beginning of the decade.

Partly to respond to the increasing scale of their customers and partly to take advantage of scale economies through reduction of redundant activities, food manufacturers had also engaged in a fury of mergers and acquisitions. Among the more notable acquisitions, Kraft (Atria) acquired Nabisco, Kellogg's acquired Keebler, General Mills acquired Pillsbury, and PepsiCo acquired Quaker Oats. Hundreds of other mergers took place in the late 1990s and into 2000, along with a great deal of "brand swapping" as companies looked to streamline and focus on core brands.

More recently, consolidation has slowed, partly as firms digest their acquisitions and partly because there are fewer acquisition targets. The Food Institute tracked 415 mergers and acquisitions in the food sector in 2003. By comparison, there were 204 mergers and acquisitions in the first quarter of 1998. There were 37 deals among diversified food processors and 35 among supermarket operators. Among the notable activity in 2004 were the acquisition of J. M. Smucker by International Multifoods and the acquisition of Veryfine Products by Kraft. On the retail side, Kroger purchased Spartan Stores and Penn Traffic Company, and Fleming sold a number of retail units to Albertsons.

Consolidation activity has led to an industry with fewer and larger players at all levels of the industry. There is an economic theory of consolidation known as matching the scale of forward players, and this explains the fact that when retailers consolidated, so did manufacturers, as well as companies that supply manufacturers (the food ingredients sector) as well as other agencies and organizations such as brokers. The most notable result of consolidation has been emphasis on cost reduction and stock keeping unit (SKU) rationalization, as packaged food companies seek to drive lower costs and higher profits through more efficient food product development and distribution.

FOOD CONSUMERS IN THE UNITED STATES

Today's food consumers, driven by changing demographics and lifestyles, have adopted different patterns of food consumption, in turn leading to differences in menu planning, food acquisition, and food preparation. Increasingly, with their demand that food be available when and where they want, and in the quantities and varieties, consumers want to support their more mobile and more diversified lifestyles. The task of the food system is changing from "bringing the consumer to the food" to "bringing the food to the consumer."

As it has for at least two decades, the changing role of women in society, specifically the increase in the percentage of adult women working, has been the single most important factor impacting food preparation and consumption patterns. The percentage of women over the age of 16 who work outside the home has stabilized at just over 60%. According to U.S. Census data, nearly 60% of children under the age of eight have mothers who work outside the home as compared with 18% in 1960. When women, especially mothers of younger children, work outside the home, households in effect substitute money for time. There are more dual-income households but there is much less time in those households for their members to perform routine household production tasks, such as cooking and cleaning up.

Family composition is also changing with attendant effects on household activities. The "traditional family" is no more, as evidenced by the statistic that only about 8% of households today feature a working dad, a stay-at-home mom, and children (as compared to 43% of households which met this definition of "traditional" in 1960). Today, approximately 30% of all households have two working adults. About 30% of all households have a single parent and about 26% of the population lives alone. Households are becoming smaller, with the average of 2.57 persons, down from an average of 3.14 persons in 1970. The effects of these changes are far-reaching for the food industry, because households, with two working adults or led by a single parent or inhabited by a single person, are in many cases, time-poor and focused on finding more convenient ways of accomplishing household tasks.

CONVENIENCE

The desire for both convenience and simplification has continued to exert ever more influence as a significant driving force in the food industry. In dual-income and single parent households especially, consumers are interested in convenience and willing to pay for it. Life is much more diversified, and people today participate in lots of activities both outside and inside the home. Family members are constantly on the go. Virtually any suburban mom or dad is familiar with the feeling that her or his primary parental role seems to be providing transportation to kids who always have to be somewhere to do something. Given that mom is working and that everyone in the family is doing so many things, it is not surprising that the result is that mom is no longer in the kitchen spending 1½ to 2 h cooking dinner everyday, as my mom did back in the 1950s and 1960s, and is no longer available to teach daughter (or son) to cook.

Consumers with busy and active lives do not sit down for three "square meals" with the family, as was the tradition, or at least we remember it that way. Meals mean different things to different people at different times. Sometimes, meals have much social significance and represent opportunities for family or friends to gather and share. At other times, meals are strictly fuel stops with emphasis on finding something quickly that tastes good and satisfies one's hunger. Because of our hectic schedules, we have less opportunity to sit down with our families, and as a result, few "all-family" meals are eaten than in the past. It is not unusual to have three or four different dinner times in a five-person household, because all of the people living in the household participate in different activates and have different obligations. Much of our eating now takes the form of "grazing," which means that we grab small amounts of food several times during the day. The new dining rooms of America increasingly are the car ("dashboard dining") and the office ("desktop dining").

Information from *Parade Magazine*, NPD Group, A.C. Nielsen, and others who track food industry trends can be summarized as follows:

- Many consumers report they are too tired to cook and are looking to simplify the task of feeding themselves and their families.
- Less time is spent planning meals; it has been noted that relatively few consumers know what they will eat for dinner as late as 4 p.m. on a given day, and even fewer plan meals more than one day in advance.
- Individuals spend less time preparing meals and are looking for shortcuts to meals that still taste good and provide adequate nutrition.
- Because there are fewer moms in the kitchen and less time spent preparing meals, cooking skills are not being passed from generation to generation as they have for hundreds of years.
- Cooking from scratch has diminished significantly; in fact, cooking has become a process of assembly.
- Meals are much simpler; the use of side dishes has dropped dramatically, while the use of kettle and crock-pot meals, cooking kits, and frozen prepared meals has increased.
- The number of meals at home has actually increased from the mid-1980s to 2005, but these meals are assembled, prepared, and sourced much differently than before.

The food industry has witnessed significant growth in eating as a secondary activity while doing something else, which leads to eating more frequently, at non-traditional times and in smaller portions. This trend has given rise to a need for portability in food and the growth of "serious snacks," such as nutrition bars and drinkable yogurt, which provide quick nutrition as well as appetite satisfaction. One example of such a product is Nouriche from Yoplait (General Mills), which is a drinkable yogurt specifically formulated to meet a nutritional profile that appeals to women.

To spend less time and effort preparing meals, consumers are increasingly "outsourcing" meal preparation. Consumers are taking greater advantage of restaurants, which are increasingly becoming "prepared foods supermarkets" and do well as supermarket food service. Consumers are not eating out more, but carrying out more, as restaurants are becoming the standard of quality for prepared foods. Supermarket operators such as Wegman's and Ukrop's have successfully tapped into this demand for outsourcing with prepared foods programs in the store.

Consumer behavior research indicates that the definition of "family" is no longer restricted to the traditional definition. A variety of relationships and numbers of people constitute families today. With a greater number of women in the workforce and no substitute at home to prepare meals, family and individual eating habits are changing. In short, consumers are seeking products that do not require much time, effort, or thought to prepare. Prepared foods are an essential resource for time-pressed consumers.

ETHNIC DIVERSITY

According to Bob Messenger, a long-time food industry expert, we have become "a nation of tribes." What this means is that contrary to the longstanding cultural homogenization of America ("the melting pot"), today various ethnicities and subcultures wish to retain their cultural identity, including food and dietary habits. Our country continues to grow more diverse, as the share of Hispanics and Asian-Americans continues to grow dramatically, and the share of whites continues to drop. Currently, Hispanics represent 10% of American households, but 20% of births. Census estimates project a 56% increase in the Hispanic population and a 55% rise in the Asian population between 2000 and 2020. By the end of the current decade, Hispanics, Asians, and African-Americans will represent one-third of the U.S. population, as compared to just 30 years earlier in 1980, when whites comprised 80% of the population.

So, we are no longer a melting pot; instead, we are a "stir fry," a metaphor that implies that we mix different groups together, but they retain their cultural integrity. What does this mean for food processors? It means that consumers will increasingly demand authentic ethnic cuisines, creative and innovative food concepts with unique and different flavor profiles, and foods that are consistent with ethnic and cultural values and folkways.

AGING OF THE AMERICAN POPULATION

The "graying of America" is a major influence on many industries, such as healthcare and leisure products, as well as on the food business. The baby boom generation is aging and life expectancies continue to increase. The two largest age groups in the American population are the 15–24 and over 55 groups, both of which are growing and have increasingly different needs. According to Willard Bishop Consulting, there are over 88 million "Gen-Y-ers" who are web savvy, expect services on demand, have distinct preferences for organic, natural, ethnic, gourmet, and prepared foods, and are primarily interested in speed and convenience. The American Association of Retired Persons (AARP) 55-plus group, the largest population segment, seeks health and wellness products and is much more focused on customer service.

According to U.S. Census Bureau statistics and projections, one out of every ten people in the United States is now 60 years and above; by 2050, one out of five will be 60 years or older and by 2150, one out of three people will be 60 years or older. These figures imply that foods that are rich in certain nutrients and can slow the aging process will continue to grow in importance. Older Americans also spend proportionately more on food that they consume at home, according to a report published by Packaged Facts. People between the ages of 65 and 74 allocate 8.9% of their expenditures to food at home, compared to 9.2% for those 75 and up. Both of these figures are significantly higher than the U.S. population at large, which allocates around 7.6%.

Older consumers have special food and nutritional needs. The food industry must be ready to meet the needs of those with special diets and unique preferences. As we live longer and health problems related to aging pose new challenges, people will demand goods that help in maintaining health. Nutritional information and labeling will have to be clear, concise, and meaningful. Aging also often brings a dampening of the senses. Smell and taste decline. Therefore, food marketers should consider the importance of enhancing flavors in products developed for older individuals. Visual acuity diminishes as well, which has implications for package label design and the need for larger print. Muscle mass and strength also deteriorate as we get older, thereby creating a need for innovative packaging structure. Furthermore, although enhanced food products with unique labeling and packaging may be more expensive to manufacture, marketers would do well to remember that older consumers will likely be willing to pay a slightly higher price as individuals over age 50 hold more than three-quarters of the nation's financial assets.

HEALTH AND FOOD

The food industry historically has been driven by three dominant consumer interests: flavor, convenience, and health. Now, in the twenty-first century, the industry finds that the bar on all three of these factors continues to rise. Consumers want more and more sensory satisfaction and more and more convenience and are increasingly driven by concerns about nutrition and health. Growing obesity and related medical conditions such as adult-onset diabetes, hypertension, heart disease, and various types of cancer, not to mention concerns over appearance and everyday issues such as fitting into an airline seat, have led to increasing consumer interest in foods with health benefits. The sad fact is that we do not eat healthily, and at the same time, the amount of physical activity has declined.

The aging consumer wants to live longer, healthier, and more productively. By the dawn of the twenty-first century, Americans in general had become more health conscious. There is now general recognition of the impact of dietary habits on physical condition and overall health.

However, despite the professed desire to eat healthier, it is not a battle easily won. Much of the food that the industry produces is highly processed and calorie-dense. But consumers have consistently voted with their food choices for food that tastes good, and so the challenge is clear. Find ways to produce food that surpasses sensory hurdles but eliminate "the bad stuff" like saturated fat, trans fatty acids, and refined sugar, and produce food rich in important nutrients such as antioxidants and omega-3 fatty acids. Companies such as Kraft and Frito-Lay (PepsiCo) have been working to eliminate trans fats from their snack items in anticipation of labeling requirements that went into effect in 2006.

The year 2004 was the year of "low-carb" frenzy, as diets such as Atkins and South Beach suddenly became all the rage for people looking for quick weight loss. Data from IRI showed that low-carb products reached sales of over $1 billion in less than 2 years, and were up 5.8% from June 2003 to June 2004. The interest in low carbohydrate foods and the number of low-carb product introductions peaked in the second and third quarters of 2004, but by the fourth quarter, interest was beginning to wane as fewer consumers were following low-carb diets. By the fourth quarter, many of the low-carb items were languishing on store shelves. What remains is a general recognition that complex carbohydrates are better than simple carbohydrates, and there will be only a small core segment of consumers who will follow the "low-carb lifestyle."

Interest has recently spurted in soy products, fiber, and whole grains. Soymilk sales have grown significantly and many other products containing soy protein, including veggie burgers, yogurt, and others, have been introduced. According to *Productscan Online*, products making high fiber claims in North America grew from around 2.3% in 2000 to 3% in 2003 to 4.2% in 2004. This trend should continue as new dietary guidelines advise promoting the consumption of increased dietary fiber by both children and adults. Similarly, increasing recognition of the health benefits of whole grains has led companies such as General Mills Inc. to react. In January 2005, General Mills, Inc. launched a complete line of Ready-to-Eat (RTE) cereals that have been reformulated so that they are made from whole grains. Many bakers have been doing the same thing with bread products.

ORGANICS AND NUTRACEUTICALS

The market for organic foods continues to increase. USDA standards for labeling organic products have helped to clarify the meaning of organic for consumers. Organic foods have a "halo of healthfulness." People feel that organic products are better for them and better for the environment, because the use of chemical fertilizers, pesticides, and fungicides is avoided. Retailers such as Whole Foods offer mainly products that are organic and all-natural. While organics represent a small piece of the food business, the category is growing at annual rates approaching 20%. Sales exceeded $15 billion in 2004 and are expected to reach $32 billion by 2009, according to MSNBC.com.

Nutraceuticals, foods that provide "medical" benefits such as lowering cholesterol or aiding digestion, represent a growth opportunity for food companies facing mature categories and stagnant sales growth. Another term that is used in the industry is functional foods. Mintel International Group defines functional foods as products that make a distinct, written health claim enhanced with added ingredients or through the act of processing. With its strict definition, Mintel estimates the 2003 functional food market was approximately $4.6 billion, with annual growth rates exceeding 10% since 1998. Mintel predicts a 39% increase in sales of functional foods between 2003 and 2008.

One of the categories of functional foods growing most rapidly is energy and nutrition bars. Nestle had been very active in this category with products such as Nestival, a breakfast bar containing carbohydrates that are absorbed more slowly and make people feel full more quickly. Nestle has also developed a type of milk protein that can help fight cavities and a chocolate component that limits the absorption of "bad" cholesterol.

It is clear that as consumers take a turn toward more active maintenance of wellness, health will become an even greater driving force in the choices consumers make about which foods to eat. Attentive marketers have already begun segmenting consumers based on health condition instead of age. Expect nutritional individualization with regard to health condition to be a key marketing emphasis down the road. And the bar will continue to be raised regarding the information consumers demand about the foods they eat.

FOOD SAFETY AND SECURITY

The tragic events of September 11, 2001 created another key food industry concern, that of food security. While food safety has long been a key issue in food processing and distribution and in food service operations, it is only recently that we have become acutely aware of the potential for agro- and bio-terrorism, intentional adulteration of the food supply for the purpose of harming people, creating economic and political instability, and reducing the trust we have in our food system. FDA has passed regulations that impact food processors and distributors and will help to both monitor food industry activity and provide for quick reaction should a terrorist event take place. The industry generally has been put on alert.

The food safety issue's greatest concern is that of reducing the risk of food-borne illness. Each year, 5000 deaths and up to 100 million outbreaks of sickness are attributed to food poisoning, according to the Centers for Disease Control. This issue is attracting greater attention and assuming greater importance for a number of reasons. First, interest in home meal replacements (HMRs) and minimally processed foods is increasing because of better quality and convenience. Although the cases of food-borne poisoning may not change because of this fact, liability may become a greater issue as lawsuits against food processor marketers, retail grocers, and HRI outlets increase. Second, the creation of a global food supply implies greater reliance on imported produce. Imported food is subject to different and many-times-less-stringent handling and processing regulations than in the United States, thereby increasing the likelihood of food-borne pathogenic microorganisms. Finally, growth in markets for organic, minimally processed, and no- or less-preservative foods further increases the risk of food-borne-illness.

A coordinated approach to food safety systems is needed. A single uniform and consistent set of standards among federal and state agencies would be a great improvement over the current policy structure. In addition, consumer education about food handling and about where food problems are likely to occur must be improved. Adequate consumer education calls for partnerships among industry, government, and the media.

OTHER KEY SOCIAL AND ECONOMIC ISSUES AFFECTING THE FOOD INDUSTRY

An issue that cannot be overlooked has to do with the critical social changes taking place in regard to income distribution. Only a small percentage of the U.S. population is gaining financially, leading to an even greater concentration of wealth among a very few. The middle class is shrinking and a bimodal income distribution is emerging. As the rich become richer, the poor are becoming poorer and hungrier. Many lower-income consumers do not have access to the full food-retailing spectrum. Finding ways to meet the food and nutritional needs of the hungry is a challenge we in the food industry must accept.

For food businesses to thrive, economic development incentives must be initiated. Tax policies and developmental incentives could be a part of this effort, as well as labor force education and training programs. Infrastructure development is another area where government can play an important role in improving the viability of food businesses. Reasonably priced access to energy and adequate transportation and communication networks are essential to the food industry. Furthermore, streamlined and relevant laws and regulations that are enforced can make a real difference to the bottom line and operating efficiency of companies in the food industry.

CONCLUSION

This opening chapter has provided an overview of some of the most important issues facing the food industry at the beginning of the twenty-first century. Perhaps the issue that the food processing/marketing industry itself is best designed to play the biggest role is that of the changing consumer. Developing convenient, user-friendly, novel, and healthy food products that meet the needs and wants of an increasingly diverse and demanding population of consumers is a challenge we will confront. Using modern technology to implement changes in products as well as improve food safety and retain quality will be essential to long-term industry sustainability. In the drive to meet the needs of the changing consumer, we must be aware of the challenges and benefits of food processing/marketing industry consolidation. Further, as suppliers of food, we have a responsibility to play a role in abating the national and world hunger problem. We must communicate our financial, transportation, communication, and energy needs to government so that decisions makers understand better how to serve the food processing/marketing industry.

Finally, we must continue to deliver new, safer, better, and more convenient foods to a dynamic population, which translates into the need for effective and efficient food product development systems.

BIBLIOGRAPHY

Anonymous 2004. A.C. Nielsen Consumers Pre*View Study Finds Consumers Too Ties to Took or Clean. Schaumburg, IL: A.C. Nielsen.

Anonymous 2004. Acquisitions and Brand-Swapping Return, www.foodprocessing.com, accessed May 2004.

Anonymous 2004. *The Food Institute's Food Industry Review 2004*. Fairlawn, NJ: The Food Institute.

Anonymous 2004. *Trends in the United States*, 2004. Washington, D.C.: Food Marketing Institute, p. 24.

Anonymous 2005. *Company Overview*: *Supervalu,* www.computerwire.com/companies, accessed January 12, 2005.

Anonymous 2005. Good News for Grocers: Older Americans Spending Proportionately More on Foods at Home, According to New Report, www.prnewswire.com, accessed January 2005.

Anonymous 2005. Safeway.com Plans Enhancements for Its Online Grocery Service, www.foodinstitute.com, accessed January 13, 2005.

Ball, D. 2004. With Food Sales Flat, Nestle Stakes Future on Healthier Fare; High Tech Additives in "Phood" Help Munchers Keep Fit; So Far, Slow to Catch On; Candies that Fight Plaque, *Wall Street Journal*, March 18, p. A1.

Berner, R., D. Brady, and W. Zellner. 2004. There Goes the Rainbow Nut Crunch: To Stay Competitive, Food Companies Are Weeding You Their Slow-Selling Products, *Business Week*, July 19, 2004, Issue 3892, p. 38.

Bishop, W. 2005. The World According to Shoppers, Presentation at Saint Joseph's University, Philadelphia, PA, April 14, 2005.

Connor, J. M. and W. Schick. 1997. *Food Processing: An Industrial Powerhouse in Transition*, 2nd ed. New York: John Wiley & Sons.

Hansen, N. 2004. Organic food sales see healthy growth: Mainstream food companies promote natural brands, http://msnbc.msn.com, accessed December 7, 2004.

Hertel, J. and P. L. Weitzel. 2004 The Changing U.S. Consumer Meets the Store of the Future, *Competitive Edge.* Barrington, IL: Willard Bishop Consulting.

Kaufman, P. R. 2002. *Food Market Structures*: *Food Retailing*, http://www.ers.usda.gov, accessed January 4, 2005.

Little, P. 2005. "Channel Blurring Redefines the Grocery Market," *Competitive Edge.* Barrington, IL: Willard Bishop Consulting.

McKinsey & Company, Inc. United States. 1995. *Foodservice 2005—Satisfying America's Changing Appetite.* Washington, D.C.: International Foodservice Distributors Association.

2 Product Policy and Goals

John B. Lord

CONTENTS

INTRODUCTION

Food processors looking to create successful new products in the current retail and consumer environment have faced increasingly difficult challenges. These include increasing competition from both existing and new types of competitors, heightened pressure on both wholesale and retail shelf space, pressure to shorten development cycles and get to market more quickly, increasingly complex technical challenges (e.g., foods that provide both "health" and good taste), and increasing regulatory complexity in areas such as labeling, food preservation, and so on. Despite these challenges, food processors seeking organic growth, as opposed to growth through acquisition, look to new products, and the frenetic pace of new food product development seems to defy the odds.

During the late 1990s, new product activity had dropped somewhat. Several reasons were cited for this. Companies recognized that there is less risk and a potentially faster return from leveraging equity in core brands than in investing in innovative new food products. The Efficient Consumer Response (ECR) initiative of the 1990s, including a move toward category management, led to a reduction in me-too types of food product launches. Market saturation in many product categories has made it increasingly difficult for small and mid-sized companies to get their products on retail shelves. While population in the United States is growing at less than 2% per year, the average supermarket stocks more than 30,000 items today, up from 6,000 just a generation ago. Food companies responded to a history of product proliferation and an increasingly strident retail sector by simplifying product lines and making stock keeping units (SKU) elimination a priority. Further, the pace of consolidation and the emphasis on productivity, exerted a significant influence on consumer packaged goods (CPG) companies, which responded by launching fewer items.

In what seems like a classic contradiction, packaged goods companies are introducing more new items than ever, but at the same time are pruning their total product lines, removing slow movers and duplicate items. According to *Productscan Online*, more than 14,826 new food items were launched during 2004 (Dorfman, 2005), compared to about 13,000 in 2000. This figure represents a significant increase from the mid-1990s. This frenetic new food product activity is taking place against the backdrop of a strategy of "SKU rationalization." Companies such as General Mills, Hershey Foods, and H.J. Heinz are weeding out slow movers to reduce costs in response to significant increases in the costs of commodities such as soybean oil and cheese. These food product line reductions also represent a reaction to Wal-Mart, which is increasingly focusing its merchandising strategy on fast-moving items, and other major food retailers, which not only focus on ways to get close to Wal-Mart on price, but are also placing more emphasis on store brands (Berner et al., 2004), primarily as a means of competing with Wal-Mart and each other. However, as long as traditional supermarkets charge new item or slotting fees, the retailers will be motivated to take on new items. As long as manufacturers have the promotional funds to get on the shelf, they will launch new items. And as long as a lot of me-too new items are launched, there will be many slow movers to take out of the product line.

While food companies launch many new items, only a few are truly new, and only a small percentage of new items make a significant impact on the market, as measured by sales volume. *Productscan Online* rates new products on whether or not they are "innovative" with only 6.7% of new food product launches earned an "Innovation" Rating, compared to 8.6% of CPG launches in 2003. Information Resources Inc. (IRI) Pacesetters 2004 reported that more than three-quarters of all new brands achieve "Pacesetter" status, which requires year one-one sales of at least $7.5 million; only 2% exceed $50 million in first year sales. Of these Pacesetters, over 90% are line or brand extensions and fewer than 10% represent new brand names.

Nevertheless, new food products are very important to the success of food manufacturers. According to IRI, in the early 1990s, one-third of CPG sales came from new products, defined as those introduced during the previous 5 years. Ten years later, that figure had risen to approximately half. A recent IRI analysis of 25 categories with the greatest 5-year sales growth showed that new brands accounted for 47% of total brands and 60% of those categories' sales growth. The real challenge is to introduce food products that add a unique difference to the category from a consumer and retailer perspective. Suppliers who have taken category management seriously carefully review both existing lines and new product projects with a more stringent emphasis on adding category value. The most successful retailers today are assigning category management partners to work with them in developing optimum approaches to managing specific aspects of new product launches, including distribution, shelf space/location, promotion planning, and pricing.

The challenge for the retailer is to be able to separate those brands that add category value from those that do not and collaborate with manufacturers operating according to that discipline. Manufacturers must eliminate unproductive SKUs and focus new item activity on a category building perspective that will likely lead to fewer but more significant (i.e., with greater potential for leading real category growth) launches. However, manufacturers, especially those who are publicly held, operate under volume and profit pressure, so they must accomplish this goal carefully to avoid short-term volume hits.

THE IMPORTANCE OF FOOD PRODUCT DEVELOPMENT

New products represent one of the few growth avenues left to packaged goods companies pressured by Wall Street to increase profits. Downsizing and productivity improvement programs in the late 1990s cut costs for many big CPG manufacturers. Cutting costs can help improve the bottom line, and can also free up funds to invest in marketing and new products, but there is a practical limit to how far costs can be cut. While there are some categories that grow well above the food industry as a whole, for example, organic foods, food companies cannot typically count on category growth to drive revenue growth, nor can they rely on price increases. Many companies have been forced to take price increases due to rising costs of both ingredients and health benefits for employees, but these increases are not taken lightly, mainly because store brands represent a viable alternative to branded items in many categories. Thus, new products and new SKUs are still the best way to drive top-line growth. Brand and line extensions are still a good way to leverage consumer awareness and reduce risk and entry cost, but innovative new food products potentially reap more rewards.

In their excellent book entitled "Leading Product Development," Wheelwright and Clark (1995) argue that "the consequences of product development have a direct impact on competitiveness. They mean the difference between falling behind a leading competitor in the marketplace and being the competitor who provides leadership, compelling others to meet similar standards." The major focus of a firm, and the bulk of its assets, is tied up in how it delivers value to its customers. When a company is saddled with old products, the wrong products, or even the right products at the wrong time, the value it provides to customers, and therefore the value of the company itself, is severely limited. Thus, Wheelwright and Clark conclude, "if the firm does product development badly, its assets—particularly its equity with its customers—will wither and erode."

Product development is fundamental to the success of the organization. Wheelwright and Clark (1995) point out that "the development of a new product is the development of every aspect of the business that the product needs to be successful. And consistently, successful new products need every aspect of the business working in harmony." Moreover, product development is the means by which a company builds the competencies and capabilities that set the stage for its future. In order to provide the value its customers will seek in the future, the firm must possess the human, financial, and structural capabilities to meet future customer requirements. These capabilities are created, enhanced, and renewed via product development activities. Product development is therefore central to future success of the firm.

It is not sufficient to have development capabilities. These capabilities must be organized and activities carried out in such a way that firms minimize product development cycle time. Today more than ever, success in new products means achieving speed to market with products that solve a customer problem in a demonstrably superior way. By bringing a product to market quickly, a company can reap several potential benefits. You begin to sell sooner, leading to fewer lost sales. The operational learning curve begins to take effect sooner, leading to lower costs. Firms can preempt competition and develop a reputation for market leadership. Being first in the market helps a firm to attract customers, because there is less advertising and promotional clutter. Quick-to-market helps to create customer loyalty and allows the firm to target most attractive market segments. First-mover advantages also include the opportunity to create barriers to entry by forging contracts or good relationships with customers. Finally, getting to market quickly can reduce risk both by incorporating most recently developed technology and by gaining first access to scarce resources.

FOOD PRODUCT DEVELOPMENT AND BUSINESS STRATEGY

It is important to understand the role of new food product development in the larger context of business unit strategy. New food product development activity, first and foremost, must be consistent with and flow from the strategic direction of the firm or the business unit. The strategic plan will chart a course for the business, specify key targets and objectives, including the time frame in which these targets are to be met, and the direction or directions for achieving growth.

Strategic marketing decisions are made at four levels, although for smaller and very focused companies with only one "business" the first two levels are the same. The four levels are:

1. The organizational level
2. The business unit level
3. The product line level
4. The brand (item, SKU) level

At the organizational level, the major issue is determining what business or businesses the organization wishes to compete in, and how organizational resources are allocated across these businesses. A company like Kraft or Procter and Gamble is in multiple businesses that may differ in terms of customers served, competitors, suppliers, and so on. A company like Tasty Baking, a sweet snack food manufacturer located (and a tradition, along with cheese steaks and soft pretzels) in Philadelphia, really has only one business—sweet snacks and desserts—and for Tasty Baking, the business and the organizational levels are identical.

At the business unit level, the major issues are choosing, first, the market or markets they will attempt to serve, and, second, the product lines they will employ to serve these markets. In addition, they need to decide how to allocate the resources of the business unit across those product lines and customers. Portfolio theory, including commonly cited approaches from the Boston Consulting Group and GE-McKinsey, suggests that the most attractive markets and the product lines with the greatest growth potential, which leverage our most significant business strengths, receive the lion's share of our resources. PepsiCo, like most other companies which make snack foods and beverages, has shifted emphasis away from full fat and sugar-laden products to those with a "healthier" profile, which it terms its "better for you" and "good for you" food products, such as Gatorade and Quaker Oatmeal. On the other hand, weaker product-market combinations receive that level of investment necessary to maintain position, or, in extreme cases, exit the industry, which can be accomplished slowly, through a harvesting strategy, or more quickly, by either liquidating or selling assets.

At the product line level, the major decisions involve which specific items in the line should be carried, how broad or narrow and how deep or shallow the line should be, and how production, distribution, and marketing dollars should be distributed among the different items. Finally, the item or brand level involves tactical decisions in terms of positioning, advertising and promotion, pricing, packaging, and distribution for that particular product, using the resources allocated to the item at the product line level of strategy.

The strategic business plan specifies the objectives the company has set for the business, typically in financial terms such as return on invested assets, earnings before income taxes, and so forth These financial objectives lead to marketing objectives, typically stated in terms of revenue and market share growth. Almost inevitably there will be a gap ("the growth gap") between where we want to be and where we will be at the end of the planning period without some changes in our strategy. Therefore, the strategic plan also specifies avenue(s) for growth over the planning horizon. Typically in food companies, a significant portion of this growth will be achieved by launching new products.

It is important to note here that not all performance improvement strategies necessarily involve growth. If you start with the most basic model for measuring performance, which is:

Revenues less (expenses/invested assets)

Financial performance can be improved in one or more of three ways:

1. Increasing revenues via one or more growth initiatives
2. Reducing expenses
3. Reducing the asset base

As the industry has both consolidated and been subjected to extreme pressure on costs, firms have looked to build shareholder value and please the investment community by reducing expenses, many times involving downsizing or "right sizing," and perhaps reductions in marketing support, reducing the asset base by selling off, or dissolving divisions, brands, and manufacturing operations.

GROWTH STRATEGIES

New product development is part of business unit strategy. Traditional marketing management texts (e.g., Aaker, 1998) typically discuss new product development as one of the strategic alternatives in the growth matrix. This model posits that the two major elements of strategy are products and markets and that all products and markets can be divided into those that are current and those that are new. Using a two by two matrix with four cells, four different growth strategies can be identified:

- *Market penetration*, which involves achieving growth by selling more of our existing products in existing markets
- *Market development* (or market extension), which involves achieving growth by taking existing products into new markets or market segments
- *Product development*, the creation of new products for markets we currently serve
- *Diversification*, which involves new products for new markets

Thus, firms have several options for growth, which, of course, are not mutually exclusive. They can grow using current products in current markets through a strategy of market penetration, which might involve changing or increasing advertising and promotion, temporary price reductions, promoting new uses (e.g., demonstrating uses for soups as sauces and ingredients in casseroles), and so on. Alternatively, companies can grow by taking existing product to new markets using a market development or market extension strategy. One seemingly obvious option for market development is to take a product into a new geographical market. Facing a saturated and declining at-home soup business, Campbell's has attempted to expand their penetration in the away-from-home sector via prepared soup programs for both food service operators and food retailers, mainly convenience stores and restaurants such as Subway. Similarly, Tasty Baking has tried for years to find a way to expand their fresh sweet baked goods from its core markets in the Mid-Atlantic States to other cities and regions.

Companies can choose to focus on markets in a variety of ways. One interesting strategy for CPG companies in a "graying" America is to focus increasingly on the aging boomers as they reach empty nester status. For example, Lee (2004) discussed Pillsbury's strategy of marketing to empty nesters. He highlighted a commercial created for Pillsbury's oven-baked biscuits and rolls, featuring a middle-aged, presumably empty nester couple, with a somewhat risque slant not typical of a company that for years has featured a giggling Doughboy. Pillsbury is now part of General Mills Inc., a company that has designed products meant to reflect the lifestyles of the aging boomer, from microwaveable cinnamon roles to feed unannounced guests to

Perfect Portions biscuits, made for smaller households, and has collaborated with the American Association of Retired Persons, advertising in the *AARP* magazine.

Another market development strategy frequently employed in the food industry is transfer. A food service supplier who packages their product for sale in supermarkets, or, conversely, a food manufacturer who develops a food service package exemplifies this. Brands such as TGI Friday and Boston Market have been on supermarket shelves for some time now. Of course, firms can grow by creating new food products for existing, or at least related, markets. This is what we call new food product development. Judging from the pace of new food product activity over the past two decades, product development is clearly the growth strategy of choice for many food companies.

Authors such as Aaker (1998) have expanded the original two-by-two growth matrix by adding another dimension, based on Michael Porter's value chain, and thereby defining expansion along the value added channel, otherwise known as *vertical integration*, as a growth strategy. Firms can vertically integrate upstream by acquiring or developing its own sources of supply or downstream by acquiring or developing its own distributors. The two major motives driving vertical integration are (1) control and (2) operational efficiencies. A food manufacturer who owns or otherwise controls farm production is vertically integrated, as is a supermarket that owns its own private label manufacturing operations or its own wholesale distribution operation.

Going outside the existing scope of our current business involves a strategy of diversification (new products aimed at new markets). This diversification can be *related*, in terms of the type of customer, technology employed, production process, distribution channels, and so on, or *unrelated*, meaning entry into a business that is entirely new to the company. Diversification can be accomplished via internal development, through establishing partnerships or alliances, licensing, or via acquisition. An example of related diversification is the decision announced by General Mills Inc. in early 2005 that the company plans to sell vitamins under its familiar brand umbrellas that resonate health, Wheaties and Total. An example of unrelated diversification is Kodak buying Sterling Drug Company (which Kodak later divested.) The fact that Kodak eventually sold Sterling Drug points out the primary challenges involved in conglomerate diversification, which include an organization having to acquire or develop different core competencies to run the new business, plus the difficulties involved in merging two disparate company cultures.

Any interested observer of the food industry will readily conclude that food-processing companies have and continue to use all of these strategies for growth. One of the most common avenues for growth employed by food companies during the past decade has been expansion via acquisition and licensing of brands. In fact, it is not too far-fetched to say that one needs a "scorecard" to keep up with which company owns which brands, given the very large number of brands which have been bought and sold during the past decade.

The focal point of this book is new food products, developed internally. We do not discuss mergers and acquisitions that bring companies completely new businesses and stables of brands, such as the purchase of the Chex Cereals line by General Mills

Inc. from Ralcorp. We, however, discuss development of new food products that use brand names licensed from other companies either individually or as cobrands, such as General Mills Inc. Reese's Peanut Butter Puffs cereal. Licensing an established brand name or character offers significant potential advantages, including instant brand recognition and awareness, established equity, consumer expectation of product benefits, strong trial vehicle, and trade acceptance. Consumers must be able to make a simple and direct link between the licensed brand and the new product or category, so there are practical limits on how and in which categories brand names and characters can be leveraged.

Strategic New Product Development

According to Gill et al. (1996), there are seven Steps to Strategic New Product Development. Step one spells out where we want to go while steps two and three indicate where we are now. Steps four and five narrow down the range of options for getting where we want to go, and steps six and seven specify how we will go about completing the tasks that need to be accomplished.

The first step involves setting new product development targets and creating a product development portfolio. This is the collection of new product concepts that are (1) within our ability to develop, (2) are most attractive to our customers, (3) deliver short-term and long-term corporate objectives, and (4) help to spread risk and diversify investments. The output at the end of step one is a clearly articulated target indicating which and how new products will contribute to overall business goals.

The second step is the situation analysis, involving both the identification of key strategic elements and environmental variables, and the sources of information needed to yield planning premises. The output at the end of step two is a compilation of current information that gives a picture of the customer, the competition, and our competencies and capacities.

The third step is opportunity analysis, by which we map the strategic geography, understand the structure of markets, and delineate market gaps that might be filled. The output of this process will be a map of the opportunities. The fourth step involves identifying potential new product options that fit the strategic geography delineated in step three. The output of step four is a complete list of all new product options in easily comparable format.

The fifth step involves establishing threshold criteria that will provide minimum acceptable performance targets. Portfolio criteria allow a business to create balance and diversity in the product line. At the conclusion of step five, we have a set of both threshold and portfolio criteria that decision-makers have agreed to use in selecting the new product portfolio.

The sixth step involves creating the portfolio. This involves operationalizing the strategy determined in steps one through five by making specific line-item decisions. At the end of step six, the output is a portfolio of new product options that fulfills the new product target, addresses key concerns in the customer and competitive domains, maintains and grows corporate capabilities, and can be developed within set budget limits. The seventh step is management of the portfolio.

PRODUCT DEVELOPMENT OBJECTIVES

Fuller (1994) maintains that growth initiatives are only one of the forces driving new product activity by food companies. Companies create new products to revitalize product lines as older products reach the maturity or decline stages of their life cycle. This generally means that changing consumer tastes, changing technology, competitive dynamics, new regulations, or changes in public policy have had a negative impact on overall market position of current products and created potential opportunity for new items. Cooper (1993) identifies some additional driving forces for new product development. These include the desire to grow into a new geographic market, the goal of gaining greater local and regional market penetration, the desire to reduce dependence on what may have become commodity items and provide more added value, and the need to expand our product and business base.

Companies introduce new products as part of both offensive and defensive strategies. When market leader Gillette introduced the "next generation" shaving system, the Mach3, in 1998, Schick responded with a preemptive but defensive strategy and employed a niche marketing approach targeted to younger shavers and African Americans with skin problems who have concerns about safety and comfort more so than close shaves. Gillette's continuing research and development efforts, leading to new generations of shaving technology, such as Mach3 Turbo and more recently M3 Power, represent an attempt to keep Gillette on the "cutting edge" (pun intended) with shaving technology that creates an unsurpassed shaving experience. This is the essence of offensive strategy. Schick cannot be content to sit back and wait for Gillette to beat it to market with every new innovation, and launched the Quattro four-bladed shaving system in 2003. Innovative new products not only create a better profit mix in the product line through higher margins, but also serve to preempt other manufacturers from gaining an advantageous market position.

Food companies frequently look to expand their market position and the range of offerings available to customers. With a dominant share in snack cakes and pies in its core market areas, Tasty Baking added donuts to its product line that gave the company better positioning for the breakfast occasion. Struggling to increase volume in a saturated and competitive market, Tasty Baking has also added cookies, snack bars, and related items, but always under the umbrella of sweet baked goods.

New food products, or at least packaging variations, are created to meet the specific needs of distributors and retailers. Many manufacturers have created club-packs and variety packs to meet the needs of club stores. Tasty Baking rolled out a line of cupcakes, called Tropical Delights, in 1998 as a response to their Puerto Rican distributor who told the company he could sell snack cakes with tropical fruit flavors such as papaya and coconut. While the sales of this line did not justify retaining it in the long run, initial sales were very strong, even in some "Anglo" markets. More recently, facing an extremely negative environment due to the obesity crisis and concerns about sugar and fat consumption, Tasty launched "Sensables," a line of sugar-free snacks. Coming late in the cycle of low-carbohydrate products that were all the rage for much of late 2003 into mid-2004, Tasty Baking decided to target the product to diabetics.

Some firms add new products to reduce costs by more efficiently utilizing existing production facilities, research and development resources, and so on. This is particularly true for food processors that engage in processing and packing on a contract basis, for either distributors selling own label products or manufacturers who want to use a copacker.

Ultimately, companies create new products for a multitude of reasons that boil down to two primary reasons, which are (1) to enhance short-term earnings, and (2) to enhance the long-term value of the brand and of the organization. There are numerous motives for looking to increase short-term earnings. The first motive is the pressure managers typically have to meet quarterly (or monthly or annual) revenue and profit targets. To the extent that rational managers are being evaluated and rewarded on the basis of meeting these short-term goals, they will behave in such a manner to do so, and a line extension or a copycat or cloned product is sometimes a good way to proceed. Top managers of publicly held corporations are under obvious pressure from shareholders to enhance the value of the organization as measured by the stock price (and by dividends). These can also be improved by new product activity. Wall Street loves companies that are innovative and active in new products, which set off streams of revenue growth.

The Role of Top Management in Setting New Product Strategy

Wheelwright and Clark (1995) identify three principal ways in which senior management should be involved in new product development:

1. Set direction and get people in the organization aligned; establish, and articulate a vision
2. Select, train, and develop people capable of realizing the vision
3. Create, shape, and influence how work gets done in order to ensure that it gets done in the best way possible

Senior managers select the core team for the project; serve as source of energy to sustain the project; serve as commitment managers, influencing, guiding, facilitating, and reviewing commitments; and play the roles of sponsor, coach, and process improver. The roles of senior management are identified as:

- Direction setter, which involves envisioning the future of the business
- Product line architect, which involves determining what should be developed and in what sequence, and what should be the connections between various products that effectively build brands and business identity
- Project portfolio manager
- process owner/creator, which involves managing the development process
- Team launcher, which involves developing and approving charters for individual projects.

These charters have been designed to realize specific objectives established by the product line architecture and the business strategy; they set out the project's business purpose and provide the framework within which the project team will operate.

The Product Innovation Charter

According to Crawford et al. (2006), successfully innovative companies are characterized by (1) having clearly defined missions, (2) seeking future customer needs, (3) building organizations dedicated to accomplishing sharply focused goals, and (4) partnering with their customers in the product development process. Successfully innovative companies focus their new product efforts with a product innovation charter (PIC), a strategy statement that guides new product development by establishing the agenda and direction for the organization's new product development activities.

Wheelwright and Clark (1995) note that senior management must "lay out the product line's evolution in terms of product types, their relationship to current and future offerings, and the timing of their introduction. Moreover, they must do so in a way that fits the marketplace, the competitive environment, and the firm's resource realities." This can be accomplished with a PIC. The four issues that must be addressed as part of establishing the future of the product line, according to Wheelwright and Clark (1995), include:

- *Position*. Deciding where in its lineup and price/performance set the business will need new products. Attention is given to opportunities in the market, possible competitive moves, gaps in the product mix, and the breadth of line needed to accomplish the business objectives.
- *Type*. Deciding the types of products that will be most effective in meeting specific customer and consumer needs. How the firm chooses to meet these needs, whether through innovative new products that create technical platforms, products that build on existing technical or brand platforms, or competitive clones will have significant impact on the economics of the business and its ability to respond to contingencies.
- *Timing*. Deciding when new products should be introduced. Timing is predicated on the pace of technological change, price/performance economics, and the evolution of the market and changes in customer tastes, plus the ability of customers to learn new habits. In addition, the ability of the firm to fund development as a sustainable pace must be considered in making decisions about how to time new product entries.
- *Relationships*. Deciding how new products will be related, that is, building on the same technical platform, using common equipment or processes, distribution channels, advertising and promotional vehicles targeting the same customers.

The product development charter specifies direction for these issues and others.

According to Dimancescu and Dwenger (1996), the charter spells out the project or program constraints and empowers a team to act around explicit expectations. When a project team is assembled, one of the empowering documents is a charter prepared by senior management. The charter outlines objectives for the proposed project and specifies the constraints within which the new product team will operate.

Elements of the Charter

Six elements constitute a full-blown charter (Dimancescu and Dwenger, 1996):

1. Rationale for the project and its links to the firm's vision and strategy
2. Specific project goals expressed strategically and in financial terms
3. The context of the project by market segments being targeted, relevant competitors who need to be accounted for, technologies, and other external factors that will affect the project
4. Project process ground rules for management review at milestone checkpoints or gates
5. A precise definition of expected deliverables
6. Constraints that may affect the project, such as management assumptions, resource availability, and budgetary limits.

An effective charter is one that provides clearly articulated expectations and objectives for the project team. This will reduce the chance of a project that is not consistent with the strategic direction of the business. Additionally, "the charter is a significant element of a team-based reward system because it spells out precise objectives against which actual performance can be measured."

Similarly, Wheelwright and Clark (1995) identify six elements of a charter:

1. Reasons for a project and its links to the firm's vision and strategy
2. Specific project goals
3. Context of the project: markets to be targeted, competitors to be attacked, technologies, external factors that will affect the project
4. Ground rules for management review at checkpoints/gates
5. Precise definition of expected deliverables
6. Constraints—management assumptions, resource availability, budgetary limits

Charters are generally focused on market opportunities as opposed to specific products, and are designed to bring common focus to opportunities. Crawford et al. (2006) identifies two major dimensions with which charters are concerned: technology and the market. Opportunities are defined in terms of these two dimensions, and any opportunity has both technology drivers and market drivers. Technology drivers for food companies would include research and development, for example, fat replacement or other ingredient technology; process technology; supply chain and logistics technology; distribution and order handling technology. Market drivers include customer groups or segments, functions performed by the product, customer benefits, and usage or consumption occasions. The charter should provide specifics regarding all of these dimensions.

Synergy

In this context, Cooper (1996) pointed out the critical role of synergy that is vital to new product success. "Step-out projects" have much higher rates of failure than projects that build on synergy, of which two types are relevant:

- Technological synergy—the degree to which the project builds on in-house development technology, utilizes inside engineering skills, and can use existing manufacturing resources and skills.
- Marketing synergy—a strong project/company fit in terms of sales force, distribution channels, customer service resources, advertising and promotion, and marketing intelligence skills and resources.

Other PIC Specifications

The product innovation charter also specifies the company's posture with respect to both innovativeness and timing. Firms can choose to be pioneers, that is, first to market with innovative products. This approach promises great potential return that is fraught with significant risk. Alternatively, companies can adopt an approach of adaptation, which means early entry into the market with the intention of improving on a competitive product. Finally, a firm may choose to take the least risky approach of imitation, that is, focusing efforts on coming into the market late with copycat products, and attempting to compete on the basis of price, location, or some other variable other than a specific product leadership advantage.

The editor of *Stagnito's New Products Magazine*, Joan Holleran, pointed out in the January 2005 edition that often the best new products are built upon the platform of tried-and-true existing products. This category offers the new product developer the best chance of success. Most successful new products are adaptations of existing products that have already established a consumer franchise. Vanilla Coke, Michelob Ultra and Tide with Downy, among the most successful product launches in 2003 and 2004, all fit in this category. These are all examples of successful launches by behemoths in the industry, but strategically, one of the ways by which a small company can compete with larger competitors is by adapting a product, because the small competitor can typically react faster. One strategy is to find a category that is dominated by two or three companies and out-flank them against segments they are missing. Unless the category grows enormously, the giants will usually ignore the small competitor. When a large company enters a category with an aggressive spending plan, awareness and consumer interest rises for the category as a whole, causing the category to grow. For example, until the recent launch of the Quattro razor discussed above, Schick had been following Gillette new product introductions for years, which allow them to save the enormous costs of advertising and R & D. The downside is that adapters (followers) are almost always number two or three, or worse in the category. When you adapt a product, the change has to be both readily apparent and meaningful to the customer.

Innovation and breakthrough products

Real innovation that leads to "new-to-the-world" breakthrough products requires a firm to go beyond its current boundaries in technology to create new product and package design, new formulations, new positioning, and new benefits. This process is both very costly and laden with uncertainty. As a result, a majority of food firms are "drilling down" to a core brand emphasis. The food industry has seen a significant shift in focus to leveraging and building on flagship brands. Kraft Foods, following its acquisition of Nabisco, continued a strategy started by Nabisco to extend the franchise of Oreos, one of the classic core brands, really an icon, in the U.S. food industryfrom "Double Stuff", to seasonal colored frostings, to Oreo Cooke Barz, to mini Oreos packed in a standup flexible pouch, to mint, chocolate and peanut butter fillings, to a "white" (vanilla) cookie, Oreo has expanded from a few to several dozen SKUs over the past two decades. Similarly, over the past several years, General Mills has expanded the Cheerios franchise with Honey Nut, Frosted, Apple Cinnamon, MultiGrain, Team, and Berry Burst versions.

On the other hand, gaining access to crowded retail shelves is getting more and more difficult as it gets ever more expensive. Retailers see no gain from new items that merely cannibalize the sales of existing items, unless margins are significantly higher or there are many promotional dollars available from the manufacturer, both of which enhance the retail bottom line. Increasingly, as the retail landscape changes and increasing pressure is felt from Wal-Mart, club stores, and nontraditional formats, retailers are more carefully selecting products on their shelves to better target specific audiences. Because manufacturers have access to data on new product activity—sales, price, assortment, promotions, and so on—and the ability to respond quickly, retailers may see an increased focus on truly innovative products, rather than a continued proliferation of line extensions.

The Importance of Platform Projects

Platform projects offer significant competitive advantage, because platforms can be leveraged into an entire family of products and are viable in the marketplace for several years. The technical platform that Nabisco created in the 1990s with its research into fat replacers not only provided the ability to create Snackwells but also enabled at the same time to create reduced-fat versions of many of their flagship cookie and cracker brands such as Oreo, Ritz, Lorna Doone, and others. Hormel's "Always Tender" brand has been extended from pork to other proteins including beef, chicken, and pork. Using a proprietary injection formula which binds water after cooking to retain moisture in the meat, Hormel has created a technical and brand platform for products that meet consumer needs for convenient meals that require little or no culinary knowledge. There are several major classes of platforms: technical platforms, that is, ingredient or process technology, packaging platforms, brand platforms, and distribution platforms.

Wheelwright and Clark (1995) discuss how Coca Cola and the Coke brand name represented a single platform. When Diet Coke was introduced, it represented, in a sense, a "half platform"; it was clearly a different formula, but it leveraged advertising and brand identification from the parent brand. Then, responding to changes in the market and advances by "better tasting" Pepsi Cola, Coca-Cola introduced the much-heralded New Coke, which the company thought would be a new platform, a new-generation offering. When a major segment of the market indicated that it still wanted the old platform, the company brought back Coke Classic, the original formula, while keeping the new Coke (at least for a while). From a market leverage point of view, the company has a brand platform (Coke) that now comes in multiple formulas and variations—Coke Classic, Diet Coke, caffeine-free Coke, caffeine-free Diet Coke, Cherry Coke, Vanilla Coke, C2, Coke Zero, Coca Cola Blak,™ and so on.

Project Portfolios

New product strategy also involves managing a portfolio of projects. Medium and large companies typically will have multiple new product projects proceeding simultaneously. These projects are related to the extent that they compete for the same organizational resources and are linked with respect to timing. Portfolio theory suggests that firms will have projects at different stages of completion and featuring different levels of innovation, for purposes of balance as well as achieving both short-term and longer-term returns.

We have already seen that there is a range of innovation opportunities from continuous innovation (product reformulations or improvements) to restaging or replacing brands to new products in current categories (line extensions) to new-to-the-company categories (e.g., through category extensions) to fundamental innovation (new to the world categories.) Using the innovation spectrum, we need to consider the portfolio of existing projects with the goal of diversification, that is, not all projects of the same type. Therefore, a company such as Gillette may be investing heavily in a fundamental new shaving technology leading to the Mach3 razor and then the MP3, while at the same time reformulating certain products and launching line extensions in other product categories with the intention of achieving some quick and low-risk market hits.

Project Selection

Which ideas deserve review? Which projects should we initiate? Given limited resources plus the pressure to generate new revenues, how do we determine which projects deserve investment capital and assignment of a team. There are several criteria that should be applied. These are listed in Table 2.1.

Strategic Approaches in the Food Industry

We can identify several examples of the impact of having some type of strategic guidance for new product development. Herr Foods is a regional producer of salty snacks located in southeastern Pennsylvania. Despite the fact that Herr's must compete with Frito-Lay, and despite the trend for smaller regional suppliers to be acquired by the big national players, Herr's sticks to a strategy of building the Herr's brand, with aggressive development of new items with unique and unusual flavors such as Mexican

TABLE 2.1
Criteria for Investment in New Product Projects

(1) Does the project have a sponsor or champion?
(2) Do we have the technology, processes, finance, equipment, and human resources to support the project, or can these be acquired via licensing or partnership?
(3) Is the project consistent with overall strategic objectives of the company or division?
(4) Does the project proposal achieve key financial targets and marketing objectives? For example, our product innovation charter may state that a new product must achieve a number one or number two position in the market, or we have a pretax ROI target of 15% within 3 years of launch.
(5) Do we have both the distribution capabilities and the trade relationships necessary to get successfully the product to market?
(6) Have we defined our target customers properly?
(7) Is there a rollout plan, which includes a timetable and measurement of both retailer and consumer response?

Cheddar, Steak, and Worcestershire and Mesquite BBQ. Another regional company in the salted snack business, Wise Delicatessen Company (also in Pennsylvania), responding to all of the hysteria about health and wellness, launched a line of chips and crisps under the Wise Choice label. Some of the items are baked instead of fried; other items contain soy protein.

The "cola wars" between Coke and Pepsi have been ongoing for decades, and the "fizzing out" of carbonated beverages in favor of bottled water, isotonic beverages, fruit drinks and juices, and flavored teas, has created additional pressure on these beverage giants in the new product arena. Coca Cola has launched products such as C2 (with half sugar and half artificial sweetener), plus Coke flavored with lemon and then lime, all of which are designed "for the long haul," and a new diet beverage, Coke Zero. Pepsi has taken a different approach with a string of limited time offerings designed to create a buzz among the members of its main target audience—youths. These limited time offerings included Orange-flavored Mountain Dew Live Wire during the summer of 2004, Mountain Dew Pitch Black in the fall of 2004 and then Pepsi Holiday Spice, designed for the Christmas season of 2004. Pepsi's approach is intended to get a product out of the market before consumers tire of it and create continual news in the category. Also, limited time launches limit both the financial commitment and the risk (Leith, 2004).

Campbell's Soup is a company with a dominant share of a category which overall is flat, and is growing only in convenience types of products. Campbell's also faces a significant challenge from the Progresso Brand, which has been revitalized since General Mills Inc. acquired Pillsbury. As a result, Campbell's has created a strategy of innovation that involves several elements including seeking to understand future trends and the forces of cultural and behavioral change, testing futuristic ideas and platforms with consumers, and questioning conventional wisdom and challenging the way they have "always done things." Condensed soup will always be at the very heart of Campbell's Soup Company, but Campbell's has been forced to extend not only to soups that can be eaten on-the-go or at your desk in the office, such as Soup At Hand

and Chunky in microwaveable bowls, the company has also tapped into the "simple meals" arena with Chunky Chili, launched just prior to the 2004 National Football League season. Another new product launched as part of this strategy of innovation is Invigor8, an energy drink designed to tap into the fast-growing energy drink market while playing off the heritage of V8.

Hershey Foods represents a good example of a company, like Tasty Baking, that has to compete with bigger and financially stronger players, such as Nestle and Mars, which have substantially greater global reach in an industry that is losing share to items such as energy bars and cookie bars. As a result, Hershey has decided to move beyond candy and gain a share of the faster-growing snack market. Hershey's Swoops, kind of a hybrid between a confectionary and a snack, was one of the most successful launches of 2004. Additional new products will leverage existing brands, which include Almond Joy and York Peppermint Patties, and could include cookies, crackers, and snack bars.

New Food Product Opportunity Analysis

Strategic planning for new food product development must involve an established and organized process for identifying new product opportunities. Opportunity identification and analysis involves ongoing environmental scanning to identify emerging consumer trends and changes in consumer habits, which drive opportunities, plus a systematic procedure for evaluating whether the opportunities fit our objectives, strategic direction, and resource base at that particular point in time. New food-product opportunity analysis flows from the business plan. Like other steps in the development process, opportunity analysis must be disciplined and directed, with specific deliverables and a procedure for evaluation. Opportunity analysis also incorporates monitoring new ingredient, processing, packaging, and distribution technologies. Ingredients like Splenda have created completely new opportunities for products such as baked goods, which retain flavor profile while containing fewer calories. Breakthroughs in processing and packaging, such as the aluminum foil-lined stand-up pouch, not only give rise to new varieties of portable snack foods, but also have redefined categories such as shelf stable tuna fish.

Some major opportunity areas in the food and beverage industry circa 2005 include (see IRI's *Times and Trends Report*, May 2004):

- Better-for-you beverages
- Nutrition and energy bars
- Portable products (e.g., drinkable yogurt, squeezable peanut butter, and portable soup)
- Fortified products (e.g., mayonnaise with Omega-3 fatty acids and orange juice with calcium)
- Dinner solutions
- Soy products
- Natural and organic products
- Heart-healthy beverages such as smoothies
- Products with new and unique flavors or flavor innovations

Opportunity analysis logically should proceed from categories with which the company is most familiar and experienced to those that represent new and uncharted territory. The reason is obvious—the risk of launching a new product in a familiar category is lessened by our ability to leverage existing brand names, distribution channels, technology and production facilities, plus our knowledge of consumers and the trade. However, launching new products in present categories typically results in reformulations or line extensions that provide low risk but minimal potential return. To the extent that the category is already crowded, contested, fragmented, and saturated, the prospects for anything more than a short-term burst of new revenues are limited. In addition, it is now generally accepted that product proliferation has a downside. It creates higher production, distribution and marketing costs, out of stocks at the wholesale and retail level, and customer confusion. Recent evidence from P&G demonstrates that market share and sales revenue can actually increase by reducing the number of SKUs in a category.

Breakthrough products and substantial new revenues require locating emerging needs that the company can serve and emerging technologies that allow us to meet consumer needs significantly better than extant competitors. To find these emerging needs and technologies, and to find market areas which offer real growth potential, we often have to expand the scope of our category scanning beyond those categories in which we currently operate. The trade-off is potentially higher investment and greater risk for higher returns. If we can find categories that allow us to use our competencies in research and development, production, distribution, and marketing, we will both limit risk exposure and enhance prospects for success. Therefore, it is vital that we focus our environmental scanning on categories that meet the criterion of fit with organizational competencies.

One possible approach to new food-product search is built on the notion that competition for a brand, which is defined by what consumers perceive to be viable substitutes in a particular consumption or usage situation, ranges from very direct to much less direct. For instance, Grape Nuts competes with Corn Flakes for breakfast, but for some people and in some instances, Grape Nuts competes with toaster pastries or even a box of raisins. Broadening the search means looking for new, novel, and emerging ways to satisfy important needs. Kellogg's launched Nutrigrain Bars, because this type of convenient, "healthful" product fits breakfast demands for a particular group of target consumers.

Indicators

What factors do marketers consider when looking for new product opportunities? What causes marketers to consider a category attractive? We need to assess several factors and conditions as listed in Table 2.2.

The first key to uncovering new product opportunities is to understand the consumer, and what benefits consumers are seeking. Health and wellness have been consumer drivers for decades, but the amount of information available to consumers about the impact of diet on health has increased dramatically over the past few years. In addition, we have finally realized that there is a global crisis with obesity. Thus, many of the successful new products during the past decade have built on the emerging consumer demand for foods that have a particular health benefit, such as those

TABLE 2.2
Factors to Consider in Evaluating Market Opportunities

- Increasing interest is new product benefits
- Category size, as measured in sales
- Category growth rate, as measured by the rate of increase in sales
- Category margins
- Number and rate of new buyers in the category
- Degree of brand switching (a measure of customer loyalty and product appeal)
- Stage of the category life cycle, indicated by rate of growth in sales and nature and extent of competition
- Amount of product/brand differentiation
- Recent innovation in the category
- Innovations in related fields that have potential application in the category
- Number of competitors
- Market shares of competitors
- Technology position of competitors
- Barriers to entry and barriers to exit: access to raw materials, capital requirements, proprietary technology, economies of scale, and customer-switching costs
- Advertising and promotional expenses to sales ratios in the category
- Trade channels and access to consumers
- Seasonality of sales

containing antioxidants or other healthful ingredients, while in the early to mid-1990s, the opportunity area was low fat. Suddenly, the industry is focused on eliminating trans-fats, adding Omega-3 and Omega-6 fatty acids to our diet, minimizing the consumption of refined sugar and simple carbohydrates, and significantly adding to the amount of fiber in the diet. These are all "hot buttons" for new food product development. Hellman's recently launched mayonnaise with canola oil, which not only eliminates the trans-fat, but also adds Omega-3 to the product for a significant health benefit.

Categories that exhibit high growth and high margins represent opportunity areas. Categories in the early stage of the life cycle, with many new buyers and relatively few competitors represent opportunity areas. Categories in which the amount of product and brand differentiation is minimal, in which few recent product innovations have taken place, and in which the advertising and promotional expenses represent a low percentage of sales also offer potential opportunities. Categories without entrenched and dominant competition and without competitors with leading edge and proprietary technical advances, may represent opportunities, provided other favorable conditions exist. Categories in which it may be possible to exploit technical innovations from other categories may be attractive candidates for new product search. Generally, it is best to avoid categories with high barriers to entry or exit, unless you can leverage a significant technical or cost advantage. Access to trade channels is important, but a company lacking a particular type of distribution competence, such as direct store delivery (DSD), may still be able to collaborate with a company that possesses such a competence.

PepsiCo acquired the Quaker Oats company primarily for Gatorade, the "800 pound gorilla" of sports drinks. In so doing, PepsiCo also acquired the Quaker Oats products, including healthy snack bars made from whole grain oats. This has created a significant opportunity for Quaker nutrition bars because of the market coverage now possible with their DSD channels.

SUMMARY

Economic and competitive circumstances in the food industry make it both necessary and extremely difficult to launch successful new products. Firms must understand the role of new products in business strategy, ensure that new product projects are consistent with, and emerge from goals and strategy for a business. Having a consistent approach to new products, including appropriate organizational support, a charter to guide new product development efforts, management of a portfolio of development projects, and specific project selection criteria enhances new product success. A critical step in new product strategy is opportunity analysis, which uncovers new product opportunity areas that fit the firm's competencies and offer strong prospects for growth.

BIBLIOGRAPHY

1. Aaker, D. 1998. *Strategic Market Management*, 5th ed. New York: John Wiley & Sons.
2. Anonymous. 2004. Chipping Away, *Food Processing*, 65(7):28–33.
3. Anonymous. 2004. *Coke's 2005 Hopes Rest in New Products*, www.preparedfoods.com/ December 20, 2004, accessed January 5, 2005.
4. Anonymous. 2004. New Product Pacesetters, *IRI Times and Trends*, May 2004.
5. Anonymous. 2004. New Products and the Consumer Connection, *Stagnito's New Products Magazine*, 4(9):10.
6. Anonymous. 2004. *Product Alert*. Marketing Intelligence Service. December 27, 2004, p. 2.
7. Anonymous 2005. Build a Better Mousetrap, *Distribution Channels*, March 2005, pp. 66, 68.
8. Barrett, A. 2003. Hershey: Candy is Dandy, But…, *Business Week,* September 29, pp. 68–69.
9. Berk, C. 2004. Campbell's Profit Rises 9% as Sales Increase, *Wall Street Journal,* February 23, p. B4.
10. Berner, R, D. Brady and W. Zellner. 2004. There Goes the Rainbow Nut Crunch: To Stay Competitive, Food Companies are Weeding You Their Slow-Selling Products, *Business Week*, July 19, 2004, p. 38.
11. Brubaker, H. 2003. Whipping up innovations at Campbell, *The Philadelphia Inquirer*, November 2, p. E01.
12. Carpenter, D. 2004. *Thin Oreos? Kraft caters to low carb fad*, www.miami.com/mld/ miamiherald/business/national/September 7, 2004, accessed September 9, 2004.
13. Cooper, R.G. 1993. *Winning at New Products*, 2nd ed. Boston: Addison-Wesley.

14. Cooper, R.G. 1996. New Products: What Separates the Winners from the Losers, In *The PDMA Handbook of New Product Development.* Rosenau, M.D., A. Griffin, G.A. Castellion, and N.F. Anschuetz, eds. New York: John Wiley & Sons, pp. 3–18.
15. Crawford, C.M. and A. Di Benedetto. 2006. *New Products Management.* Boston, MA: Irwin McGraw-Hill.
16. Dahm, L. 2002. Foolproof and Flawless: Hormel's Always Tender Products Offer Consumers a Perfected Meal Solution, *Stagnito's New Products Magazine*, 2(4):18–19.
17. Dimancescu, D. and K. Dwenger. 1996. *World Class New Product Development.* New York: AMACOM.
18. Dorfman, B. 2005. Foodmakers Rush New Items, www.biz.yahoo.com/rg050226/food_ products_1.html, accessed February 26, 2005.
19. Fuller, G.W. 1994. *New Food Product Development: From Concept to Marketplace.* Boca Raton FL: CRC Press.
20. Gill, B., B. Nelson, and S. Spring. 1996. Seven Steps to Strategic New Product Development, In *the PDMA Handbook of New Products Development.* Rosenau, M.D., A. Griffin, G.A. Castellion, and N.F. Anschuetz, eds. New York: John Wiley & Sons, pp. 19–33.
21. Hill, J. 1997. The New Product Hurdle That May Cost You the Race, *New Product News*, 33(8):13.
22. Hoban, T.J. 1998. Improving the Success of New Product Development, *Food Technology*, 52(1):46–49.
23. Holleran, J. 2205. Big Brands, Benefits Spur New Product Trial, *Stagnito's New Products Magazine*, 5(1):12, 15.
24. Horovitz, B. 2005. Can't Eat Your Wheaties? Try Vitamin, www.usatoday.com/money/industries/food, March 2, 2005, accessed March 3, 2005.
25. Lee, T. 2004. Why foods fail, *Minneapolis Star Tribune*, August 30, p. 1D.
26. Leith, S. 2004. *Drinks try to sustain the buzz,* www.ajc.printhis.clickability.com/ October 29, 2004, accessed November 9, 2004.
27. McCarthy, M.J. 1997. Food Companies Hunt for a "Next Big Thing" But Few Can Find One, *Wall Street Journal*, May 6, pp. A1, A8.
28. Peterson, R. 1993. Speed Is Critical in New Product Introductions, *Marketing News*, 27(5):4.
29. Roskelly, N. 2004. Wise Choice: Traditional Snack Ventures into Healthier Options, *Stagnito's New Products Magazine*, 4(11):40–41.
30. Uhlman, M. 2005. A Trimmer Tastekake, *The Philadelphia Inquirer*. May 16, 2004, pp. E1, E4.
31. Wheelwright, S.C. and K.B. Clark. 1995. *Leading Product Development.* New York: The Free Press.

3 New Product Failure and Success

John B. Lord

CONTENTS

INTRODUCTION

New products play several roles for the organization. They help maintain growth and thereby protect the interests of investors, employees, and suppliers of the organization. New products help keep the firm competitive in a changing market (Patrick, 1997). The consequences of product development have a direct impact on competitiveness. They mean the difference between falling behind a leading competitor of the marketplace and being the competitor who provides leadership, compelling others to meet similar standards (Wheelwright and Clark, 1995). Finally, new products spread the marketing risk. The investment community values new products; new products affect the top line and therefore enhance the value of the firm and shareholder value (Patrick, 1997).

The academic and the business periodical literature is replete with detailed listings and explanations of both why new products fail and what factors are related to success. There is no shortage of guidance available to those interested in achieving the revenue growth, profit growth, and reputation for innovation and leadership associated with successful new product launches. Organizations invest many human, material, and monetary resources in new product development. In addition, much research, by both the academic and industry sectors, has been conducted regarding the factors involved in new product failure as well as success. Yet, the statistics that we frequently hear cited about product development and the rates at which new food products fail are frightening. How can these seemingly contradictory facts be reconciled?

During the 1990s, the food industry launched the Efficient Consumer Response (ECR) initiative in an attempt to find ways to increase the productivity of the food supply chain and remove excess and unnecessary costs from the system. As part of the effort to quantify and ultimately control the critical cost components of the food distribution channel, the consultants who supported ECR estimated that the excess cost to the grocery system in the product development and introduction process ranges as high as 4% of net sales. These costs include both:

- All development and introduction costs associated with failed products, including products canceled before introduction as well as products withdrawn after launching.
- Excess costs incurred in launching successful new products, principally excess manufacturing costs due to an initial massive inventory buildup needed for introductory deals and special offers, for example, free goods.

Industry data from a study commissioned by the Joint Industry Task Force on New Products and conducted by Deloitte & Touche Consulting Group in 1995 suggests that new product introductions cost the food system (manufacturers, brokers, wholesalers, and retail grocery stores) approximately $252 per stock keeping unit (SKU), per store. It is important to note that the study was conducted using 1988 data, so it is safe to assume that today's estimate, incorporating only the most modest inflation figures, would be at least 50–60% higher. And given the conventional wisdom that between 80% and 90% of new consumer packaged goods (CPGs) products do not succeed in the marketplace, it is clear that the food industry spends a great deal of money on products that are introduced but do not succeed.

GETTING THE NUMBERS STRAIGHT

One of the four key components of the ECR initiative, efficient product introduction, addresses the concern about the alarming number of new products launched each year, and the fact that most of these are line extensions. The July 1997 issue of *Progressive Grocer* (Matthews, 1997) included a supplement entitled "Efficient New Product Introduction." This report was intended to, "describe techniques for new product introduction ... advancing the understanding of distributors, brokers, and manufacturers within the grocery industry" and cited a project undertaken by Ernst & Young who provided the data cited in the report. Prime Consulting Group Inc., as part of the study, computed product introduction, success, and failure rates. While this study is somewhat dated, it appears to be the most recent comprehensive industry study of its kind.

Defining what constitutes a new product success or failure is a critical first step in computing and assessing success and failure rates. If a product concept demonstrates enough strength during early stage testing to warrant investment in product development, but fails to survive beyond product or market testing, is this a new product failure? If a new product is launched and gains retail distribution and generates revenues, but those revenues fail to meet stated targets, is that new product a failure? If a new item is launched and generates significant first-year distribution and revenues, but loses distribution and revenues after the first year, is that a new product failure? How do we account for products that are seasonal or are merely replacements for other products in our existing line?

Clearly, therefore, the industry needed to develop a consistent definition of what constitutes a new product success or failure. However, even more basic was the need to define clearly, what constitutes a new product. Is a product new, because it is new to the consumer? Most industry observers agree that a product new to the consumer is a new product. But how about a product that is new to the companyor a product that is reformulated or repackaged to create improved performance or new consumer benefits?

The *Progressive Grocer* report is notable for its specification of (1) a classification scheme for new food and allied products which differentiates new products from new Universal Product Codes (UPCs); (2) a specific definition of product failure that is used as a criterion to determine whether a new product is classified as a success or a failure; and (3) empirical data on failure and success rates. Previously, with the exception of data provided by Information Resources Incorporated (IRI) in their annual "New Product Pacesetters" reports, little actual data has been provided to either confirm or contend the conventional wisdom that four out of five (or worse) of all new products fail.

NEW PRODUCT CLASSIFICATION

The Ernst & Young project group's first task was to develop a consistent new product classification process. The basic premises (Matthews, 1997) were that "(a) the classification should reflect a consumer perspective and (b) that all new items are not ... new products ... that a new UPC is not the same thing as a new item." A new UPC is not necessarily a new product, and not all products new to a company are new to the consumer. New UPCs can be divided into six classifications as follows (see *Progressive Grocer*, 1997):

- Classically innovative—"new to the world" products
- Equity transfer—apply an existing brand name to a new item in a different category
- Line extension—a new flavor or form of an existing brand
- Temporary, seasonal, and special packages—introduced for short time period only
- Competitive entry or clones—"me too" items
- Conversion or flow through, for example, replacing the 8-ounce yogurt with a 6-ounce package

Only the first three of these classes are considered to be new products, and only about 6% of the new UPCs qualify as new products, of which 22% are new brands or segments and 78% are line extensions. Thus, the number of "really" new products may be in the range of 1500–2000 per year instead of the much higher numbers reported by organizations such as Mintel who track new UPCs. According to Datamonitor's *Productscan Online* database of new products (www.productscan.com), of the 33,000 plus new food, beverage, health and beauty, household, and pet products introduced in the United States and Canada in 2004, only 6.7% earned an Innovation Rating from *Productscan Online*. This means that fewer than 7% of all the new CPG items launched in 2004 offered at least one breakthrough benefit in formulation, positioning, packaging, technology, merchandising, or creating a new market (*Distribution Channels*, 2005).

According to the *Progressive Grocer* study, of the items introduced to the trade each year, 56% represent UPC changes on existing items; 28% are test market items and new price look-up (PLU) items; 10% are seasonal items; and only 6% represent true new product introductions. Similarly, a presentation made by A.C. Nielsen at the 1997 Prepared Foods New Product Conference included an estimate that just 7% of the new products introduced in 1996 represented innovations in formulation, positioning, or technology. There are no new data to suggest that the numbers 10 years later are dramatically different.

SUCCESS AND FAILURE DEFINED AND MEASURED

Conventional wisdom suggests that 80–90% of all new CPG items fail. According to IRI, the failure rate for new products and line extensions is about 50%, but is closer to 75% if counted by SKUs instead of brands (IRI, 2003). A study by A.C. Nielsen and Ernst and Young fixed the failure rate in the United States at about 95% and 90% in Europe (Clancy and Kreig, 2003). Kraft claims a success rate of 50%, which they indicate is one of the highest in the industry (*Advertising Age*, 2004).

Why the disparity in these reported rates of success and failure? It all depends on the definition of "failure" and of "success." This problem is not easily solved. Obviously, a new item's sales cannot be tracked unless the product is launched. IRI, for instance, does not even measure new product performance unless that new brand has met a criterion of at least 30% All Commodity Volume (ACV) distribution. The task of measuring the number of ideas, concepts, prototypes, and other forms of new products, which never see the light of the market, in order to establish a more theoretically appealing rate of new product failure, is daunting if not impossible. In the light of this challenge, the project team for the *Progressive Grocer* report established the following definition of success:

"…a new product is considered a success if it achieved at least 80% of 26-week sales per distributing store after two years" (Matthews, 1997).

Performance data are displayed in Figure 3.1. These data were derived by applying the above-referenced definition of success to both new products and line extensions, and using information provided by Efficient Marketing Services Inc., which measured sales, distribution, and sales per distributing store. Charts 5 and 6 show that sales patterns for successful products are different from those that fail. Chart 7 illustrates success rates as measured by Prime Consulting. While these data reflect what happened a decade ago, it is interesting to note that IRI, the providers of the data for the study, continues to quote the same success and failure rates.

One of the caveats mentioned in the study is that while distribution of new products has historically been driven to achieve overall ACV penetration, this may not be appropriate for new items that are targeted to specific market segments. This problem takes on more importance in an era of more and more precise target-marketing, with product variations created to serve very specific consumer (and even trade) niches in search of very specific product benefits.

NEW PRODUCT FAILURE

Hoban (1998) conducted a survey of food and allied products manufacturers and distributors to ascertain both positive and negative influences on new product success. Respondents were asked how much impact each of 11 different barriers or negative influences had on the success of their firm's new product development efforts. Results indicated that lack of strategic focus, limited understanding of the market, priorities not set or communicated, lack of financial resources, and focus on short-term profitability were rated as having the highest negative impact. Poor product quality, limited creativity or vision, and lack of support for risk-taking were also rated as high impact by more than half of the respondents.

Hoban's survey respondents were also asked to state in their own words why some new products are not successful. Answers were coded into eight main categories. The top four reasons listed by manufacturers were, insufficient product marketing, duplication or lack of innovation, lack of support during rollout, and lack of a compelling consumer benefit. The top four reasons listed by distributors included: lack of a consumer benefit, duplication or lack of innovation, insufficient product marketing, and inadequate market research (Hoban, 1998).

Lots of food products get to market, obtain early distribution, and fail to sustain it. Before too long, distribution is waning, and the product is being "swapped out" for another new item. Some notable failures from big companies during the past few years include Nabisco Ooey Gooey Warm 'N Chewy Chips Ahoy microwaveable cookies,

(a) What classifies a product as new?
New item classifications (New UPC # new item)

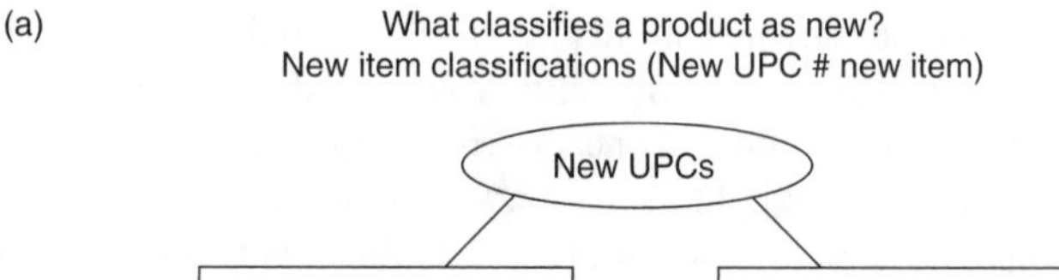

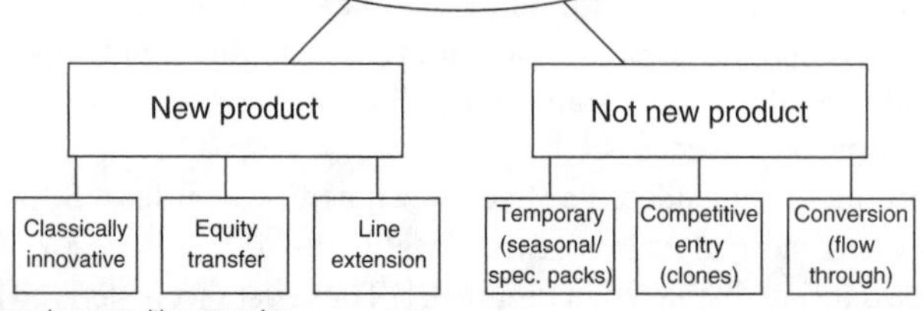

Source: *prime consulting group Inc.*

The first step in the research process was to define a new product. Surprisingly, the study participants were unable to discover any commanly agreed to industry definition of new products. So the group's first task was to develop a consistent new product classification process. Two basic premises guided that effort: The classification should reflect a consumer perspective; and all new items were not, infact, new products–in other words, a new UPC is not necessarily the ssame thing as a new procut, The Classifications that did qualify as true new products were: classically innovative products, equity transfer products, equity transfer products and line extensions. Those that did not qualify as new products were: temporary items (such as seasonal products and special packs) competitive entry items ("clone products") and conversion items (flow through products/substitution items) for more on the classifications, see "Efficient new product introduction shattering the myths."

(b) How many new prodcuts are introduced?

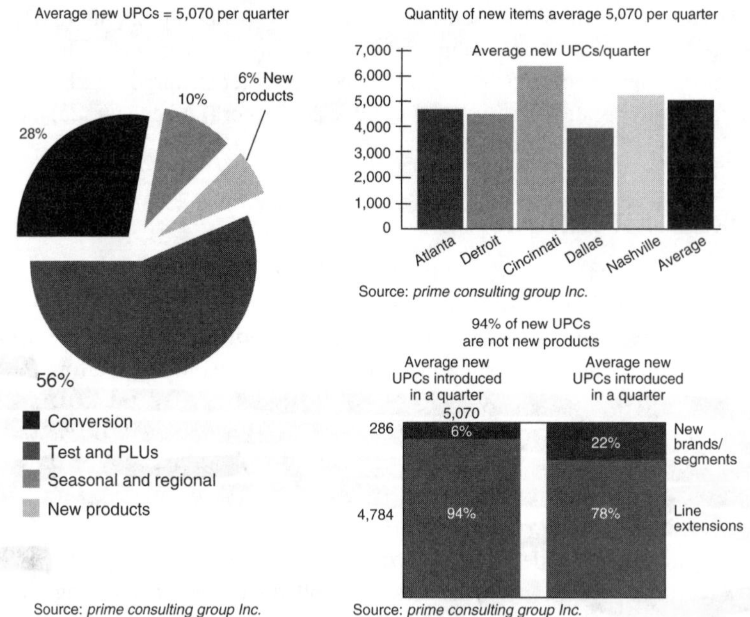

Source: *prime consulting group Inc.* Source: *prime consulting group Inc.*

To get a bandle on exactly how many new products are introduced, all new UPCs were tracked by quarter over a two-and-half-year period in 500 and-half-year period in 500 Kroger stores in five markets (Atlanta, Detroit, Cincinnati, Dallas and Nashville). The numbers ranged from 4,000 to 6,000 new UPCs, 56% were conversion items, 28% were test-market products and new PLU items, 10% were seasonal and regional products, and only 6% were true new products. Of those new products 22% represented new brands/segments and 78% wwere live extensions.

FIGURE 3.1 (a) What classifies a product as new? New item classifications (newUPC # new item) (From Anonymous, In *Efficient New Product, Introduction*. Supplement to *Progressive Grocer*, 1997, 44–46. Reproduced with permission from Progressive Grocer, July 1997, pp. 44.); (b) How many food products are introduced? (From Anonymous, In *Efficient New Product, Introduction*. Supplement to *Progressive Grocer*, 1997, pp. 44–46. Reproduced with permission from Progressive Grocer, July 1997, pp. 45.)

Snackwells Cereal Bars, Doritos Thins, Cottonelle Fresh Rollwipes, and "America's first and only disposable, premoistened wipe on a roll" from Kimberly-Clark. Why did these, and numerous other products fail? Crawford et al. (2006) has identified a fairly complete set of reasons including poor planning, poor management, poor concept, and poor execution. Virtually every reason for failure falls into one of these categories.

Poor Planning

Poor planning incorporates issues such as developing a product that does not fit a company's strategy, competencies, and distribution strength. Bic makes great disposable pens and razors, but not perfume. Anheuser-Busch felt it could make Eagle Snacks a success because snacks are complementary to beer and can be distributed using the same channels, but snack manufacturing and distribution relies on different competencies than beer brewing and distribution. Poor planning also includes failure to analyze the market properly to understand whether and what type of opportunity may exist in a category and what specific unsolved problems consumers have. Also included in this category of reasons for failure is not understanding the stakes of the game, that is, the cost to enter and sustain a position in a given product category. Finally, products can fail because firms did not perform due diligence in discovering patent and copyright issues or understanding the potential impact of regulations such as those governing the use of certain types of advertising claims or labeling information. These failures are all attributable to poor planning.

Poor Management

Poor management is all about the organizational culture, support and resources for new product process, management's expectations and focus, the process used to develop products, and the process of deciding upon which projects are funded and which are not. New product development inherently involves investment, risk, and rewards that may not be immediately evident. Firms that do not encourage entrepreneurial behavior will not observe any. Firms that discourage or penalize taking risks will find that no one in the firm is willing to take risks. Firms that do not provide adequate investment capital can never achieve big winners. Firms that do not put the best people, in effective cross-functional teams, on new products will succeed only rarely. Firms that demand immediate returns will have only limited successes with line extensions but no innovation. Projects which lack the individual—the new product champion—willing to see the project through and are willing to fight the battles to obtain necessary resources will wither for lack of support. Organizations lacking clear goals and direction for new product efforts will find a real difficulty in a achieving a coherent and coordinated effort which is both effective and efficient. The bottom line is that firms that do not invest in innovation would not experience any innovation.

Poor Concept

A poor product concept lacks a compelling consumer benefit, is a simple me-too item with no real and relevant difference from items already available, does not have a defined market target with adequate sales potential, or is introduced at the wrong time. Products lacking a significant consumer benefit as well as competitive clone items simply do not provide consumers a reason to buy. The trade may take on such an item if the up-front financial incentives are adequate to pay for the shelf and display space, but such products cannot sustain their market position and will quickly lose distribution.

Poor concepts are also those that lack appropriate positioning. A product's position is the place it holds in the consumer's mind, relative to the positions of other brands in the category, and based on the characteristics and benefits that consumers associate with the brand. What the marketer says about the product, how it is priced, packaged, merchandised, and how it performs create a product's position. A poor position is one which associates benefits with a product that are not relevant nor important to the consumer, or alternatively one for which the consumer is unable to perceive a meaningful difference between brands. In either case, the consumer sees no reason to buy the product.

One of the big issues in food product distribution is that of variety versus duplication. A new product possessing unique characteristics and benefits adds to the variety offered by a retailer and available to the consumer. This uniqueness adds to the appeal of the product to both the trade and the household consumer. On the other hand, me-too products, which merely duplicate products already on the shelf, add to shelf clutter and to consumer confusion and serve merely to further fragment the product line. Some have argued that too much choice in a category actually reduces consumer economic utility by making it more difficult to ascertain and acquire the correct item. Moreover, fragmentation of the product line adds operating costs in the areas of production, distribution, and inventory control, and can threaten the existing product base by increasing potential stock-outs and reducing on-shelf visibility.

Poor Execution

Poor execution covers a multitude of sins. Included in this category are products which fail to deliver on the promise, that is, they simply do not perform to customer expectations as well as products that simply do not work as intended or expected. Food products that lack flavor or texture, packages that are difficult to open or do not dispense properly, and products which deteriorate before going out of code are all examples of products which do not execute correctly.

Poor execution extends to all areas of the marketing program. Failure to properly execute the sales plan and achieve targeted levels of distribution can be fatal to a new product, as can failure to achieve appropriate display and retailer support. Advertising, which is of insufficient weight to build awareness and trial according to plan, is poorly scheduled relative to seasonal sales factors or when the product is available, fails to reach the target consumer, or fails to communicate clearly the new product's brand name, benefits, and availability are all examples of poor execution. Lack of proper promotions to incent consumers to try the new item will lead to failure. Products that are mispriced or have a poor price–value relationship, as perceived by the target customer, are likely to fail.

Products that are introduced too late, either because they have been preempted by competitors or because the market has moved on to something else, are poorly executed. Responding to the mania surrounding Atkins and South Beach diets and low-carbohydrate foods that seemed to be all the rage, several food processors launched new low-carb products in 2004 that, by the time they got to market, were too late to take advantage of what turned out to be mostly a fad. Products that are introduced before their time are poorly executed. A classic example is the late 1990s launch of Intelligent Cuisine by the Campbell's Soup Company. "IQ" was a product line with a very strong concept and rationale, but the idea of a medically prescribed, fully prepared diet designed to remedy specific medical conditions such as high cholesterol and hypertension, proved to be ahead of its time. Of course, another execution issue with Intelligent Cuisine is that Campbell lacked the competency to market through physicians who were supposed to have recommended the IQ diet to their patients.

ACHIEVING SUCCESSFUL NEW PRODUCT LAUNCHES

Matthews (1998) reported on a food industry task force that was commissioned in 1997 to study new products and how they function in the industry. The work of the task force has resulted in some key findings. Only one in 20 U.S. brands experience increasing sales in their second and third year in the market. There is a direct relationship of consumer trial to sales volume. Sales volume for new brands seems to be almost completely dependent on sustained advertising, and a reduction of advertising after the introductory period almost invariably leads to dramatic sales declines. Study results indicated that 69% of new products have sales declines after the first year. As a result of their research, the task force concluded that there are four basic consumer truths associated with new products:

- The product needs to deliver on the concept promise. Products with high consumer acceptance, as measured by after-use purchase intent, are likely to succeed. Those with low acceptance are likely to fail.
- Advertising quantity and quality matter. Advertising drives awareness, which drives trial, which drives sales.
- Distribution drives sales.
- New brands need long-term support if they are going to succeed.

Understanding what truly drives new product success requires evaluating new product activity from the point of view of the entire food distribution channel. From this vantage point, other "truths" about new products are: (1) a new item or line needs to add incremental dollars and profit for both the category and the brand; (2) new products should enhance the manufacturer's and the retailer's competitive position in the market; (3) launches should be carried out with minimum disruption to the distribution system as a whole; (4) a launch should be accompanied by both a product service plan and a consumer target plan defined by both the manufacturer and distributor; and (5) there should be an identification of activity-level costs (Matthews, 1998).

TABLE 3.1
New Product Success Requirements

(1) An appropriate organizational environment and top management support
(2) Use of a disciplined new product development process
(3) Dedicated development teams plus the willingness and ability to partner and outsource
(4) Product development activities which start with and flow from business unit strategy
(5) Understanding the environment and identifying the opportunities
(6) Identification and specification of what is driving the consumer and what the consumer wants
(7) Processes and techniques for keeping the pipeline filled with a wide variety of new product ideas
(8) Clear and focused product definition early in the process before development work begins
(9) A superior and differentiated product and package
(10) Use of research to measure reaction to the product and all elements of the program throughout the development process
(11) Use of category management philosophy to align manufacturer and retailer focus on the consumer
(12) A well-executed launch from procurement and production to distribution and selling to advertising and promotion to monitoring in-market results
(13) Ability to adapt, grow, and improve as market and competitive conditions evolve

While useful to the industry, these findings are incomplete, because the focus is on the back-end of the development process, specifically, new product launch. Cooper (1996) reported the results of various studies that concluded that the most important discriminators between new product winners and losers were, in rank order:

1. Understanding of user's needs
2. Attention to marketing and launch publicity
3. Efficiency of development
4. Effective use of outside technology and external scientific communication
5. Seniority and authority of responsible managers

These factors have much more to do with the front-end of the process. Wheelwright and Clark (1995) maintain that the entire range of functions and activities, every dimension, of the organization drives the success or failure of product development. The development of a new product is the development of every aspect of the business that the product needs to be successful. And consistently successful new products need every aspect of the business working in harmony.

Thus, it may be instructive to construct a listing of new product success requirements that integrates the findings of numerous industry experts and observers and will provide guidance to firms looking to enhance their success in new product development. These success requirements can be developed into a 13-point listing, which is shown in Table 3.1.

An Appropriate Organizational Environment and Top Management Support

An organizational environment that is right for successful new product development is one in which there is a nurturing of innovative thinking, such as the case with a leading innovator like 3M, and a culture that rewards creativity. Firms must also support risk-taking and take a long-term perspective, which involves supporting new products through the challenging "early years." Firms must recognize that while there are corporate hurdles for any investment, really new products often show losses during the first year to third year because of high introductory marketing costs, and this effort must be sustained. Organizations with a strong track record in new products invest in continued innovation. Organizations must clearly communicate that it is acceptable to take risks and even make mistakes or else they will clearly wind up with a "nothing ventured, nothing gained" type of situation. Finally, firms should commit to putting their best people on new products.

Cooper (1996) discussed the importance of top management support in driving projects to market, that is, seeing the project through to completion. Top management's main role in supporting innovation is to set the stage: commit to a game plan and make available the right resources. Top management must define project boundaries, missions, and charters. Top management must pick the right kind of teams and create job opportunities and career paths that result in qualified team leaders and qualified core team members, and that support individuals when, where, and in quantities needed. Top management must also ensure that teams are not only effective in achieving their individual project goals, but also fit well with the overall strategy of the business (Wheelwright and Clark, 1995). Good managers use objectivity to deal with failures, correct problems, and celebrate successes.

One of the most important characteristics of successful product development is achieving speed to market, but with quality of execution. The right management approach to new products can help achieve this. For example, Pillsbury launched Totino's Stuffed Nachos in only a few months, as opposed to the projected one year. Management realized that in order to be in full distribution prior to January, the top snacking month, the product had to be ready for shipping by late summer, and they accomplished this by telling the development team for Stuffed Nachos that the management group would assume all risks associated with the accelerated timetable, allowing the development team to focus on the task at hand. The team delivered (Dwyer, 1997). Tasty Baking took Sensables to market in 6 months, because top management made the commitment to support the project, accept risk, use outside ingredient suppliers, and let the team run with the project (Uhlman, 2004).

Use of a Disciplined New Product Development Process

Successful product launches involve a number of different activities and an increasing amount of investment as the project proceeds closer to launch. A disciplined process with defined stages and gates such as the Arthur D. Little process (Rudolph, 1995) (Chapter 4) provides the basis for objectivity in decision making, because deliverables and metrics are defined for each stage. Such a process helps to ensure that investment capital will be expended wisely and on the best projects. Cooper (1996), the originator of the stage-gate process, notes the importance of the quality of execution of technological activities such as preliminary technical assessment, product development, in house product or prototype testing, trial or pilot production, and production start up. There is a need for completeness, consistency, and proficiency, which is much more likely with a well-planned and disciplined process for which all constituent activities are defined and sequenced at the front end.

A disciplined process should contain realistic but aggressive timetables. Being quick to market with a product that does not deliver on the promise is perhaps worse than getting to market after a key competitor. While there are definite advantages, chronicled above, in reducing development cycle times, launching a product that does not perform to the company's or the consumer's expectations is surely a path to nowhere.

Dedicated Development Teams Plus the Willingness and Ability to Partner and Outsource (see Chapter 7)

The use of cross-functional teams encourages communication across functions and enhances speed to market. Interaction among team members who are colocated and share a common set of objectives facilitates the implementation of projects. Dean Foods fundamentally changed and revitalized the milk category with the launch of Milk Chugs, which led to the launch of flavored milk in upscale and novel packaging by a number of dairies throughout the United States. The firm's ability to mobilize into teams allowed it to scale high hurdles during the two-year development cycle. A flat organization such as Dean Foods uses cross-functional teams so that individuals can contribute from their own perspective, then cross over and contribute from the perspective of other functions. The team approach allowed engineers to come up with the best and economical choice for machinery, with finance determining how to pay for equipment and marketing to create programs so that volume projections could be met.

Cross-functional teams make it easier to marshall necessary resources from the many disparate functions required to launch a new product successfully. At Nabisco, now a part of Kraft, a cross-functional new products team generally consists of 8–12 people from commercialization, marketing, product development, engineering, quality assurance/quality control, packaging, sensory services, logistics, and finance, and may include a representative from a supplier in the very beginning of the project. The team is almost certain to interface with the company's legal, regulatory, and public relations personnel as well as with top management. In addition, every new product project needs a champion and a sponsor to compete for scarce organizational resources and see the project through to completion. At Nabisco, each team has one person assigned in the following roles: sponsor, team leader, phase leader to motivate and manage the project.

The experience that Schwan Food Company had taking Red Baron Pizza Slices to market represents a good example of cross-functional teams at work. The unique attributes of the product, including proprietary crust, triangular shape, and microwave susceptor were the result of cutting edge technology developed by cross-functional teams in regular communication with each other, plus strong relationships with Schwan's packaging and equipment suppliers. The cross-functional team included members from R&D, process engineering, sensor, regulatory affairs, culinary, and applied research. Other disciplines including sales and marketing, manufacturing, engineering, and packaging services were also involved. The team assembled early in the process and discussed the concept and opportunity openly and in detail, so that each team members was on board, up to date, and aware of design criteria and his or her respective responsibilities in delivering against project expectations. (Roberts, 2003).

As a strategic response to intensifying global competition, food processors are reassessing and redefining core competencies, then downsizing and restructuring to focus on these competencies that add value to the business. Core competencies increasingly are being focused on consumer marketing and product innovation, not plant and process design. As a result, while most medium to large-sized companies have at least the minimal in-house capabilities to generate new products, few possess all of the necessary skills to develop the new products efficiently, which today's market demands. This situation creates the need to outsource certain new product development activities.

Most food processors lack the scale to incorporate every type of function and expertise necessary to successfully create, develop, and launch new items, so it is the norm for many food companies to see to take advantage of supplier expertise in the realms of ingredients, equipment, and research/laboratory facilities. Services provided by suppliers may involve one or more of chemistry, microbiology, food technology, processing (including scale-up), or sensory evaluation/market research (Morris, 1996b). Some manufacturers are partnering or joint venturing with each other and with suppliers to retain core competencies, pool resources so they can access new markets or new technologies, cut costs, leverage manufacturing resources, and optimize assets and gain a competitive advantage (Morris, 1996a).

According to Ferrante (1997), there are four basic reasons to turn to copacking:

- When there is a short term need for emergency or additional capacity
- To leverage technology and expertise that the company does not own internally
- To gain speed to market; to expedite a product line to market with low risk by using a third party manufacturer
- To reduce up-front costs and investment (capital investment, fixed and variable costs, transportation of ingredients, supplies, finished product, and so on)

Similarly, Berne (1996) maintains that outsourcing is useful in certain circumstances. These include when firms:

- Do not have the internal resources or time to accomplish all necessary development work
- Do not have the needed technology in-house
- Have a blue sky idea so far out they do not know where to start
- Have had core competencies weakened due to downsizing
- Need some expert advice

According to Demetrakakes (1996), the biggest reason for the increased reliance on outside help is money. The food industry clearly lags other industries in spending on research and development. A study conducted by PRTM Consulting and reported in *Prepared Foods Magazine* (June 2004) indicated that R&D spending typically represents only 1% to 2% of revenues, far less than other industries. As a result, firms do not have all of the research nor the development capabilities they need, and outsourcing becomes key to new product development. During the past several years, for example, firms who provide flavorings and ingredients have found that they must be consultants as much as they are suppliers to food processing companies, providing basic food science and technology support.

There are four basic reasons why Kraft turns to copacking: (1) a short term need for additional/emergency capacity, which may result from a seasonal spike in sales or an internal manufacturing issue; (2) to leverage technology and experience which Kraft does not own internally; (3) risk management and speed to market (to avoid the risk of investing in new equipment when someone else already has the needed equipment and the market outlook is uncertain and to reduce cycle time; (4) cost, because a copacker can occasionally produce for less (Ferrante, 1997).

PRODUCT DEVELOPMENT ACTIVITIES MUST START WITH AND FLOW FROM BUSINESS UNIT STRATEGY

At the business unit level, strategy involves choices of products and markets, and how product-market combinations will be exploited for achieving business objectives. Product development cannot be isolated from overall business unit strategy; on the contrary, it is essential that new product efforts emanate from a clearly articulated set of objectives and strategy for business growth. Consistent with the notion of building on and leveraging core competencies, this "rule" stipulates that our new product efforts must be clearly focused and must start with those business processes and assets that are the bases for the firm's competitive advantage.

Hawkins (1996) states that one of the "eight ways to win in the marketplace" is building new products on the company's strongest assets, which might be a technology platform, a particular production process, or brand equity. According to Cooper (1996), firms should attack markets from a position of technological strength, and elements of technological synergy, including fit with research and development, engineering, and production skills should be critical criteria in the prioritization and evaluation of new product projects. Some basic rules for new product success include developing a sound strategy that leverages core capabilities and clearly differentiating new products from those of competition through carefully conceived, developed, and designed brand identities.

Cooper (1996) also notes the need for marketing synergies, which means a strong fit between the needs of the project and a firm's marketing competencies and assets: brands, research and market intelligence capabilities, distribution channels, sales force, advertising resources, and so on. When firms launch products that sell into markets that are familiar to the firm, and which utilize a firm's existing distribution channels and sales force, prospects for success are heightened considerably.

We must also consider financial and human resources since both are quite obviously essential aspects of new product development. Projects must be both adequately funded and properly sourced with key people from a variety of disciplines. The number, type, and timing of projects must be set with reference to available financial and human resources.

UNDERSTANDING THE ENVIRONMENT AND IDENTIFYING THE OPPORTUNITIES

One of the most important guidelines for new product development voiced by both experts and practitioners is the need to consider the attractiveness of the market. Opportunities for successful new products arise when firms can leverage their technical, marketing, financial, and human resources against marketplace gaps. These gaps represent combinations of benefits sought by groups of consumers who can be identified, reached selectively, represent a segment of adequate size to represent market potential, and who have not found an existing product that provides these benefits. What creates a gap? A gap is created by changing consumer tastes, which are in turn driven by changes in the marketing environment. Changes that have significance for marketers are primarily demographic, socioeconomic, and lifestyle changes, which drive changes in the way consumers eat and the types of products that fit their consumption system. A good example is how the recognition of a global crisis in obesity, coupled with our desire for a weight loss "magic bullet," led to the low-carb hysteria of 2003–2004, as seemingly overnight millions of people wanted to reap the weight loss benefits of eating mainly protein and fat. Only a year later, the emphasis has turned to consumption of whole grains and nutrients such as antioxidants, soy protein, and omega 3 fatty acids, and tracking the glycemic index.

In addition, advances in ingredient, process, and packaging technology, and changes in the regulatory environment affect both the manner and the economics of delivering relevant consumer benefits, and provide food processors with capabilities of delivering taste, convenience and health benefits in ways previously not practical or even possible. New product success rates improve significantly when companies monitor the environment and become aware early on of changes that create marketplace gaps and define attractive markets.

Market attractiveness is a strategic variable of increasing importance. Attractive markets have at least some of the characteristics listed in Table 3.2.

Identification and Specification of Consumer Drivers and Consumer Wants

Unsatisfied consumer needs represent potential new product opportunities. Thus, effective and successful new product development has to start with consumers. Having identified the market opportunity, the next step involves studying and understanding the dynamics of the consumers in the category, using extensive market research and working closely with customers/users to identify customer needs, wants, and preferences, and to define what the customer sees as a "better product." According one prominent industry executive, "effective new product development means extensive research to understand the consumer plus intuition. You must understand why people

TABLE 3.2
Characteristics of Attractive Markets

- Sufficient size (bigger is better ... other things being equal)
- High growth rates (more for everyone, so less competition)
- Few barriers to entry, such as patents, very high advertising to sales ratios, restricted access to raw materials, and so on
- Few barriers to exit, such as restrictive service contracts with customers or fixed assets with no alternative uses
- Few recent product or technical developments in the category (meaning that the category may be ripe for something new and different)
- Technical breakthroughs that might be leveraged to create significant new benefits in the category (stand up foil packages from beverages to sweet and salty snacks to tuna to rice)
- Categories that are not dominated by a strong competitor or set of competitors (unless you want to be different, like Jones Soda and not compete with the big guns)
- New emerging needs driven by demographic and lifestyle changes.

do what they do and say what they say. This is where data and consumer research ends and intuition begins." (O'Donnell, 1997).

Consistent and long-term success in the marketplace cannot be achieved unless you truly understand your consumers—their problems, wants, needs, preferences, and even their aspirations. This requires classic market research, which will yield data and insights about the market in the aggregate. However, astute marketers recognize that truly understanding your customers requires that you go beyond traditional surveys and data collection and engage your customer on an individual basis. This can be done in a variety of ways, including depth interviews, focus groups, and listening to and analyzing customer 800 number or hot line calls.

Traditional research is useful for uncovering and defining customer's underlying needs and motivations, preferences, likes and dislikes for current products, and changing tastes and habits. Quantitative research is also useful in quantifying market opportunities by providing marketers with such information as market size, market structure, and awareness and usage patterns in the category. Quantitative research also can be used to develop demographic and lifestyle profiles plus media usage patterns that can be correlated with product usage to yield valuable market insights. However, this is no longer sufficient; you have to get out in the field and talk to your customers as they shop and in their homes.

Ethnography

Increasingly, food companies, like other businesses, recognize the value of ethnographic research and are incorporating this approach into more traditional marketing research methodologies. (The following description of ethnography is drawn primarily from www.ethno-insight.com, the Website of Ethnographic Insight Company.) Ethnography is a holistic technique that goes beyond the limitations of focus groups by taking marketing research outside the laboratory setting. Ethnographers observe, interview, and videotape people in the context of their everyday lives: where they work, live, shop, and play.

Ethnography is the study of people in their natural or "native" environments and requires a well-trained researcher, skillful in immersing him or herself in diverse environments, cultures, and populations; in establishing rapport with people in these social contexts and in interacting with them through participation, observation, and dialogue to uncover their attitudes, beliefs, perceptions, and values, as well as the unspoken cultural patterns that shape behavior.

For food marketing research, ethnography involves the study of individual consumers or members of a household or other small group of subjects in their homes and as they shop—in their own environment. Rather than looking at a small set of variables and a large number of subjects ("the big picture"), the ethnographer attempts to get a detailed understanding of the circumstances of the few subjects being studied. Ethnographic accounts are descriptive and interpretive; descriptive, because detail is so crucial, and interpretive, because the ethnographer must determine the significance of what he or she observes without gathering broad, statistical information. Observational research yields surprising insights into how people use products and services and what improvements or innovations are needed.

In a marketing research context, ethnography is used to uncover, interpret, and understand the consumer point-of-view and the hidden rules of environments. Whereas focus groups and surveys rely on self-reporting and memory out of context, ethnography provides a holistic view of consumers in the context of their daily lives. Consumers do not interact with products in isolation; they are affected by changing family patterns, unseen cultural factors, and other products and objects in the proximate environment. Ethnographic research is the best means for getting at these unspoken cultural and social patterns that shape consumer behavior. Ethnography can be used as a stand-alone technique or can be used in conjunction with other qualitative and quantitative marketing research techniques to provide a real-world understanding of consumer preferences, motivations, and needs. Such insights translate into strategic business opportunities, including improved customer loyalty and increased competitive advantage.

A great example of a new product that was clearly driven by an understanding of the customer is packaged precut (fresh cut) salads. Consumers want to eat healthy, at least some of the time, but lack both the time and the energy to wash, peel, cut, core, dice, and slice several different vegetables and fruits to make up that healthy lunch or dinner. Understanding the consumer helped to create a product idea that revolutionized the produce category. The same technique has been applied to fruit. The revised United States Department of Agriculture (USDA) food pyramid, which greatly increases the recommended intake of fresh fruits and vegetables, should enhance the prospects for selling even more fresh produce, as long as it is convenient and tastes good.

Keeping the Pipeline Filled with a Wide Variety of Product Ideas

While there is typically no shortage of new product ideas, they do not appear magically. Firms must have a systematic process and set of techniques for uncovering new product ideas. Ideation cannot occur only periodically or on a pure project-by-project basis. Through competitive intelligence, market research, and creative techniques, firms need to consistently identify and screen a large multitude of new product ideas.

At Campbell Soup Company, ideas may come from several sources, ranging from global symposiums looking at culinary trends and "flavor potential opportunities" to what the company calls its "grass roots" idea campaign, an Internet site where employees can put ideas into what is known as the "sand box," a company initiative that gives employees the opportunity to test their ideas to see if they fit any consumer needs. (see http://knowledge.wharton.upenn.edu.) Most good new product ideas come from consumer insights, the understanding of what is driving consumer needs and choices in the marketplace.

Del Monte invested over $20 million in research and development in fiscal 2004, mainly to maintain a steady supply of fresh new product ideas that will help deliver the kind of growth that the investment community demands (Lindeman, 2005). Del Monte employs "innovation tables" across all of its product lines. This technique allows researchers to get free time on a periodic basis to develop their own ideas and build prototypes, similar to the "bootlegging" technique at 3M Company, and both cultivates and encourages innovation. Ideas that get sufficient support lead to concept development research and economic analysis. For really big ideas, management has created a new ventures team, which ultimately recommends whether Del Monte will invest in the project (Lindeman, 2005).

One of the keys to success for Kraft is the process of exploration, whereby Kraft identifies literally hundreds of new product ideas. For every 50 new product ideas identified, approximately 6 survive initial screening. For every 25–30 concepts screened, 5 concepts are refined, 3 developed, 2 go to in-home use and 1 to test market. These numbers emphasize the need to be very productive in the ideation process, which is a function of developing a superior understanding of the consumer. Ideation also requires both a good invention system that encourages and rewards creativity as well as excellent market intelligence to uncover what is happening in the marketplace. Great ideas require breaking away from the traditional and letting wild ideas fly without succumbing to the temptation to evaluate or judge.

Many sources of new product ideas

Beyond the traditional and not-so-traditional ideation techniques of problem detection, brainstorming, and others, firms must also keep in mind that successful new products can often come from other parts of the food industry, including foodservice and niches such as gourmet and health foods. Many food companies now have a research chef on staff, whose responsibilities include monitoring trends in the culinary arts, translating consumer insights into "truly wondrous dishes with unsurpassed eye, nose, and mouth appeal" and creating and refining products that meets or exceeds target consumer expectations. (Thomas and Brody, 2004) (Chapter 13), Successful new products also come from creating new categories by shooting the gaps between products. Combination products that solve consumer problems include combination glass and household cleaner, analgesics and cold medicines in the same pill or liquid, and so on. New categories can also be created through value-added bundling, for example, Healthy Choice Salad Bar Select. Sometimes ideas previously rejected might merit consideration if market or competitive conditions have changed. According to Dornblaser (1997), "Although there seem to be almost no truly new product ideas, sometimes products hit the market at the wrong time, marketed the wrong way, or without the needed support. That's why a failed product concept from the past may be a success in the present." It is interesting to note that while Pepsi and Coke launched Edge and C2 (both containing) half-sugar and half-artificial sweetener to take advantage of the low-carb trend in 2004, Pepsi had a product called Pepsi Light for a short time in the early 1970s with exactly the same attribute.

Clear and Focused Product Definition Early in the Process Prior to Development Work

Cooper (1996) notes that a pivotal step in the new product process is early definition, prior to the development phase, of the target market, product concept and positioning, and the consumer benefits to be delivered. A fully articulated product concept represents a customer "wish list," which is in turn a protocol for product development and gives the development team a set of requirements that guides technical development, package development, process development, equipment specifications, and preparation of the marketing program. Without this focused and agreed-to definition, the prospects for an efficient and effective process that minimizes time to market are adversely affected.

Pillsbury is now part of General Mills. Its motto (Dwyer, 1997) is to "test early, fail early, and eliminate options that won't work." Initial screening, preliminary market and technical assessments, a detailed market study, and an early stage business analysis are included among predevelopment activities that are pivotal to new product success and must be built into the process. Similarly, Kraft (Fusaro, 1995) includes among its principles for new product success early development work on positioning, which includes specification of the target audience, key benefit(s), point of difference, and source of business.

A superior and differentiated product and package

Ultimately, we have to give the consumer a product that promises and delivers meaningful, unique, and superior benefits. Kraft achieves new product success by developing a product bundle that solves an important customer problem in a demonstrably superior way. The product itself—its design, features, advantages, and benefits to customer—is the leading edge of new product strategy, according to many experts. According to Cooper (1996), six items comprise a superior product: innovative, possessing unique features, meeting customers needs better, delivering higher relative product quality, solving the customer problem better (than competitive products), and reducing total customer costs. The message is that products succeed in spite of the competitive situation—because they are superior, well defined, and executed well, and has certain synergies with the firm. Cooper (1996) argues that perhaps there is too much preoccupation in today's firms with competitive analysis and not enough focus on delighting the customer.

Innovation is important to new item success and gaining the interest of retailers. According to Mark Baum of Grocery Manufacturers of America, "in the past few years consolidation in the retail industry has resulted in a shorter trial period for new products to prove their viability. Where a new product used to have 9–12 months to secure an acceptable threshold of distribution, now a new item may be given

3–6 months, and true product innovation and products that are radical departures from what is standard can be the key to securing shelf space with given retailers" (Dahm, 2002b).

Some recent examples

The success of Frito-Lay's Baked Lay's salted snacks (Matthews, 1997) highlights the ability of a new brand with a significant and meaningful point of difference to drive top-line growth. Baked Lay's built upon a consumer trend toward eating "better-for-you" foods and delivered a key benefit, lower fat, without sacrificing taste. An aging and increasingly health-conscious population, but one that wanted taste sensation and an ability to engage in "low guilt indulgence" was willing to pay more for the product, because it delivered on the promised benefit. The objective was to develop a line that would grow the category and offer higher margins for retailers through providing consumer value in the way of variety and taste satisfaction. Early versions of the product did not meet taste expectations, requiring breakthroughs in both formulation and processing. Sustained consumer testing and in-market testing was necessary to insure that the significant investment in process and equipment was wise. Ultimately, Baked Lay's delivered the highest trial rates in the Frito-Lay history and a brand that was the largest-selling new item in the 1990s.

Minute Maid Premium Heartwise is an orange juice proven to lower cholesterol. The product concept was based on a clear consumer opportunity, in that consumer studies from organizations such as Gallup showed that lower cholesterol levels are among the most desired benefits from functional foods, and that concern over cholesterol in the diet is growing at all age levels. Key consumer insights guided the development process: consumers want health solutions that are integrated into their daily life and routines but will not sacrifice taste, do not want to pay a premium, even for value-added benefits, and need an easily understood health benefit. Heart Wise contains plant sterols that are clinically proven to help lower total and low-density lipoprotein (LDL) cholesterol by inhibiting the absorption of cholesterol. Once the nutritional science team had identified plant sterols as a lead solution for a cholesterol reduction benefit, R&D and consumer research indicated strong potential for a beverage application. Development work led to a product that delivers as per consumers' taste and value expectations and created a major innovation in the category. According to information released by Coca-Cola, Heartwise, launched in the fourth quarter of 2003, achieved 90% ACV by March 2004.

Another example of a new product with a true novelty is Barilla Restaurant Creations pasta sauce, a product that is packaged in a distinctive hourglass-shaped package that has two 17-ounce jars within it. The top jar is the "chef's recipe" made with ingredients including olive oil, cheeses, herbs, and wine. The bottom jar is the base sauce made with tomatoes and olive oil. A home chef can create pasta sauce by mixing the contents of the two jars, heating, and mixing in a pound of pasta. Each recipe was inspired by the distinctive and authentic Italian flavors of a particular region. For instance, Sugo all Romana, made with Romano and Ricotta cheeses, olive oil, and herbs is inspired by the cooking of the Lazio region. The result is a

TABLE 3.3
Barb Stuckey's Top 10 New Product Introductions of 2004

- *Uncle Ben's Ready Rice*, a ready-to-heat, retorted rice in a pouch followed their frozen bowl meals as products designed for the time-starved consumer who wants to put a good meal on the table. From regular rice to Minute Rice to Ready Rice, Uncle Ben's reflects the evolution of meal preparation in the United States.
- *Kraft 100 Calorie Packs*, an on-the-spot response to our mania over obesity and weight management, offering consumers the familiarity of brands like Oreos and Chips Ahoy with the convenience of a calorie-controlled and counted snack.
- *Forkless Gourmet*, an alternative to Hot Pockets that adults actually eat. Forkless Gourmet offers high quality and ethnic flavors in a convenient hand-held microwaveable mini meal form.
- *Ritz Chips*, the "best line extension of the year," in that they will help to minimize cannibalization, because they are in chip form. These crispy snacks satisfy a potato/tortilla chip craving with a wholesome "baked' halo.
- *Francis Coppola's Sofia*, a slim-line energy drink in a can a la Red Bull. The single serve Blanc de Blanc sparkling wine comes with a petite straw for sipping. It is sporty, portable, feminine, and makes the consumption of sparkling wine as convenient as that of bottled beer.
- *Splenda Sugar Blend*, an easy, no-sacrifice way to cut in half the sugar in coffee, tea, and even baked goods. A great product even for those not on a low-carb diet.
- *Starbucks Frappucino Light* lets customers enjoy their made-to-order treat without sacrifice. It feels and tastes indulgent but without being really high in calories.
- *Fresco Style at Taco Bell.* A restaurant menu option that allows the customer to substitute chunky salsa for the higher fat cheese and sauce, leading to a low-fat, low-calorie item without having to expand the menu dramatically.
- *Hormel & Stagg's Chili in a Box.* There was a lot of activity in shelf stable chili with Campbell's Chunky Chili in a can and Bush's chili in a glass jar. What makes this product so different is that it is packed in a paperboard "juice box" type of Tetra packaging.
- *Jennie-O's Oven Ready Turkey*, a product that makes cooking the Thanksgiving turkey almost foolproof, and eliminates the messy and unsafe handling of a raw turkey. Cooking time is faster and the shopping bag package adds utility by making a twenty plus pound bird much easier to maneuver.

Source: From Bob Messenger's Morning Cup, January 2, 2005. Used with the author's permission.

superior-quality, authentic Italian product based on a combination of culinary and technical expertise (Fusaro and Toops, 2004) Table 3.3.

Use Research to Measure Reaction to the Product and All Elements of the Program Throughout the Development Process

Listening to the customer does not end with the birth of the product idea. Successful product development must involve talking to the consumer early and often. The product development process is essentially a series of screens and evaluations, starting with preliminary market assessment to determine market size, market potential, and the competitive situation. Detailed market studies are used to determine customer needs, wants, and preferences that serve as input to product definition and product design. Pushing the numbers early and often is the key; quantitative research must be employed to gauge market size, growth, and trends. Early stage volumetric estimates can be generated from concept tests (see Chapter 6) that help to measure likely market acceptance of the product and provide an intent-to-purchase score that can be normed to provide estimates of trial. It is vitally important to establish the correctness of the proposition with the consumer early in the process: engage the consumer to help define and refine concepts.

Throughout the development process, the concept must be continually refined and the physical product evaluated to insure that sensory attributes and nutritional characteristics match the product concept. Rigorous testing of the product, including shelf life, quality and safety elements must be undertaken under normal and extreme customer conditions. Successful firms make sure the customer is exposed to the product as it takes shape, and throughout the entire new-product project. All elements of the program must be evaluated as they are created: packaging (structure and graphics), labels and advertising (copy and media plan), promotions, display, and shelf sets. Even after the product and package survive testing during the early and intermediate stages of development, and after all elements of the marketing program have been identified, refined, and tested, the reaction of both the trade and the target household customer must be assessed. This assessment typically involves some type of market testing or field trial. Ultimately, companies that use disciplined processes that feature regular and rigorous evaluation have better overall new product performance than companies lacking such evaluative processes.

Use of Category Management Philosophy to Align Manufacturer and Retailer Focus on the Consumer

Ernst & Young has identified eight critical success factors in new product launches that can be applied to both the manufacturer and the retailer. The key is partnering with a focus on the consumer and on achieving objectives for the category. These success factors are shown in Table 3.4.

The basic idea is simple, and goes to the root of ECR, a term no longer used in the United States but still very much a part of the food industry lexicon in Europe. Create a true partnership between the manufacturer and the retailer that focuses on creating value for the consumer through the best possible retail assortments, specific to the demographics and lifestyles of shoppers in a specific store trading area that in turn creates value for both the retailer and the supplier via profitable sales and share. Thus, new items must do more than generate new item allowances that add to the retail bottom line or create movement only because of promotional incentives, because items that cannot sustain themselves on the market for lack of consumer interest will ultimately create waste from the retailer all the way back up the supply chain. The key is consumer insight to help manufacturers launch products with a compelling consumer benefit that adds to instead of merely cannibalizes sales in the category. In addition, pricing decisions are integrated across the category and promotions are designed to maximize return on investment of trade funds. Together, the supplier and the retailer must do their homework as partners and base decisions on shared category and consumer data.

TABLE 3.4
Critical Success Factors in New Product Launches

Manufacturers Does the Manufacturer…	**NPI** Critical Success Factors	**Retailers** Does the retailer…
• Practice or plan to implement ECR, category management, and efficient assortment concepts • Involve the retailer in test marketing and collecting and analyzing data at the SKU level when determining the optimal number of SKUs to introduce • Focusing to bring real value and innovation to the category	Category Management	• Have a category strategy that details the Category's role Relative size of the category Optimal number of category SKUs Full P&L product evaluations for each product line in a category Category teams, with well-defined goals that are empowered to make decisions on all aspects of marketing the category
Strive to match technology innovations with unmet customer needs	Match Technology with Unmet Consumer Needs	
• Evaluate the number of UPC changes that occur within the category • Perform extensive consumer research, with early retail involvement, which drives a formal business case evaluation, including volume projections	Fact-Based Analysis and Planning (doing your homework)	• Have a fact-based evaluation process as a baseline for all introduction evaluations regardless of category • Have a documented monitoring process to ensure successful introduction, where problems are investigated to determine underlying cause and appropriate corrective actions
• Have a documented New Product Innovation (NPI) process with formalized review points and identified critical-path model of the process	A Formal NPI Process with Pre-Established Criteria	• Participate with the manufacturer in the development and execution of promotions and retailer programs • Get involved early in the NPI process to improve greatly odds at execution-level success
• Plan and develop effective programs with respect to consumer marketing/promotions, strong advertising, and retail programs	A Strong Implementation Program A Strong Advertising Plan	• Work with manufacturer to optimize introduction dates to retailer schedules
• Ensure early retailer and consumer involvement to improve greatly the odds of execution-level success • Work with the retailers tying the critical-path model and planned introduction time to largest cluster(s) of customer re-planogram time(s)	Interaction, Coordination, and Communication	• Work with the manufacturer in evaluation category performance so the manufacturer can suggest new assortments • Partner with key manufacturers to improve process, share data and category insights, and help test-market products
• Plan for and support execution activities in conjunction with the retailers early in the process	Store Support	• Support introductions with adequate inventory, facings, and effective joint promotions • Employ computer based planograms and stocking level/facing optimization programs to improve execution
• Evaluate category-level rewards and compensation based on profitability of the category, along with individual recognition	Aligned Performance Metrics and Rewards	• Have fact-based performance evaluations of buyers on volume projections, contribution, traffic, growth, and profit, depending on category role • Rotate key employees through the various roles in stores, merchandising, and buying to ensure a broad perspective
• Have detailed cost and target pricing policies in the NPI process	Profitability	• Have NPI evaluations based on category growth, sales volume, and long-term profit potential

A Well-Executed Launch

The launch is where "the rubber hits the road" and is all about execution and proper timing. The launch consists of setting specific goals and targets for market performance, running initial production, distribution and selling in to the trade, implementing the advertising and promotional programs we have planned for the trade and for the household consumer, and tracking all of these activities relative to goals and targets. Success is measured by performance in four key areas:

- Production and distribution—having the product available in good condition in the correct quantities and at the right times and places
- Sell-in—executing the sales activities to the trade and achieving the targeted levels of distribution penetration, including the targeted shelf sets, displays and numbers of facings, plus the correct on-shelf and in-ad price
- Being on the shelf, which means not only gaining retail authorization but also making sure the product moves from the warehouse, to the store, and onto the shelf and on display
- Consumer reaction, as measured by awareness, trial, repeat purchase, and customer satisfaction
- Overall performance, as measured by sales revenue and sales growth, market share and share trends, and source of volume—cannibalization vs. category growth vs. taking sales from competition

One of the key challenges of the launch is coordinating sourcing and production, and availability of the product, with both the sales presentations to the trade and the breaking of the initial consumer advertising and promotional activity. We discuss this process in more depth in Chapter 20. Product must be in the store before advertising breaks and coupons are dropped. When planning the launch, seasonal sales patterns must be taken into account. It should seem obvious that seasonal items cannot afford to miss the major selling season. The firm needs to set the launch date and then back up from there. It must be determined, primarily from experience, when the product needs to be in wholesale warehouses, when it has to be available to ship, when production must be run, and ultimately when the ingredients and raw materials need to be acquired. For consumers to buy the new product it must be on the shelf and we must build awareness through advertising and motivate purchases through promotion.

Finally, we need to use a combination of syndicated data and market research to track the launch. We need to obtain data on retail distribution and overall market penetration (e.g., using Market Scan (MS) data). We need to gather sales and share data at the retail level (typically through IRI or Nielsen scanner data). We need to monitor purchase dynamics including patterns of substitution and the incentives associated with purchase (household panel data), and we need to track awareness, trial, repeat purchase, and customer satisfaction through both household panel data and consumer surveys.

Ability to Adapt, Grow and Improve as Market and Competitive Conditions Evolve

Proper execution of the development process and the launch does not guarantee the long-run success of a new product. We do not exist in a competitive vacuum; there is a need to adapt and grow as we adjust to realities of the marketplace and competition. In most food product categories, new items, including those that feature a technical advancement, can be easily duplicated. A competitive difference lasts but a short time if competitors note that the new product has a significant impact on consumer purchase patterns. Moreover, the trade behaves impatiently with respect to new products, and if sales objectives are not achieved, there is always another new item to fill the shelf. Thus, it is essential that firms plan to: (1) ensure market success via contingency planning and (2) to continuously improve the product bundle.

Contingency planning involves establishing alternative scenarios of market and competitive reaction to the new product launch, setting up a monitoring system to track market and competitive reaction, and having alternative strategies to implement for different market and competitive conditions. For example, what do we do if the reaction to our new product launch is for a competitor to lower price? We need to be proactive instead of purely reactive to minimize the potentially negative impact of a competitive countermove.

Kraft is one practitioner of the model of continuous improvement. A key part of Kraft's new product strategy is to stay on the offensive by value engineering to improve continuously, to add value to products, to delight the customer, and to drive down costs. Another aspect of staying on the offensive is to flank the new brand with line extensions and to meet the needs of a market with needs and wants that become more refined and more differentiated over time. The original Kraft Oscar Mayer Lunchables brand has been modified several times during its highly successful life. First, the container was changed to lower unit costs. Then, items such as branded drinks and desserts were added to create additional customer value. Later, Kraft rolled out versions such as nachos and pizza targeted specifically at kids.

FINAL THOUGHTS

The Leo Burnett Company New Product Planning Group recommends that new product marketers follow the principles listed in Table 3.5, while "10 Ways to be More Successful in Beating the New Product Odds" from IRI are in Table 3.6.

All of this boils down to the reality that new products must be in the store and available, and consumers must become aware of the new item, be incentivized to try it, and have a reason to keep buying it past the first purchase occasion. If any of this does not happen, the product will not be a success, plain and simple.

TABLE 3.5
Leo Burnett's 15 Principles to Guide New Product Success (1995)

(1) Distinguish your product from competition in a consumer-relevant way
(2) Capitalize on key corporate competencies and brand strengths
(3) Develop and market products to people's needs and habits
(4) Market to long-term trends, not fads
(5) Don't ignore research, but don't be paralyzed by it
(6) Make sure your timing is right
(7) Be a marketing leader, not a distant follower
(8) Offer a real value to consumers
(9) Determine a product's short and long-term sales potential
(10) Gain legitimacy and momentum for the brand
(11) Give the trade as good a deal as the consumer
(12) Clearly define, understand, and talk to your target
(13) Develop and communicate a distinctive and appealing brand character and stick to it
(14) Spend competitively and efficiently, behind a relevant proposition
(15) Make sure the consumer is satisfied … and stays that way

TABLE 3.6
Ten Ways to be More Successful in Beating the New Product Odds

- Follow a disciplined product development process that includes rigorous consumer testing of product, packaging, and advertising.
- Offer unique benefits and superior value (IRI data show that highly innovative brands have year-one sales 50% higher than products offering no new benefits.
- Focus on big brands and big categories. The biggest new products are in the big categories.
- Be in sync with consumer trends.
- Manage and plan for variety. IRI's "Pacesetters" tend to offer a large variety of items and refresh their line new SKUs in year two and year three.
- Extend your brand wisely. Focus on the core equity of your brand. Do not venture into unfamiliar territory alone; find a strong partner.
- Be the first to market, but only after you have gotten all of these other points right.
- Advertise in year-one to generate awareness and well into year-two to cement the brand's foundation for long-term success. Media weight must be accompanied by copy that is both attention-getting and persuasive.
- Get the product into distribution quickly to build momentum for the start of advertising.
- Study the strategies that worked for highly successful brands in recent years.

Source: From Information Resources Inc. 2004.

BIBLIOGRAPHY

1. Anonymous. 1997. Getting the Numbers Straight, In *Efficient New Product, Introduction*. Supplement to *Progressive Grocer*, pp. 44–46.

2. Anonymous 2004. How Aggressive New Product Strategies Drive Success at the Hain Celestial Group, Inc., *Stagnito's New Products Magazine*, 4(1): 23–25.
3. Anonymous. 2005. An Elusive Goal: Identifying New Products That Consumers Actually Want, http://knowledge.wharton.edu/index.cfm, accessed March 15, 2005.
4. Anonymous. 2005. Build a Better Mousetrap, *Distribution Channels*, March 2005, p. 66.
5. Anonymous. 2005. www.ethno-insight.com, accessed March 6, 2005.
6. Anschuetz, N. F. 1996. Evaluating Ideas and Concepts for New Consumer Products, In *The PDMA Handbook of New Product Development*, Rosenau, M.D., A. Griffin, G. A. Castellion, and N. F. Anschuetz, eds. New York: John Wiley & Sons, pp. 195–206.
7. Bargman, T. and R. Pomponi. 2004. "Product Development Benchmarked," *Prepared Foods*, 173(6): 52–57.
8. Berne, S. 1996. Options in Outsourcing, *Prepared Foods*, 165(5): 24–26.
9. Clancy, K. J. and P. C. Krieg. 2003. Surviving Innovation, *Marketing Management*. 12(2): 1061–3846.
10. Cooper, R. G. 1993. *Winning at New Products*, 2nd ed. Boston, MA: Addison-Wesley.
11. Cooper, R. G. 1996. "New Products: What Separates the Winners from the Losers," In *The PDMA Handbook of New Product Development*, Rosenau, M. D., A. Griffin, G. A. Castellion, and N. F. Anschuetz, eds. New York: John Wiley & Sons, pp. 3–18.
12. Cooper, R. G. and E. J. Kleinschmidt. 1990. *New Products: The Key Factors in Success*. Chicago: American Marketing Association.
13. Crawford, C. M. and A. Di Benedetto. 2006. *New Products Management*, 7th ed. Boston, MA: Irwin McGraw-Hill.
14. Dahm, L. 2002a. Go Team! How Multi-Functional Team Strategies Can Result In New Product Success, *Stagnito's New Products Magazine*, 2(2): 22.
15. Dahm, L. 2002b. The New Product Game, *Stagnito's New Products Magazine*, 2(4): 24–28.
16. Davis, Ann. 2001. Multi-Functional Launch Teams: An Idea that Actually Works, *Stagnito's New Products Magazine*, 1(4): 22–26.
17. Demetrakakes, P. 1996. Take Out Technology, *Food Processing*, 57(6): 53–62.
18. Dimancescu, D. and K. Dwenger. 1996. *World Class New Product Development*. New York: AMACOM.
19. Dornblaser, L. 1997. This Glass is Half Full, *New Product News*, 33(11): 3.
20. Dwyer, S. 1997. Hey, What's the Big Idea? Pillsbury Knows, *Prepared Foods*, 166(5): 8–20.
21. Ferrante, M. A. 1997. Outsourcing: Gaining the Manufacturing Edge, *Food Engineering*, 69(4):87–93.
22. *Foods*, 166(10):26–30.
23. Fusaro, D. 1995. Kraft Foods: Mainstream Muscle, *Prepared Foods*, 164(5): 8–12.
24. Fusaro, D. and D. Troops. 2004. "2004 Innovation Awards—A Pageant of New Products," *Food Processing*, 65(12): 21–24.
25. Hawkins, C. 1996. Rookies of the Year, *Prepared Foods*, 165(11): 13–17.
26. Hoban, T. J. 1998. Improving the Success of New Product Development, *Food Technology*, 52(1): 46–49.
27. Holleran, J. 2004. Best New Products of 2004, *Stagnito's New Products Magazine*, 4(12): 36–42.
28. Hustad, T. P. 1996. Reviewing Current Practices in Innovation Management and a Summary of Selected Best Practices, In *The PDMA Handbook of New Product*

Development, Rosenau, M. D., A. Griffin, G. A. Castellion, and N. F. Anschuetz, eds. New York: John Wiley & Sons, pp. 489–510.

29. Leo Burnett U.S.A. 1994. 15 Principles, Presented at the *Prepared Foods Annual New Products Conference.*
30. Lindeman, T. F. 2005. Del Monte Gives Investors Ideas to Chew On, *Pittsburgh Post-Gazette*, March 22, p. E23.
31. Maremont, M. 1998a. Gillette Finally Reveals its Vision of the Future, and it has 3 Blades, *Wall Street Journal*, April 14, pp. A1, A10.
32. Maremont, M. 1998b. How Gillette Brought its MACH3 to Market, *Wall Street Journal*, April 15, pp. B1, B10.
33. Matthews, R. 1997. Efficient New Product Introduction, July 1997 Supplement to *Progressive Grocer*, pp. 8–12.
34. Matthews, R. 1998. What's New in New Products? *Grocery Headquarters*, 64(8):18–19.
35. Morris, C. 1996a. Beyond the Box with New Manufacturing Alliances, *Food Engineering*, 68(4): 63–70.
36. Morris, C. 1996b. Cut to the Core: Food Companies Reassess Engineering Competencies, *Food Engineering*, (68)11: 57–67.
37. Moss, R. 2004 *Of Pacesetters and Transformational News*, www.retailwire.com, accessed May 7, 2004.
38. O'Donnell, C. D. 1997. Campbell's R&D Cozies Up to the Consumer, *Prepared Foods*, 167(1): 34.
39. Parker-Pope, T. 1998. The Tricky Business of Rolling Out a New Toilet Paper, *Wall Street Journal*, January 12, pp. B1, B8.
40. Patrick, J. 1997. *How to Develop Successful New Products.* Lincolnwood, IL: NTC Business Books.
41. Roberts, W. A. Jr. 2003. *Schan Food Company: Red Baron Pizza Slices*, www.preparedfoods.com accessed January 26, 2003.
42. Rudolph, M. T (1995). The Food Product Development Process, *British Food Journal.* 97(3): 3–11 and Chapter 4.
43. Stein, J. 1998. The Men Who Broke Mach3, *Time*, 151(16): 4.
44. Stinson, W. S., Jr. 1996. Consumer Packaged Goods (Branded Food Goods), In *The PDMA Handbook of New Product Development,* Rosenau, M. D., A. Griffin, G. A. Castellion, and N. F. Anschestz, eds. New York: John Wiley & Sons, pp. 297–312.
45. Symonds, W. C. 1998. Would You Spend $1.50 for a Razor Blade? *Business Week*, April 27, p. 46.
46. Symonds, W. C. and C. Matlack. 1998. Gillette's Edge, *Business Week*, January 19, pp. 70–77.
47. Thomas, M. and A. Brody. 2004. Let's Work Together to Deliver a Culinary Gold Standard, *Food Technology*, 58(5): 16.
48. Thompson, S. 2004. Kraft Pasta Extensions Leave Retailers Unimpressed, *Advertising Age*, April 19, 2004, pp. 4, 60.
49. Uhlman, M. 2004. A Trimmer Tastekake, *The Philadelphia Inquirer*, May 16, 2004, pp. E1, E4.
50. Von Bergen, J. M. 1998. Newest Tasty Treats Have Tropical Flavor, *Philadelphia Inquirer*, May 12, 1998, p. C1.
51. Wheelwright, S. C. and K. B. Clark. 1995. *Leading Product Development.* New York: The Free Press.

4 The Food Product Development Process

Marvin J. Rudolph

CONTENTS

Although no guarantees can ever be offered for a new food product success, the implementation of carefully orchestrated plans significantly increases probabilities of success. Planning is a series of well-considered steps to be taken from gap analysis through concept generation and evaluation to prototyping and assessment, positioning through mapping, optimization, market testing, and scale-up for launch. The emerging discipline of quality function deployment may have an integral role to play in establishing the phases and benchmarks before products are actually formulated.

INTRODUCTION

The current process for developing new food products is seriously flawed. Of the 8077 new food products (stock keeping units or SKUs) introduced to United States retail markets in 1993, only about one-quarter of them were novel–not simply line extensions (Morris, 1993). Although there are no published data on successes and failures of new food products, it is estimated that 80–90% of them fail within 1 year of introduction. These are just the products that made the retail cut; consider all the products whose efforts fell short and the retail introduction never took place. This means that about 200 novel food products introduced in 1993 will make it after their first year on the U.S. retail shelves in 1994. The failure cost to the U.S. food industry is estimated at $20 billion (Morris, 1993). This cost results from missed sales targets, lost revenues, and postponed profits in addition to wasted development resources.

Being complex and iterative, the food product development process has proved difficult to define and model. It begins with a concept and ends with either the entry of the product in the market or the maintenance of the product in the marketplace, depending on whose model is studied. Barclay's review of research work into the process of product development (not only food) and the way in which it has progressed over 40 years shows that much of the week studying the new product development process is unknown to product development managers (Barclay, 1992a). After investigating current practices in new product development at 149 United Kingdom-based companies, only 78 were found to have some form of a new product guide to help manage the process, and 65 of the 78 stated their guide had originated through experience. Only one company based its new product development process on literature describing process models.

Traditional approaches to managing the product development process often fail, because they result in unbalanced milestone structures. Typically, there is no preplanned structure of milestones that identifies the deliverable for each functional group at each step of the program. Project managers apply their often-limited experience to develop a program plan that emphasizes their area of expertise. Often, milestones for following product performance are absent. For example, production data are not fed back to the team members who developed the product. Furthermore, management traditionally concentrates on the process only when authorization for large amounts of money is requested. Management attention therefore is focused typically on the purchase of equipment, not on concepts or brainwork.

We believe that Arthur D. Little Inc. has developed a comprehensive philosophy to guide food product development activities. It is based on establishing clear, consistent milestones for the entire development process and identifying the required deliverables by each of the functional elements contributing to product development with the firm. Milestones are viewed as an opportunity to monitor progress against a planned set of goals, to review the next tasks and anticipate problems, and to initiate program changes. This approach to milestones is analogous to the use of bivouac site by a mountain climbing team to regroup and make adjustments before proceeding on the journey. Figure 4.1 depicts the milestone structure graphically, highlighting the natural and effective shifting in the level of activity for each functional area.

While the peak activities shift, it is critically important that each group provide input to all of the milestones throughout the process. As Barclay states, "The [food product] development process needs to be linked with the corporate objectives and to the external environment to allow new ideas into the organization" (Barclay, 1992b).

The Arthur D. Little (ADL) milestone-driven product development process recognizes that a good process is flexible and continuously evolving (Figure 4.1). The many

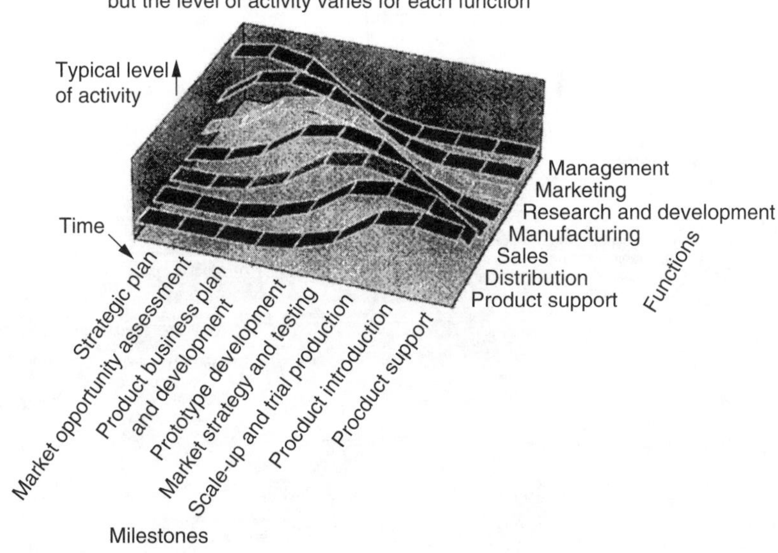

FIGURE 4.1 Organizational involvement in the product development process.

advantages of this process for project leaders, team members, and management are based on the following features:

- Use of common, project-specified vocabulary facilitates communication among the case-team, management, and project reviewers.
- Development of a standard framework of milestone deliverables reduces project start-up time.
- A consistent definition of milestone structure allows for internal benchmarking.
- A proven methodology allows for more accurate project planning, including allocating resources, establishing budgets, and scheduling tasks.

The total milestone-driven food-product development process under investigation is illustrated in Figure 4.2. It can be envisioned as a skier racing in a three-phase giant slalom, the phases defined as product definition, product implementation, and product introduction.

PHASE I: PRODUCT DEFINITION

STRATEGIC PLAN

The skier (product developer) begins the race by proceeding form the starting gate under a strategic plan implemented by what we at ADL refer to as "third generation R&D" (Roussel et al., 1991). Third generation R&D is simply a holistic linking of business and technology goals.

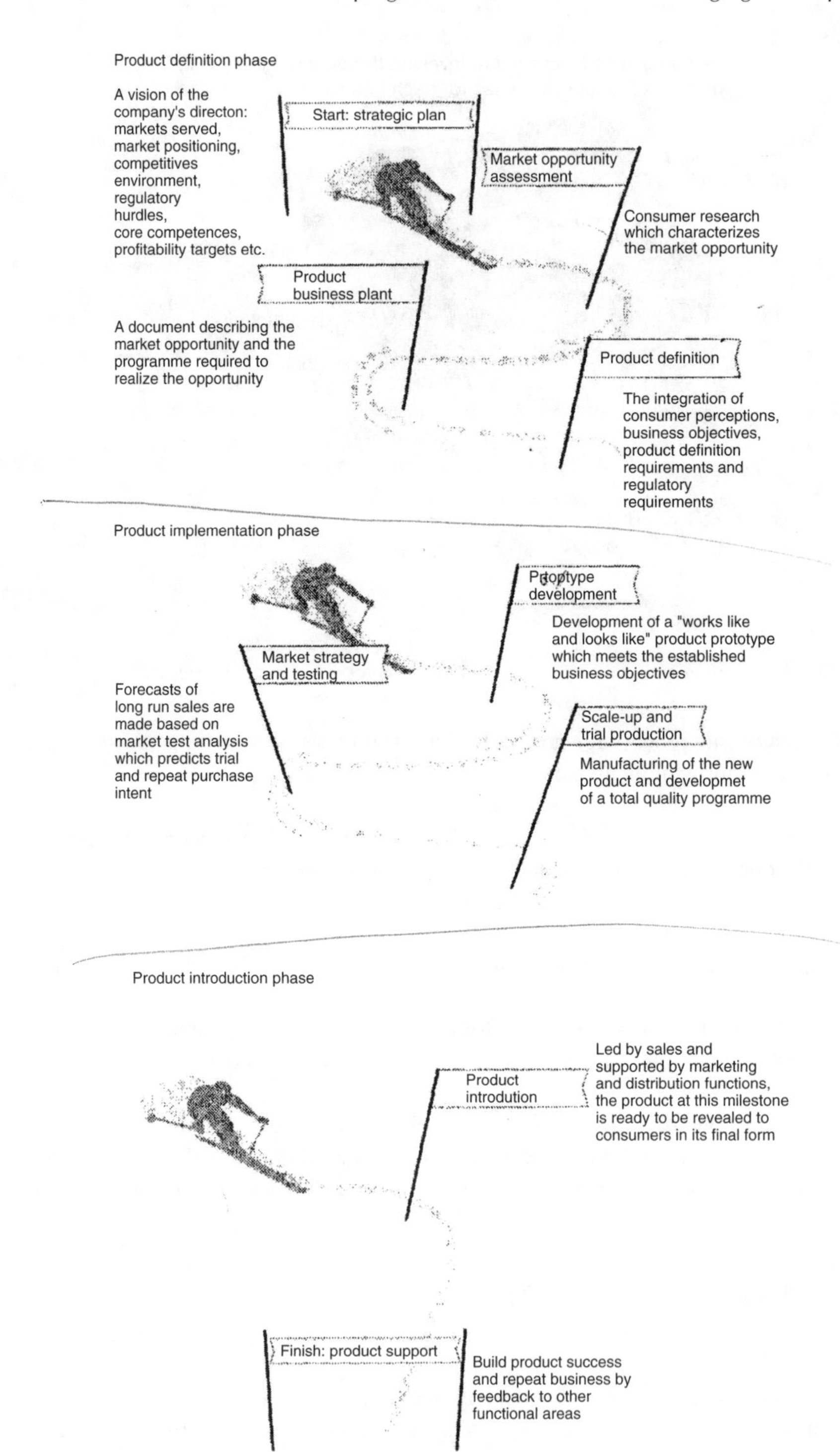

FIGURE 4.2 Milestone structure.

Prior to the early 1960s, there was first generation R&D, which can be defined as the strategy of hope. A research center was established, usually separated in distance from the main activities of the corporation, whose mission was to come up with interesting phenomenon. Work was self-directed, with no explicit link to business strategy. R&D activities were a line item in the corporate budget. If something came out of the effort, so much the better. However, the attitude was fatalistic—expectations usually were nonspecific.

From the early 1960s to today, we have second generation R&D, which is best characterized as "hope for a strategy."

It is characterized by a mutual commitment to goals by upper management and R&D, consideration of the corporate strategy when setting these goals, and the implementation of a control system to track progress towards the goals. There is, however, no corporate-wide integration of R&D activities.

The future belongs to third generation R&D, emergent in leading companies. It articulates the issues that the firm must consider as it decides how to define overall technology strategy, set project goals and priorities, allocate resources among R&D efforts, balance the R&D portfolio, measure results, and evaluate progress. The crucial principle is that corporate management of business and R&D must act as one to integrate corporate business and R&D plans into a single action plan that optimally serves the short-, mid-, and long-term strategies of the company. A major output of third generation R&D is a vision of the company's direction; it characterizes the markets served and that competitive environment details regulatory hurdles and identifies the company's market positioning, core competencies, and profitability targets.

MARKET OPPORTUNITY ASSESSMENT

Once the strategic plan is in place, the second gate to traverse is the characterization of the market opportunity. This means that market requirements must be defined. In the food industry, this usually means consumer research. Frequently, focus groups (real-time knowledge elicitation) are conducted to identify potential opportunities for new products. The appeal of focus groups is in the "free format"—usually resulting in qualitative (anecdotal) comments that may be misinterpreted when a strong observer bias exists. Lack of a rigorous method for sifting through consumers' verbal comments may result in a misunderstanding of the real trade-offs that consumers are making or a failure to uncover an opportunity because of insufficient probing.

On the other hand, quantitative methods, like conjoint analysis (Rosenau, 1990), are criticized because of the limited and unrealistic nature of the options that can be presented and evaluated by the consumer.

The ADL approach to conducting real-time knowledge elicitation combines the advantage of the unrestricted choices offered by free-choice profiling with the powerful collection of quantitative data for statistical analysis. We use a software tool for real-time data capture and analysis during the focus group. The online analysis provides the moderator with feedback on how well the consumer descriptions differentiate the samples. The moderator can then pursue further probing to uncover subtle yet important differences among sample. Subsequent statistical analysis is used to reduce the tension set to those sensory characteristics that account for the majority of the variance.

The method is effective in explaining the full range of sensory dimensions where individuals are able to discriminate effectively between items. As alternative features are defined, these consumer opportunities can then be translated into product concepts for further testing during consumer real-time knowledge elicitation activities. This can improve the success rate of the concepts as they move further through the milestone gates.

THE BUSINESS PLAN

The output of consumer real-time knowledge elicitation is the identification of new consumer needs and product concepts that can be incorporated in a product business plan (Hopkins, 1985), a document that describes a market opportunity and the program required to realize the opportunity. The business plan, usually written for a 12-month period, does the following:

- Defines the business situation—past, present, and future
- Defines the opportunities and problems facing the business
- Establishes specific business objectives
- Defines marketing strategy and programs needed to accomplish the objectives
- Designates responsibility for program execution
- Establishes timetables and tracking mechanisms for program execution
- Translates objectives and programs into forecasts and budgets for planning by other functional areas within the company.

PRODUCT DEFINITION

The last step in this phase is product definition. The key to product definition is the integration of multiple and often conflicting objectives. The integration of consumer requirements, business objectives, product delivery requirements, and regulatory requirements is illustrated in Figure 4.3.

ADL has pioneered a structured approach to integrating research of consumer needs and descriptions of the competitive environment with technical realities into a unique product specification. We use a product definition process based on quality functional deployment (QFD) to help us combine and translate consumer requirements into product specifications. (See also Chapter 14 for the application of QFD for package development.) QFD (Sword, 1994) was developed in Japan as a tool to provide designers with an opportunity to consider the qualities of a product early in the design process. QFD is a method that allows us to consider the qualities of a

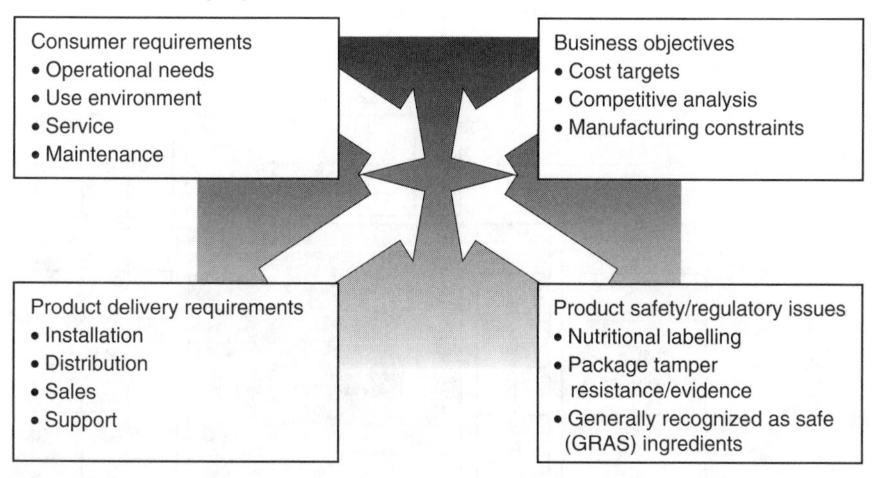

FIGURE 4.3 Integration of objectives.

product, process, or service. It helps us to focus our activities on meeting the needs of the customer:

- Who are the customers?
- What is it that they want?
- How will our product address those wants?

Using QFD leads to a better understanding of customer needs that the product must meet to exceed competition.

QFD methodology evolves around the "house of quality" (see Figure 4.4), a graphical representation of the inter-relationships between customer wants and associated product characteristics. It maps product requirements, helps to identify and understand requirement trade-offs, and predicts the impact of specific product features. Additionally, it provides a team-building tool for interdisciplinary product planning and communication. It is a method that is an important part of the process to develop successful products that fit the strategic and tactical needs of the business.

PHASE II: PRODUCT IMPLEMENTATION

PROTOTYPE DEVELOPMENT

Once the food product is defined, a "works like, looks like, tastes like" product prototype is constructed or formulated (see Chapter 12). To demonstrate that the

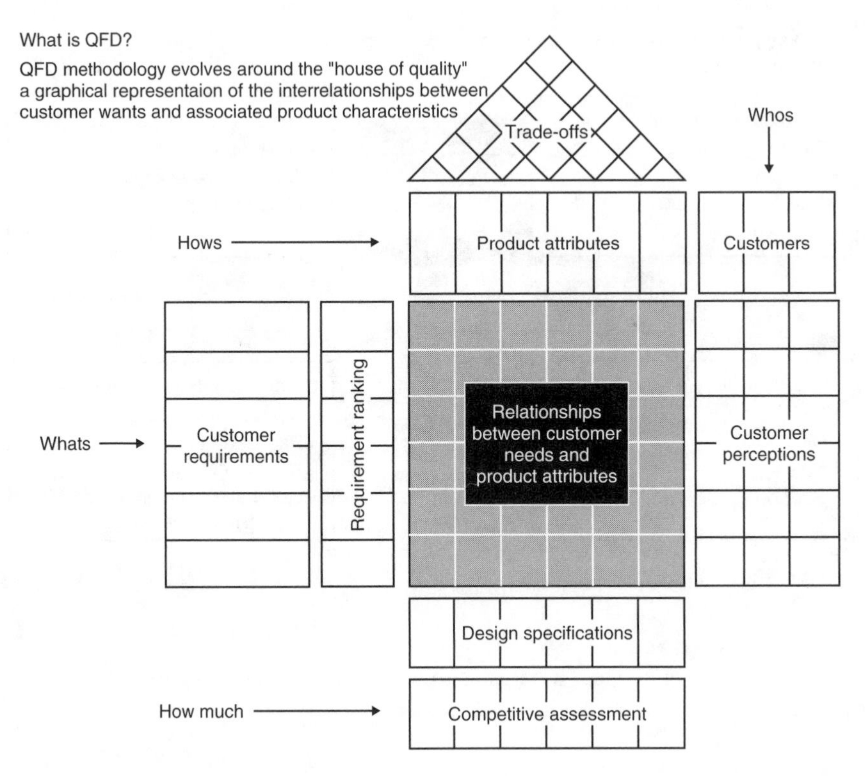

FIGURE 4.4 The house of quality.

product prototype in its conceptualized final form will meet the technical and business objectives established, ADL staff use their profile attribute analysis (PAA) method (Hanson et al., 1983). PAA is an objective method of sensory analysis that uses an experienced and extensively trained panel to describe numerically the attributes of the complete sensory experience of a product. These attributes are a limited set of characteristics that provide a complete description of the sensory characteristics of a sample. When properly selected and defined, little descriptive information is lost. PAA is a cost-effective tool for product development that takes advantage of the use of powerful statistical techniques, such as Analysis of Variance.

PAA is used in product prototype development in two ways; competitive product evaluations (benchmarking) and product optimization (see Chapter 13).

BENCHMARKING

Competitive product evaluations provide formulators with objective information regarding the flavor quality of competitive products and the areas of flavor opportunity. The following example demonstrates a competitive product evaluation of several nationally branded oatmeal cookies. The products sampled (four full-fat oatmeal

TABLE 4.1
Profit Attribute Analysis (PAA): Competitive Product Evaluation

Full-Fat Cookies	Archway Oatmeal Cookies
	Nabisco oatmeal cookies
	Pepperidge farms oatmeal cookies
	Pepperidge farms Santa Fe cookies
Reduced-fat cookies	Pepperidge farms wholesome choice oatmeal cookies
	Nabisco SnackWell's reduced-fat oatmeal cookies
No-fat cookies	Entenmann's no-fat oatmeal cookies

Note: The sensory analysis evaluated two replications of each of three lots of each oatmeal cookie product. The samples were presented in a random order with no visible identification of the brand.

cookies, two reduced fat cookies, and a no-fat oatmeal cookie) are identified in Table 4.1.

The evaluation was conducted to determine whether reduced fat or no fat products provide sensory characteristics similar to those of their full-fat counterparts.

- A panel of trained sensory analysts evaluated two replications of each of three lots of each type of cookie. The samples were presented in a random order with no visible identification of the brand.
- The panel used PAA to evaluate the products against the 13 sensory attributes shown in Figure 4.5.
- Summary indices for texture and flavor were developed using principal components analysis to summarize the attribute data and were interpreted as shown in Table 4.2.
- The resultant map of the flavor and texture indices (see Figure 4.6) shows that the full-fat Pepperidge Farms Santa Fe oatmeal cookie is not significantly different from the Nabisco SnackWell's reduced fat oatmeal cookie.

The balance and flavor indices are shown in the flavor map (see Figure 4.7); the reduced and no fat products, as well as two out of the four full-fat cookies (Pepperidge Farms and Pepperidge Farms Santa Fe), exhibit a thinner and less blended flavor. The Archway and Nabisco full-fat products had a fuller, more blended flavor but differed significantly in texture. Archway is a soft, fragile type of cookie and Nabisco is a moister, chewier cookie.

The results of this study indicate formulation improvements to the flavor of the reduced and no-fat cookies should be focused on improving the balance and fullness of flavor.

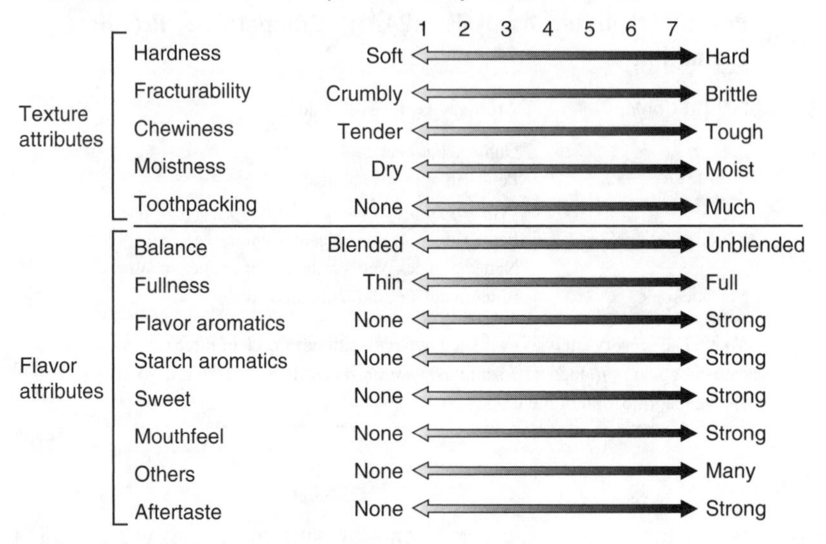

FIGURE 4.5 Sensory attributes of oatmeal cookies.

TABLE 4.2
Summary Indices for Oatmeal Cookies

	Higher Scores	Lower Scores
Texture index	Less crumbly	More crumbly
	Harder	Softer
	More moist	More dry
	More chewy	More tender
Flavor index	More starch aromatics	Less starch aromatics
	More others	Fewer others
	Less blended	More blended
	Thinner flavor	Fuller flavor
	More mouthfeel	Less mouthfeel

Note: Summary indices for texture and flavor were developed using principal components analysis to summarize the attribute data.

PRODUCT OPTIMIZATION (SEE CHAPTER 15)

Response surface methodology (RSM) can be used to achieve product optimization. In experimental food product formulations with multicomponent mixtures, the measured response surface, usually a flavor attribute, can reveal the "best" formulation(s) that will maximize (or minimize) the attribute (Cornell, 1990). An RSM experimental design for optimizing a ketchup formulation is illustrated in Figure 4.8.

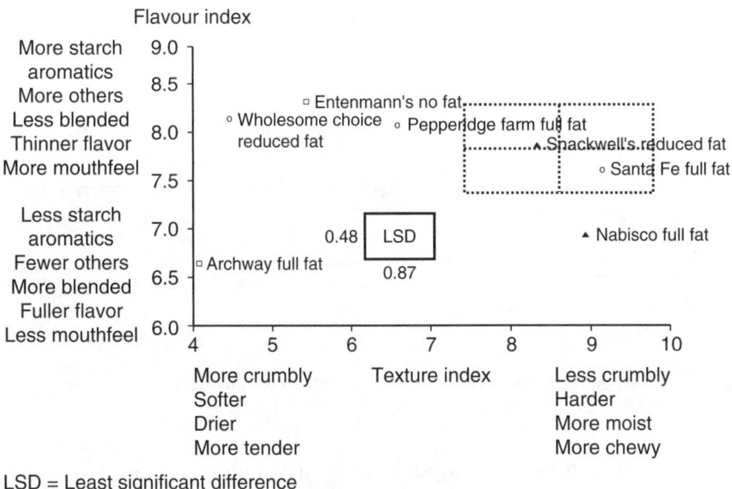

FIGURE 4.6 Texture index versus flavor index, oatmeal cookies.

The reduced and no fat products as well as two out of the four full-fat cookies (Pepperidge farms. Pepperidge farms santa fe) exhibited a thinner, less blended flavor improvements to the flavour of the reduced and no fat cookies should be focused on improving the balance and fullness of flavour

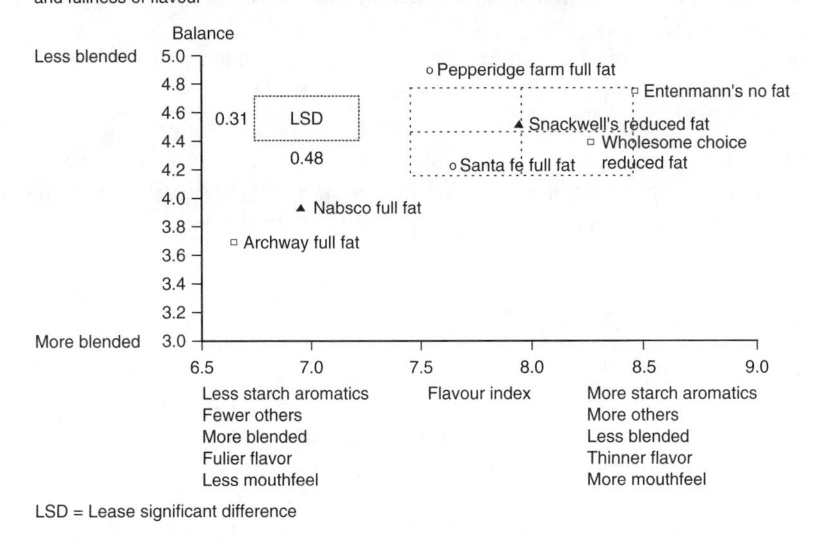

FIGURE 4.7 Flavor index versus balance, oatmeal cookies.

Professional panelists characterized the initial market samples form five plants with six code dates and the subsequent reformulated experimental products. The three most important variables were identified as percentages of salt, acid, and high-fructose corn syrup. The experimental design called for the manufacture and sensory analysis

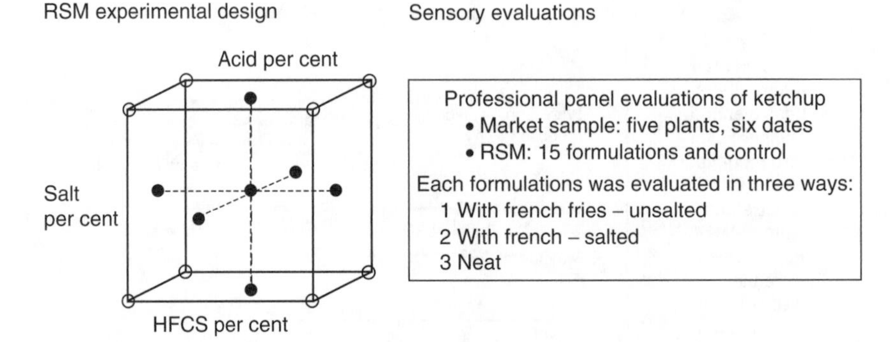

FIGURE 4.8 Response surface methodology experimental design.

of 15 ketchup formulations and a control. Each formulation was evaluated three ways: with unsalted french fries; with salted french fries; and neat. The RSM results revealed that our client's original evaluation design (without including different "carriers") was misleading.

- Ketchup evaluated neat has different flavor characteristics from ketchup evaluated with french fries.
- The original "optimized" formulation was acceptable in limited use situations.
- The resultant optimized formulations were acceptable in all ketchup uses when a standard manufacturing process was followed.

When RSM is utilized, product optimization time is greatly reduced from traditional "cook and look" optimization techniques that depend on subjective formulation and evaluation procedures and often stop short of fully realized product improvements.

MARKET STRATEGY AND TESTING

At this point in the product development process, the organization has invested time and money in developing a new product from the initial concept to product optimization. If marketing forecasts look good, the temptation is to prepare for a full-scale launch. However, the product definition phase is based on models that are reasonably accurate representations of market response, but not of reality (Chapter 2). Things can still go wrong, as witnessed by the failed national introductions of "New Coke", Milky Way II candy bars (25% fewer calories and 50% fewer calories from fat), and ConAgra's Life Choice Entrees (low-fat diet regimen).

Long-run sales are based on two types of consumer behavior, product trial, and repeat purchase. Forecasts of long-run sales can be made if market test analysis can predict the percentage of consumers who become repeat users as well as those who will try the product. Numerous models have been developed that present the new product to consumers in a reasonably realistic setting and take direct consumer measures leading to the forecasting of cumulative trial and repeat purchases. Although

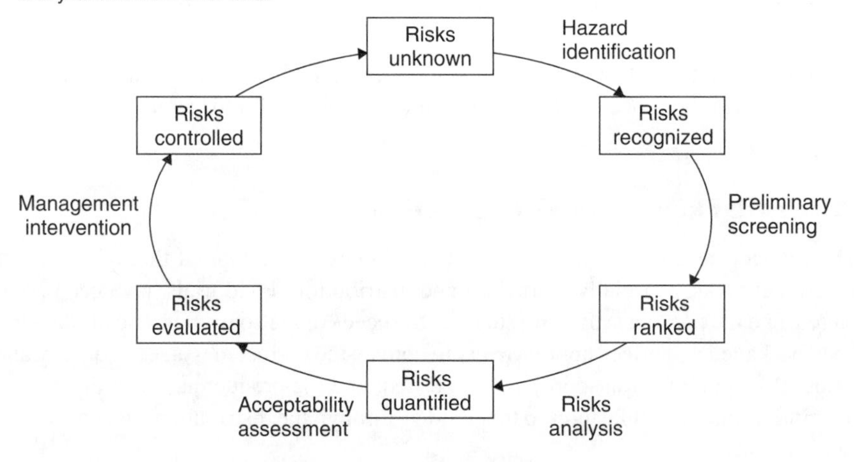

FIGURE 4.9 Food industry risk controlling process.

they will not be discussed in detail here, some of these models are based on previous purchasing experience, the so-called stochastic or random models, and a combination of trial/repeat and attitude models (Urban and Hauser, 1980). New services are constantly being developed commercially, and it is clear that technically strong models and measurement systems will be widely available to forecast sales of new packaged food products.

SCALE-UP AND TRIAL PRODUCTION

Ultimately, the new food product has to be manufactured to meet the needs of the consumer. Early involvement of the manufacturing function in the product development process helps to avoid problems that invariably surface when consumer expectations conflict with engineering constraints. The product's success is often linked to the level of compromise that is reached between the R&D and manufacturing functions (Chapter 14).

Implicit in the scale-up and trial production of the new food product is a total quality program that continuously identifies, analyses, and controls risk. The risk controlling process, as shown schematically in Figure 4.9, begins with the identification of all potential hazards and proceeds through the screening, analysis ranking, quantification and evaluation stages, and ultimately to the controlling of the risks.

A hazard analysis critical control point (HACCP) matrix (Chapter 8) is a useful tool for identifying and prioritizing hazards that may affect food product quality. Such a matrix has the following elements:

- Identification of critical control point
- Evaluation hazard potential
- Assignment of degree of concern (low, medium, and high)
- Development of criteria for hazard control

- Preparation of monitoring/verification procedures
- Designation of corrective action alternatives that may be required

Since food safety is always of paramount concern, new products often linger or die at this point in the process if the issues cannot be resolved satisfactorily.

PHASE III: PRODUCT INTRODUCTION

The product introduction milestone is led by sales but supported through all other functional areas, especially marketing and distribution. Field trials have been completed, and the product is designed to meet the needs of the consumer. The product has been packaged and priced appropriately to convey the correct messages of quality and value. Packaging for transport has been tested, and the product has been distributed in a timely and correct fashion so that it flows through the distribution system without impediments.

This phase is perhaps the most exciting and anxious, where customers see the product for what it is. Their initial response generally reveals the potential for success or failure of the product.

PRODUCT SUPPORT

Product support is a complementary milestone that builds product success and repeat business, because it feeds back valuable information to other functional areas that can lead the process for line extensions, product upgrades, and the creation of all new opportunities. Product support is the "infantry" for the battle at the retail shelf, the first line of communication from the point-of-sale back to the organization.

CONCLUSIONS

It is worth assessing how effectively your company controls the basic product development process. Symptoms of a "broken" process are:

- Longer development time than competitors
- Missed targeted introduction dates
- Significant number of "crash projects"
- A succession of stop/go decisions.

If your company exhibits these symptoms, you may need to overcome a common misconception about control: milestones are not an administrative burden imposed by senior managers eternally worried about the team's capacity to deliver. Rather, they are a self-help tool without which a team can feel hopelessly lost. Ultimately, the mountain-climbing team and the base camp are jointly responsible for the success of a climb (Vantrappen and Collins, 1993).

BIBLIOGRAPHY

Barclay, I. 1992a. The New Product Development Process: Past Evidence and Future Practical Applications, Part I, *R&D Management*, vol. 22, no. 3, pp. 255–63.

Barclay, I. 1992b. The New Product Development Process: Part 2. Improving the Process of New Product Development, *R&D Management*, vol. 22, no. 4, pp. 307–17.

Cornell, J. A. 1990. *Experiments with Mixtures.* New York: John Wiley & Sons. p. 8.

Hanson, J. E., D. A. Kendall, and N. F. Smith. 1983. The Missing Link, *Beverage World.* pp. 1–5.

Hopkins, D. S. 1985. *The Marketing Plan.* New York: The Conference Board Inc.

Morris, C. E. 1993. Why New Products Fail, *Food Engineering*, vol. 65, no. 6, pp. 132–6.

Rosenau, M. D. 1990. *Faster New Product Development.* New York: American Management Association, p. 239.

Roussel, P., K. N. Saad, and T. J. Erickson. 1991. *Third Generation R&D.* Boston, MA: Harvard Business School Press.

Sword, R. 1994. Stop Wasting Time on the Wrong Product!, *Innovation.* Winter, pp. 42–5.

Urban, G. L., and J. R. Hauser. 1980. *Design and Marketing of New Products.* Englewood Cliffs, NJ: Prentice-Hall.

Vantrappen, H., and J. Collins. 1993. Controlling the Product Creation Process, *Prism.* Second Quarter, pp. 59–73.

5 Food Product Concepts and Concept Testing

John B. Lord

CONTENTS

INTRODUCTION

The new food product development process flows from strategic business planning and starts with opportunity analysis, followed by idea generation. Having established both a strategic direction and identified a category in which there seems to be a gap that offers opportunities for a new product or a new product line, the objective of the ideation phase is to create, discover, and uncover as many new product ideas as possible. During ideation, we suspend evaluation so as not to hamstring the creative process. Whatever the number of new product ideas put forth, many will not be feasible. In addition, the firm, regardless of size, has limited resources for development. Thus, the next step after ideation involves screening ideas with respect to technical, manufacturing, distribution, marketing, and financial feasibility, and passing along to the next stage only those few ideas with the best prospects for success on all of these dimensions. The next stage involves turning ideas into concepts and then using marketing research to elicit consumer reaction to and the level of interest in concepts before proceeding to prototype development.

Some of the most necessary and most effective research for food product development can be conducted at the concept development and testing stage. Early in the development process, we need to prioritize concepts according to which are the developments most worthy. Successful new food products come from winning ideas and the firm must test propositions with consumers to minimize the chance of proceeding with marginal ideas and wrongly positioning a winning new product idea. Because the firm must allocate always-scarce resources to competing projects, we need to measure consumer purchase interest and intent in order to generate early estimates of revenues. Separating the potential losers from the potential winners early in the process also lessens the risk associated with the later stages of product development.

In this chapter, discussion will center on product concepts and concept testing. We will discuss both the theory and practice of concept testing, discuss how concept testing has evolved in a world of broadband Internet, and provide examples of testing methodologies from research suppliers and clients.

The Hierachy of Effects Models

One of the most commonly accepted theoretical models in marketing is called the "hierarchy of effects model." This model centers on the stages consumers go through in responding to marketing communications, providing a basis for planning, executing, and evaluating marketing strategies, particularly promotional strategies, and also providing the theoretical underpinning for the estimation of first-year sales volume for new consumer packaged goods. There are alternative versions of the hierarchy of effects model. One has the acronym AIDA, which stands for awareness-interest-desire-action. We will discuss the A-T-R model, which stands for awareness, trial purchase, and repeat purchase.

The hierarchy of effects involves the following process. At the start, consumers have no awareness of a brand, particularly a new brand. Through a combination of advertising, promotion, selling, publicity, and word-of-mouth, including the new phenomenon of "blogging," consumers become aware of the brand, and then develop some beliefs about the brand, or what is termed "comprehension" of the brand's attributes and benefits. The first two stages are termed cognitive. The next stage is termed affective, meaning that the consumer forms some type of positive or negative associations, or degree of liking for the brand, normally based on some evaluation of what the brand is perceived to offer versus what the consumer is seeking. If that liking is strong enough, the consumer may form a preference for the brand. Ultimately, if the consumer thinks the brand will solve a problem, if the brand fits within the consumer's lifestyle and budget, or even satisfies some need for novelty, the consumer may form an intention to purchase, meaning that he or she has made a judgment that the brand offers potentially want-satisfying benefits. The consumer may act on this purchase intention and decide to try the product, provided that it is available and other situational variables, such as an out-of-stock, do not interfere with the purchase decision. Finally, the consumer evaluates his or her experience with the product and assesses whether or not the product provided the expected benefits. If the consumer is satisfied, he or she may decide to buy the brand again. The latter stages of this process are called behavioral stages.

The concepts of awareness, trial, and repeat are commonly used both to set objectives for and measure the effectiveness of a new product launch program. We develop a marketing communication program to create a specified level of awareness. In order to become aware of something, we must be exposed to the message and be attentive to the external stimulus. For example, for a specific television commercial to have any potential impact, we must not only be watching television at a specific time on a specific network, we must also not be out of the room, or distracted by attending to some other stimulus, such as a book or a conversation with someone else in the room. Awareness is a function primarily of (1) a consumer's interest in the category, (2) the positioning of the product, specifically in terms of benefits associated with the product, (3) the amount of communication about the product to which the consumer is exposed, and (4) the strength of the impression created by the communications medium. Both the number of impressions and the quality and relevance of the message to the consumer will have a significant impact on creating awareness.

Trial is primarily a function of consumer usage of the category, the consumer's exposure to messages about the product, the extent of distribution, and incentives, such as samples or cents-off coupons. Trial consumers, if satisfied with their initial purchase and if the new product fits their lifestyle and consumption patterns, may be induced to buy the product again. Besides satisfaction with the product, which is a function of product quality, repeat purchase is related to perceived price-value. The number of repeat purchases will depend upon these factors, plus the extent to which the product has been adopted as a regular part of the consumer's lifestyle and consumption system.

PRODUCT CONCEPTS

A product concept is a stated, printed, pictorial, or mocked-up representation and description of a new product. The concept provides a device for communicating to both consumers and the development team the nature of the new product, how it will work, the product's features and characteristics, benefits, reason for being, and what problems it will solve for the user. A concept also represents a protocol that provides the development team a specific set of expectations as well as direction for both the development process and the launch strategy. The concept provides the basis for positioning strategy and product formulation, plus packaging, advertising, pricing, and distribution.

Concepts evolve during product development. At the earliest stages, a concept may be a simple standard description using a sentence or phrase. A more elaborate concept will include major benefits, features, package size, and price; a simple drawing may accompany the description. At later stages, the concept is enhanced and may resemble a rough print ad with an interesting headline focusing on the key consumer benefit, attractive graphics, and copy with style and personality.

CONCEPTS CAN BE PRESENTED IN DIFFERENT FORMATS

Depending on both the relative newness of the concept and the stage of development, there are multiple formats in which concepts can be presented. These include the following:

- Simple verbal descriptions
- More elaborate verbal descriptions incorporating advertising language
- Storyboards including product descriptions and typically an illustration of the package or product
- Images which can be viewed on a computer screen, both static and streaming
- A mock-up of the package
- A product prototype

Since concept testing, like the development process itself, differs in complexity and depth across different products, the format for the concept becomes more elaborate as more learning takes place, allowing for more detailed information from concept tests.

The development team must take into account the fact that as we add more detail, particularly creative execution to the concept, we risk testing reaction to the advertising instead of reaction to the concept. This will confound the feedback we are gaining from consumers and may result in both unreliable and invalid data on market reaction to the concept. As Patrick (1997) advises, because we use concept boards that include advertising language to garner reaction from consumers during interview situations, both care and creativity must go into the development of a concept board so we avoid displaying pure advertisement, as opposed to the features and benefits of the proposed new product.

The concept statement must communicate the main point, the benefit, right up front, and simply stated, so the respondent knows immediately *what's in it for me.* Any graphic or illustration must support the main product idea. The results of research in the social sciences indicate that any type of concept is communicated more effectively if presented in an environment that is familiar, and the graphics should help create that familiar environment. Make certain respondents can easily recognize the main benefit. Remember to make sure that concept statements are clear and understandable as well as realistic.

Core Idea Concepts and Positioning Concepts

Schwartz (1985) differentiates core idea concepts and positioning concepts. A *core idea concept statement* is usually quite short, just a few sentences, or brief paragraphs. The core idea concept focuses directly on the product's main benefits, puts little emphasis on secondary features, and avoids persuasive communication. Core idea concepts can be supported by rough artwork to help communicate how the product might look or function. Core idea concepts are used as building blocks. In other words, the development team creates alternative concepts that can be screened, elaborated, and fleshed out on the way to becoming a positioning concept. From the Gibson Consulting Group (Schwartz, 1985), we have these examples of core idea concepts:

> Here's a new line of snack dips for vegetables from French's that comes in convenient dry mix form.
>
> Here's a new product from French's that does for your hot vegetable side dishes what salad dressings do for salad vegetables.
>
> Here's a new product from French's that adds new taste variety to your everyday chicken dinners.

A *positioning concept* statement is longer and more detailed than a core idea concept. Positioning concepts can run several paragraphs, and attempt to communicate all of the product's main as well as secondary benefits. The positioning concept incorporates language that compares the advantages of the new product with those of other products, and is often supported with a high-quality photograph or illustration. Positioning concepts typically use both visual *and* verbal stimuli to describe the product, its benefits, and its end-use applications. Following on our earlier example from Gibson Consulting, the full positioning concept for the new French's chicken flavoring product is as follows:

> French's Introduces Chicken Dippin' Sauces.
>
> Chicken is chicken, right? Broiled, fried, or baked, it all tastes the same, right?
>
> Not any more!
>
> Now French's brings you new Chicken Dippin' Sauces for delicious tasting, moist chicken, right at the table.
>
> Chicken Dippin' Sauces come in two tasty varieties that complement the taste of chicken: there's Mild—a light fruity flavor; and, Zesty—a blend of rich robust flavors.
>
> Regardless of how each member of your family eats chicken—with their fingers or a fork—every bite will be flavorful and moist with new Chicken Dippin' Sauces.

> New Chicken Dippin' Sauces from French's … adds flavor and moistness to chicken right at the table.

Product concepts, which typically include visualizations now that most concept testing takes place on the Internet, provide the basis for gaining feedback from consumers to help measure the strength, viability, and sales potential of a new product. A poorly worded concept can lead to inaccurate data, which in turn can lead to development mistakes and wasted investment capital, not to mention the waste of the development team's time and effort. Good concepts clearly communicate the nuances of the product idea but are realistic. Realism is often a major deficiency in writing concepts. Concept statements must reflect reality, both in terms of the capabilities of the product and in terms of the realities of the competitive environment.

Some other issues with writing concepts include clarity of language, focusing on what the consumer needs to know, proper length, and appropriate degree of emotion. Frequent errors in writing concepts include placing too much emphasis on the headline and illustration at the expense of poor body copy, making the concept too long, and overstatement of what should be obvious points. Writing a concept is an art, not a science.

Concepts Evolve

In many instances, concepts evolve and become more complete as feedback is gathered and the effort becomes more focused. A good example of this is seen in the evolution of the concept of a product from Hershey Foods, Reese's Peanut Butter and Cookie Cups illustrated in Figure 5.1. You can see how the concept developed from version 1 to version 3 as illustrated in the figure.

Not only has the concept evolved, it is also clear that product formulation has evolved, and there has been at least initial work on package design.

Concept Screening

The idea generation phase of new product development is designed to generate large numbers of ideas. Since resources must be expended to write concepts and create concept boards, and because development resources and time are severely limited, the new product group must quickly reduce a large number of ideas to a much smaller number for which concepts will be formalized. This requires a preliminary screening process that must be accomplished quickly and in a very short period of time. In many instances, screening is a process that occurs internally, using very basic criteria, and relatively unsophisticated methodology basically the following: The key criteria are: (i) can we make it? (ii) will someone buy it? (iii) can we make adequate margin on it? A host of technical, market, and competitive issues are represented by these three basic questions. Use of a simple weighted rating scale on these and other criteria management wishes to employ, can yield a set of scores (sum the weight times the rating for each criterion) that can be used as a simple assessment tool. The lowest scoring concepts drop out in this iteration. Those that meet a minimum number and achieve the higher rankings stay in.

Introducing
intrigue

The new candy
bar from
reese's
that gives
you a rush
of intriguing
flavors and texture:

Two chocolate cups are filled with reese's peanut butter sandwiched between two chocolate cookies. This candy bar is available in a standard size (2 cups per package–1.4 oz.) and in a convenient slx-pack size (6 standard size candy bars).

Introducing
Reese's
cookie cups

We've put
two of your
favorites together–
chocolate cookies
and peanut butter–
in the unique
reese cup shape

A chocolate cup filled with reese's peanut butter sandwiched between two chocolate cookies, each package will contain two cups.

Introducing
Reese's
Peanut butter and cookie cups

Now you can enjoy America's favorite peanut butter cup with an added crunch.
Rees's peant butter and cookie cups combine the delicious taste of peanut butter with the crisp crunch of a chocolate cooke. Each milk chocolate cup contains a chocolate cllkie water and real reese's peanut butter.
Look for Reese's peanut butter and cookies cups in the candy section of your favorite store. They come packaged 2 ways: a 1.44 oz. signle-serve bar containing 2 cups and a convenient 6-pack. Both are priced the same as other candy bars.

FIGURE 5.1 Evolution of concept statement for peanut butter and cookie cups.

The ability to reach a large number of potential target customers by tapping into an Internet panel allows product developers to screen a relatively large number of ideas quickly and relatively inexpensively. BASES *SnapShot* is a relatively new system that gives packaged goods companies an option for a less expensive concept test. *SnapShot* can help identify winning concepts and prevent costly mistakes such as taking a mediocre concept through a costly development process by serving as a sophisticated screening device. *SnapShot* was developed in response to the demands of clients who wanted a faster, less expensive service that can be used to test dozens of new product ideas and whittle the list down to a manageable number.

BASES *SnapShot* is "a simple, low cost, and highly standardized concept screening device ... designed to reduce the cost of the test..." (www.BASES.com). The test uses a standardized questionnaire and standard set of marketing inputs, including a very brief overview of launch support, provided by the client. The deliverable is a one page summary of results, which includes most of the key concept indicators, including purchase intent, overall liking, perceived value for the price, perceived uniqueness, claimed number of trial units, claimed annual purchase frequency, a BASES concept potential score, and simple diagnostic analysis, such as an analysis of concept performance on several standard attributes.

Concepts Testing

Concept testing is defined as a marketing research technique that is used to evaluate a concept's market potential and provide information useful in strengthening the concept and developing introductory marketing strategy. Testing a concept involves exposing a product idea to consumers and getting their reaction to it, using a predetermined series of questions designed to measure various reactions, feelings, and opinions.

Concept testing has several important purposes. First, a concept test should enhance the efficiency of the development process, because only concepts judged to have sufficient competitive strength and market potential pass this particular gate and go on to the next stage, thus conserving product development resources. Second, the fully elaborated concept that emerges from the test provides direction in terms of a specific protocol for the development team to follow through subsequent development stages. Third, the concept test helps the development team to gain understanding of consumer reaction to the concept and its components. Fourth, concept testing allows an estimate of purchase intent, and thus provides data for early-stage sales volume estimation. Fifth, concept test results provide diagnostic data as well as guidance for positioning strategy.

Concept testing can use either qualitative or quantitative techniques. Qualitative techniques, such as in-depth interviews and focus groups, can provide a great deal of useful knowledge regarding consumer reaction to the concept, the concept's strengths and weaknesses, and recommendations for improvement. Qualitative research, however, cannot provide data necessary to make early sales volume estimates.

The most important purpose of a positioning-concept is to present the product idea realistically to learn whether consumers will eventually buy the product. The emphasis should be on clear and accurate product-description communications rather than persuasion so as to portray the product as it will eventually be presented to the market. Sometimes copywriters can get carried away with a concept description; we must remember to test the concept and not the advertising.

A concept test can yield numerous important classes of information, which are listed in the following section. Not all of this information will be yielded by any one test, and what information the test provides will be dependent on the research methodology, specifically the types of questions asked, as the proper execution of the test. The types of data provided by a concept test are listed in Table 5.1.

Virtually every quantitative concept test will contain an intention to buy or purchase interest question. Most will attempt to measure uniqueness and price-value, and most will attempt to elicit consumer reaction not only to the concept as a whole but also to the elements or characteristics that comprise the new product.

TABLE 5.1
Concept Test Data

- Consumer attitudes and usage patterns in the product category
- *The competitive setting or market structure in the category*
- Size of the potential market
- Segments and characteristics of the consumers making up each segment
- The main idea being communicated by the concept
- The importance of the main idea to consumers
- The concept's relevance to consumer needs
- Consumer reaction to product attributes and features, and the relationship between consumer reaction to product attributes and overall concept rating
- Purchase interest and intent (at a given price point)
- Reasons consumers give for purchase interest
- Major strengths and weaknesses of the concept
- Strength of the concept versus other concepts tested
- Perceived advantages and disadvantages relative to competitive products
- Uniqueness—a measure of competitive insulation and a strong indicator of concept strength if combined with high purchase interest
- Believability, the extent to which consumers feel that the concept can deliver the benefits promised
- Perceived price-value—consumers expect good value so they are not necessarily turned on by concepts which provide a good value, but a poor price-value relationship can dramatically depress overall appeal
- Usage situations and frequency of usage
- The expected frequency of purchase
- Potential source of volume (incremental category sales; from competitors; cannibalization of the developing firm's product line)

According to Schwartz (1985), a properly planned, executed, and analyzed concept test can provide several important pieces of information:

- An early read on sales potential. If potential is adequate, test results should aid in identifying the segments of the market that are potentially most responsive to the new product. If results are below expectations, the test results should permit determination of any potential targets of opportunity if the market population were to be segmented into different or narrower demographic, behavioral, or attitudinal groups. Alternately, we may find simply that the concept lacks sufficient appeal to be a viable entry into the category.
- The degree to which the message that consumers receive from the concept (in both objective and subjective dimensions) is consistent with the message intended. This is diagnostic element that lets the development team know if the concept's level of success or failure is attributable to the attractiveness and strength of the new product idea, or is compromised in some way by a biased or inaccurate execution. This is especially important in the case of

a poor-performing concept when a decision must be made to either kill the idea or make another attempt at positioning and describing the concept.

- Identification of the individual strengths and weaknesses of the concept, providing insight into the relationship between consumers' overall evaluation of the concept and the role of each of the concept elements in that evaluation. The test should relate which elements of the concept contributed to positive evaluation and which contributed nothing or to negative evaluation. This insight aids in strengthening the concept via heavier emphasis on positive characteristics and elimination or downplaying of neutral and negative elements.

METHODOLOGY

In its most basic form, concept testing involves communicating one or more concepts to potential survey respondents who are typically representative of the target market for the product and measuring reaction to the concept. One of the first decisions a firm makes about concept testing is whether to outsource. Many commercial research suppliers offer concept testing, and the options for food companies to outsource at least the fieldwork portion of concept testing have increased dramatically with the development of the Internet as a major means of administering consumer research. A.C. Nielsen BASES has several alternative testing designs and is a frequent research supplier to the food industry.

BASES

(Author's note: Most of the material in this section is derived from two sources, BASES.com and HBS Case Services case number 595-035 Nestle Refrigerated Foods: Contadina Pasta & Pizza (A) Rev. 1/97] A.C. Nielsen BASES (formerly the BASES Group) specializes in estimating and analyzing sales potential for new products, line extensions, and restaged brands, using market simulation models and analytical tools. BASES is the industry's leading simulated test market product and has been around for over 25 years. Most companies use BASES tests as a screener for products to be test-marketed or rolled out. BASES tests, like most concept testing methodologies, attempt to help decision-makers reduce risks by providing accurate estimates of consumer sales volume at the earliest possible stages of product development, well before major resources are invested in production, packaging, marketing, or management time. According to the company, "the BASES system yields reliable, accurate forecasts and actionable recommendations that improve concept and product appeal, increase the productivity of marketing efforts, and raise the odds for in-market success" (see http://www.bases.com/services/bases1.html). Consumers are exposed to information about the concept in a format similar to their likely exposure in market (for example, a television commercial or print ad, a visual of the package, and the retail price). Respondents are questioned about interest in the concept, likelihood of purchase,frequency of purchase, quantity purchased at each purchase event (transaction size), and perceptions of price-value. If the product is available, those consumers who exhibited the highest level of interest, the likely triers, are given an opportunity to try the product. After some time, triers are asked about their experiences with the product and given an opportunity to buy it again.

A.C. Nielsen BASES offers alternative testing products, including:

- *Pre-BASES*: combines survey measures and assumptions about marketing support to give an early read on sales potential. Pre-BASES integrates multiple volumetric consumer measures, including measures reflecting purchase interest, value perception, transaction size, and purchase frequency.
- *BASES I*: concept test with volume estimates generally thought to be within a 25% accuracy range. The methodology employed attempts to assess the current level of awareness and usage in the category, and gain insight into consumer perception of alternate possible positioning statements with respect to competitive brands. A BASES I test provides an estimate of first-year trial volume, simulates total Year-1 sales volume, and provides some understanding of the likely effect of alternate positioning.
- *BASES II*: concept test in combination with a product taste test, with first-year sales volume estimates reliable within 20% of actual. In addition, BASES II tests, measure claimed source of volume.
- *BASES II Line Extension*: test for new products that are part of a pre-existing product line.

A BASES test starts with consumer attitudinal data and reactions to concept boards, including purchase intent, which are adjusted in order to provide accurate predictions of consumer behavior, specifically trial of a new product. Repeat purchase rate is a function of product satisfaction. Different repeat purchase rates can be projected for products that prove to be, mediocre, average, and excellent.

These adjusted consumer data are combined with client-provided assumptions of distribution, media plans, and trade and consumer promotional events in order to estimate trial and repeat purchase potential for the new brand. Given their extensive product category experience, with results of tests of close to 6000 new concepts and 3000 new products during the past 10 years, BASES has in-market databases, which help ensure the reasonableness of inputs and outputs to the model. BASES' clients routinely share their in-market experience with BASES for products that were tested in the BASES system and subsequently introduced in-market, and BASES buys IRI panel data. These data allow BASES models to be adjusted and fine-tuned so that about 90% of its forecasts accuracy falls within plus or minus 20% of actual market results.

The outcomes of a BASES test include a forecast of first-year sales volume, comparison of the target concept to other concepts and products in the BASES database, evaluation of the concept's positioning and overall appeal, evaluation of the product's performance (BASES II) and its effectiveness in meeting triers' expectations, suggestions for improving the concept and product, and specification of which improvements will have the most significant impact on potential sales volume, and analysis of planned marketing activities and recommendations for increasing their effectiveness.

Major factors contribute to the reliability of a BASES test. These include the sample size, the similarity of the estimated marketing plan to the actual plan used to launch the product, and the BASES experience with similar product categories. A BASES II test also employs in-home product usage tests to derive more accurate assessments of product satisfaction and repeat purchase.

OUTSOURCING CONCEPT TESTING

In an era of outsourcing, it is not surprising that many packaged goods companies outsource a substantial portion of their research. Everything but the stimulus, that is, the concept, is outsourced to a research supplier or suppliers; specifications regarding the target audience are given to the research supplier who then plans and executes the fieldwork.

There are two main reasons why concept tests are outsourced. The first is that most of the research suppliers have recruited Internet panels that give rapid and economical access to a large sample of potential respondents. The second is that these suppliers, by conducting multiple tests over time, have assembled a large database of test results that allows for fine-tuning their methods and models, yielding potentially more accurate sales forecasts.

Companies conducting their own tests typically employ central location tests (CLT) such as mall intercepts that use a combination of convenience and quota sampling, or use mail or Internet surveys sent to households or individuals who are members of a preselected and demographically balanced panel.

ADMINISTERING THE TEST

Concept tests traditionally have been administered in person, typically via the use of central location tests or focus groups; by the use of mail questionnaires; and by the use of telephone surveys. The relative advantages and disadvantages of these methods of gathering data are well-documented in marketing research texts. In-person interviews allow a great deal of range and flexibility in the types of questions researchers can ask and the visual stimuli that can be employed, and typically allow for greater amounts and depth of data to be gathered. In-person interviews are expensive and require trained interviewers as well as a location in which the test is administered. CLTs require careful screening of potential respondents to ensure sample representativeness. However, concept tests in many cases use closed-end questions and scaling questions, which can be effectively and efficiently handled via telephone or by the use of mail surveys.

Telephone interviews allow a great deal of control, access to a greater range of respondents in a shorter period of time than that of CLTs, and, through the use of random digit dialing, can be used in a random sampling situation. Limitations include the ease with which respondents can refuse to be questioned or terminate the interview before completion as well as the lack of any visual stimuli. Telephone interviews can be accomplished very quickly, but the range and depth of data gathered is somewhat limited. Telephone interviews are most useful when the objective is to gather some basic consumer response data in a short period of time, which is typical of new product development projects.

Mail surveys tend to be the least expensive, provide an opportunity for the respondent to ponder questions and even gather information before answering questions, allow for visual stimuli to be presented, and allow for careful targeting of households or at least neighborhoods. Response rates tend to be lowest for this type of questionnaire administration; however, this disadvantage is negated if we have access to a panel of households who have been recruited and screened.

CONCEPT TESTING MOVES TO THE INTERNET

As the twenty-first century dawned, advances in information technology and increasing household access to the Internet have fundamentally changed how CPG companies conduct research. Over the past few years, use of the Internet to administer surveys and elicit feedback to product concepts has increased dramatically. Companies such as Hershey and General Mills have shifted the bulk of their quantitative research to the Internet, in place of more traditional telephone or shopping mall methodology. As early as 2002, General Mills reported that 74% of its quantitative research was being conducted over the Internet. Hershey Foods had moved new product testing online in 1999 and 2000. Similarly, Campbell Soup Company has moved most of their consumer testing to the Web.

The primary advantage of replacing typical mail and CLT testing with Internet testing is the time frame in which research results are available. For example, Arnold (2002) reported that Hershey Foods found that the typical 6-week time frame for mail concept testing was cut to 2 weeks by putting the surveys online. The impact on the new product development cycle is obvious. In an era where speed to market is all-important, the opportunity to reduce one key phase in the development process from several weeks to a few days represents a potentially huge advantage.

An article on e-research by Ray and Tabor (2003) indicated another significant advantage of Internet research: "the Web also can locate hard-to-find respondents who themselves 'find' the researcher by visiting the company's Web site relevant to their interests, and build online panels of prerecruited individuals." Much Internet research uses online panels, and the prerecruitment of such panels by both companies and research suppliers allows for quick online access to a sample that can be tailored to the interests of the survey and closely reflect the target audience for the research. A concomitant disadvantage of Web-based surveys, however, is that error can result from the incomplete representation of all households in the potential sample, because, while most households in the United States have access to the Internet, the figure is well below 100%. It is also very difficult to use probability-sampling techniques on the Web.

Internet research creates another advantage over more traditional methods of administration. Researchers can use multimedia applications with both audio and video presentations of products and product comparisons. Respondents, for instance, can rotate a package to view nutritional information, usage instructions, and other informational cues. Researchers can manipulate product characteristics to determine an optimal product (Ray and Tabor, 2003). In many cases, researchers have moved to less text, that is, shorter concept descriptions, with more visuals including lifestyle visuals. For instance, when testing a product like Soup At Hand, Campbell Soup Company might show a person sipping the soup in a car or at a desk. The intention is to replicate what a print ad might communicate, while keeping in mind that the purpose of the test is to elicit reaction to the concept, not the ad itself. Overall, there is substantially more emphasis on visuals with Internet concept testing.

When a company moves concept-testing administration from traditional CLT and mail surveys to the Internet, one of the keys is to conduct validation testing, in order to make sure that historical data and norms can still be used. Hershey, working with NFO Worldwide, did online and mail panel tests concurrently in 1999 and found comparable results with correlation between the two sets of tests achieving 0.9, an extremely high degree of correlation (Arnold, 2002). Likewise, Campbell Soup ran a number of parallel tests, both Internet surveys and mail surveys, in order to determine how test results differ systematically as a result of the different technology, and to calibrate the results, that is, to adjust historic scores so that comparisons could still be made across different tests. Since concept-test scores can only be useful when compared to norms, this type of historical comparison is necessary.

BASES e-Panel is a data collection methodology that allows BASES to gather consumer reaction to concepts using the Internet with no compromise in database comparability or forecast accuracy (www.bases.com). As of May 2004, the BASES e-Panel stood at about 90,000 households, or approximately 265,000 individual consumers. While the first BASES e-Panel was built in the United States, new panels have been built in other global regions with similarly high household penetration of the Internet. Globally, the BASES e-Panel operates in Canada, the United Kingdom, France, and Germany.

Members of the e-Panel have been profiled on demographic characteristics, purchasing habits, media usage, and consumer attitudes. As a result, clients can use the e-Panel to examine the characteristics of consumers who are most interested in their products. The ability to define key target groups with criteria related to their behavior as consumers, beyond simple demographics, gives marketers the ability to do very exact and specific target marketing.

The major advantages of the BASES e-Panel include:

- Concepts can be tested as concept boards or videos
- Security, confidentiality, and data quality
- Cost savings of approximately 20% over traditional mall-intercept testing
- Greater interviewing capacity, often improving the timing of the study
- Efficient means for reaching select groups such as pet owners, presence of children, and so forth
- Greater convenience for panelists, leading to higher participation rates and compliance

A Sampling of Commercial Concept Testing Services

In addition to ACNielsen BASES, other companies offer proprietary methodologies for evaluating concepts. MRSI offers a product, which allows clients to screen concepts, refine ideas into core concepts, evaluate concept appeal, confirm concept and product performance, and forecast initial sales volume (prism.mrsi.com/stage2.html). MRSI screens concepts quickly among target consumers, using an opt-in panel. Concepts are brief, 3–4 sentences, and may include a simple visual. Each respondent is exposed to multiple concepts in groups of 3–5 that the respondents rate on key measures such as purchase intent, uniqueness, and believability.

MRSI tests concepts using their *Online-iN-Site* interviewing system that offers full graphical capabilities, using either images or video. MRSI also offers clients the ability to determine whether a product delivers against the expectations established by the concept by giving respondents the concept test prototype product to use, and then gathering purchase intent and other evaluative data after use. This type of research can be accomplished in the traditional way using mall intercepts and follow-up phone calls or entirely via the Internet so long as the product can be mailed to the respondent's home for home use.

Decision Analyst (www.decisionanalyst.com/Services/concept.asp) offers a set of three Internet-based concept testing services. *ConceptScreen* provides clients with the capability to test 10–20 early stage concepts together as a group. Qualified respondents are invited to Decision Analyst's website to participate in the survey. The order of the concepts is randomized from respondent to respondent to eliminate positional and sequencing bias. Each respondent sees all of the concepts twice, first reviewing but not rating the concepts, then the second time through respondents are asked to answer four questions about each concept. This test provides a rating of the better concepts.

Concept Check is an Internet-based system that provides diagnostic feedback for early stage individual concepts, and includes verbatim "why" responses. Each concept is tested monadically. A total of 50–75 target consumers are asked to review and evaluate the concept. The test employs both standard closed questions and open-end questions used to elicit information as to how the concept can be improved.

Conceptor uses the Internet to test a new product concept among 150 target consumers who view the concept and complete a battery of questions and diagnostic ratings. Each concept is tested monadically. A predictive model calculates an overall *SuccessScore* for each concept, which is compared to norms to determine if the concept warrants further development. The *SuccessScore* is based on trial interest, purchase intent, expected frequency of purchase, uniqueness, reaction to price, perceived value, and category breadth.

Ipsos-Insight Inc. (www.ipsos-insight.com) offers a set of research tools for concept screening and evaluation. *GOProspect* is a method for prioritizing early-stage concepts that are most likely to succeed as fully articulated concepts, and is useful for narrowing a list of potential concepts to the most promising ones before committing substantial development resources. *GOProspect* uses the Ipsos U.S. online panel. Panelists are sent emails containing a URL to link them to the survey site, and are asked to rate a series of new product ideas in the form of a short phrase or sentence. Using a choice modeling methodology, *Concept Evaluator* delivers a full assessment of a concept, including customers' perceptions of the underlying themes, a visual depiction of which attributes or features increase or decrease consumer appeal, and full analysis of the data in total and by subgroups. *Concept Evaluator* identifies key words and phrases in a concept that drive consumer interest and addresses key concept communication issues such as brand name, taglines, positioning statements, packaging options, emotional benefits, pricing, and variety lineups.

Ipson *Concept/Tester.online* uses the Web to have respondents evaluate a single concept in a 10 min interview with 20–25 question, providing data comparable to a traditional mall intercept test in about half the time. A statistically representative sample of 300 respondents is drawn, customized for each study. *Ipsos Concept/Forecaster.online* provides validated early sales volume estimates, using a model built by calibrating online concept survey data to actual sales data for new products that have hit the market.

NPD Foodworld offers a research product called *Concept Checks* (www.npdfoodworld.com/foodServlet?nextpage=proprietary_concept.html) that offers concept testing capability to the foodservice industry. Clients can choose from among three Internet-based options, including *Concept Sort*, for testing several ideas shown as simple text for early idea prioritization; *Sequential Monadic Concept Screen*, which tests several, more fully finished concepts displayed as artwork, simple photos or simple descriptions, providing economical evaluation of consumer interest during the early development stage; and *Concept Check*, which uses traditional monadic concept testing that provides thorough evaluation of advertising and sales potential displaying concepts as photos or artwork with full descriptive text. NPD Foodworld claims that Concept Checks helps its clients to increase new product success rates and focus resources on the plans and ideas with the highest potential, with both reliability and time-savings.

Kellogg's partnered with an Internet research company called BuzzBack Market Research to gather information prior to launch of Pop-Tarts Yogurt Blasts. Kellogg's wanted to create a compelling and memorable product name, and wanted to understand why one name would be preferred to another in order to develop effective communication about the new brand. Literally, over a weekend, Buzzback surveyed 175 mothers and their children (respondents under the age of 13 must have parental permission to participate in online research). Survey respondents reviewed name alternatives, package design ideas, and messaging statements about the yogurt ingredients and nutritional value. The results of the survey provided the Kellogg's team with quantitative data as well as insights into moms' and kids' perceptions of the name, the packaging, and the nutritional value of the product (Little, 2004).

Sample Size

There are no hard and fast rules governing sample size. As the population becomes more heterogeneous, achieving a more representative sample requires using a larger sample. Convention dictates that concept tests use samples consisting of 300–500 potential respondents. This number is necessary to provide adequate size of each cell when doing cross-tabulations. Smaller sample sizes can be used for relatively

TABLE 5.2
Dimensions of Concept Tests

- Uniqueness
- Believability
- Level of consumer interest
- Whether the new product will solve the consumer's problem
- Practicality and usefulness of the new product
- An affective or relative liking dimension
- Reaction to price or a price-value rating
- Strengths and weaknesses
- Problems with the concept

homogeneous populations. However, as we have seen above, most commercial concept tests conducted on the Internet use samples in the 75–150 range.

Designing the Questionnaire

Just as there are no hard and fast rules governing sample size, there is no single approach to the number, type, and sequence of questions in a concept test. However, a typical concept test questionnaire would begin with an assessment of the respondent's current experience and practice in the product category, demonstrate the concept, elicit reaction to the concept, measure intent to purchase, and then ask for relevant demographic and other data useful in classifying respondents. Table 5.2 lists several dimensions on which reaction to the concept can be assessed.

Purchase Intent

The single most important question in concept test is the purchase intent question; this is the tool most frequently used to measure concept success, both as a decision-making variable and as a key element in various volume-prediction models. By asking this question, we not only measure the overall strength of the concept, but we also can pinpoint the type(s) of individuals most likely to buy the product. A typical statement of the purchase intent question is as follows:

> If (this product) were currently available in your local store, how likely is it that you would buy/try this product?

Since most available normative data are based on a five-point scale, the purchase intent question usually employs such a scale with the following scale points:

- Would definitely buy
- Would probably buy

- Might or might not buy
- Would probably not buy
- Would definitely not buy

The percentage of respondents who respond "definitely would buy" is called the *top box score*. The percentage of respondents who respond "definitely would buy" plus the percentage that respond "probably would buy" is a measure of positive purchase intent and is called the *top two-box score.*

While gauging purchase intent is critically important to a concept, the development team must consider several key issues. First, not every respondent who indicates "definitely will buy," will buy. Over time, marketers develop norms for a category whereby we can hypothesize a certain numerical relationship between positive purchase intent and projected trial rate. When you analyze the results of a concept test, it is helpful to have scores available from previous concept tests to act as a point of reference or benchmark, especially when these scores have been related to actual in-market performance. In effect, these data serve as a historical perspective against which the current concept's result can be compared. Accurate forecasts of trial and first-year sales volume are completely dependent on these norms.

Second, many marketers simply place too much value on purchase intent. New product ideas often live or die on the basis of one score. Blind reliance on any single question is dangerous, and purchase intent is no exception. Third, if the primary objective is to predict actual volume, once the product is introduced to the marketplace, make sure to tell respondents the price *before* asking about their interest in purchasing the product. This is critically important to the accuracy of the forecast of sales volume. Fourth, because the reasons consumers give for having high or low levels of purchase interest often provide valuable insights about a concept's strengths and weaknesses, we should assess reasons for purchase interest. This can aid in spotting winners and losers and in understanding the factors responsible for their performance.

Other Key Concept Test Criteria

Uniqueness represents an important concept test dimension for an obvious reason: successful new products must be meaningfully different from competitive products. Two elements of uniqueness—of the overall product and of individual product characteristics—should be studied. Uniqueness is an excellent diagnostic tool to help pinpoint how to improve a deficient concept. Cross-tabulating uniqueness by purchase interest can help pinpoint factors that are responsible for weak concept ratings.

Determining whether and to what extent consumers perceive a concept as delivering important benefits or satisfying recognized needs is particularly important in understanding whether consumers might adopt a new product. Often, we are willing to buy a new product once or twice, because novelty is an important motivation, but also because new products often come with significant incentives that lower the effective price. However, one or two purchases do not sustain a new product. Unless we are specifically planning to launch a "fad" item, such as a ready-to-eat breakfast cereal using a licensed character, or a seasonal item, such as a Christmas snack cake, we need consumers to make the product part of their regular lifestyle and consumption system. This can only happen if consumers perceive the item as having the ability to meet ongoing and important needs.

Virtually every concept-test questionnaire includes a question that asks respondents what they consider to be the main idea of the product. Purchase intent and attribute ratings can only be valid if consumers understand the concept. There is also value in discovering how consumers evaluate the importance of the key benefit. This can be particularly valuable when testing a concept for a breakthrough product—something completely different from products currently available. Early stage consumer evaluation of a "new to the world" product presents certain challenges. It is crucial to determine the importance of the concept to the consumer to measure the strength of breakthrough concepts properly. This type of concept, even if it has high appeal, may score low on purchase intent due mainly to the fact that consumers may not believe that the product can really deliver the promised benefit.

Purchase intent provides a basis for estimating trial. In order to estimate overall sales volume, we need also to assess potential purchase frequency and make some projections about quantities purchased per purchase occasion. Of course, products that have high purchase frequency may represent a high-volume opportunity. Schwartz (1985) points out that questions that measure intended purchase frequency are often omitted from concept tests, because the data may prove to be unreliable—consumers often significantly overstate their intended purchase or usage frequency. However, eliminating the question is not the solution. It is better to ask the question and use some care in interpreting the result, assessing data in a relative sense by using norms from previous introductions in the category or a similar category. We can also employ consumer panel data to deflate responses to purchase frequency questions to bring them into line with actual purchase patterns. It is insightful, with a sufficiently large sample size, to cross-tabulate intended purchase-frequency with different population characteristics such as household size, household income, and so on. This will be particularly helpful in identifying the heavy user segment if one exists.

Other useful findings from concept tests include identifying the user and the usage occasion(s), and how the concept might be improved. The user is the type of person that respondents associate with your product. This information is important primarily in planning our advertising, since user imagery is a key element of creative strategy, particularly in categories such as beverages, cosmetics and fragrances, and tobacco products. Usage occasions not only help to identify the competitive set, which may and often does differ according to usage occasion, but also provide clues for our positioning and marketing program. Consumers often can provide very good insight into how to improve a concept. Simply include the question, "In what ways, if any, could this product be changed or improved?"

ABSOLUTE VERSUS RELATIVE CONCEPT TESTS

Concepts can be tested in two ways: either individually or with control concepts. In a monadic concept test, a respondent evaluates only one concept or idea. In a monadic sequential test, a respondent is exposed to several ideas, one at a time. Respondents see and evaluate the first idea, then repeat the procedure across other ideas, with the order of presentation rotated across respondents. The alternatives are compared by the same sample of respondents.

Relative concept tests involve testing concepts against other concepts or existing products in a competitive setting. While this type of test introduces a more realistic aspect to the process, most experts prefer absolute or monadic tests, because absolute measures are less biased and more amenable to sales volume estimation.

Choice Modeling

Many packaged goods manufacturers have adopted choice modeling in order to develop a detailed understanding of how specific product attributes or features and the perceived levels of attributes present in new product concepts impact consumer evaluation of concepts and intent to buy. Product developers need to understand what drives consumer reactions, and what types of trade-offs consumers are willing to make among different product features. This can be accomplished during concept testing with the use of customer-choice modeling methodology.

An in-depth discussion of customer-choice modeling and conjoint analysis is discussed by Moskowitz et al. in Chapter 10. A substantial literature on each of these topics exists. One exposition of the use of conjoint analysis in concept testing is by Green et al. (1997). In a conjoint analysis study, product concepts are represented by the attributes or features that comprise them, where each attribute can have multiple possible levels. The goal of the methodology is to determine which attribute and attribute levels customers prefer and how much customers value the attributes (Dahan and Hauser, 2001). Product development teams can learn what consumers want and value from their reactions even if the consumers cannot articulate those wants directly (Moskowitz et al., 2004.) According to Green and his colleagues, "conjoint studies of new product concepts are particularly useful when product class histories or norms are not available; when the researcher is not certain of final product design features; when the researcher is interested in alternative price/demand relationships, marketing positioning, and buyer segmentation; and when the new product is technologically complex and requires investigation of consumer learning, where such learning depends on the nature of the product's features."

Customer-choice modeling, according to Verma and Plaschka (2003) is a series of related methodologies that allow researchers, using data from a variety of industry and consumer sources, to develop a list of core market-choice drivers, that is, purchase attributes that influence perceptions of value and purchase intent for a product or service. For new product research, the goal is to find the optimal combination of benefits and levels of these benefits that increase both the value perception and the intent to purchase by target consumers. Because only a limited number of attributes, each with a limited number of levels, create potentially thousands of potential choice configurations, technologies such as fractionated factorial design are used to limit the number of configurations that must be tested while preserving the ability to estimate all attribute-level main effects.

In customer-choice experiments, respondents are asked to choose among alternatives presented in choice sets. Advances in information technology, including broadband Internet connections, digital imaging, and streaming video allow researchers to develop realistic experiments that provide realistic and easy-to-use formats that lead to a high level of respondent interest and involvement. Verma and Plaschka (2003) believe that "choice modeling can yield valuable insights for market-driven strategy development by revealing customer clusters, suggesting the potential effects of changing the levels of value drivers, assessing overall brand equity, and identifying customers' switching barriers."

In order to obtain a quick read on the product features people wanted most in low-fat yogurts, Dannon USA contacted a research supplier named Affinnova, which e-mailed 40,000 men and women and asked them to click on a link to "help create a new product and get a shot at winning $10,000." (Wells, 2005) The survey achieved a response rate of 11% and 705 people qualified—as diet conscious yogurt eaters who were responsible for grocery shopping in their household—to take the test. Affinova presented online a series of yogurt containers in such a way that respondents could evaluate different combinations of name, package design, nutritional labeling, and size. While there were 11,268 possible combinations, each respondent was asked to choose from among a limited number. The algorithm recombined features to highlight frequently selected elements. The test changed as more people completed it, and the program determined the most popular combinations of characteristics.

All the Internet activity required 6 days, and by the end of the project, Affinnova had concluded that the new yogurt would be sold in a red container, in a four-pack of 4-ounce cups, highlight the benefits "80% less sugar" and "3 g of carbs" and be called Carb Control. The new product hit supermarket shelves 6 months after the test was concluded, and 1-year sales hit over $70 million in late 2004 (Wells, 2005).

Using a similar methodology, Taco Bell found over 1100 participants on the Web to help them design a healthy burrito. Respondents chose from among 10 categories of fixings, including 3 kinds of chicken and 11 sauces, and watched as an animated program assembled and cooked their concoctions. While the research showed Taco Bell which combinations consumers preferred, the company was surprised to learn that, instead of a low calorie item, consumers wanted a more indulgent burrito with three cheeses and were willing to pay a premium to get it.

Concept Test Validation

We have previously discussed the notion of norming concept test results, particularly purchase intentions. This is one type of validation procedure. The major premise of any concept testing procedure is that consumers' reactions to a concept are valid and reliable indicators of their likely future behavior in the market. To establish the validity of purchase intent scores as predictors of trial requires the analysis of a significant number of cases wherein we can compare purchase intent scores with subsequent performance of the product in actual market conditions, either a sell-in test market or after launch. Without this information, we can have little confidence in the validity of our purchase intent data and the ability to predict trial. One technique used by some analysts is to develop a trial estimate based on a weighted average of our purchase intent scores. For example, the "top box" score is weighted by 80, the second box score is weighted by 60 and so on so that the percentage of respondents who checked "definitely would not buy" is weighted by zero. It is important to remember in this context that high scoring concepts MAY be successful and low scoring concepts will most likely not be successful.

Volume Estimation

One of the most critical outcomes of the concept testing phase is an estimation of first-year sales volume. Estimating sales volume is predicated on the A-T-R model and relationships we discussed earlier in this chapter. Sales volume estimates require the data shown in Table 5.3.

Market size estimates are based on category data that measure the number of households purchasing items in the category plus a projection of the geographic scope of the launch. For instance, canned soup has close to 100% household penetration and there are approximately 110 million households in the United States. Thus, Campbell Soup Company might identify the potential market size to be 55 million households in a launch that will reach about 50% of the population during the introductory period.

Awareness levels are projections based upon relationships between the strength of the introductory marketing campaign and the percentage of households in the target market that will become aware of the new item. Introductory marketing campaigns, for purposes of these projections, typically consist of several different elements. Media advertising campaigns help build awareness. The strength of an advertising campaign or program is usually measured by the number of gross impressions delivered (that is, the number of exposures of the message to consumers, including multiple exposures to the same consumer), or by gross rating points (GRPs). GRPs represent a measure

TABLE 5.3
Data Requirements for Estimating Sales Volume for a New Product

- Estimated market size, usually stated in terms of number of households in the target market
- Projected level of awareness of the new item, which will be created primarily through consumer advertising plus consumer and trade promotions
- Projected level of distribution penetration, typically stated in terms of a percentage of ACV distribution to be attained within a given period after launch and based on both the strength of the concept and the strength of the trade program
- Estimated percentage of households that will try the product plus the average trial volume per household
- Estimated percentage of households that will buy the product at least once after trial (repeat purchase) plus the projected number of purchase cycles and the average volume per purchase during the introductory period
- The price of the product to the trade.

of the weight of an advertising campaign. GRPs are calculated by multiplying the net unduplicated reach of the campaign, stated in terms of a percentage, by the frequency, or average number of times each household is exposed to at least one message. Thus, a campaign that reaches 80% of the target audience an average of 3 times generates 240 GRPs. Historical category norms allow product developers to estimate with reasonable precision the relationship between GRPs and awareness level.

Introductory marketing campaigns also include consumer promotions, such as coupons, trade promotions, and monetary incentives provided to the trade to encourage the stocking and displaying the new item. In many cases, the retail trade will pass at least some of the savings along to consumers in the form of lower introductory prices. Again, there is no way other than conducting a controlled experiment to determine the relationship between the strength of the introductory marketing campaign and the awareness levels created without reference to historical launch data. One of the major reasons many food processors outsource this type of research is because of the large number of cases research companies, such as AC Nielsen BASES, have developed that provide reasonably accurate and dependable mathematical relationships.

Projections of distribution penetration rely on the same type of historical database identified above. Wholesalers and retailers will "cut-in" new items if there is money to be made. Thus, the ability of the concept to sell to consumers, trade margins, and trade incentives will impact distribution penetration. We need reference to historical data to make accurate projections.

Trial can be measured empirically through the use of a simulated test market such as BASES II, which provides a sample of consumers an opportunity to purchase the product under simulated market conditions. Alternatively, trial can be projected from purchase intent scores. This is where norms and validation of these scores comes heavily into play. Again, historical data that allows us to determine some sort of probability relationship between "top box" or "top two box" scores and trial rates, are necessary components of the process. The average quantity purchased at trial may simply be projected as one unit. However, if we are launching a line of items simultaneously, and we incentivize the consumer to buy more than one item at a time, we have to adjust the quantity estimate accordingly.

How does one estimate repeat purchase? Repeat volume is a function of how many times a household purchases a new item after trial and the average volume at each purchase occasion. Repeat purchase can be measured empirically using tests that allow respondents an opportunity to repurchase the item after initial trial in a simulated shopping environment. Historical data from launches of products in the same category as well as the specifics of both the category the launch program must be considered in estimation of purchase frequency and the average quantity purchased.

Repeat purchase is a function of satisfaction with the product, so that we can use ratings of concept strength, in the absence of any empirical data, as a surrogate for satisfaction, and estimate repeat purchase in a manner similar to that with which we estimated trial from purchase intent data. Data on purchase frequency and amount purchased can be gleaned directly from questions included in the concept test, adjusted in some manner. The reason for the adjustment is that consumers usually cannot accurately provide volume and frequency information; they tend to overstate their intended consumption. Adjustment factors are calculated by examining historical relationships. One technique is to correlate responses to similar questions in previous concept surveys versus actual diary panel, or store scanner data for those same products. The repeat purchase estimate is a weak link in the predictive process, regardless of how it is derived. You never know for sure if your concept will stimulate more or less than average repeat buying until it reaches the market.

The price of the product to the trade comes from the launch team. We must consider both normal and introductory prices and trade margins. We also need to consider both seasonal sales patterns and the timing and pattern during the introductory period of volume build, which more sophisticated models allow. A procedure for estimating first-year sales volume is shown in Table 5.4.

These computations will be extremely sensitive to the values we assign to the parameters and variables, and so realistic numbers must be assigned to awareness, distribution penetration, average quantities purchased, and number of purchase events. Moreover, the estimates of trial and repeat, if overstated, can mislead companies into spending money to develop losers. Finally, one should note that we could perform sensitivity analyses with models of this type. If projections do not meet critical levels, we can adjust the strength of our consumer or trade marketing campaigns to create greater awareness or distribution penetration, to make the numbers move. Clearly, we can also adjust components of trial and repeat purchase estimates.

TABLE 5.4
A Procedure for Estimating First-Year Sales Volume

- Start with the number of households in the target audience.
- Multiply by the level of awareness. The product of these two numbers is the percentage of target households who become aware of the existence of the new item.
- Multiply the aware households by distribution penetration, because one cannot purchase the product if it is not available in stores they shop.
- Multiply the number of aware households with access to the product by the projected trial rate. You have now computed the number of trial households.
- Multiply the number of trial households by the average volume purchased at trial. You now have trial volume in units.
- Multiply the number of trial households by the percentage of repeat purchasers. You now have the number of repeat purchase households.
- Multiply the number of repeat purchase households by the average number of purchase events, based upon the repeat purchase cycle, which is the average amount of time between purchases and the average number of units purchased per purchase event. You now have the repeat purchase volume in units.
- Add trial volume plus repeat volume, multiply by the price at which the product is sold to the trade, and you have projected first year sales volume in dollars (manufacturer sales).

TABLE 5.5
Sample Simulation

Assumptions	Low Spend	High Spend
• Distribution	80%	90%
• FSIs	2	4
• GRPs	1500	2500
• Sampling	No	Yes
Output/forecast		
• YR 1 trial	4.2%	9.5%
• YR 1 repeat	41%	43%
• YR 1 volume (000 cases)	1700	4000

Table 5.5 provides an example of a simulation leading to a sales forecast. Two treatments are simulated: a low spending introductory plan and a high spending introductory plan, yielding dramatically different sales forecasts:

Caveats

Fitzpatrick (1996) notes several important caveats to keep in mind regarding concept testing:

- The concept statement is a description written to explore potential consumer interest; include some "sell," that is, how the product is different, makes life easier, and so forth
- Include some sort of price context to avoid major misunderstanding
- Do not make materials too elaborate; retain flexibility in presentation
- Be prepared to modify the benefits as you get consumer reaction
- Consider "layering" different options on the basic concept
- Elicit individual as well as group opinion
- Make sure you understand the difference between lack of consumer understanding (that is, they understand the concept differently from what we understand) and consumer confusion (they do not know what we are talking about)

Improving Concept Testing

Clancy and Kreig (2003) suggest some modifications to improve concept testing accuracy and projectability. First, they suggest larger samples, between 300 and 500 respondents, to insure that the results are more reliable. Second, achieve a balance between different methods of administration, particularly telephone and Internet data collection, mailing a concept description and scoring scale to some respondents before contacting them by phone, doing some one-on-one interviews to balance the results from an Internet survey, and making sure that the firm conducting the Web survey is using a demographically balanced, representative panel. Third, use full concept descriptions, complete with name, positioning, packaging, features, and price. Present the concept in a competitive context. Fourth, measure purchase probability using a scale superior to the traditional 5-point purchase intent scale; the authors have tested alternative scales and concluded that an 11-point scale better predicts real world behavior, especially for high involvement decisions. Keep in mind that any scale will yield purchase intent that is overstated and needs to be adjusted using historical category norms. Finally, a modified multiple trade-off analysis using either conjoint measurement or choice modeling methodologies can predict consumer behavior and sales for a range of alternative concepts, and allow the researcher to understand which elements of the concept drive consumer perceptions and decisions.

Another method is to display concept scores graphically in two relevant dimensions. In order to understand why concepts are perceived as acceptable or unacceptable, strong or weak, and so on, we can ask respondents to rate the concept on specific product attributes, including the importance of a given attribute. Ratings on individual product attributes represent a "disaggregation" of the overall concept rating. Plotting the concept ratings on each attribute and the importance ratings of each attribute on a on a two-dimensional scale provides a clear visual presentation that highlights the strengths and weaknesses of the concept.

The importance ratings are plotted on the vertical axis, highest at the top and lowest at the bottom. The concept ratings are plotted on the horizontal axis, highest on the left and lowest on the right. The upper left quadrant contains attributes that are important to the respondent and that the test concept scores well on. The upper right quadrant contains attributes that are important but low-rated. This provides direct and dramatic insight for the development team regarding how the concept can be improved by reconfiguring the mix of attributes that comprise the new product concept, and thus the positioning.

SUMMARY

The A-T-R model provides a theoretical underpinning to the process of launching a new product as well as the estimation of first year sales volume, which is a key outcome of the concept testing process. Product concepts are verbal or pictorial representations of products. Concept statements provide a description of a new product's attributes and benefits, as well as a stimulus to which consumers can respond to assess the overall strength of the concept and the level of consumer interest. Concept statements also provide a protocol to guide the work of the project team as the concept is developed into a physical product. Concept tests employ a variety of research methodologies to measure consumer reaction to the concept, determine the concept's strengths and weaknesses, and measure purchase intent, and serve as a gate through which only high-scoring concepts pass to the next stage of development. Purchase intent, along with other data, is used to create early stage sales volume estimates. Purchase intent scores must be validated, usually through comparison with previous test scores combined with actual market performance. This requires that a substantial database of concept test results be maintained. A.C. Nielsen BASES and M/A/R/C offer concept and simulated test market products frequently used by food processors.

BIBLIOGRAPHY

Anonymous. 1997. HBS Case Services, case number 595-035, Nestle Refrigerated Foods: Contadina Pasta & Pizza (A) Rev. 1/97.

Anonymous. 1999. *The CBC System for Choice-Based Conjoint Analysis*. Sequin, WA: Sawtooth Software, Inc.

Anonymous. 2005. *Concept Evaluator*. www.ipsos-insight.com, accessed April 9, 2005.

Anonymous. 2005. *Concept Testing*. www.decisionanalyst.com/services/concept.asp, accessed January 24, 2005.

Anonymous. 2005. *GoProspect*. www.ipsos-insight.com, accessed January 24, 2005.

Anonymous. 2005. *Proprietary Research*. www.npdfoodworld.com, accessed January 24, 2005.

Anonymous. 2005. *Research to Guide Product Development: PRISM Stage 2 Concept Development*, http://prism.mrsi.com, accessed April 9, 2005.

Anonymous. 2005. *Sorenson In-Store Sales Forecast*. www.sorensen-associates.com, accessed April 9, 2005.

Arnold, C. 2002. Hershey research sees net gain, *Marketing News,* 36(24):17.

Clancy, K. J. and P. C. Krieg. 2003. Surviving Innovation, *Marketing Management*, March 1, 12(2):1061–3846.

Clancy, K. J., R. S. Shulman, and M. M. Wolf. 1994. *Simulated Test Marketing.* New York: Lexington Books.

Crawford, C. M. and A. DiBenedetto 2006 *New Products Management.* Boston, MA: Irwin McGraw-Hill.

Dahan, E. and J. Hauser. 2002. "The Virtual Customer," *The Journal of Product Innovation Management,* 19:332–353.

Dahan, E. and V. Srinivasan. 2000. The Predictive Power of Internet-Based Product Concept Testing Using Visual and Animation, *Journal of Product Innovation Management*, 17:99–109.

Dahan, E. and H. Mendelson. 2001. "An Extreme-Value Model of Concept Testing," *Management Science*, 47(1):102–116.

Fitzpatrick, L. 1996. "Qualitative Concepts Testing Tells Us What We Don't Know," *Marketing News,* 30(20):11.

Green, P., A. M. Krieger and T. G. Vavra. 1997. Evaluating New Products, *Marketing Research*, 9(4):12–21.

Light, B. 2004. Kellogg's Goes Online for Consumer Research, *Packaging Digest*, 41(7):40.

Moskowitz, H. R., R. Katz, J. Beckley and H. Ashman. 2004. Understanding Conjoint Analysis, *Food Technology*, 58(1):35–38.

Patrick, J. 1997. *How to Develop Successful New Products.* Lincolnwood, IL: NTC Business Books.

Ray, N. and S. Tabor. 2003. "Several Issues Affect e-Research Validity," *Marketing News,* 37(19):50–53.

Rubin, J. 2000. Online Marketing Research Comes of Age, *Brandweek*, 41(42):26–27.

Schwartz, D. 1985. *Concept Testing.* New York: AMACOM.

Verma, R. and G. Plaschka. 2003. The Art and Science of Customer-Choice Modeling: Reflections, Advances, and Managerial Implications, *Cornell Hotel and Restaurant Administration Quarterly*, 44(5/6):156–165.

Wells, M. 2005. *Have It Your Way*, www.forbes.com, accessed April 9, 2005.

6 Consumer-Driven Product Design of Foods and Beverages: Methods, Mind-Sets, Metrics

Howard R. Moskowitz, Matthias Silcher, Jeffrey Ewald, and Jacqueline Beckley

CONTENTS

This chapter is about us, the consumer. We dictate winners and losers by purchasing some items and rejecting others. Because of this understanding the algebra of the consumer mind is a major contribution to successful product development and marketing. The objective of this chapter is to show new ways of using consumer insights to make "bottom up" product development more efficient and in the end more successful.

This chapter presents an integrated approach to new product design, beginning with the study of consumer demands, moving on to concept development and creating databases, and finishing with product design. It presents a synthesis of approaches that meld together traditional sensory evaluation, consumer research, product optimization, and marketing.

At the beginning, we describe methods that help the food product developer to understand the current marketplace and what drives consumers to demand certain products. The chapter then describes the applications of online conjoint analysis and their possible use in consumer driven product design. It deals in depth with the creating of cross-section and longitudinal databases that help the product developer to see bigger consumer patterns.

When writing about product designing, it is tempting to concentrate on the research design and analysis, while leaving the actual data acquisition un-discussed. This chapter deals profoundly with data acquisition; with questionnaires and scales it addresses the respondent's capacity to evaluate products and different types of test venue.

In addition, the chapter deals with the use of consumers, interactively, to provide feedback on product innovation at the very early stages. Several case studies illustrated within this chapter show effective and successful practical applications.

PART 1—THE IMPORTANCE OF DESIGN IN TODAY'S COMPETITIVE WORLD

Folk wisdom in the food business holds that a vast majority of food products introduced by a company will fail in the marketplace even though many of these introductions have enjoyed some level of research and are being introduced with the power of knowledge behind them. Anyone looking at the success rate of products must wonder occasionally whether the marketing knowledge has been correctly applied, and even whether product developer's design was appropriately implemented to guide marketing. In this context, product development design comprises the different information-gathering steps, from knowledge of trends to new concepts, new product formulations, and new packages. Are we, in fact, seeing the advance of methodology in consumer insight and a simultaneous meltdown in the ultimate product for which that insight was formalized and professionalized in the first place? If so, it is time to revisit the approaches for early stage product design, especially those that have gone under the rubric of *consumer insights.*

This section deals with the fundamentals of product design from the consumer point of view. It bridges the gap between two large fields, psychophysics/sensory analysis with its hoary history and practices and the newly emerging fields of *quali/quant* consumer research. We will get a good idea of how these methods work when we trace problems from their scientific aspects, through our base of scientific knowledge and onto these newer methods. The chapter also presents the approaches in the light of today's ever-increasing speed of design and development (Gleick, 1999) and the need to accelerate knowledge building and insights in order to remain competitive. The chapter differs from the traditional approaches of sensory analysis first codified by Amerine et al. (1965) and later extended by the researcher. Instead, the chapter presents a synthesis of approaches that meld together traditional sensory evaluation, consumer research, product optimization, and marketing.

Consumers are notoriously fickle about the products they select; they become bored with products; their sensory systems pick up discrepancies between what a product promises versus what it delivers, and pick up sensory signals about product quality. Just as importantly, consumers are hard-wired in the chemical senses, taste, and smell, to provide an initial reaction of accept/reject, and only afterwards to assess the characteristics of the product. Finally, the reader should always keep in mind that the consumer we speak of in this chapter is us, not some disembodied organism whose job it is to evaluate test stimulus. The *us* we speak about eats and drinks to live and to enjoy, spending hard earned money in doing so.

With these issues squarely in front of us, how do the product developer and the marketer understand the mind of the consumer, what is desired, and use that information to translate consumer desires into products? What do we know about the practical aspects of our sensory systems—not the esoteric and arcane facts of sensory information processing, but rather what happens when our senses meet the physical stimuli that we know as food and drink? And finally, ever the pragmatist, how do we apply this sensory knowledge to food design?

More importantly than even sensory function is an understanding of hedonics—likes and dislikes. What do we know about the measurement of what people like? Can we engineer liking in a scientific fashion in order to increase the probability that a food will be accepted. What do we know about the influence of expectations on food and the need to fit a product to a concept, so that the consumer is given something that he will eat?

And finally, with all of this knowledge, how can we increase the chances of success by developing a knowledge-based system? What disciplines do we invoke, beyond the usual suspects of "sensory evaluation," and its scientific cousin, psychophysical measurement? Is there room for a new approach in this field, grounded in sensory analysis principles, but incorporating ethnographies of how we interact with food, assessment of the language we use about food to identify core ideas, rapid development of ideas and concepts, and finally powerful methods for developing new products that fit these concepts?

The Critical Role of Consumer Requirements and Inputs

The growing competition in the food and beverage industry has catapulted the consumer to new levels of influence. One could, of course, argue that it is consumers who have driven product development and marketing all through the years, and for the most part that is a truism. A product cannot succeed without consumer acceptance, unless the manufacturer is willing to pay enormous sums of money to give the product away. This strategy eventually fails as the company runs out of patience and money. With the accelerating competition faced by companies, however, and with the limited consumer budgets and shelf space at the market, all companies cannot win in the marketplace, even if they spend the money and give the product away. Consumers are willing to spend only a limited amount of money on food and beverage. Even more importantly, consumers will not even accept for free products that they do not like. With increasing numbers of market entries, some must necessarily lose out. It is the consumer who dictates winners and losers by purchasing some items and rejecting others. Consumer requirements, what the consumer wants and will accept, become increasingly important in this "supply economy," where it is supply, not demand, that is in abundance.

Beyond accepting the need to acknowledge consumer requirements, the corporate product designer needs to establish those requirements in an actionable way. We know that consumers like some products and reject others. We know that there are health trends (Health United States, 2002) that foods enjoy, fads, those ideas about one's state of being drive purchases. A good product design program combines knowledge about consumer preferences with knowledge about trends and with knowledge about fundamental consumer nutrition. These consumer requirements join together to provide the structure in which product development should take place.

This chapter presents an integrated approach to new product design, beginning with the study of consumer demands, moving on to concept development, and finishing with product design and package graphics. The objective is to show how new ways of thinking about consumer insights are making product development more efficient and business focused.

PART 2—COMPETITIVE ANALYSIS AND CATEGORY APPRAISAL

Developers and researchers will enjoy more success more frequently if they do homework before designing the product. By homework is meant an analysis of the competitive frame. Certainly, there are those occasions when an idea for a product emerges *deus ex machina*, like Venus from the head of Jupiter. Everyone loves to hear stories of the emergence, almost by magic, of new ideas that revolutionize the industry and enjoy unexpected commercial success. Such stories are wonderful, but they do not portray the real situation. Developers will be successful if they understand what it is about the current marketplace that drives consumers to demand products, what it is about the products themselves that drive repeat purchase, and the long-term trends in which the industry finds itself. These insights are not difficult to achieve. They require a disciplined evaluation of the environment, a disciplined evaluation of current product communications by corporations, and assessment of existing products. In the patterns that emerge will lay the answer.

SCANNING TRENDS/MACROENVIRONMENT

Trends are long-term shifts in what is happening in the world around us. Examples of trends include the focus on health, the focus on food as a delivery mechanism for physical features that promote health, and the emphasis on convenience. Trends can be technical, such as trends in ingredients, or they can be popular, such as trends in consumer demands. The literature on trends is diffuse, with articles appearing in trade magazines as well as in the popular press. We will not deal with trends here other than to mention that they are sources of ideas about product features. For example, as the trend continues toward improved knowledge about the nutritional properties of foods we may expect consumers to become interested in the ingredients of their products, beyond the typical ones of sugar, salt, and fat. Healthful components such as phytoestrogens, leutein, and so forth, may become important, and serve as a basis for a food concept. By scanning trends on a regular basis, the food designer creates a framework in which to design foods that may satisfy consumer desires.

Deconstructing Competitor Messaging

Competing companies cannot be particularly secretive about their products. In the food industry, there are patents to protect intellectual property, but the consumption of food is generally dictated by consumer demands. Consumers select what they want to eat, often on the basis of what they read and what they hear. Consequently, product designers can learn a great deal about what consumers want, based upon deconstructing the current messaging and getting consumer responses to those messages. For example, Table 6.1 shows best versus worst performing concept elements on the basis of the rank order of their utilities or "ability" to drive interest in a concept that contains them. These elements come from the current competitive frame of "good for you foods." The researcher must guard against closing his mind by testing too few elements from the competitive messaging. In this case, the more the elements tested the better the data set will be, because it will comprise more information, and give a fuller picture of the different messages.

TABLE 6.1
Ranking of Elements for Good for You Foods, and the Source of Information (Web Site)

Element	Source	Rank
Foods with active ingredients that provide health benefits	healthwellexchange.com	1
Food that may prevent disease or promote health	ificinfo.health.org	2
Disease prevention through food	healthwellexchange.com	3
A food that does more than just provide good nutrition	healthwellexchange.com	4
Contains natural antioxidants	bestlifeint.com	5
Helps prevent and fight cancer	bestlifeint.com	6
Fresh vegetables	healthwellexchange.com	7
Food that provides a health benefit beyond the traditional nutrients it contains	healthwellexchange.com	8
Fresh fruits	healthwellexchange.com	9
Made with Beta-carotene	richmond.k2.ga.us	10
:	:	:
:	:	:
A refrigerated functional beverage	healthwellexchange.com	120
Helps ease the symptoms of menopause	whitewave.com	121
Effective in helping women during menopause	bestlifeint.com	122
All ingredients are certified organically grown	turtlemountain.com	123
Specializing in natural health food	soland.com	124

Note: The rank goes from 1 (best performing element among consumers, based on the utility values) to 124 (worst performing element among consumers).

Deconstructing Competitor Products

Another very good way to derive direction for design and development looks at the actual products offered by competitors. These products often span a reasonably wide sensory range, principally because the competitors try to outdo each other in pleasing customers, while at the same time searching for a niche that they can occupy alone. Like the well-known development of niches in evolution, products tend to fall into niches of different sensory profiles. The competitors in a product category may or may not know the nature of the consumer sensory preferences, but with sufficient trial and error they eventually discover the different locations in the sensory world that allow them to survive.

Deconstructing the competitor products requires two pieces of information, evaluation of the sensory characteristics of products and evaluation of acceptance, or commercial success. These are three separate pieces of information, requiring different decisions rules. Acceptance and commercial success are not the same. A person can accept a product but not buy it because it is too expensive. A person cannot accept a product but buy it because it is so cheap to be considered a "bargain not to be missed." The objective in this type of deconstruction is to create a model or relation between sensory characteristics and acceptance that the product designer can use to create new products. Although some practitioners aver that they can get good direction by having the respondents profile the "ideal" product, or indicate how far "off-target" a product may be in sensory characteristics, the weight of evidence is clear that most success comes from developing sensory-liking relations using many products.

The output of competitive product analysis comprises two things—a table of attributes by products, and curves relating sensory attribute level to attribute liking or overall liking (so-called drivers of liking analysis). The tabular data is not particularly revealing by itself, because it comprises simply a set of numbers, without relations between variables. By looking at the data the researcher can get a sense of the sensory range achieved by the different competitors, and the range of overall liking. This type of information can be useful to identify, within the existing competitive framework, what are the attribute levels that have been achieved by competitors, and how acceptable are the competitive products (see Table 6.2, where the numbers are ratings on anchored 0–100 point scales). If the researcher tests the competitors both unbranded and branded (i.e., identified as to brand), then it becomes possible to determine the value of the brand, and the degree to which the different brands affect overall liking.

A lot more insight emerges when the researcher plots a scatter gram relating sensory attribute level on the abscissa to liking on the ordinate. The typical type of curve shown on the left side of Figure 6.1 is "noisy." That is, at first glance there is no discernible pattern. This lack of pattern is to be expected, because overall liking is determined by several attributes interacting with each other, not by one attribute alone. Nonetheless, if the researcher fits a quadratic function through this noisy data, it may be possible to identify the approximate sensory level at which overall liking maximizes (Moskowitz, 1981). Figure 6.1 shows the scattergram before and after a curve is fitted to the data. The approach is meant as a heuristic, to aid the developer,

TABLE 6.2
Competitive Analysis Table for Eight Commercially Available Bologna Products Currently Being Marketed to Consumers

Test Condition	Rating Attribute	Statistical Summary			The Eight Competitor Bologna Products								Sensory-Liking Curve*
		Min	Max	Range	A	B	C	D	E	F	G	H	Optimum
Liking attributes													
B	Overall	42	59	17	54	42	56	50	59	52	53	58	
I		49	69	20	58	53	54	49	68	50	69	63	
B	Appearance	34	52	18	45	34	41	42	52	41	49	50	
I		44	68	24	55	54	44	46	66	51	68	57	
B	Aroma	47	62	15	52	47	58	49	60	51	52	62	
		53	66	13	60	58	61	53	66	57	66	63	
B	Taste	41	62	21	54	41	57	51	59	53	51	62	
I		49	69	20	59	55	58	49	66	52	69	63	
B	Texture	38	63	25	55	38	55	52	61	56	53	63	
I		49	68	19	61	49	62	50	68	54	64	60	
Sensory attributes													
B	Lightness	55	52	3	46	52	48	59	56	44	59	15	15
		60	66	6	50	66	52	73	58	49	73	24	
B	Aroma	48	56	8	48	51	54	56	51	48	54	50	51
I		50	59	9	53	57	55	55	59	55	59	50	
B	Taste	47	60	13	49	47	58	55	60	56	56	51	58
I		53	68	15	53	54	60	59	68	55	66	55	
B	Firm	46	59	13	46	59	47	52	54	50	55	55	51
I		48	61	13	50	55	53	56	61	48	54	58	
B	Juicy	49	60	11	49	51	58	55	60	55	53	51	60
I		51	64	13	58	56	55	54	64	51	62	56	
B	Chewy	36	55	19	36	55	41	46	42	39	41	46	42
I		39	52	13	39	52	45	46	46	42	45	48	

B = Blind test, I = Branded/Identified Test. Optimum sensory level is estimated from the quadratic plot of attribute liking (ordinate) plotted against sensory intensity (abscissa). All numbers are expressed in average ratings across respondents from anchored 0–100 point rating scales.

* The optimum score identifies the "On-Target-Area." Products with a number far below the optimum score have "not enough" and products with a score much higher than the optimum score have "too much" and can be considered as "Off-Target."

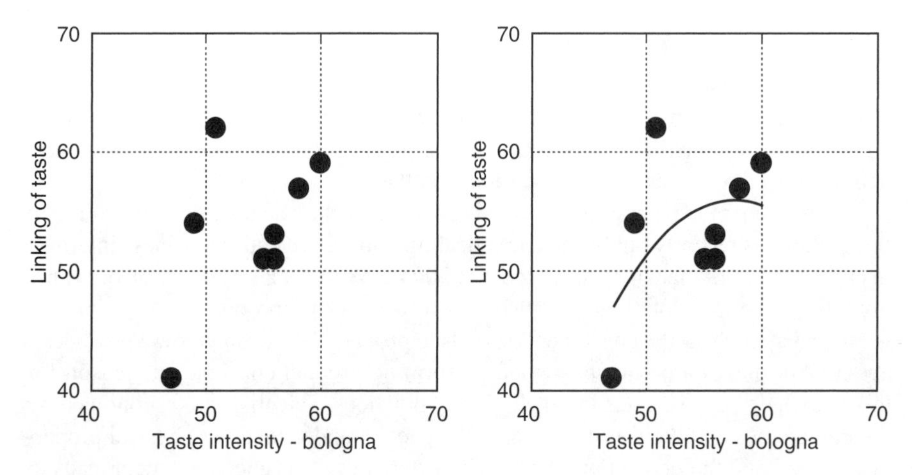

FIGURE 6.1 Sensory-linking curve for bologna before and after curve fitting.

rather than setting the optimal level as a fixed point. As the sensory intensity of taste increases for the bologna product, the liking of taste increases as well. However, there is a point, beyond which additional increases in taste intensity do not lead to increasing the liking of taste. Table 6.2 shows the optimal levels that might be estimated from this approach, keeping in mind that acceptance is driven by the interaction of several attributes.

PART 3: CONCEPTS—THE "BLUEPRINT" OF THE PRODUCT

In product design very few developers will start out without a product concept. Concepts come in two types—product concepts and positioning concepts. *The product concept tells the developer what the product should be. The positioning concepts tell the developer (but more typically the marketer) why the consumer should buy the product.* A sense of the difference between these two types of concepts can be seen from the example below. The product concept is more concrete—it allows the respondent to visualize what the product will be like. The positioning concept, in contrast, is more emotional, more oriented toward motivation. In practice, the distinction between the product and the positioning concept is often blurred, but for early stage product design it is important that we work with concepts that lead to specific development action. Product concepts prescribe the product, and these concepts act as a guide for his desired action.

We can get a sense of the differences between the concepts by comparing the product and the positioning concepts for yogurt. The differences between the two types of concepts are not always clear. For the most part the product concept tries to describe the product, without trying to sell the benefits. In practice, there is a lot of cross-over between product and positioning concepts, but for the purposes of product design, we will try to focus on product concepts that tell us the features of the product.

What Goes into a Concept?

Product concepts do not emerge fully formed from the development laboratory. Matters are not so direct. A product emerges to fit a need, whether this need is simply a new flavor of an existing product, or some physical product that solves a clearly defined need (e.g., a sandwich to be eaten "on the go").

There is not much literature on product concepts in the scientific domain, although one author has recently published a text dealing with the scientific/business interface in concept development (Moskowitz et al., 2004). We know a lot about sensory functioning, we know a lot about the physical properties of a product, and we know a great deal about how the consumer responds to product. We do not know a great deal, about what makes a particularly good performing product concept. The reason for this knowledge can be traced to the industry practices. Scientists are encouraged by the universities and their professional colleagues to understand the physical properties of food and the perception of food characteristics. This encouragement leads to publications about sensory perception in the public, refereed scientific literature, and to conferences. In contrast, companies do not talk about what makes a good product concept. Students in the university do not learn about how to construct good product concepts. The university emphasis is on measuring responses to product concepts and to product concept performance, in a technical way, without much soul. Finally, the companies test many product concepts, embellishing and refining the ones that look promising until they become fully fleshed concepts that turn into products. The companies discard the poor performing concepts. No one really looks at why the poor performers are poor performers—everyone is interested in the winning ideas.

Measuring Fully Formed Concepts

Not every concept is well accepted by consumers. Although many product developers, marketers, and innovation specialists feel that they intuitively "know" the consumer market, rarely does the corporation trust these intuitive "gut" feelings. Rather, the product and positioning concepts are generally put to the test. The concepts may be evaluated by an internal group, with winning concepts put to focus groups or to concept screening. In focus groups, a small group of relevant consumers discuss the concept, arguing the pros and cons of one or several concepts put out for discussion and analysis. Occasionally, the participants in the focus group may vote on the concept. The enterprising researcher might run several of these focus groups and tally the results. Some researchers have gone so far as to recruit relatively large numbers of respondents to evaluate a large number of concepts, using "dials" to capture data live, and computer analysis to present the results to the client audience in attendance (MS Interactive, 2004).

The most general way to measure concepts consists of creating a number of these concepts, testing them among consumers in a concept screen, and reporting the data. The respondent may be intercepted in a mall, sent concepts by mail, or sent email invitations to participate in a concept screen. No matter which method is selected, from personal to phone to computer interview, the objective remains the same—to rank these ideas in terms of acceptance, and if possible estimate potential of these concepts for the market. Quite often researchers and marketers make use of norms to estimate the likely market success using concept scores as part of the input.

Concepts can be rated on a variety of scales, to assess the different dimensions of the concept. Of course, the most basic dimension is acceptance. Acceptance might be as vague as overall acceptance or as ostensibly specific as expected frequency of use and nature of use (e.g., will it replace current products or add to the current array). There are many different scales on which one can rate these concepts. The most typical is purchase intent, ranging from definitely not purchase to definitely purchase.

Consumers rate concepts on category scales, usually the 5-point purchase intent scale or the 9-point liking scale. Often the researcher transforms the data so that the ratings become binary (accept or reject). For this transformation into binary, the analysis shows how the elements drive the conditional probability of accept or reject the concept. Market researchers, interested as they are in the performance of groups of consumers, typically work with this type of binary data, such as the percentage of respondents who assign a concept rating that represent moderate to high "interest" in the concept. For purchase intent, this is called the "top two box" percentage, corresponding to the percentage of respondents who say that they would definitely or probably buy the product, based on the concept. For 9-point liking ratings, this is the "top three box" percentage, corresponding to the percentage of respondents who rate the concept 7, 8, or 9. This paper will deal with 9-point liking ratings for concepts and present data based upon transformation to the "top three box" percentage.

How Does the Concept Get Developed—Traditional Top Down Methods

If there is no public literature on concepts, then how do these concepts manage to get developed? Certainly, in business, researchers test thousands of concepts in focus groups, sifting through this mass of hopeful product description until the focus group moderators, researchers, brand managers, and product developers come upon that one piece of gold that can constitute a new product idea.

The reality of concept development is not as disciplined as one would like. For the most part early stage concept development comes out of the mind of the marketer, or just as frequently out of the mind of a specialist "creative" hired to produce these concepts. The creative can be a freelancer whose specialty is the creation of product concepts, or an employee of an advertising agency that hopes to be selected to create advertising for this new product.

Most commercial concept creation works from the "top down." That is, there is a briefing about the product needs, and eventually there emerge one or several concepts. The concepts are generally fully formed, because the creative group avers that the pieces of the concept go together, and cannot be separated from each other. More frequently, than one might wish to admit, the concept is "suggested" by someone of high authority in the corporation, at which time the concept is considered to be finished, blessed by this higher authority, and then, if necessary, submitted to a test.

The standard outputs of the "top down" method are scores from concept tests (also called concept sorts or concept screens). The researcher tests one or several concepts among consumers, instructing the consumer to rate each concept on a set of attributes. The attributes include acceptance (e.g., purchase intent or overall liking), believability, uniqueness, value for the money, and so forth. These attributes provide a signature of the concept on a set of key evaluative criteria, used by the researcher to estimate likelihood of success. Whether the concepts are fully formed with pictures, simple text concepts, or even single ideas, the objective is the same—identify a winner, and determine the reason for the concept win. Corporations have literally tens of thousands of these studies in their collective repositories, but for the most part, the concepts are disconnected sets of product ideas, and not related to each other. That is, from most of the concept information owned by a single company the product designer still cannot determine the rules for creating a winning new idea. The concept tests are mainly report cards or "beauty contests," to identify winners, not rules for winning.

Back to Basics—Developing Product Concepts from the Bottom Up

During the past 30 years, marketers have espoused the method of *conjoint analysis* to create concepts from the "bottom up" (Green and Srinivasan, 1980; Gustaffson et al., 2001). Rather than relying on some external authority or inspired moment to create a concept, these marketers have suggested that the concept be developed by mixing together components into new combinations. The act of creation turns into an exercise of creating raw materials (elements), mixing them (design), testing combinations (field world), and estimating utilities (statistical regression analysis). This disciplined approach, while seeming overly rational, has a greater chance of success than the more conventional inspiration methods, simply because it requires the designer to do homework. Anything that promotes the increased understanding of consumer's wants and needs will increase the probability that the concepts will score well.

Conjoint analysis constitutes one of today's most popular methods for developing concepts from the "bottom up." The combinations mentioned in Figure 6.2 are examples of the test concept. The scale can be purchase intent, fit to an end use, or any of a variety of different types of ratings, such as uniqueness, believability, and so forth. The key to the approach is the systematic combination of the concept elements into different concepts, such that each element appears as a free agent. Experimental

A product concept for a health yogurt
with active cultures
a low fat yogurt
available in plain, fruit, and tropical flavors

A 'mostly' positioning concept for a health yogurt
its active cultures improve digestion
specially formulated so you don't need to feel guilty
the array of flavors is specially designed for your sensory delight

FIGURE 6.2 Comparison of a product concept and a positioning concept for a health yogurt.

design has been dealt with in various statistics books and is now well accepted (Box et al., 1978). At the end of the exercise, the researcher can identify which of the elements drives the rating, even if the respondent is unable to articulate the reason underlying his ratings. The fact that the respondent need not have to articulate the reasons for his choice of concepts means that the researcher can rely on ordinary consumers to drive concept development, rather than having to select specific types of consumers such as those who are unusually articulate.

Let us consider the use of conjoint analysis to develop concepts for a coffee beverage. In the simplest of cases, the research begins with a series of elements, such as those shown in Table 6.3, combines them according to an experimental design such as the one shown in Table 6.4, and creates test concepts, such as those shown in Figure 6.3. Each test concept is really a product description in miniature. The respondent, looking at the concept, is generally unaware that the elements have been systematically varied, and simply reacts to the concepts, one at a time. Table 6.3 shows utility values or the ability of the concept element to drive interest in the concept. More technically, the utility value is the conditional probability that the concept will be rated 7–9 on the 9-point scale if the element is present. The utility can be positive, which means it increases the odds of the concept being rated highly. An example of this is element E02 (*A coffee that's guaranteed to wake you up*). The utility is 3, meaning that an additional 3% of the respondents will rate the concept as interesting (7–9) if the phrase E02 is put into the concept. Conversely, some concept elements decrease the conditional probability. (e.g., element E01: *A lively decaffeinated coffee that won't weigh you down*). The utility is −11, meaning that 11% fewer respondents will rate the concept as 7–9 if that element is used in the concept. The additive constant is the utility value if no elements are present in the concept.

The analysis of conjoint data is made straightforward by regression modeling, available on personal computer programs. The researcher knows what elements appear in each concept (see Table 6.4) and knows the rating assigned to the concept, or in the case of yes/no whether the respondent said he would be interested or not interested in the product. Using ordinary least-squares regression analysis (OLS), the researcher estimates the part-worth contribution of the elements. OLS regression modeling, whether at the individual respondent or at the group respondent level, immediately reveals which elements work and which do not. The regression model is expressed quite simply as:

$$\text{Rating} = k_0 + k_1(\text{Element \#1}) + k_2(\text{Element \#2})..k_n(\text{Element \#}n) \qquad (6.1)$$

An example of the regression results appear in Table 6.3 (right hand column, labeled utility). The bottom of Table 6.3 shows "norms" for these utilities, to allow an interpretation of what they mean. Using these norms the product designer can identify what features appear to make a difference, at least at the concept level. The additive constant, k_0, is the estimated utility value that the concept would achieve if there were no elements. The constant is an estimated parameter from the regression. It can be used to gauge "basic interest" in the idea, without the presence of the elements, which themselves add or subtract to this basic idea. The coefficients $k_1 \ldots k_n$ are the utility values, which, as explained above, are the additive conditional probabilities of

TABLE 6.3
Elements for Coffee Beverage and Their Utilities. Results Based on a Study with 312 Respondents

	Element Category and Text	Utility
	Category 1 = richness, appropriate use	
E01	A lively decaffeinated coffee that won't weigh you down	−11
E02	A coffee that's guaranteed to wake you up	3
E03	100% organic coffee . . . healthy for you and the planet	−1
E04	Dark fancy house blend . . . an extremely rich cup of coffee	2
E05	A distinctive, well rounded cup of coffee . . . the ideal way to start a busy day	5
E06	A slightly caffeinated iced coffee drink . . . to help you get through your day	−13
E07	A jolt of caffeine to awaken your senses	−1
E08	Iced to the max . . . for those hot summer days	−13
E09	The mini-drink six pack . . . iced drinks for people on the go	−16
	Category 2 = flavor	
E10	Invigorate your senses with Cinnamon Apple Spice & French Caramel	−18
E11	A unique flavor, sure to delight . . . sweet and smooth rich cream complement this delectable treat	−3
E12	White chocolate mousse, and wild raspberry . . . a melt in your mouth dessert in a flavored coffee	−9
E13	New classic combination . . . pistachio & maple walnut . . . unleash the nutty side in you	−20
E14	Chocolate and cognac give this coffee a flair . . . try it once and you'll come back for more	−7
E15	Enjoy the taste of toffee in a light cream . . . a new summer favorite	−7
E16	Vanilla and chocolate fudge combined . . . a unique flavor that is sure to please	−4
E17	Mocha and spicy Java create a one of a kind chocolate fantasy	−7
E18	Thrilling burst of vanilla flavor and sweet, crisp taste . . . gives you "more to go wild for"	1
	Category 3 = promise, quality	
E19	The freshest cup of coffee possible	2
E20	A masterful combination of carefully chosen coffee from each year's harvest	1

E21	Highly aromatic, rich in taste with smoky overtones	−1
E22	Wonderfully smooth with deep tones	−1
E23	Its unique aroma will appeal to your senses	0
E24	Tangy taste, rich body and pleasing aroma	−3
E25	Exceptional aroma and a deep mellow body	0
E26	Spicy aroma, medium body and clean flavor make this coffee stand out	0
E27	Aroma, body and flavor . . . perfectly balanced	−1
	Category 4 = heritage	
E28	Made from exotic Jamaican beans, experience the magic of another world	1
E29	Premier espresso made from the finest beans	−2
E30	Made from a select combination of African and Central American beans	−2
E31	A dynamic blend of washed Arabian coffee	−2
E32	An Italian favorite . . . cappuccino with a flair	−2
E33	What you always wanted . . . café Americano with all the works	−1
E34	One of a kind coffee developed by top quality growers	2
E35	Dark exotic taste . . . a superb Turkish brew	−3
E36	A robust strong coffee blend . . . made from dark roasted Brazilian beans	0

Norms for utilities

>15 Superb performer, should break through clutter in a concept and drive interest

10–15 Excellent, will draw attention, positive contributor to interest

6–10 Acceptable, reasonably good contributor to interest, but probably not breakthrough

0–5 Not really relevant, probably has minor effect, but not detrimental

< 0 Detrimental to acceptance, with higher negative numbers being more detrimental

TABLE 6.4
Example of an Experimental Design*.

Concept	Category 1 Richness, Appropriate Use	Category 2 Flavor	Category 3 Promise, Quality	Category 4 Heritage
1	Element #8	0	Element #4	Element #6
2	0	0	Element #3	Element #2
3	Element #1	Element #6	0	Element #3
4	Element #4	Element #4	Element #9	Element #1
5	Element #6	0	Element #2	Element #5
6	Element #2	Element #1	Element #6	Element #4
7	Element #7	Element #3	0	Element #2
8	Element #3	Element #7	Element #5	0
9	0	Element #8	Element #9	Element #3

Note: Each Row Corresponds to a Concept, with Nine of 36 Shown. The Design Comprises Four Categories, Each Category Comprises Nine Elements.

* The experimental design is set up so that each concept may be either complete with exactly one element from each category, or incomplete with a concept lacking a category. This format of complete/incomplete combinations allows the ratings to be analyzed by ordinary least squares using a dummy variable format, which in turn allows us to estimate the true utility value of each element relative to a baseline of that element being "absent."

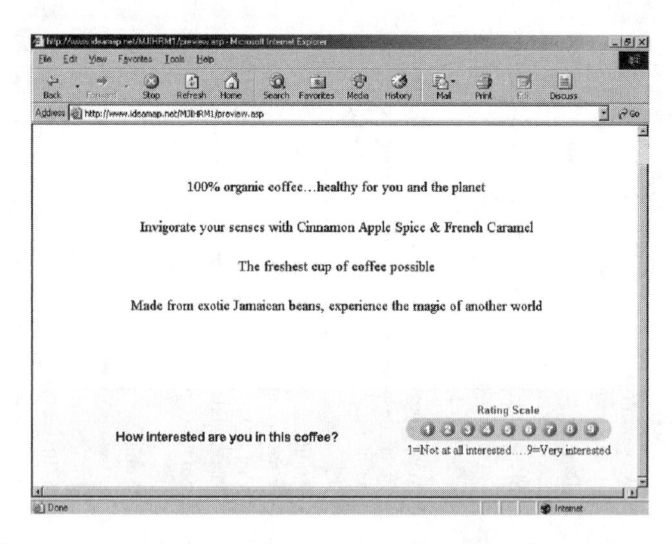

FIGURE 6.3 A test concept for a coffee beverage and the nine-point rating scale.

a concept achieving a rating of 7–9 on a 9-point scale, if the element is put into the concept.

The final use of conjoint analysis is to create new and presumably better concepts. The researcher knows how each of the elements performed. It becomes a

TABLE 6.5
Best Versus Worst Performing Concepts for a Coffee Beverage, Based on the Conjoint Analysis Results

	Additive constant	46
	Best performing concept: utility = 56	
E05	A distinctive, well rounded cup of coffee . . . the ideal way to start a busy day	5
E18	Thrilling burst of vanilla flavor and sweet, crisp taste . . . gives you "more to go wild for"	1
E19	The freshest cup of coffee possible	2
E34	One of a kind coffee developed by top quality growers	2
	Poorest performing concept: utility = −8	
E06	A slightly caffeinated iced coffee drink . . . to help you get through your day	−13
E13	New classic combination . . . pistachio & maple walnut . . . unleash the nutty side in you	−20
E24	Tangy taste, rich body and pleasing aroma	−3
E30	Made from a select combination of African and Central American beans	−2

straightforward exercise to recombine the elements into new and better combinations, by choosing winning elements. Of course, the elements have to go together, and have to be physically realizable in a product, but the conjoint analysis gives the designer a head start toward identifying what works. Table 6.5 shows two concepts—the best performing and the poorest performing, concepts, created from the elements. Table 6.5 shows the utilities of the elements, which were explained above. Table 6.5 also shows the additive constant. The total utility of the concept is the sum of the additive constant and the individual utilities. The total utility shows the conditional probability that the concept as a whole would be rated 7–9 on a 9-point scale.

SEGMENTATION—DO ALL RESPONDENTS SHARE A SIMILAR MIND-SET?

We know that people have different points of view about food. An old Latin proverb brings this out clearly and succinctly: *de gustibus non disuptandem est*—of taste one does not dispute. The person-to-person variation is especially marked in the chemical senses (Ekman and Akesson, 1964; Pangborn, 1970) and pervades foods in general (Meiselman, 2000). We all know that there is a plethora of food products in the marketplace. Does this plethora reflect a plethora of different mind-sets?

The answer to the foregoing question is yes, but with some anomalous behavior that continues to appear from study to study. We know that people, when asked to profile their optimal cup of coffee, all want the coffee to be full, rich, and flavorful. They all think they want the same thing, but in actuality, they have rather different patterns of sensory preference. So, at least when it comes to coffee, and when we deal with the emotional, not clearly descriptive statement about coffee, we are not so sure. On the other hand, if we ask these respondents to tell us about the flavors of coffee they like, whether they like the coffee with milk, with sweetener, very hot or iced, then they have an easier time of it. They can describe what they like fairly well, and when they evaluate concepts describing these characteristics, they show different patterns.

We are dealing here with what marketers call segmentation. The designer can segment people in many different ways, using all sorts of criteria. Criteria include age, income, brand used most often, method for preparing the product, additives used in conjunction with the product, and so forth (Hall and Winchester, 2000). The possibilities for segmentation are limitless. The segments have both general implications for understanding the differences among people (Meiselman, 1978), and for creating specific products targeted to the segments. Dividing people is easy; finding groups of people that differ meaningfully on their responses to product concepts is both harder but ultimately extraordinarily valuable for the business practitioner (Green and Krieger, 1991).

Fortunately, for us, methods for to identifying segments in the population have been well established by statisticians (Systat, 1997). The segmentation that we will find most important for product design is the segmentation of people into like-minded groups, based upon what they find acceptable in a product or in a product concept. By limiting our focus to product features, we ensure that the segments we uncover differ in ways meaningful to product design.

Our segmentation is called concept-response segmentation (Moskowitz 1996a, b). The segmentation divides people so that their patterns of responses to the product features are similar within a concept, but differ across concepts. Methods such as k-means segmentation are well established in this respect (Hartigan and Wong, 1979), and have been to establish groups of individuals with different mind-sets about a product. Each respondent generates a vector or set of utilities denoting that person's response to the concept elements. We saw the average utilities for 36 elements across 312 respondents in Table 6.3. Let us look at what happens when we divide the respondents by the pattern of their utilities, rather than looking at the total panel. We see the results in Table 6.6, which shows three segments. The table shows the winning elements for each segment. Based upon these segments the product designer might then opt to develop different products, one product tailored for each segment. The waker-upper is looking for a jolt of caffeine, in a well-rounded and dark fancy *house-blended* cup of coffee. Other considerations would enter into the decision as well, such as the size of the segment (percentage of respondents in the population belonging to the segment), their readiness to accept a new coffee beverage, and the economic value of their consumption as shown by the frequency with which they consume coffee beverages.

The particular segmentation method is worth a short mention, because it provides the reader with a sense of how the statistical analysis generates the responses. Each respondent participated in the study on coffee, with 36 elements combined into 60 combinations. Each respondent generated 36 utilities, one for each concept element. Each person is similar or different to every other person in the study, based upon the pattern of utilities. This degree of similarity is indexed by the statistic Pearson R. R is at its lowest at -1.00 when two people show a perfectly inverse pattern; that is, what one person likes the other person dislikes, and vice versa. R is at its highest value of 1.0 if whatever one person likes the other person likes to the same relative degree. The distance between two people is then defined as the number (1-R). The distance

TABLE 6.6
The Three Different Mind-Set Segments for the Coffee Beverage

		Total	Seg1	Seg2	Seg3
	Additive constant	46	50	47	43
	Segment 1—waker upper				
E07	A jolt of caffeine to awaken your senses	−1	12	−1	9
E05	A distinctive, well rounded cup of coffee . . . the ideal way to start a busy day	5	10	5	1
E02	A coffee that's guaranteed to wake you up	3	10	6	−4
E04	Dark fancy houseblend . . . an extremely rich cup of coffee	2	8	5	−6
E18	Thrilling burst of vanilla flavor and sweet, crisp taste . . . gives you "more to go wild for"	1	8	−18	8
	Segment 2—the gourmet				
E19	The freshest cup of coffee possible	2	4	7	−3
E20	A masterful combination of carefully chosen coffee from each year's harvest	1	−1	6	−1
E02	A coffee that's guaranteed to wake you up	3	10	6	−4
	Segment 3—the flavor seeker				
E16	Vanilla and chocolate fudge combined . . . a unique flavor that is sure to please	−4	−1	−32	12
E12	White chocolate mousse, and wild raspberry . . . a melt in your mouth dessert in a flavored coffee	−9	−8	−41	11
E18	Thrilling burst of vanilla flavor and sweet, crisp taste . . . gives you "more to go wild for"	1	8	−18	8

Note: The table shows the winning elements for each segment, and how those elements perform for the total panel and for the other two segments.

between the two people, in turn, can be used to divide them. The clustering program puts the different people into segments such that the distances between people in a segment is small, whereas the average distance between segments is large.

Applying the Bottom Up Approach by Using Appropriateness for End Use

A continuing problem in innovation is the need to create new and meaningful product ideas. The bottom up approach using conjoint analysis can work by putting the consumer into a situation or mind-set and then asking the respondent to rate concepts as fitting the specific situation. The exercise generates a database of elements that fit the situation (Moskowitz, 1998).

Let us continue with the example of a coffee beverage, only this time expand the end uses or mind-sets to venues (e.g., Burger King) to day-parts (e.g., breakfast), to age (e.g., teenagers), or to emotional states (e.g., happy). We can see the beginnings of the

Welcome to the new beverage study!

Please click on one of the yellow buttons below to take one of our surveys. You are always welcome to come back and take another one, but you can only take each one once. Feel free to forward the link to this study to your family, friends, and colleagues.

Disclaimer: If you are under age 18, please get a parent's permission before completing this survey.

- ❍ Coffee
- ❍ Burger King
- ❍ Starbucks
- ❍ Dunkin Donuts
- ❍ Maxwell House
- ❍ Folgers
- ❍ Taster's Choice
- ❍ Before breakfast
- ❍ Breakfast coffee
- ❍ Mid-morning coffee
- ❍ Lunch-time coffee
- ❍ Afternoon coffee
- ❍ Early evening coffee
- ❍ Late evening coffee
- ❍ Relaxation coffee
- ❍ Energizing coffee
- ❍ Coffee (teenagers)
- ❍ Coffee (ages 20–30)
- ❍ Coffee (ages 31–50)
- ❍ Coffee (ages 51+)
- ❍ Coffee (friends)
- ❍ Coffee (family)
- ❍ Coffee (happy)
- ❍ Coffee (sad)
- ❍ Coffee (restless)
- ❍ Coffee (tired)
- ❍ Coffee (social)

FIGURE 6.4 The "wall" for the coffee study. The respondent is led to the wall by clicking a link in the e-mail invitation.

mind-set in Figure 6.4, which shows a "wall." The wall is a screen that the respondent sees and introduces different study topics, all related to one particular project—here a coffee beverage. The respondent receives an email invitation to participate in a study on a new coffee beverage. The email contains a "link" to a Website. When the respondent clicks on the link, the respondent is shown the "wall" in Figure 6.4, and then chooses a particular study. The respondent then evaluates concepts with the mind-set of the particular study (e.g., a coffee for breakfast). Unbeknownst to the respondent, all of the studies comprise the same 36 coffee beverage elements shown in Table 6.3.

Let us now see what happens when the respondent's mind-set is tuned to a specific day-part. The key issue is what particular product features fit a day-part, which will help the product designer create a product concept. Some of the results from the conjoint analysis appear in Table 6.7, which shows the best performing elements for the morning day-parts versus the afternoon/evening day-parts.

A product designer looking at these results can now ask and answer the following two questions, leading to the new product concept:

1. *Do the features of the coffee beverage differ by day-part?* The results show that they do. Morning day-parts involve coffee as a stimulant. They do not involve coffee flavors. Thus, the new coffee beverage is probably going to be more like coffee than like a flavored beverage with a coffee base. Furthermore, there are strong winning elements for the morning day-parts. To the product designer this means that the consumers have an idea of what should be a morning beverage. To the degree that the coffee beverage

TABLE 6.7
Winning Concept Elements for 6 Day-Parts (Morning Group vs. Afternoon/Evening Group)

Morning coffee drinks	Pre Breakfast	Breakfast	Mid Morning	Average
A coffee that's guaranteed to wake you up	11	13	8	10
A distinctive, well rounded cup of coffee . . . the ideal way to start a busy day	10	11	5	8
A robust strong coffee blend . . . made from dark roasted Brazilian beans	2	1	0	1
Tangy taste, rich body and pleasing aroma	1	1	0	0
Afternoon and evening coffee drinks	**Afternoon**	**Early Evening**	**Late Evening**	**Average**
Vanilla and chocolate fudge combined . . . a unique flavor that is sure to please	−9	0	1	−3
Chocolate and cognac give this coffee a flair . . . try it once and you'll come back for more	−10	0	1	−3
White chocolate mousse, and wild raspberry . . . a melt in your mouth dessert in a flavored coffee	−12	3	3	−2

Note: These concept elements provide the designer with an idea of what consumers expect in a new coffee based beverage.

departs from traditional "coffee," the product concept is likely to be less accepted.

2. *Do respondents show similar desires for afternoon versus for evening?* The data suggests that they do not. Afternoon coffee beverages are more like traditional coffees. In contrast, the evening coffee beverage may have a different set of flavorings and forms. The difference in afternoon and evening coffee beverages suggests that a product designer might consider two beverages—one, more a traditional coffee based beverage, and the other, a more fancified beverage with a coffee base.

BEYOND SINGLE CONCEPT STUDIES TO INTEGRATED DATABASES OF CONCEPTS—THE IT!® PARADIGM

Most commercial research with concepts comprises one-off studies that deal with immediate problems. Whether the problem involves a single product idea, a day-part, or even a group of related products, the focus of concept research is always on the immediacy of the data, its applicability to the particular problem, and its focus on specifics rather than on general patterns. One of the recurrent weaknesses of this targeted focus is that there is no body of knowledge in the concept world by which to understand general patterns across products. There is no systematic knowledge or database available to understand how consumers respond to different brand names, features of products, emotional aspects of products, or aspects of the buying situation. For example, what are the features that people want in different beverages or foods? Are there segments or respondents that transcend specific foods and can act as an organizing principle for product designers? Do consumers have similar needs and wants across foods, and are these needs/wants reflected in their response to concepts?

Creating a cross-section *and* longitudinal database to understand the "algebra of the consumer mind" is a major contribution to product development, marketing, and consumer sciences, respectively. The author and colleagues (J. Beckley, H. Ashman) have created unique databases to profile the consumer mind in a set of related product areas, with these databases using conjoint analysis, and a common classification questionnaire (Beckley and Moskowitz, 2002). A database (e.g., Crave It!) comprises 30 smaller studies (e.g., french fried potatoes, hamburgers, and the like). Each smaller study, in turn, comprises 36 concept elements whose utilities are quantified by the method of conjoint analysis, implemented on the Internet. What is important here is that each of the 36 concept elements has a raison d'être, such as nature of product feature, emotional benefit, brand, usage, and so forth. As much as possible the elements are parallel across the different product studies, to facilitate discovery of patterns that transcend specific products (see Table 6.8). It is straightforward to compare the results of one food product to the results of the other food products, since the structure of the concept elements is parallel across the 30 studies. Each smaller study also has the same classification questionnaire at the end of the conjoint measurement, allowing for meta-analyses.

The studies on food (Crave It!), beverages (Drink It!), health and functional foods (Healthy You!), fast food dining (It's Convenient), and the buying situation (Buy It!) provide unique, actionable databases comprising utility values for the different elements in the 30+ product categories. Target database users are product developers who need to determine what are the features of a product, marketers who needs to know what to say, and trend watchers who needs to know what aspects of products are emerging and on their way to becoming "hot."

A sense of the results that can be obtained from the It! Databases. For example, the 2001 Crave It! ® database revealed that:

1. Most of the high positive utility values lay in the first silo, dealing with product descriptions. People want to know about the product itself. Respondents did not respond to emotional statements, at least with the simple phrasing that could be done in the concept formats. This means that emotions may be relevant, but not easy to uncover in concept research.
2. Respondents did not respond to brand names at all, with the exception of the Coca Cola® brand in the study dealing with cola (data not shown here).
3. Running the same study with two different groups of respondents (e.g., adults and teens) shows the products and statements across which the two groups differ most strongly. For example, one study was run across 20 products with both adults and teen (Ashman et al., 2002). Teens and adults

TABLE 6.8
Organizing Principle for the Crave It!® Database

EL	Category	Rationale	Hamburgers	Peanut Butter	Cola
E01	Primary attributes	Basic physical attributes	Fresh grilled hamburger	Smooth brown peanut butter, sticks to the roof of your mouth	Classic cola
E02	Primary attributes	** (Continuum: basic to complex/detailed physical attributes) in some cases . . . "healthy"	A char-grilled hamburger with a taste you can't duplicate	Smooth peanut butter with a hint of honey blended in for sweetness	Cola . . . . carbonated and sparkling, just the right amount of taste and bubbles
E03	Primary attributes	** (Continuum: basic to complex/detailed physical attributes)	A grilled aroma that surrounds a thick burger on a toasted bun	Peanut butter with a real peanutty taste	A perfect beverage: cola for breakfast, lunch, a break, or dinner
E04	Primary attributes	** (Continuum: basic to complex/detailed physical attributes) in some cases . . . "real"	Moist bites of bun, burger, and onion	Real peanut butter, no preservatives added.. separates and everything	Cola . . . . The dark brown color, faint smell of vanilla, and bubbles tell you, you have real cola
E05	Primary attributes	** (Continuum: basic to complex/detailed physical attributes)	Juicy burger with the crunch of lettuce and tomato	Crunchy peanut butter, with real nuggets of peanuts mixed in	Cola—all the taste but only one calorie
E06	Primary attributes	** (Continuum: basic to complex/detailed physical attributes)	Gooey grilled burger with rich sauce and fresh lettuce and tomato	Peanut butter blended with ribbons of jelly . . . . Tasty and so convenient	Diet cola with a slice of lemon . . . the world's most perfect drink! . . .
E07	Primary attributes	** (Continuum: basic to complex/detailed physical attributes)	Layers of burger, sauce, pickles, and lettuce on a moist sourdough sesame seed bun	Peanut butter and chocolate hazelnut cream in blended ribbons for a dreamy taste	A thick slushy of cola and ice

Continued

TABLE 6.8
(Continued)

EL	Category	Rationale	Hamburgers	Peanut Butter	Cola
E08	Primary attributes	** (Continuum: basic to complex/detailed physical attributes)	Lots of crispy bacon and cheese on a juicy grilled hamburger on a lightly toasted bun	Scoops of peanut butter with ribbons of marshmallow throughout	Cola—the perfect mixer for everything you drink
E09	Primary attributes	Complex physical attributes; details	Burger smothered in onions and cheese	Sliced bananas and peanut butter layered for a special taste	An ice cream float—cola, ice cream . . . chilled and tasty
E10	Secondary attributes/mood	Party pleaser/inviting	Burgers are a party pleaser	When it's cold outside peanut butter is cozy and inviting	Cola is a party pleaser
E11	Secondary attributes/mood	Beverages	With a chilled glass of water . . . or carbonated beverage	With a chilled glass of milk . . . or carbonated beverage	With a warm burger and french fries
E12	Secondary attributes/mood	With . . .	With great tasting french fries . . . and that special sauce	Straight from the spoon . . . nothing else added	With twice the jolt from caffeine . . . gives you just the added energy you need
E13	Secondary attributes/mood	Premium quality/classic taste	Premium quality . . . that great classic taste, like it used to be	Premium quality . . . that great classic taste, like it used to be	Premium quality . . . that great classic taste, like it used to be
E14	Secondary attributes/mood	Savor it . . .	You can just savor it when you think about it during work and school	You can just savor it when you think about it during work and school	You can just savor it when you think about it during work and school
E15	Secondary attributes/mood	All natural/changing flavors	100% natural . . . a real beef burger!	100% natural . . . only the best peanuts	New choices every month to keep you tantalized
E16	Secondary attributes/mood	With all the extras you want . . .	With all the toppings and sides you want . . . pickles, relish, jalapenos . . . lettuce, tomato, chips . . . .whatever	With great tasting white bread and fruit spreads in any flavor you want	In your choice of sizes, bottles, or cans . . . . whatever you're looking for

E17	Secondary attributes/mood	Imagine the taste . . .	You can imagine the taste as you walk in the door	You can imagine the taste as you walk in the door	You can imagine the taste as you walk in the door
E18	Secondary attributes/mood	Lick your lips twice . . .	So tasty and juicy you practically have to lick your lips twice after each bite	So good . . . you practically have to lick your lips and the top of your mouth twice after each bite	So refreshing . . . you practically have to drink another can
E19	Emotional	Quick/fun/alone	Quick and fun . . . eating alone doesn't have to be ordinary	Quick and fun . . . eating alone doesn't have to be ordinary	Quick and fun . . . drinking alone doesn't have to be ordinary
E20	Emotional	Have to have it . . . can't stop	When you think about it, you have to have it . . . and after you have it, you can't stop eating it	When you think about it, you have to have it . . . and after you have it, you can't stop eating it	When you think about it, you have to have it . . . and after you have it, you can't stop drinking it
E21	Emotional	Fills that empty spot . . .	Fills that empty spot in you . . . just when you want it	Fills that empty spot in you . . . just when you want it	Fills that empty spot in you . . . just when you want it
E22	Emotional	Cheers you up . . .	When you're sad, it makes you glad	When you're sad, it makes you glad	When you're sad, it makes you glad
E23	Emotional	Escape routine/celebrations	Now you can escape the routine . . . a way to celebrate special occasions	Now you can escape the routine . . . a way to celebrate special occasions	Now you can escape the routine . . . a way to celebrate special occasions
E24	Emotional	Multidimensional sensory experience	A joy for your senses . . . seeing, smelling, tasting	A joy for your senses . . . seeing, smelling, tasting	A joy for your senses..seeing, smelling, tasting
E25	Emotional	With family and friends	An outrageous experience . . . shared with family and friends	An outrageous experience . . . shared with family and friends	An outrageous experience . . . shared with family and friends

Continued

TABLE 6.8
(Continued)

EL	Category	Rationale	Hamburgers	Peanut Butter	Cola
E26	Emotional	Ecstasy . . .	Pure ecstasy	Pure ecstasy	Pure ecstasy
E27	Emotional	Satisfies HUNGER . . .	It feeds THE HUNGER	It feeds THE HUNGER	It feeds THE THIRST
E28	Brand or benefit	Basic brands/experiences	At White Castle	From Reese's	From Shasta
E29	Brand or benefit	** (Continuum: basic to premium brands)	At Jack-in-the box	From Jif	From Tab
E30	Brand or benefit	** (Continuum: basic to premium brands)	At McDonald's	From Peter Pan	From Dr. Pepper
E31	Brand or benefit	** (Continuum: basic to premium brands)	At Wendy's	From Health Valley	From Pepsi
E32	Brand or benefit	** (Continuum: basic to premium brands)	At Burger King	From Smuckers	From Coca-Cola
E33	Brand or benefit	Premium brands/experiences	At Fuddruckers	From Skippy	Dispensed fresh from the fountain, especially for you
E34	Brand or benefit	Fresh . . . for you . . . by you	Fresh from the grill, especially for you . . . by you	Fresh and natural . . . especially for you	Kept chilled . . . especially for you
E35	Brand or benefit	Best in world . . .	Simply the best burger in the whole wide world	Simply the best peanut butter in the whole wide world	Simply the best soda in the whole wide world
E36	Brand or benefit	Safety . . .	With the safety, care and cleanliness that makes you trust it and love it all the more	With the safety, care and cleanliness that makes you trust it and love it all the more	With the safety, care and cleanliness that makes you trust it and love it all the more

Note: The categories and rationales for the different products are as similar as possible and identical when feasible and meaningful.

TABLE 6.9
Comparison of Winning and Losing Elements, and Elements Showing the Largest Teen vs. Adult Differences

	Average Utility	Teen Utility	Adult Utility	
Winning elements—average of teens and adults				
Chicken	21	22	21	Plump, juicy chicken breast, marinated in a special sauce and cooked over an open-fire for a smoky grilled taste
Brownie	21	20	21	Fudgy, chewy brownies with a sweet buttery vanilla taste and a thick layer of rich and creamy chocolate frosting on top
CheeCak	20	19	21	Cheesecake with swirls of raspberry, chunks of white chocolate, baked in a crunchy crust and garnished with pecans
Tacos	19	19	19	Homemade soft taco shells wrapped around warm simmered meat and topped with chunks of tomato and shreds of lettuce and cheese
Losing elements—average of teens and adults				
Donuts	−8	−10	−6	Cakey donuts with thick, cinnamon apple chunks spilling out of the dough
Coffee	−13	−6	−19	Decaffeinated whole bean coffee for those who want all the taste and none of the caffeine
Olives	−13	−6	−19	Dark wrinkled olives marinated with hot pepper flakes
Pizza	−13	−13	−14	Pizza with feta cheese, spinach, and a thick chewy crust
Elements showing biggest difference between teens and adults				
	Difference (teen-adult)	Teen	Adult	
Nuts	24	−1	23	Large cashews, brazil nuts, pecans, with just the right amount of salt
Nuts	−20	6	26	Fresh mixed nuts like pecans and cashews, not a peanut anywhere
Coffee	22	20	−2	Brewed coffee blended with ice cream and caramel then topped with heavy whipped cream
Coffee	14	17	3	A hearty cup of Cappuccino, frothy with foam and the rich taste of espresso

Source: Data from the Crave It!® database, for teens and adults.

were quite similar in most products, except for nuts and for coffee. This in-depth insight, show in part in Table 6.9, helps the product designer understand both specific issues in a food designed for different age groups. The larger array of elements across products may actually help the designer to see bigger patterns that transcend individual foods, leading to greater long-term understanding.

TABLE 6.10
Comparison of Total Panel (Tot) and the Three Concept-Response Segments (C1, C2, C3) for Hamburger Elements, from the 2001 Crave It!

Hamburger	Tot	C1	C2	C3
Additive constant	30	52	9	47
Segment 1—classic				
Lots of crispy bacon and cheese on a juicy grilled hamburger on a lightly toasted bun	17	11	34	–22
Segment 2—elaborate				
Lots of crispy bacon and cheese on a juicy grilled hamburger on a lightly toasted bun	17	11	34	–22
With all the toppings and sides you want . . . pickles, relish, jalapenos . . . lettuce, tomato, chips . . . .whatever	10	2	19	3
Burger smothered in onions and cheese	5	–7	18	–5
A grilled aroma that surrounds a thick burger on a toasted bun	10	3	17	4
Layers of burger, sauce, pickles, and lettuce on a moist sourdough sesame seed bun	7	6	17	–21
So tasty and juicy you practically have to lick your lips twice after each bite	8	4	14	1
A char-grilled hamburger with a taste you can't duplicate	7	3	14	–2
Juicy burger with the crunch of lettuce and tomato	5	–3	13	2
Premium quality . . . that great classic taste, like it used to be	7	5	12	–3
Segment 3—imaginer				
Fresh from the grill, especially for you . . . by you	5	–1	7	13
You can imagine the taste as you walk in the door	7	6	6	12

Database: Numbers in the body of the table are the utility values for the elements, for each group.

4. There were three segments that continued to appear from study to study; the elaborates who responded to fancified descriptions of food; the classics who responded to the traditional aspects of the product and disliked novel ideas; the imaginers who responded to other concept statements besides food descriptions, such as ambience (Beckley and Moskowitz, 2002). Table 6.10 shows how different these segments are in terms of their response to the features of a hamburger. The segmentation will play a big role when it comes to developing new product. Similar types of segments emerged for other large-scale databases run the same way, such as those for beverages (Hughson et al., 2004).

Two Issues with Concepts as the Drivers of Product Design

Despite the importance of concepts as the first stage in product, it is important to surface two recurring issues that should give one pause and reason to ponder. These two issues are the limiting range of concepts, and that concepts may drive the design of impossible-to-achieve products.

1. *Concepts limit the wider potential product range:* People are often biased by their experience. When they hear a product described, they may react more negatively than would be the case were they to sample the product without the description. That is, concepts rely on language and the mental evocation of a product. All too often, our previous experience biases us. We limit the range of what products we want to hear, at least at the concept stage. If, in turn, it is the concept stage that provides the platform for product development, then product developers will be hindered. They will fail to create products that will delight the consumer, because at the concept stage these products will sound unappealing. We could summarize this problem by saying that the product designer can produce more than the consumer is apparently willing to tolerate. The practical solution to this problem is to reverse the development cycle, first producing prototypes, then selecting the winning prototypes according to consumer responses, and only then creating the product concept around the prototype. The product concept thus plays very little role in guiding development and only serves as a summary of the product. This reversal of the conventional sequence deserves more consideration, especially in the light of the narrowness of one's imagination in contrast to the wide range of blind product acceptance.
2. *Concepts occasionally describe an impossible to produce prototype:* Designers face the inverse problem as well. Sometimes, the consumers want things that cannot be delivered together. In the worst of these situations, the consumers may want abstract features of products that simply cannot coexist, such as good taste, low fat, low carbohydrates, and high healthfulness. In product concepts, this type of general concept would not be even considered. However, a more likely candidate is an orange beverage, artificially sweetened, with a taste like that imparted by sucrose, and with a real mandarin orange flavor. Given the sweetener that one is working with, it may be impossible to create that type of flavor sensation. This issue of impossible to create prototypes based on concepts also needs study, for how does then the researcher know what part of the concept to believe and deliver, versus what part of the concept should be ignored as not feasible?

PART 4: DESIGNING A PRODUCT TO FIT THE CONCEPT

What does the product developer do to create the prototype? Do ideas for the products come out fully formed? We saw the need for homework when the designer develops product concepts. The same happens for product prototypes. For every product that reaches consumer evaluation, there are literally dozens that fail to make it off the developer's bench, or even if they make it that far, they may prove unstable in storage, unacceptable to the consumer, or infeasible to produce on a commercial scale.

When writing about product design it is tempting to concentrate on the design and analysis, while leaving the actual data acquisition undiscussed, or at least referred off handedly to those who specialize in product testing (and concept testing).

Nothing could be further from practical truth. It is important to know the limits of product and concept testing; what are reasonable approaches to the acquisition of information, what can be done, and what cannot be done. The issue becomes even more relevant when we realize that a great deal of applied product testing comes from the practices of professionals, not from textbook prescriptions implemented by professionals.

Data Acquisition

We will look at a few of the more practical issues and see some of the alternatives and their pros and cons. The issues divide into three groups—those dealing with the questionnaire (1, 2), those dealing the respondent's capacity to evaluate many products (3, 4), and those dealing with test venue (5, 6, 7, 8).

1. *Should the respondents be trained to understand the meaning of attributes, beyond a simple explanation?* We dealt with this above in the language of food and drink. For the most part, there is no need to train the respondents in terms of knowing the attributes, except for a short explanation of terms that might not be familiar. The objective with consumer respondent is to get a measure of their perception, not to train them into being instruments with high reliability. That reliability is desirable, but not necessary.

2. *Can we mix evaluative and descriptive attributes?* At first glance, this may seem like an irrelevant question. Certainly, we can instruct respondents to evaluate products on many attributes and respondents will willingly do so. The real question is whether introducing descriptive attributes in an evaluation forces the respondent to pay attention to characteristics that otherwise would be irrelevant. We are not sure of the answer. We do not know whether we would make a different decision about the products if we were to introduce the overall rating attribute at the beginning of the evaluation, in the middle of the evaluation, or at the end of the evaluation. Often there is no effect of other attributes on the overall rating attribute, other than perhaps fatiguing the respondent, who has to assign many ratings for attributes that will not be used for decision making.

3. *How many products should a respondent assess?* This is probably the most common issue to be raised. There is no clear answer. Sometimes the practitioners state strongly that the respondent should test only one product, because, the reasoning goes, in the so-called *real world*, the consumer eats or drinks one alternative at a time of the product, not more. This evaluation of one product is called *pure monadic evaluation*. The problem with pure monadic testing is the loss of discrimination. If each respondent tries one product, and there are two or more products to be evaluated, then each respondent tries only a fraction of the products. Studies show that the discrimination among the products is reduced with this one-product strategy and that even when the test stimuli are sounds of different loudness it takes many more respondents to create a dose-response curve than would be the case when each respondent evaluates several stimuli (Stevens, 1956).The reason for this problem is simple—each judgment from a respondent comprises two parts—the actual intensity and the variability from the individual respondent. If, however, the respondent evaluates all or many

TABLE 6.11
Maximum Number of Products That a Respondent Should Test in an Extended Test Session and Recommended Waiting Time in Minutes between Samples

Product	Maximum Number of Samples	Minimum Waiting time between Samples	Comments
Carbonated soft drink (regular)	15	7	Little adaptation or residual exists and there is no fat to drive satiety
Pickles	14	10	Garlic and pepper aftertaste build up
Juice	12	10	Watch out for sugar overload
Yogurt	12	10	
Coffee	12	10	
Bread	12	4	
Cheese	10	15	Fatty residue on tongue
Carbonated soft drink (diet)	10	10	After taste can linger
French Fries	10	10	Fatty residue on tongue
Cereal (cold)	10	10	
Milk based beverage	10	10	
Soup	10	10	Can become filling
Sausages	8	15	Fat leads to satiety
Hamburgers	8	15	Amount ingested has to be watched
Candy—chocolate	8	15	
Croissants	8	10	Fat leads to satiety
Salsa	8	10	Longer waits necessary for hot salsas
Cereal (hot)	8	10	Amount ingested has to be watched
Ice cream	8	10	Fat leads to satiety
Mousse	6	15	Combination of fat + sugar drives satiety
Lasagna	6	10	Amount ingested has to be watched. Satiety rapidly emerges if a full sample is consumed.

Note: Times are based upon actual studies run by Haward R. Moskowitz from 1975–2004, and represent observations about *what has worked in practice.*

of the products, many of the individual differences are eliminated. This is called a within-subjects design. Table 6.11 shows recommended base sizes for different foods and beverages, assuming that the researcher can work with the respondent for an extended period (e.g., 3–4 h).

4. *How many attributes can a respondent actually evaluate for a specific product without becoming fatigued or bored?* This question is also important in the practical evaluation of products. Researchers tend to hoard attributes and rarely eliminate attributes. The questionnaire can end up looking like a laundry list of attributes. The underlying reason for this accretive growth is the reluctance to give up attributes,

because of the fear that the research will be somehow "missing something." Most of the time, this fear is unfounded.

5. *What are the key aspects of the different test venues (central location vs. home use)?* Much of the traditional early stage flavor and texture work was done in an unstructured, informal fashion in the development laboratory, using one's colleagues as the panelists. Before the field of testing grew into a profession with standards, it was common to "test" a new formulation by tasting it oneself, giving some to one's laboratory associates, or to one's clerical staff, soliciting opinions about direction (viz., "was the product and thus the developer on target"?), and then moving on to make the requisite changes. Developers did not concern themselves with formalized instructions, did not worry about the validity of what they were doing, and generally did not worry that their informal tests "at the bench" were anything but useful to guide them. This informal "benchtop" testing has evolved into three distinct and formalized ways to test products.

Central location intercept: The interviewer sets up the test facility in a well-trafficked area (usually an indoor shopping mall). The interviewer intercepts potential panelists, screens them for appropriateness for the test, and then invites the person to participate. The intercept test must be very short, permitting the panelist to evaluate two or at most three products. For more products, the interviewer must pay the panelist or use more panelists with each panelist testing a partial set of products. Intercept tests are very popular, because they are fairly cheap. They, however, do not exercise the best quality control of the interview, because the quality of the interviewing staff varies from market to market, the turnover in interviewers is high, and the interviewer is rushed for time in each interview in order to make quota. Panelists do not want to participate because they are not paid, and because the interview often interferes with other tasks that they must perform when they go to the shopping mall. The interview becomes more of a nuisance than anything else.

Central location prerecruit or "hall" test: For hall tests, the researcher recruits a consumer to participate, gives the consumer several products to test in some type of randomized order, and then obtains ratings of the product(s). The hall test can be conducted with a fairly high degree of control, the interview can be observed, and the results come in quite quickly. The panelist can test many products, not just one or two as typical with the central location intercept test. In the most advanced hall tests, the consumers use computers linked together by a local area network, so that the results are instantaneously computed after the last panelist has finished.

Home use testing or normal venue of consumption: The consumer is either called afterwards by phone to report their ratings, or invited back to a central location for a personalized interview. The home use test can accommodate several products, as long as the researcher takes care to ensure that adequate controls are maintained. Controls include separating the products by a day or more, dictating the order of products to be tested at home, and creating a questionnaire that the consumer understands. Home use tests are often used for products that the consumer has to prepare, where consumption is relevant rather than just taste alone, and when there is the possibility that with repeated consumption the consumer may become bored with the product. Repeated consumption of the same product at home can ferret out the boredom factor.

TABLE 6.12
Advantages vs. Disadvantages of Central Location Versus Home Use Tests

Aspect	Central Location	Home Use Test
Control over ingestion and evaluation	High	Low
Test site versus typical consumption	Unnatural	Natural
Amount of product to be evaluated	Limited	Unlimited
Number of different products tested	Many	Few
Measure satiation and wear-out	No	Yes
Mix many concepts and many products	Yes, easy	No, hard
Number of panelists/product for stability	20–50	50–100

6. *Comparison of central location test and home testing.* A brief comparison of the pros and the cons for hall testing vs. home use testing appears in Table 6.12. It is left up to the reader, faced with the actual real world problem, to decide. For the most part, hall use tests and central location tests will yield the same or similar answers—winning products will remain winners, losing products will remain losers. There are, however, certain products that may appear to be winners in a short test but may become less acceptable when the panelist is forced to live with them on a daily basis. In these cases, it is vital for the test to be home use because a central location hall test may provide incorrect answers. If time and resources permit, an alternative strategy uses hall tests to screen down to a few products, and then follows with a home use test of the winning (and perhaps modified) products in a home use test.

7. *Data biases due to test venue.* The test venue can affect the ratings of products, and, therefore, the decision about what to do in terms of product design. These differences due to venue can become quite serious, because they affect the scores received by the products, and thus affect the decisions made when directing product development. Home use tests may inflate the scores of products, whereas hall tests may deflate these scores. Often products that score well at home score poorly in central location. A decision to "go with a product" because it scores well in home use tests may mislead the developer, who always looks for positive results. Part of this difference between home use tests and hall tests can be traced to the number of products tested, and the frames of reference brought to bear by each test venue. In home use tests, the researcher tests only one or a few products. Consumers do not have a frame of reference provided by other products, and tend to uproot products, especially the product tried first. If there is one product in the home use test, then 100% of the products are tried first. If there are two products, then 50% of the products are tried first. In hall tests, however, where consumers have many products, they may up-rate the first product, but accurately rate the second and other products. Consequently, in hall tests, only a portion of the products tested are subject to the up-rating bias. If the hall test comprises eight products, then only 12.5% of the products will be tested in the first position.

8. *Which of the three test venues is best*? In early stage, testing there is no right or wrong location for the test. Whichever test venue does the job is best, but the researcher should always focus on efficiency and the cost/benefit ratio. Under the most normal of circumstances, home use tests provide more valid data, but the data is most expensive, comes in more slowly, and may not enable each panelist to test every product. From the statistical viewpoint, it is ideal to have each consumer test every product. The power of testing products in their natural consumption state (viz., at home) is offset by the weakened statistical power of the home use test, since for large scale tests the panelist cannot evaluate all of the products.

The Language of Product Sensory Attributes

To understand modern methods of product design we first have to go to the basics. The most fundamental of these is the language of food and drink. In order for the product designer to create a product, it is important to be able to describe the sensory characteristics. A couple of generalizations are relevant here.

1. *Consumers are capable but limited.* Consumers can do fairly well in recognizing that a product fits a concept, but have a harder time describing the specific characteristics of the product that they want. Even if they knew the product intimately, consumers might have a difficult time describing its characteristics in a way that the food designer could use as a product blueprint.

2. *Categories of sensory attributes are not created equal.* Consumers can describe appearance characteristics fairly well in the abstract and can also do a pretty good job of describing the texture characteristics of products that they want. Much of the consumer capability in description comes from the fact that consumers need not refer to existing products to describe color or appearance or texture. If the consumer wants a reddish orange product, if the consumer wants the surface to be somewhat mottled, if the consumer wants a slightly crunchy product, then the food designer has a pretty good idea about what the consumer says he or she wants. Whether the consumer is actually describing a product that will be successful is something entirely different. We will deal with this issue of ideal versus real products below. Nevertheless, the language of visual and textural perception is fairly clearly, pretty unambiguous, and easy to deal with.

3. *Taste is a fairly straightforward, not particularly complicated sense.* Consumers can describe the taste characteristics of products, using the four basic tastes (salty, sweet, sour, or bitter). Often consumers confuse bitter and sour, especially at low levels, but not salty with sweet, or with bitter.

4. *Aroma (olfactory-driven) is often described incorrectly as "taste."* At the same time, however, consumers bring in words that more properly belong to aroma. In the consumer's mind, taste and flavor are interchangeable. They are not. Taste comes from the sensory stimulation of the tongue receptors, and is limited to four tastes (or a fifth taste, *umami*). Flavor is in part taste, but in part the odor perception coming from the back of the mouth, as the volatiles of the food travel up to the nose, inside the buccal cavity. This is called *retronasal stimulation* and is responsible for the enormously complex sensory perception that extends far beyond the so-called taste. When, for example, we experience meat flavor certain taste sensations emerge, for example

salty. However, at the same time there is the exquisite interplay of many volatiles in the mouth, appearing and disappearing in different times as we chew the meat. The retronasal stimulation, really smell from the back of the mouth, is integrated with the taste perception to produce this flavor. Consumers, by and large, have a hard time breaking down this complex flavor perception of its components, although they can often give it a label, and it often evokes quite vivid memories.

5. *Flavor terms are object-related.* Continuing the previous point, we know that the consumer language of flavor generally makes use of specific objects. Unlike sweet, salty, hard, mottled, and like "general" words from the sense of taste and texture, the language for flavor makes use of specific words that point to an object.

6. *Consumers can help the designer with some attributes, but not with others.* The difficulty of actually using consumers to describe the product sensory profile has not been lost on product designers. Although consumers have definite sensory preferences, it is difficult for them to express the precise nature of these preferences in a language that the designer can work with. It is reasonable to use consumers to rate degree of liking. Furthermore, a consumer can tell the product designer what he or she wants for many sensory attributes, but not for all attributes. Consumers are good in describing simple, straightforward attributes (e.g., appearance, texture; Moskowitz et al., 1979), but have a harder time describing more subtle attributes (e.g., some flavor attributes).

7. *Experts play a key role in product design for attributes the more subtle attributes.* Experts are better in describing the nuances of a product, especially with language that the consumer does not know. Consumers are excellent in reporting degree of liking, and using their own consumer language to describe the simpler, more concrete sensory attributes of a product.

8. *Consumers, not experts, are the purchasers of the food and beverage.* Despite the great efforts involved in working with experts, it is ultimately the consumer who makes the difference. Yet, as suggested above, the role of the consumer in the early stages of product evaluation is marked by a history of controversy. There are some practitioners who proclaim that the consumer is actually incapable of doing much more beyond simply saying that he likes or dislikes the product. There is little evidence to support this radical position. There are others who proclaim with equal vehemence that the consumer can act as the profiler of sensory characteristics, the judge of sensory magnitude, and the acceptor of the product. The former group, who feel that the consumer is merely a registrant of acceptance trace their intellectual history back to those who dealt with the profiling of wines, perfumes, beers, and fragrances. In these fields, there is a rich vocabulary, not necessarily available to the consumer. A lot of training needs to be done to achieve proficiency. The latter group, who feel that the consumer can do far more, comprise individuals with a leaning towards psychophysics. Psychophysics is the study of the relation between the measured aspects of well-defined physical stimuli and the subjective sensory or hedonic response. Psychophysicists do rather elegant experiments with consumers, such as varying the composition of mixtures of odorants or tastants, instructing the respondent to profile the intensity of the mixture on attributes, and developing quantitative relationships between the stimulus magnitudes and the consumer responses. The typical respondent in a psychophysical experiment is not an expert, but may

have received a short instruction on which to look for in the stimulus, and how to use the scale. The thousands of scientific papers in hundreds of journals attest to the fact that the well-instructed consumer respondent can do remarkably well in these studies. The only failing may be that the consumer judge does not know the mean of a rating scale—meaning that is inculcated into the expert panelist as part of his training.

9. *The big opportunity for product designers is to create a system that relates sensory ratings, whether by consumers or experts, to product acceptance*. The big problem comes when the corporation uses experts to profile products on characteristics, but does not use this information in a meaningful way to drive product design. If the company simply uses experts to create profile after profile, that activity is self-indulgent. If the company uses the expert profiles of products in a systematic, demonstrable, public way to drive development, then the exercise is meaningful.

Examples of Descriptor Systems Used by Experts to Describe Sensory Perceptions

Researchers at Arthur D. Little Inc. in Cambridge, Mass., appear to be the first to recognize that in order to guide the development properly, the researcher had to understand what the consumer was experiencing, the nature of the sensory characteristics, their magnitude, and their order of appearance. The Flavor Profile® is a process by means of which a panel is trained to recognize the different sensory aspects of a product and then choose reference stimuli to act as examples of these attributes (Cairncross and Sjöström, 1950; Caul, 1957). The pioneering work of the Arthur D. Little Inc. group was later followed by similar, albeit modified descriptive methods, such as the Quantitative Descriptive Analysis© (QDA) (Stone et al., 1974) and the Sensory Spectrum© (Meilgaard et al., 1987). All three approaches are *process-oriented*. They guide and enable the developer or sensory researcher to create appropriate descriptor system for the product being studied. They do not provide a fixed list of terms per se, primarily because in the description of flavors a myriad of attributes often emerge, many of which are unique to the particular product.

Communicating descriptive analyses can be difficult, because the product designer may be faced with a wall of numbers in the table. Creative ways have been developed to plot the descriptive analysis results, and by so doing communicate these results in an easy to understand way. Figure 6.5 shows an example of the "spider-plot" for a product, and a modified formulation. The plot provides an easy way to represent attribute data for two or more products. The attributes correspond to vectors originating from the center, and the length of the vector corresponds to the intensity of the attribute. The plot visually highlights the differences between products on attributes.

The complexity of odor perception primarily, but of texture perception as well, led to a number of systems by which to classify odors. For example, Dravnieks and his colleagues (Dravnieks, 1974) attempted to create a standardized thesaurus for describing flavor, which appears in Table 6.13. There are many other schemes

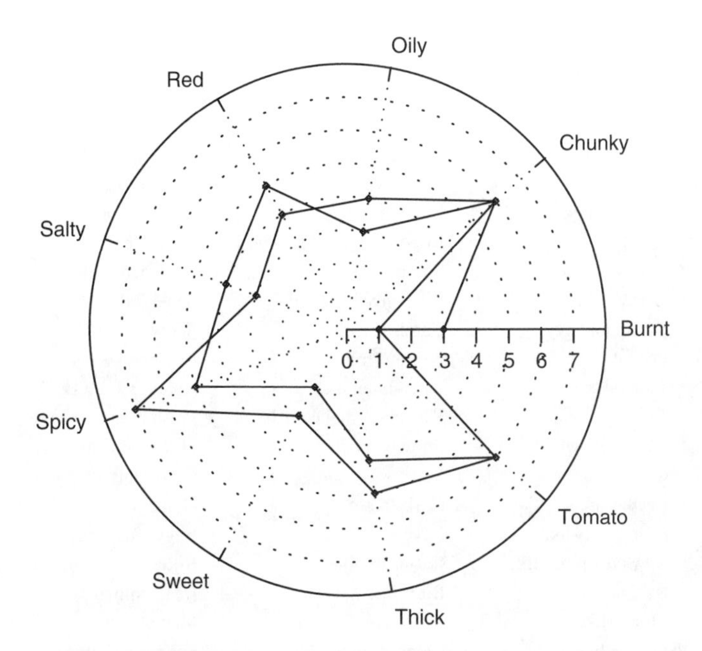

FIGURE 6.5 Polar coordinate diagram of two products rated on nine attributes along a 7-point scale. The products are plotted so that the length of the vector is proportional to the intensity of the attribute. The attributes are arrayed around a circle, separated by vectors with equal angles between them. The angular separation has no meaning other than making the display more readable.

to classify flavor, with these schemes deriving from specific industries. Table 6.14 shows a system for beer (Clapperton et al., 1975). Table 6.15 shows a system for fragrance provided by Haarman & Reimer Inc.(now Symrise Inc.), an ingredient and flavor/fragrance supplier serving the food and personal products industries.

Texture, in contrast to flavor, has had an easier time, perhaps because there simply are fewer attributes to describe texture. Brody, 1957 and later Szczesniak and her colleagues at the General Foods Corporation in White Plains New York proposed a list of terms to describe texture (Szczesniak et al., 1963). These appear in Table 6.16. These terms deal with the mechanical and the geometrical aspects of food. There are fewer nuances in texture, so the challenge for Szczesniak was to create the basic list first and then provide reference samples to show gradations of each texture attribute.

When all is said and done, the most important thing to keep in mind about any descriptor system is that it is an attempt to boil down the rich experience of a product into a usable set of terms and scale values. Description by itself does not tell the developer what to do. It simply shows the developer what the panelist is experiencing for the particular product. It is the developer's job to translate descriptions of the characteristics of products into operationally meaningful activities that alter the product in the proper, consumer driven, direction.

TABLE 6.13
Dravnieks' Classification Scheme for Odor

Eucalyptus	Beer-Like	Wet-Paper Like
Buttery	Cedar wood-like	Coffee-like
Like burnt paper	Coconut-like	Peach (fruit)
Cologne	Rope-like	Laurel leaves
Caraway	Sperm-like	Scorched milk
Orange (fruit)	Like cleaning fluid	Sewer odor
Household gas	Cardboard-like	Sooty
Peanut utter	Lemon (fruit)	Crushed leaves
Violets	Dirty-linen like	Rubbery (new)
Tea leaf like	Kippery (smoked fish)	Fresh bread
Wet wool, wet dog	Strawberry-like	Oak-wood, cognac
Chalky	Stale	Grapefruit
Leather like	Cork-like	Grape juice
Pear (fruit)-like	Lavender	Eggy (fresh eggs)
Raw cucumber-like	Cat-urine-like	Bitter
Raw potato	Bark-like	Dead animal
Mouse-like	Rose-like	Maple syrup
Pepper-like	Celery	Seasoning - meat
Bean-like	Burnt candle	Apple (fruit)
Banana-like	Mushroom-like	Soup
Burnt rubber-like	Fish cigarette smoke	Fried fat
Geranium leaves	Nutty (walnut)	Urine

THE LANGUAGE AND METRICS OF PRODUCT ACCEPTANCE

If a product does not taste good then the odds are that it will not succeed in the marketplace. The goal of development is to create a product that truly does well in the market. To this end, the key evaluative criterion that designers use is liking. The higher the liking rating the more likely it will be that the product will succeed, all other factors held equal. The important thing about scaling liking is not the scale used, but rather whether the scale has norms associated with it. Norms provide a context to evaluate what the liking (i.e., acceptance) rating actually means.

Consumers, not experts, are the primary source of measurement about product liking. For liking, experts *may*, with some caution, replace the consumer when it comes to two uses:

1. Evaluating a product as "good" or "poor," at least in the early design stages.
2. Later on, when the product characteristics have been specified, one can use experts or machines to determine whether a particular batch of product lies within the specifications or departs from the specification.

When it comes to measuring degree of liking, pronouncements and dogma become even more severe than they are for descriptive analysis. Some researchers insist that in order to avoid bias the acceptability scales comprise an even number of categories,

TABLE 6.14
Classification System for the Sensory Characteristics of Beer

	First Tier Term	Second Tier Term
1	Spicy	
2	Alcoholic	Warming, vinous
3	Solvent like	Plastic like, can liner, acetone
4	Estery	Isoamyl acetate, ethyl hexanoate, ethyl acetate
5	Fruity	Citrus, apple, banana, black currant, melon, pear, raspberry, strawberry
6	Floral	Phenyl ethanol, geraniol, perfumy, vanilla
7	Acetaldehyde	
8	Nutty	Walnut, coconut, beany, almond
9	Resinous	Woody
10	Hoppy	
11	Grassy	
12	Straw-like	
13	Grainy	Corn grits, mealy
14	Malty	
15	Worty	
16	Caramel	Primings, syrupy, molasses
17	Burnt	Licorice, bread-crust, roast-barley, smoky
18	Medicinal	Carbolic, chlorophenol, iodoform, tarry, bakelite
19	Diacetyl	Buttery
20	Fatty acid	Soapy/fatty, caprylic, cheesy, isovaleric, buytric
21	Oily	Vegetable oil, mineral oil
22	Rancid	Rancid oil
23	Fishy	Amine, shellfish
24	Sulfidic	H2S, mercaptan, garlic, lightstruck, autolyzed, burnt rubber
25	Cooked vegetable	Parsnip/celery, dimethyl sulfide, cooked cabbage, cooked sweet-corn, tomato ketchup, cooked onion
26	Yeasty	Meaty
27	Ribest	Black currant leaves, catty,
28	Papery	
29	Leathery	
30	Moldy	Earthy, musty
31	Sweet	Honey, jammy, over-sweet
32	Salty	
33	Acidic	Acetic, sour
34	Bitter	
35	Metallic	
36	Astringent	
37	Powdery	
38	Carbonation	Flat, gassy
39	Body	Watery, characterless, satiating

Note: The second tier terms become important when the first tier term has many alternatives of different sensory character.
Source: Table from Clapperton, Dagliesh & Meilgaard, 1975.

TABLE 6.15
Fragrance Classification System According to the Perfumers of Haarman and Reimer, Inc (Holzminden, Germany)

	General Class	Sub-Classes
1	Citrus	Classic
		Modern
2	Green	Fresh
		Balsam
		Floral
3	Floral	Fresh
		Floral
		Heady
		Sweet
4	Aldehydic	Fresh
		Floral
		Sweet
		Dry
5	Oriental	Spicy
		Sweet
6	Chypre	Fresh
		Floral-Animal
		Sweet
7	Fougere	Classic
		Modern
8	Woody	Dry
		Warm
9	Tobacco/leather	
10	Musk	

TABLE 6.16
Descriptive Attributes for Texture Properties

Textural Properties	Geometrical—Related to Particle Size and Shape	Geometrical—Related to Orientation
Hardness	Powdery	Flaky
Fracturability	Chalky	Fibrous
Chewiness	Grainy	Pulpy
Adhesiveness	Gritty	Cellular
Viscosity	Coarse	Aerated
	Lumpy	Puffy
	Beady	Crystalline

Source: After Szczesniak, A.S., Brandt, J.A., and Friedman, H.H, *Journal Of Food Science*; 28, 39–403, 1963.

with the same number of points allocated to "good" as "poor." Other researchers feel that in a commercial setting, most of the products will be acceptable, because they are designed for consumption. Furthermore, these researchers feel that it is more important to differentiate among the different levels of good products by means of more scale points than it is to balance the scale. Finally, some researchers insist that the scale must have a point for "neutral" (viz., a middle category), whereas other researchers insist that the scale should have no neutral point at all, thus forcing the panelist to decide whether he likes or dislikes the product. Table 6.17 shows scales that have been proposed over the years.

Labels constitute another area of contention in acceptance scaling. The label assigned to the particular category makes a great deal of difference. Scales vary in their language as Table 6.17 clearly shows. Some practitioners prefer that the scales have discrete points, and that each point be labeled. Indeed, a great deal of controversy surrounds the number of points and he appropriate label for each point. Issues in acceptance scaling surround other decisions about the scale, such as whether the scale should deal with liking or some other action, such as readiness to consume (so-called Food Action (FACT) or Food Action scale, proposed by Schutz in 1965).

TABLE 6.17
Category Scales—Liking

2-Point	**5-Point**	**9-Point Hedonic Scale**
Like	Very good	Like extremely
Dislike	Good	Like very much
	Moderate	Like moderately
3-point	Tolerate	Like slightly
Acceptable	Never tried	Neither like nor dislike
Dislike		Dislike slightly
Never Tried	Very good	Dislike moderately
	Good	Dislike very much
Like a lot	Moderate	Dislike extremely
Dislike	Dislike	
Do not know	Tolerate	
Well liked	FACT scale	Purchase intent scale
Indifferent	Eat every opportunity	Definitely would buy
Disliked	Eat very often	Probably would buy
	Frequently eat	Might/might not buy
100-point scale (Moskowitz Jacobs Inc. Scale)	Eat now and then	Probably not buy
0 = hate	Eat if available	Definitely not buy
100 = love	Don't like—eat on occasion	
	Hardly ever eat	
	Eat if no other choice	
	Eat if forced	

Source: After Meiselman, H.L, In: *Encyclopedia of Food Science*; Petersen, M.S and Johnson. A.H., eds.; AVI, Westport, CT, 675–678, 1978.

Today's two predominant scales are the 9-point hedonic scale and the 5-point purchase intent scale. Both scales appear in Table 6.17. The 9-point hedonic scale was developed as the major "workhorse" scale by the U.S. Army Food and Container Institute in Chicago, and published in 1957 (Peryam and Pilgrim, 1957). The hedonic scale appeared during the time that scaling was becoming very important in academic psychophysics, and measurement was becoming very important in application. The hedonic scale comprises four levels of acceptance, four levels of rejection, and an intermediate level of indifference. The hedonic scale has seen use in studies of food procurement by the U.S. military, large-scale food preference surveys, and industrial and scientific applications. The scale is typically analyzed by parametric statistics, such as the average and standard deviation, and processed by inferential statistics.

More recently, in keeping with the growing interest in psychophysics to use labeled magnitude scales, a labeled magnitude scale of liking was developed at the U.S. Army Natick Laboratories (Schutz and Cardello, 2001). This scale, which was derived from quantitative assessments of the semantic meaning of a wide range of verbal labels that can be used to describe liking and disliking, places the terms "greatest imaginable liking" and "greatest imaginable disliking" at either ends of a visual analogue scale, with "neither like nor dislike" in the middle. The other verbal labels chosen for placement on the scale include those used in the original 9-point hedonic scale. However, these labels are placed along the line such that the distances among the labels reflect the *psychological* magnitudes of the differences in their semantic meaning along the like—dislike dimension. The resultant scale produces ratio level, rather than interval level data. In a series of studies, this scale was found to be as easy to use by consumers as the 9-point hedonic scale and to be significantly less difficult than magnitude estimation. It was also found to have equal reliability and sensitivity to the 9-point hedonic scale, while providing greater discrimination among highly liked foods.

The 5-point purchase intent scale represents a different type of scale. The scale is also balanced, labeled at each point, but asks the respondent to indicate what he would do with the product with respect to purchasing it. As noted above in the section on concepts, marketing researchers rather than sensory analysts work with the scale. The marketing researcher studies the market potential for the product, rather than dealing intimately with the R&D product developer. Market research usually works with the percentage of respondents who say that they would definitely or probably buy the product. This is the so-called top-two box percentage, obtained by computing the percentage of respondents who select the two top scale points on the purchase intent scale (4 and 5 on the 5-point scale). Researchers then work with these percentages, rather than with the mean. The focus of market researchers, as their clients the marketers, is directed toward the percentage of respondents who will do a certain behavior, rather than on the intensity of that behavior. In a sense this difference in analytical strategy between averages and percentages reflects the different intellectual histories of the sensory analyst and the market researcher.

No matter what scale the product designer uses, it is important to keep in mind that a scale is only a measuring instrument. Developers want to know what the scale point means. On a 1–9 point scale of liking, for instance, what does it mean to say that a product scores an average of 5.5, whereas another product scores an average of 7.0?

Is this a statistically meaningful difference? What behaviors are to be associated with products scores of 5.5 vs. 7.0, respectively? Does the consumer select products scoring 7.9 more often than products scoring 5.5? If not, then the scale differences are an artifact with no real meaning, and in fact with a great deal of potential to confuse.

More About Liking—A Practical Example of Results, Analysis and Interpretation

An example about yogurts will illustrate some of the thinking regarding acceptance testing and analysis. The panelists each tested each of seven yogurts in a randomized order, rating the degree of liking using the 9-point hedonic scale (Peryam and Pilgrim, 1957). Table 6.18(A) shows raw data from the first six panelists. Table 6.18(B) shows the mean and standard deviation from the full panel of 30 consumers, for each yogurt, respectively.

The first question that a researcher would ask is how well the yogurts perform. This is especially important in product design, because the objective is to create a highly

TABLE 6.18
Data Set for Liking, Showing How to Analyze Products to Determine How Well the Product is Liked, and Whether the Products Differ from Each Other

A: Ratings of consumers on seven yogurts, A–G (partial data set)

Panelist	A	B	C	D	E	F	G
101	7	6	6	7	6	8	3
102	6	8	5	6	3	4	2
103	8	7	7	6	4	6	4
104	6	7	4	6	5	4	4
105	4	8	8	7	7	7	4
106	7	5	3	5	5	6	5

B: Summary statistics

	A	B	C	D	E	F	G
Mean	6.0	6.1	5.9	6.1	6.3	5.1	4.8
Standard deviation	1.21	1.13	1.58	1.03	1.48	2.04	1.41

C: Analysis of variance

Source of variation	Sum of squares	Degrees of freedom (df)	Mean square	F ratio	Probability by chance
Yogurt	116.781	6	19.463	9.281	0
Error	425.7	203	2.097		

D: Probability that two yogurts (row, column) differ by chance alone

	A	B	C	D	E	F
B	1.00					
C	0.20	0.20				
D	0.50	0.50	1.00			
E	0.00	0.00	0.62	0.28		
F	0.00	0.00	0.38	0.13	1.00	
G	0.00	0.00	0.04	0.01	0.83	0.96

acceptable product that will succeed in the marketplace. Acceptance is measured by the 9-point hedonic scale. The yogurts vary in their level of acceptance. Five are fairly highly acceptable, whereas two are less acceptable.

The second question deals with a comparison of pairs of yogurts. It is not sufficient to look at the numerical differences between the yogurts and stop there, because the difference in means across the yogurts may be due to chance alone, or may actually be due to repeatable differences. A standard analysis of liking ratings looks at the ratio variability due to the variability of the seven means vs. the random variability. The ratio of these two variabilities is an index number called the F ratio, which has a defined statistical distribution. By computing the F ratio, and knowing the number of observations for the means and the number of observations for the error term, the researcher can determine the probability that the observed F ratio for the yogurts comes from chance alone, or represents a true effect (viz., that the difference is not due to chance). This method is the *analysis of variance*, Table 6.18(C) shows the analysis of variance table for these seven yogurts, and reveals that indeed there are differences among the yogurts in degree of liking that were captured by the hedonic scale. That is, by using the scale to represent overall liking panelists told us that they liked some yogurts more and other yogurts less. (We deduce that there is a significant difference among the yogurts by considering the F ratio, which is significant, meaning that it would be very rare to observe this F ratio if the products were really identical to each other, and differed only by random chance). Finally, Table 6.18(D) shows a comparison of pairs of yogurts for probability of being significantly different, using the Tukey test (Box, et al., 1978). The difference in liking ratings for pair of yogurts along with the measure of variability of means reveal whether or not this specific difference can be considered significant, or just a result of random variability. There may be an overall significant effect (viz., yogurts do differ, in the aggregate, one from the other), but on a pair-wise basis only some yogurts will differ, whereas others will be the same.

From this simple, but instructive, exercise the reader can begin to see the power brought to early stage product design by having the consumer rate liking. Rather than force the panelist to rank products, category scaling method gives the panelist a yardstick, and has the panelist "have a go" at the products. The panelist can assess one product, five products, 100 products, or more, although by the time the panelist has assessed 15 products or more fatigue sets in. The products can be rotated, and indeed each panelist need not scale all of the products, but rather need only evaluate a subset of the full product set. No memory is involved, since the panelist reacts to each product as it is presented, trying his best to locate the product on the liking scale.

STATISTICAL TESTS VS NORMATIVE INTERPRETATIONS

Researchers use two different approaches by which to assess the meaning of liking ratings. One approach uses statistical difference testing and inferential statistics, which we saw above for the yogurts. By performing the standard tests of difference the researcher discovers whether the numerical differences between products observed in the study would repeat themselves in subsequent tests, or whether the differences observed are due to chance. The statistician does not talk about the external meaning of the difference but rather concentrates on the repeatability of the results.

Researchers often talk about the statistical significance of differences, but these statements about statistical significance do not necessarily have anything to do with the performance of products in the marketplace. Small differences in ratings of liking can become highly significant (viz., not be due to chance alone) when the researcher vastly increases the number of ratings for each product. The greater the number of ratings per product the more significant will be the same numerical difference.

The analysis of liking through norms is often more germane to product development. Normative data enables the researcher to consider the numbers in light of existing products whose performance is already known. The norms provide a means by which to interpret the data. Let us look at the results from a study of nine cereals, this time using the 0–100 point liking scale (0 = hate, 100 = love). Given this scale, what then does a rating of 56 mean? On an absolute basis what do the different points on the scale mean? Does a rating of 60 on the 100 point liking scale insure success, whereas a lower rating (even slightly lower) of 58 suggests that the product will fail? The data appear in Table 6.19 (A). The norms from the 0 = 100 scale (see Table 6.19(B)) are taken from more than 2000 studies run by the author since 1981.

TABLE 6.19A
Scores for Nine Cereals on the Moskowitz Jacobs Inc. 0–100 Point Liking-Scale, and a Verbal Interpretation of the Liking-Rating

Cereal Product	Mean Liking-Rating and Interpretation
1	66 = Very good
2	62 = Very good
3	60 = Good, may need a little work
4	56 = Good, probably needs work
5	54 = Good, probably needs work
6	43 = Fair, needs a lot of improvement
7	35 = Very poor
8	31 = Very poor
9	27 = Exceptionally poor
	Least significant difference = 3.2

TABLE 6.19B
Norms From the 0 = 100 Scale Taken from More Than 2000 studies

	Norms
70+	Excellent, needs no work
61–70	Very good, can use slight improvement
51–60	Good, needs improvement
41–50	Fair, needs a lot of improvement
40–	Poor

Descriptions such as these help the developer to understand what the scores really mean, whether the product is a winner, or a real dud.

Intermediate Solution #1 to Product Design: the Curious Cases of the Just About Right (JAR) Scale and the Self-Designed Ideal

Mid-way between the sensory scale of intensity and the scale of acceptance we find two other scales; just about right (JAR) and self-designed ideals. These two scales move the respondent forward from evaluation of the product to direction given to the product designer.

Directional JAR scales require that the respondent say whether or not the product has too much or too little of a characteristic. The JAR scale attempts to do to things simultaneously—first have the respondent evaluate the intensity of a perception, and secondly, decide whether there is too much or too little of that perception for some unknown ideal level residing in the consumer's mind.

Directional JAR scales are quite popular among developers, because they appear to be easy to use. For a complex product, it is not unusual to ask the panelist to rate the product on 10–30 directional scales. From the scales, the developer can get an idea of the degree to which a product is off-target on specific characteristics. Table 6.20 shows a profile of a complex product, American noodles + beef, on 14 directional scales. Panelists find the directional scales easy to use, and at a superficial level, product developers can easily interpret the scale results. The table is easy to

TABLE 6.20
Example of Directional Scales for a Noodles + Beef Product

Directional Attribute	Directional	Comment
Amount of mushrooms	−33	Far too little
Onion flavor	−27	Far too little
Aroma strength	−16	Far too little
Meat flavor	−13	Far too little
Amount of meat pieces	−8	Too little, but not too far off target
Size of meat pieces	−7	Too little, but not too far off target
Flavor strength	−6	Too little, but not too far off target
Amount of pasta	−5	Too little, but not too far off target
Thickness of gravy	−3	On target
Saltiness	−3	On target
Darkness of gravy	−1	On target
Firmness of pasta	8	Too much, but not too far off target
Spiciness	11	Too strong
Tomato flavor	12	Too strong

Note: All attributes were rated on a 0–100 point scale, anchored at 0 = far too little, 50 = just right, 100 = far too much (+0 = too much; 0 = just right, − = too little).

understand, because the developer need only look at whether the product has too much of an attribute or too little. Of course, the scale does not tell the developer how to correct for the under-delivery or over-delivery, but the data acts as a good diagnostic of problems.

Beyond the simple-appearing data and diagnostic interpretation in Table 6.20, however, lie years of experience and accretions of different points of view.

1. *Implicit recognition of segmentation.* Some practitioners work with the distributions, rather than with the means. In their mind, the mean disguises the fact that there will be some individuals who feel that the product has far too much, and others who feel that the product has far too little of an attribute. These product designers aver that the distributions of JAR values that they obtain have to be qualified by over-delivery and under-delivery, so that they do not run the risk of making a mediocre product. Practitioners who look at the distribution of under-delivery and over-deliver implicitly recognize the existence and importance of sensory-preference segmentation, but rarely address that major issue head-on.

2. *Implicit use of norms or rules of thumb in interpretation.* No single product is ever perfectly "right" on all of the attributes. Respondents by their very nature try to comply with the implicit demands of the study, and in doing so look for departures from normal, whether or not they would do so if they were undirected. One consequence is that each of the product attributes will generally be "off-target," some to more degree, some to less degree, simply because the respondent feels that he must "do his job." It takes years of experience and some built-in rules of thumb to identify which of the JAR scale values correspond to product problems that must be solved, and which are simply off "just right," because the respondents feel compelled to identify some problem. Does the departure from just right a full scale point, a half scale point, a pattern of departures that signals a problem? The answer to this question is generally not recorded in a company's annals, but rather passed down as laboratory lore by one researcher to the next.

3. *Confounding and misdirection.* Some attributes may be confounded and mislead. In a fruit flavored beverage, one can lower the sweetener level considerably. Panelists will report that the "flavor" is too weak. Yet adding flavoring (the superficial response to this directional result) will simply produce a more tart, bitter taste, which is still too low on "flavor." Consumer reports flavor, but the key ingredient is sugar. Thus, if the panelist says that the flavor is too low, the developer may add flavoring. The panelist will continue to say that the flavor is too low, and the developer will take this direction seriously, and add more and more flavoring until the product becomes unpalatable. The real answer is to add sugar. A similar issue applies to chocolate flavor. Take away the fat and the chocolate flavor drops. Consumers will demand more chocolate flavor, but the real key is to add fat.

4. *Implicit recognition that all attributes are not created equal and that political correctness plays an important role.* Respondents have no problem when they evaluate many sensory attributes on the JAR scale, especially those dealing with appearance and with texture respectively. Respondents have a more difficult task when they evaluate certain flavor words, especially those words that are easy to understand in common parlance but have associated with them emotional baggage. Thus, it is fairly

easy to determine that a product is too dark or too light, too soft or too hard. There is no real emotional baggage. However, consider characteristics such as chocolate for a candy, or its more extreme term, *real chocolate flavor*. In the candy bar business, chocolate flavor is important. More likely than not, candies will score lower on real chocolate flavor than higher. There will be some candies that score high (over-delivery) on real chocolate flavor, but the odds are strongly in favor of under-delivering products. The same type of problem occurs for products attributes like fatty or salty, especially because these attributes have associated with them the notion of "political correctness." That is, developers and marketers are applauded for creating products that taste good, but have low fat, or low salt. In public, respondents are reluctant to say that a product has too little "fat" or "too little salt." The respondent gets publicly recognized as a good citizen for opting to accept the healthier food, low in fat, and low in salt.

Intermediate Solution #2 to Product Design: The Self-Designed Ideal Product

The self-designed ideal product is the first cousin to the JAR scale. When rating a self-designed ideal the respondent is assumed to have available two sources of information:

1. The intensity of the product(s) as it is currently perceived.
2. Knowledge in an intellectual sort of way the intensity of a hypothetical product.

The hypothetical product may be one that is considered to be "ideal" in terms of acceptability, but there is no reason to limit oneself to that type of product. One might present the respondent with some type of a set up concept or scenario, and instruct the respondent to rate the product that fits this scenario. The procedure is attractive, at least on the surface, because it links together the characteristics of a product being developed (i.e., the ideal) with actual products.

At a practical level in industry, the use of self-designed ideals is becoming increasingly popular, especially when linked with the profile of actual products. Whether or not the self designed profile is really a valid way to drive product development is an entirely different story. Again, personal observation suggests that respondents often 'know' the desired appearance and texture characteristics of a product, but they may not know the desired taste and aroma of the product. This lack of knowledge makes the self-designed ideal less than accurate.

A second problem that often arises is that the ideal profiles are remarkably similar from respondents who are known to have different sensory preferences. The case of coffee presents a wonderful example of this convergence. Two individuals, belonging to segments that like strong, bitter, bold coffee versus weak and watery coffee, generate similar sensory ideals of a rich, full-bodied coffee, with low bitterness. The self-designed profile is more similar than the actual sensory preferences would lead us to believe. This issue of similar self-designed ideals for people with radically different sensory preferences is reminiscent of the relative narrowness of direction from concepts for product development versus the wider direction from responses to actual products.

PART 5: TESTING MULTIPLE PRODUCTS TO GENERATE OPTIMAL FORMULATION

During the past 40 years, product designers have come to realize that doing one's homework in product development can be very critical to ultimate success. We saw earlier that marketers promote the use of experimentally designed concepts. Far earlier, however, than marketers were agricultural engineers who developed the principles of experimental design to optimize crop yields. At about the same time, chemical engineers in the petroleum industry worked on experimental designs to increase the yield of gasoline through better methods of cracking crude oil. About 40 years ago, experimental design methods (known as response surface procedures) were even introduced into product design (Gordon, 1965), but it would take decades to make the procedures accessible to the designer, affordable, and acceptable to the many potential users.

Applying experimental design to the product design task is somewhat harder. Historically, product designers began their work with descriptive panels, whose job it is to profile the characteristics of the product. Using trained panelists seemed to make a great deal of sense to designers, and had its origin in the world of "experts" among brewmasters, flavorists, and perfumers. It was a simple jump from the expert and his experience to the expertized consumer panelist. Looking back, however, it appears that the expert panelist called into play a set of beliefs about who was the appropriate group of people to guide product design. The expert panel as the product designer's development tool eventually gave way in the 1980s to consumer-driven approaches, and statistical methods of product optimization. The process of changeover from expert to consumer guidance is still going on and still leads to heated arguments among professionals holding opposing opinions.

The speed of business and the competitive environment dictate that for continued success, the marketer creates a system to produce consumer-acceptable, profit-making products. It is tempting to create products by "rifle shots," under the mistaken belief that the marketer, product developer, market researcher, and sensory analyst really understand the consumer. Nothing can be further from the truth. The marketplace is littered with products developed from intuition, researched incorrectly, launched with fanfare, and producing nothing. Homework, homework, homework, and disciplined development must substitute for arrogant guesswork and a belief that "speed to market" alone suffices to win. Homework becomes even more important when there are competitors lurking about, ready to steal one's customers.

The structured approach to product design boils down to a simple thing. Rather than testing single products that represent best-guesses, the so-called "rifle shots" designers have tried to learn about their product by assessing systematically varied products. This systematic approach involves the use of statistics, from the branch of statistics known as experimental design and modeling (equation fitting). In systematic experimentation, the goal is to learn relations between the variables under the developer's control and the consumer response.

Consumer acceptance need not be the only information collected. The enterprising developer might wish to create the different prototypes, and for each prototype collects data from consumers on acceptance and sensory attributes, as well as the rating of the product on image attributes and on fit to a concept. While he gathers these data from consumers, the researcher might also collect data on the profiles of these products by expert panelists, using their language and scales, as well as collect instrumental information by using machines to take readings of the products. Finally, the developer might wish to obtain information about the cost of goods for each of the product.

Creating a Product Model—Ice Creams

We can best understand the experimental design approach by a case study, which illustrates the principles. Our particular product is an ice cream product. When working with experimentally designed products, the research emphasis is to cover a wide space of sensory impressions, because thorough understanding that space the product designer begins to understand what the consumer likes and dislikes. Experimentation of this sort is the antithesis of rifle shots, because rifle shots naively assume that the developer knows what the consumer wants and the research is simply to confirm their ingoing decisions.

The design process from experimental design to test execution and analysis is fairly straightforward, and should not in anyway be daunting to the researcher. The developer identifies the physical variables (ingredients and process conditions). These are actual stimuli that are physically varied to create the actual prototypes. The variables are under external control (e.g., formulation and process). The product developer creates a set of combinations of these variables, called prototypes, which become test products to be evaluated by actual eating and drinking. The variables cover a range of different product alternatives.

The panelists evaluate the test products, rating them on the sensory, liking, and image characteristics. Executing the study is fairly straightforward, as discussed above. The researcher prepares the product, invites the respondents to participate, and collects the data. Each of the respondents evaluates some or even all of the products, rating the products on a set of scales. These scales include liking, sensory attributes, and the more complex image attributes. At the end of the evaluation, the researcher collates the data, and computes the average rating for each product across all of the respondents who evaluated that product. The data may be subjected to additional analysis in order to tease out segments of individuals with different sensory preferences (sensory segmentation), as well as further analyzed to compute the averages from key subgroups identified ahead of time (e.g., users of different products in the category). Typically, the only ratings to be analyzed in this "subgroup-level" detail are the overall liking ratings. The other attributes may be tabulated by these subgroups, but the subgroup level results are not usually considered for further analyses.

The data matrix then can be analyzed by regression to develop equations showing how the consumer-rated attributes (and other measures, such as cost of goods) change with changes in the independent variables. The equations allow for interactions among ingredients, and for nonlinearities. Equations summarize the relation between the independent variables and the dependent variables. Product developers can then identify the optimum combination of the independent variables that, in concert,

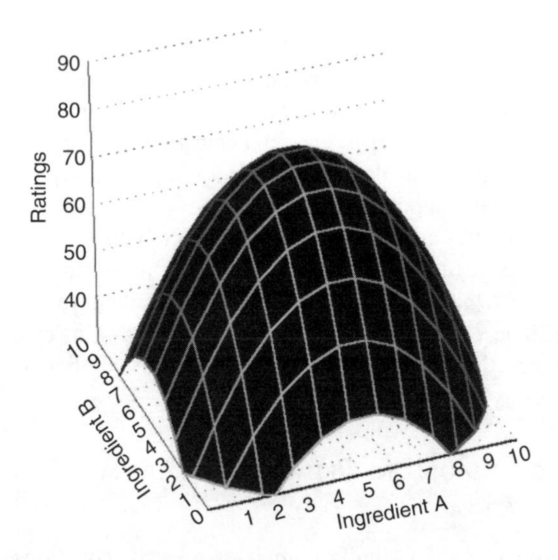

FIGURE 6.6 Schematic of a response-surface model, showing the relation between an attribute rating, and two independent (ingredient) variables.

generate a highly acceptable product, within constraints of cost, and with specific sensory characteristics. Figure 6.6 shows a schematic of the response-surface "hills" Optimization finds the highest point on that hill, subject to constraints. Research need not be limited to two variables, even though the graphical representation can only show two variables. The product model may encompass 3–10 variables, or even more. The only requirement is that there be more products tested when the number of variables increases (see Chapters 5 and 15).

Experimentally designed combinations of product ingredients and processes find their most important use in creating a product model. The model comprises a set of equations, one equation for each response attribute. Although many possible equations could fit the data, typically, the researcher uses a quadratic function. With two variables the equation is written as follows:

$$\text{Rating} = k_0 + k_1(\text{Ing A}) + k_2(\textit{Ing}\ \text{B}) + k_3(\text{Ing A}^2) + k_4(\textit{Ing}\ \text{B}^2) + k_5(\textit{Ing}\ \text{A} \times \text{B}) \quad (6.2)$$

When the researcher deals with three, four, five, or more independent variables the equation becomes slightly more complicated, but not very much. For example, in the case of six independent variables, each at three levels to capture nonlinearities and interactions, there might be 45 combinations. The equation would comprise an additive constant, six linear terms, six quadratic (square) terms, and a possible 15 pair-wise interaction terms, for a total of 28 predictors. The parameters of this 28-term equation are very easy to estimate with today's regression programs widely available on the personal computer.

The quadratic equation (Equation 6.1) allows the rating to peak at the middle level, an especially important property when the rating attribute is liking. As a formula variable increases, liking first increases, peaks, and then drops down. This inverted U shaped form does not necessarily have to occur for any particular product, but it often does. By allowing the quadrature, but not forcing it, the equation can capture a key aspect of consumer responses to the product.

Often, but not always, the ingredients interact with each other to generate acceptability or other response. One never knows ahead of time the extent of such interaction, such as which particular pairs of independent variables will interact with each other. The cross term A×B allows the equation to capture the interaction, but if the interaction is minimal then the value of the coefficient, k_5, will be very small.

Using the Product Model to Optimize an Ice Cream Product Formulation

By itself, the product model simply comprises a set of equations, without much intuitive substance. The product model becomes much more interesting when we use it to identify the appropriate formulation corresponding to specific objectives. For example, the designer might wish to identify the highest scoring product within the range tested, but also ensure that the product has an affordable cost of goods, so that it makes reasonable business sense. The product model allows the designer to explore many thousands of ingredient combinations, looking at each combination through the two objectives. Each combination has an expected liking rating and a cost of goods.

Statisticians have developed methods by which to explore the many thousands of combinations, looking for that combination that satisfies the two criteria. There may be even additional criteria placed on the product formulation (e.g., the product is very acceptable, has a specific highest cost of goods, and enjoys specific characteristics). Sometimes the constraints imposed on the product are so severe that no product can possibly emerge that satisfied all of the constraints. This should not be viewed as a problem, but rather should be welcomed, because it reveals that the objectives are impossible to satisfy. Such demonstration of impossibility is important in the applied world of product development, because it "short circuits" a process that ultimately wastes time, money, and misses other market opportunities.

Let us use the model to identify some optimal formulations. Table 6.21 shows three candidate ice cream product formulations (columns A–C) designed from the same product model. The objective of optimization was to identify the candidate formulations (within the ranges of prototypes tested) that would deliver an optimal product, subject to a constraint. There are many objectives and many constraints possible. Since the designer has done the requisite homework, it is now easy to create new and potentially more acceptable, cost-effective products by simply consulting the product model, and imposing specific constraints on the solution. This optimization "by the numbers" is based upon the empirical data set that needed to be created just one time. In the business world, the product model is often used for 5–10 years after the study, to address new issues about cost reductions, new product opportunities, and questions dealing with "what if" issues. Product modeling and optimization provides a

TABLE 6.21
Three Optimal Ice Cream Products, Formulated with the Four Ingredients Used in the Ice Cream Experimental Design (A–C), Optimized for Different Business Objectives

	A	B	C	D
	Business objectives			
Maximize	Maximize overall liking	Maximized fit to the ice cream concept	Maximize overall liking	Reverse engineer product should fit four goals
Imposed constraint (A–C) or goal to be satisfied (D)	None	None	Rating of "*Fit to the ice cream concept #1_*" had to exceed 67 on the 0–100 scale	Fit concept = 70 Older = 60 Sophisticated = 60 For evening = 60
Suggested ingredient levels based on the product model				
Ingredient A	1.00	1.95	1.43	1.83
Ingredient B	3.00	1.00	1.00	1.00
Ingredient C	2.42	2.05	2.23	2.92
Ingredient D	1.00	2.16	1.04	1.79
Expected ratings or cost based on the product model				
Expected cost and liking				
Cost of goods	257	259	200	224
Overall liking	66	58	63	56
Expected image				
Younger vs. older	57	62	61	62
Unsophisticated vs. Sophisticated	51	45	49	55
For morning vs. evening	49	55	55	57
Fits ice cream concept #1	62	71	67	62
Expected sensory				
Softness	46	51	49	44
Visible vanilla specks	31	32	33	39
Flavor Intensity	35	37	36	39
Smooth vs. grainy texture	51	56	52	58

Note: The fourth optimal product, developed by "reverse engineer" fits a desired sensory profile. The test prototypes were rated by consumers on anchored 0–100 point scales. Cost of goods was estimated by summing the unit costs of the four ingredients.

better, more integrated database than does continuing rifle-shot research, even though the effort is greater to create the prototypes up-front.

Reverse Engineering: Using the Product Model to Create a Product That Has a Specific Profile

Let us turn the design problem around 180 degrees, and specify a product opportunity. The opportunity may be a product that fits a concept, or a product with a specific sensory profile. Perhaps a scan of the marketplace reveals that there are no products with a specific sensory profile. A product with this desired sensory profile might be accepted or rejected, but it would certainly be different from the existing products in the market. How this gap in the marketplace is discovered is dealt in the sections that follow, but right now let us assume that the marketer has identified the gap, and has an idea of the product sensory profile or image profile.

How might the developer identify the formula corresponding to this particular sensory profile? One way uses the product in a novel way. The product model relates the ingredient variables to the ratings, with each attribute having its own equation. Rather than maximizing liking or some other criterion, which we saw in Table 6.21, why not search for a combination of ingredients that generates a sensory or image profile as close as possible to a goal profile that one wishes the product to have. This is an example of *reverse engineering*—the designer begins with the response profile, and identifies the formulation that would generate it. Where the response profile comes from is entirely different from how it is created with a set of ingredients. The response profile might correspond to the "gold standard" of identity for the product, with the business objective to deliver that gold standard with a new set of ingredients. The response profile might be a totally new entry into the market, with a sensory profile that can be advertised. The response profile might even be an image profile, rather than a sensory profile, as shown in column D. The designer needs only a set of equations relating formula variables and their interactions to attribute ratings. If the attributes are limited to sensory descriptions, then the reverse engineering approach can only create formulations with given sensory profiles. If, however, the researcher has used image attributes and created the requisite equations, then the image profile itself can be used as a goal.

How Does the Product Designer Optimize When There Are No Experimental Designed Products?

We have just seen how the product designer can use the power of modeling to identify optimal formulations. In many situations, however, there is no experimental design that can be used, although the researcher can test many products. We dealt with this issue earlier in the treatment of the category appraisal. Beyond simply score-cards for product performance and sensory-liking curves, however, there are many more possibilities if the researcher is prepared to follow the foregoing analytic steps that were prescribed for product modeling. This time, however, there are a few differences in the analytic approach, but the results are well worth the effort. We can summarize the bootstrapping approach in eight simple steps, which if followed,

TABLE 6.22
Steps Involved in Creating a Product Model with Category Appraisal

Objective—Locate the full set of competitor and rifle-shot (prototype) ice cream products in a geometrical space, with the property that sensorially similar products lie close together. Then, use the coordinates of the map as independent variables. Identify places on the map where there may be no products, with these locations having the property that they correspond to a relevant product, for example, highly acceptance.

Step 1—Set up the data. Identify the sensory attributes that will be used to create the sensory dimensions

Step 2—Create the map. Perform a principal components factor analysis on these attributes. This generates a factor space, whose dimensions or axes are the primary sensory factors. Principal components factor analysis is efficient, easy to do on a personal computer and computationally very fast. The factor analysis is done with the objective of using the factor scores as predictors or independent variables in the product model, rather than used in the more conventional way of reducing the number of attributes. The principal components factor analysis creates a 'basis set' of future predictor variables, with much of the sensory information embodied in that basis set.

Step 3—Locate ice creams. Compute the location of each ice cream product in this factor space

Step 4—Create the product model. Create the equation relating the location in the factor space to a specific attribute. Each attribute has its own equation. The equation is a quadratic function, written as: Attribute $= k_0 + k_1$ Factor 1 $+ k_2$ Factor 2... k_n Factor $n + k_{n+1}$Factor $1^2 + k_{n+2}$Factor 2^2. Cross terms may be used as well. A cross term would be represented by the term Factor 1* Factor 2. There may be more than two factors, depending upon the results of the principal components factor analysis.

Step 5—Optimize. Within the range of factor scores in the study, and within the sensory levels achieved by the ice creams, (convex hull) identify a factor location corresponding to the optimal product or to a product with a desired sensory or image profile

Step 6—Solve for the sensory profile corresponding to the optimum. Using the equations developed in Step 4, estimate the full sensory profile of the optimal product

Step 7—Identify target products. Referring back to the table of products vs. sensory attributes, identify the products having the requisite sensory attribute identified in Step 6. This becomes the reference or landmark product for development for the specific attribute.

Step 8—Follow the same approach for reverse-engineering. Identify a goal to be matched, determine the location in the factor space corresponding to that goal, estimate the full sensory profile generated by that location, then identify target products.

generate a map of the competitive product space, a set of equations to use with that map, and the optimization/reverse-engineering technologies to identify sensory targets (Moskowitz, 2000). Table 6.22 presents the eight steps. Let us look at this approach for the ice cream problem.

Looking at the competitive frame requires considerably less work than creating prototypes. If a "new" food product is defined as "one that is new to the consumer," only 7–25% of food products launched can be considered truly novel (Lord, 1999). Rather than experimentally varying the ice cream on different ingredients, let us have the product designer purchase the different ice creams currently being marketed. Although such a task sounds easy, there are some issues to consider, such as the sensory variation in the ice creams and the assurance that the products are "in spec." Sensory variation simply means that the ice creams cover a wide sensory range. This sensory range can be ensured, at least roughly, by having expert or consumer panelists taste the vanilla ice creams, rate them on sensory characteristics, and ensure that the products reasonably differ from each other. One can never hope to cover the entire range of sensory variation on all attributes, but limited range is one of the shortcomings that is accepted in order not to expend a lot of early-stage effort on prototype development.

In order to maximize the information the researcher might opt to follow the steps listed in Table 6.22. These steps begin with the actual ice cream profiles as provided by consumers, as well as possibly by experts and by instruments, although that latter information is not necessary. The steps continue with the creation of a basis set, or set of factors using as inputs the sensory attributes. The principal components factor analysis allows the developer to do two things; locate all of the commercially available vanilla ice creams in the geometric space created from principal components factor analysis, then build an ice cream product model, using the factor scores as the independent variables. The final objective is to optimize the acceptance, or reverse engineer the product to fit a profile, all the time ensuring that the resulting ice cream "product" has an *achievable sensory profile*. This achievability is obtained by making sure that the expected sensory attributes lie between the upper and lower limits of the ice cream products tested in the actual study. The outcome is a sensory profile of the optimal product. There is one caveat, however. The product designer did not work with the actual formulations themselves. Rather, the designer worked with in-market products. Thus, the sensory profile does not immediately prescribe a formulation. Rather, with the sensory profile in hand, the designer can identify specific products in the set that have the requisite sensory level, and can act as targets. Each sensory attribute has an optimal level, and a corresponding physical representation among the ice cream products tested.

There are both benefits and caveats to this category appraisal approach, merged with mapping and with optimization:

1. The developer does not have to do as much work. The designer merely has to buy the products "off-the-shelf." Occasionally the designer will have R&D create some products to fill in gaps.
2. The independent variables (factor scores) integrate all of the sensory data. The factor scores, based on the sensory attributes, act as substitutes for the sensory attributes, and cover the sensory space. Rather than choosing a limited set of sensory attributes, the designer can use the information in most of the sensory attributes, by availing himself of the factor scores, which comprise this sensory information, albeit in a mathematical set of primaries.

3. The designer can identify holes in the space from the mapping. The holes correspond to sensory profiles where there do not exist any products. To find a hole the designer needs to plot out the different products in the space, and looks for the holes.
4. The designers can estimate the likely acceptance rating, and the sensory profile of the product in the hole, by using the product modeling.
5. All of the recommendations come out as factor scores, which are immediately translated into sensory profiles by means of the product models.
6. The sensory profiles, in turn, are converted into target products that have the necessary sensory levels. Several products may have the requisite sensory level. No single product may have all of the requisite sensory levels. The optimum is a composite profile.
7. The results do not, however, prescribe a formulation per se, nor can they. The results can only prescribe a sensory profile, because the researcher skipped the step of converting a sensory profile into a formulation. If the researcher has collected objective physical measures of these ice creams, or expert panel data, then the recommendations for the optimal product or the reverse-engineered product can be couched in the language of these objective measures or expert panelists, respectively.

PART 6: RAPID PRODUCT DESIGN AT THE "FUZZY FRONT END"

This chapter began with the question about the way the product designer identifies the characteristics of a product, and then moved on to identifying trends, creating concepts, and finally creating products. The chapter finishes with new approaches to product design, using currently available Internet-based technologies. As the food and beverage industries are becoming increasingly competitive, there is an emerging need to accelerate the process of discovery and design. Whereas in previous years the corporation had the luxury of time and resources to understand consumer wants and then build products, today the story is about lack of time, lack of resources, an overabundance of competition, and a surfeited consuming public, not to mention stiff criteria for products placed by the trade.

One of the big market buzzwords today is the "fuzzy front end." A quick look through a search engine such as Google® reveal 4,090,000 references to innovation, and 1,270 references to the phrase "fuzzy front end" as of August 2004. When all is said and done, however, this phrase refers to the early stage of development (Stinson, 1996). For the most part, product designers under the direction of outside experts are content to do qualitative research at this stage. Market researchers are well equipped with a variety of tools to facilitate ideas, but except for the qualitative specialist, most researchers shy away from using these tools at the early stage (Hoban, 1998; Rosenau et al., 1995).

Over the past decade, the process of corporate innovation has generally been left to outside consultants and to internal/external teams specializing in the invention process (Griffin, 1997; Wheelwright and Clark, 1992). Researchers have shied away, preferring to act in the evaluative mode when ideas are created, rather than in an active creative mode. A consequence of this situation is that product designers are all-too-often neglected in this process, even though the newly emerging tools of research can accelerate innovation. It may be that the expectation of using new consumer research technology is premature, that it may take years before the talk is converted into action. What is needed is a knowledge-development and knowledge-enhancing system, operating efficiently and cost-effectively (Moskowitz, 1997). The system should have these six properties that deliver knowledge in a way consistent with the world-view of the product designer:

1. Rapid and user friendly, executed in days and weeks so the answers will really be used
2. Consumer-based, to ensure ongoing inputs from the end user
3. Knowledge-based, using data, not guesswork,
4. Reality-based, using observation to identify actual behaviors leading to these new products,
5. Iterative, to allow for changes in direction with each step of the new learning
6. Cost-effective to promote learning with little risk

With these issues in mind, let us now finish this chapter by looking at the confluence of some new technologies in product design, principally at the very early stages, the so-called "fuzzy front end." As the Internet-based technology continues to grow in power and corporate acceptance, there are emerging a number of approaches to understanding the consumer mind. We present an integrated approach to this understanding, which uses a combination of "off-the-shelf" databases, ethnographic observations, consumer interviewing with artificial intelligence programs, and finally concept development, segmentation, and optimization.

CREATING THE PRODUCT DESIGNER TOOLBOX FOR INNOVATION

A key defining aspect of product design in the past 10 years, and apparently for the next several years will be, is the use of "tools." Product designers, sensory analysts, and marketing researchers graduating from universities today are awash in technological aids to creative thought. These aids span the range from computerized interviews either on personal computers or on the web to statistical analysis methods and high-level quantitative treatment of qualitative data. One of the key phrases often heard is the "toolbox." A toolbox simply comprises a set of well-accepted research procedures that have been designated as appropriate to help solve a problem. Although there is a method and analysis toolbox for many common research problems (e.g., product testing, tracking, and so on), there is no comparable product design toolbox for innovation. There are no algorithms, procedures, or analytical strategies to deal with the fuzzy front end of development, although there are numerous practitioners who provide one or another method. The business literature is replete with these methods, be they in the form of books or journal articles, in either the popular press or the archival academic press, respectively. A product design toolbox comprising high-level data acquisition and analysis techniques for the early developmental stage would, therefore, be a very welcome addition to the product designer's armory.

Intellectual Foundations—Models from Strategies of Adaptation in Evolutionary Biology

Although there are no well-accepted algorithms, procedures, or strategies to deal with the fuzzy front end of development, there is an emerging view suggesting that organization improvisation and innovation appears to arise from the recombination of previously successful subroutines (Borko and Livingston, 1989). Models of adaptive systems as well as some evidence from evolutionary biology have shown that recombining routines provides one of the most fruitful sources of change, allowing systems to prosper and to adapt to new circumstances (Holland, 1975; Levinthal, 1991). We believe that, in the same way, a company using well-developed product design oriented subroutines can utilize the principles of adaptive planning, recombining successful routines in rapid iterative learning cycles. Each subroutine should provide a clear "piece of the puzzle" so that practitioners can easily recognize "recombinatorial" possibilities as new learning occurs. The consequence would be the research-driven ability to produce and evaluate new ideas in response to unexpected learning or market changes (Flores and Briggs, 2001).

To round out this overview of intellectual foundations we can look at some of the critical new thinking around the concept of an "adaptive enterprise" that might embrace this new paradigm. Haeckel suggested that Adaptive Enterprises are "*Sense and Respond*" organizations rather than "*Make and Sell*" organizations. The barriers to innovation are removed when the organization is in constant search for new ideas and recognizes publicly that it has a constant need to add new ideas to the conveyer belt. "Change is no longer a problem to be solved, but rather an indispensable source of energy growth and value" (Haeckel, 1999).

An Organizing Principle for the Toolbox—Ideas as a Combination of Function and Form

One way to create the product designer's toolbox involves adapting the proposition of Finke et al. (1995), that areas comprise both functions (viz., consumer needs) as well as their relation to forms (viz., solutions). They identified three types of cognitive search strategies that may be relevant to the creation of new product ideas:

1. Identify or define a function (viz., consumer need), and then perform an exploratory search for a suitable form (viz., solution)
2. Identify a form (viz., solution) followed by an exploratory search for a meaningfully related function (viz. consumer problem)
3. Generalize an already known function-form relation

Goldenberg et al. (1999) adopted this organizing principle to the context of ideation for new products, developing a classification that we can use here for the "subroutine."

1. "Need-spotting," in which need identification precedes product (viz., form) development
2. "Solution-spotting," in which either a form is identified and the inventor searchers for a suitable need, or in which both need and solution are concurrently identified, with the generalization following shortly afterward
3. "Mental invention" in which there is a decision to innovate, and afterwards both the need and the solution are developed interactively.

The foregoing organizing principle provides the theoretical groundwork underlying innovation. The primary objective is to provide a system that identifies needs efficiently, identifies solutions equally efficiently, and which provides an ongoing stream of alternative needs and solutions so that the company need not rely upon serendipity to create new products. The secondary objective is to make the system inexpensive, very rapid, easy to use, and powerful, so that the design process can become more powerful and widely used, thus increasing the chances of market success through true knowledge.

A Recommended Beginning Paradigm for Product Design

The paradigm comprises a set of discrete steps, each of which has been used in a variety of applications. The sequence and combination could provide the necessary insight for innovation and continuous development. The paradigm is grounded in the notion that innovation in the food beverage and foodservice industries should not be at the whim of chance and happenstance, but rather should be harnessed in a systematized, public fashion. The paradigm comprises steps that focus on each of the cognitive search processes relevant to ideation—specifically "need spotting," "solution spotting," and "interactive mental invention." The process encompasses two dynamically interacting opposites—innovation (new, different, unique, and perhaps radical) and systematized/public fashion (current, conventional, and same).

The paradigm comprises five specific stages: The first two stages are informational. They provide context (Foundation Study, such as the It! Database discussed earlier) and within that context identify behavior that may lead to a new product opportunity (Ethnography). In the scheme above, these are subsumed under "need spotting."

1. *Foundation study*. The foundation study comprises a body of knowledge about a product or service category, which is available prior to the need for the innovation development. The foundation study presents the landscape currently known to the developer and marketer, albeit in a way that may go deeper than conventional means. The foundation study uses well-defined methods, such as conjoint measurement, to identify what specific features of a current product or service category are attractive to consumers, versus what repel the consumers. The foundation study also provides indications about the existence and nature of segments in the consumer population (Beckley and Moskowitz, 2002; Moskowitz and Rabino, 1994). One might consider the foundation study to provide a corpus of publicly available knowledge that can be used as a background within which to interpret observations, problems, and so forth.

The importance of a foundation study cannot be overemphasized, for it comprises systematized learning.

2. *Observation* (e.g., ethnography). Observation means looking at actual behavior in the environment rather than considering reported behavior. Ethnography is becoming very popular today as a way to understand the customer in-depth. Ethnography by itself, however, simply provides snapshots of behavior. When merged with a Foundation Study, ethnographic observation puts the behavior into a context (e.g., typify a problem to be solved thus revealing an opportunity; show the way a person solves an every-day problem, and so on). All too often, it is easy to obtain ethnographic data by video camera and other recording devices, but hard to locate this behavior within a matrix that reveals the business opportunity. Detail often overwhelms, hindering insight. The foundation study enhances the potential usefulness of the ethnographic information (Abrams, 2000; Ericson and Stuff, 1998; Stewart, 1998).

The latter three stages deal with the use of consumers, interactively, to provide the innovation—viz., "solution spotting." The assumption here is that the consumers are not particularly verbal, may not express themselves, are not "lead users" (von Hippel, 1986) and in general are not highly motivated to provide award-winning insights. Rather, the assumption is that the consumers who will provide the innovation are conventional, cannot easily think "out of the box" in their ordinary lives, unless forced to and given the opportunity. However, it is far easier and more process-friendly to present consumers with specific stimuli, and let them respond with simplistic, albeit occasionally inspired answers. These last three stages, thus work with the modal type of consumer, rather than with the articulate individual or lead user. The innovation is designed to come from the "actual consumers," not from the "elite, verbally articulate, creative consumer." Although it is always desirable to have a source of inspiration among consumers, the system should depend upon that source of inspiration, because then the system depends on serendipity.

3. In-depth interviews, on computer, using artificial intelligence to analyze the language and identify key ideas. The objective here is to create an interview, on computer, similar to that done by an intelligent interviewer who probes. The in-depth probing is done using software methods rather than an interviewer (Cleveland, 2001). The software is set up to help the respondent elaborate on key themes, by selectively repeating certain key phrases in the "probe" mode. The verbatim from dozens, or perhaps even hundreds or thousands are then analyzed to identify linkages of ideas. This area of artificial intelligence has been rapidly developing over the years, and has reached the point where the computer analysis of the verbatim can isolate key ideas (Cleveland, 1986). Of course, the critic might argue that machine probing could never replace the human interviewer, which criticism is still valid. On the other hand, with rapidly acquired, inexpensively obtained, easily analyzed, dozens, hundreds, or even thousands of interviews, there is a plethora of raw material from which to extract key ideas, and sufficient data from which one or a few key, new ideas might well emerge. What the computer lacks in truly profound intelligence may be partly compensated by the ability of the computer to access the minds of hundreds of people in a cost-effective and time-effective fashion. The output of this third step is a particular problem or situation, with which the respondent

can identify, and which forms the frame of reference for concept development and solution. The availability of relatively inexpensive computer interview administration and analysis of the interview makes the in-depth interview affordable. Furthermore, the ease of administration makes the potential for iterative interviewing quite real, allowing the researcher to change the focus of the interview as more knowledge is obtained.

4. *Create raw materials for product concepts by framing a situation in the respondent's mind:* The objective here is to obtain elements for concept creation from consumers, who are placed into a specific mind-set. Step 3 (in-depth interview by computer) provides the necessary information from which to formulate a frame of reference. Step 4 presents that frame of reference to consumers, asks them to provide elements that could either expand the problem, or provide part of the solution or the benefit. The problem statement serves as a springboard, catalyzing the respondent's creativity, and focusing the respondent's output into a specific and relevant direction. Through a Delphi-like procedure (Flores, 2002; Jolson and Rossow, 1971), the respondents in a variety of locations provide the elements, and also react to elements provided by others. Through this creativity process, each individual respondent is not required to be particularly creative, but rather to provide some few elements, and to judge a few other elements provided by respondents who previously participated in the process. The outcome, however, comprises a rich set of raw materials that serve as the basis for concept development and optimization. One of the key aspects here is that hundreds of respondents participate, removing the onus of creativity from any single respondent. Another aspect is that the elements are polished and voted on as they move through the system. The output provides a ranking of the relevance of the most important to least important elements, as well as a complete list of the elements offered by the participants. Another feature of the approach is the possibility of iterative work. With relative ease, the researcher can return again and again, with new questions or mind-sets, to create more information with the consumer's help. Thus the issue is not just quantity. There is the possibility of better, more specific elements for use in concept creation.

5. *Concept evaluation and optimization using online conjoint analysis*. The final step in the design process comprises the creation and evaluation of test concepts through conjoint analysis. The conjoint measurement is set up on the Internet by the researcher, executed with hundreds of respondents, and analyzed to identify the contribution of each element to consumer interest, at the individual respondent level. The output, therefore, of this final stage is the impact or utility value of every element, and the ability to create newer, better, more impactful concepts. It is worth noting again that the Internet makes the conjoint approach quite cost-effective. The ease of set-up permits the conjoint to be iterative, a benefit to researchers that has not always been the case for the traditional conjoint approaches (Moskowitz et al., 2001).

A Case History for Innovation—The Car as a Restaurant and Food Items for the Car Trip

An easy way to understand the paradigm for innovation in product design comes from a simple case history. The case history deals with the common problem of time-pressure and its effect on eating and driving. As the demands on consumer time increases, available time decreases. This shows up in an increasing number of individuals eating in cars. Indeed, it is becoming obvious that more and more fast food restaurants are featuring drive up windows where the customer can order food, in order to eat in the car. Eating on the go is increasing, and so are the problems associated with it such as messy cars. The case history deals with the creation of a product to fit with this new lifestyle trend. We will follow the above-mentioned five steps to identify a product opportunity, and develop the design for that product at a concept level.

Step 1: Foundation Study

The 2001 Crave It!® database (Beckley and Moskowitz, 2002) dealt with 30 different food products, ranging from hamburgers to french fries to ice cream, and so forth. The goal of the foundation study was to identify what particular features of foods drove rated "craveability." As noted earlier, the foundation study revealed three radically different mind-sets of consumers, transcending the 30 different foods. These were defined as "Classics" (want food the traditional way); "Elaborates" (want lots of variation of their food, including toppings, flavors, and the like.); "Imaginers" (responsive to emotion and promiseless so to descriptions of food). These three segments appeared in all the foods, with a great many respondents falling into the 'Elaborates'. The usefulness of this segmentation as an organizing principle for development will become clearer as we proceed. Going back to Table 6.10 we see that the three segments like different types of hamburgers, and that the Elaborates want a lot of toppings on their hamburger, which could become quite messy. For example, the Elaborates want "lots of crispy bacon and cheese on a juicy grilled hamburger on a lightly toasted bun," or "with all the toppings and sides you want … pickles, relish, jalapenos … lettuce, tomato, chips … whatever." These features would lead to a problem eating in the car.

Step 2: Ethnographic Observation

By itself, the foundation study provides a corpus of information for the marketer and developer, but does not seek out problems. Ethnographic observation of people in their daily lives does reveal problems. A later project that studied fast food consumption observed consumers in cars. These consumers were fast food customers, who purchased and ate some of the food in the car. During the course of some of the interviews, it became obvious that a key issue was "cleanliness." Although this issue did not surface directly, observation of behavior revealed that customers were having problems eating the food neatly in the car.

Ethnographic observation is not survey research. The observation records behavior, but it is up to the analyst to put a structure around that behavior. Looking at the video records suggested that there might be a linkage between the category of "Elaborates" discovered in the foundation study and the messy situation in which customers found themselves when trying to eat fast food in the cars. The connection was not made at a conscious level by the respondents, who, in any case were unaware of the foundation study. The connection did occur to the researchers that perhaps those individuals eating messily in cars might belong to the same class of respondents classified as "Elaborates". With a large number of 'Elaborates' in the foundation study (exceeding 40% across all of the 30 categories), this connection of messiness and "Elaborates" took on additional meaning. The connection suggested an opportunity for new products, designed for the fast food restaurant, geared toward in-car food consumption.

It is noteworthy that this type of connection is not necessary in all cases. The connection comes from the availability of the foundation study, which sets a framework for understanding, the behavior that is observed. Were the foundation study not to be available, and thus the categorization of people into "Classics," "Elaborates," and "Imaginers" not available, the linkage with eating in the car might never have been made.

Step 3: In-Depth Interviews Using Computerized Methods Facilitated by the Internet

The recognition of eating in the car as a messy situation (ethnographic output) that might afflict the group of consumers in the elaborate segment (foundation study output) led to the need to further identify the problem. Through in-depth interviews powered with artificial intelligence and analyzed quantitatively (Cleveland, 2002), it soon became apparent that there were a number of issues involved in eating fast food, especially with children, and especially in the car. Through evaluations with several dozen respondents, and analysis of key issues, the findings showed that the foods purchased were too large, too messy, and that there was an opportunity to downsize the product in order to make it easier to eat in the car (Table 6.23).

Step 4: Internet-Facilitated Ideation and Collaborative Filtering Using Brand Delphi®

The open-ended questioning in Step 4 provides a sense of the problem and at some level one or two solutions. These solutions are provided by respondents, almost in a serendipitous way, during the depth interview. What is needed, however, is a concentrated attack on the problem, once the problem is identified. This attack is done in Step 4.

The set-up or *aufgabe* question, emerging from the in-depth interview, was phrased as follows to the respondent: "We are interested in making a new type of bread/bun for breakfast. We are looking to develop new/buns that will make 'eating on the go' easier and less messy (e.g., in cars, trains and while walking). We welcome your ideas that will help us make better designed bread/buns." This paragraph, presented at the start of each Delphi-like exercise, to each of the 480 respondents on the Internet, sets the stage for creativity. The paragraph focuses the creativity to move in the right direction, namely a specific product, but does not provide any additional ideas.

TABLE 6.23
Key Topics and Conclusions from In-Depth, Web Survey and Automatic Analysis of Interviews

Core Text List of Key Vocabulary Phrases Used in Interview

Food that works in a car
Bite or Bites (size food)
Plain or hamburgers
Barbecue (sauce container)
French or fries
Wrappers (that hold food in)
Finger or food (not fall part)
Candy and wrapper (type)
Chicken and nuggets (type food)
Key conclusions
Eating food in cars is messy for mothers. The causes of messy food are foods that are uncontrolled, too much ketchup, and so forth, or foods that fall apart or food that cannot be handled by the hands they are given to, such as "little" hands.
The fact that mothers have strict criteria of what they will allow in their cars, and what they will not allow, yet there is still a mess, says that the criteria, no matter how strict, and the criteria are not working for all mothers

1. The food they receive does not conform to what they ordered; too much ketchup.
2. The food they order comes in portions too large to easily control.
3. The food they order does not come in wrappers that keep them contained, such as drinks, shakes, and burgers. There are leaks

Small portions, controllable wrappers, and getting what they ordered are the keys to controllable and nonmessy eating in the car.
The number one idea for a nonmessy food in the car was the burger bite, a single toddler size bite.

The respondents were invited to participate by e-mail for a 10-min session. During this session they:

1. Read the set-up paragraph
2. Looked at up to eight ideas previously provided by other respondents, checked off whether they liked or disliked these ideas
3. Provided up to two new ideas or refined/polished other respondents ideas
4. Rated the ideas selected on an importance scale
5. Completed a short classification questionnaire

With several hundred respondents participating and rating different ideas as well as providing their own, the result of the exercise generates a matrix of ideas that can be ranked in terms of relevance and importance. What is important here is that the creativity is directed (by the set-up paragraph or by an image, even perhaps a video), large-scale (by virtue of hundreds of respondents), somewhat iterative (by having

TABLE 6.24
Ranked Elements from the Brand Delphi® Exercise on the Internet

Option	Rank
Four top tier elements	
Something that hold all of the contents (breakfast sandwich) without the possibility of overspill when you bite into it.	286
Whatever the product, it should not be flaky or too pastry-like as that is likely to create crumb problems all over the front seat of a vehicle. Wrappers that could catch crumbs are ideal. As is a container that one could put on their lap so it would catch the food	284
Be able to handle with one hand	274
Cannot have too many wet ingredients, like sauce that can come shooting out the bottom.	241
Four middle tier elements	
Bite-sized pieces (less mess)	202
Do it with less flaky bread and less gooey filling for less mess	199
Less oil on sandwich	160
Pocket bread (pita)	150
Lowest tier elements	
Similar to a Subway sandwich bun, very soft. I eat these often when I'm driving and the filling(whatever that may be) does not fall out.	66
Cutting a sandwich in half when it is wrapped in paper makes it much easier to eat	63
Bite size with filling, several in one Velcro type package	56

Note: Results are shown from the first 480 respondents gathered in one evening.

the respondents vote on the contributions of others), and quantifiable (by having the respondents rate the importance). Table 6.24 shows part of the results of this exercise after 6 h on the Internet.

It is important to remember that this type of creative exercise can be repeated with different set-up paragraphs, images, or videos or a combination of both, time after time, and with relatively little difficulty, until a very large number of elements have been created. Managing this abundance of information becomes a matter of reading them, and keeping the elements that are considered promising by respondents, as well as meaningful from the designer's viewpoint. Of course, there will be many redundancies with a lot of respondents participating, but this plethora can be narrowed down at the designer's leisure, after the material has been collected. It is worth noting that this approach is equivalent to a large-scale "bioassay" of the customer's mind. That is, if the Internet acquisition of the elements is continued for a week or more, with tens of thousands of elements, one can begin to count the frequency of appearance of each type of idea. This gives a sense of the state of the consumer's mind when it comes to problem solutions. What types of solutions are typically present (viz., the ones that keep coming up), what types of solutions are around but infrequent (they come up a few dozen times in thousands of interviews), and what types of solutions are truly unique (they come up once or twice). Thus the system, conducted on a regular basis, also measures the customer mind, for solutions, in a tracking-like fashion.

STEP 5: CONCEPT DEVELOPMENT BY USING CONJOINT ANALYSIS TO IDENTIFY THE WINNING ELEMENTS

The key issue is the need to craft a concept for a new product (or service), that has uniqueness and acceptance, and which answers a problem. The problem is made clear in the in-depth interviews, and in the nature of the ethnographic observations (e.g., messy cars resulting from messy eating). The business opportunity is clear from the combination of the Foundation Study, which shows the existence of the segments, and the ethnographic observation, which shows with more immediacy how individuals eat messily in their cars. The potential solutions to the problem came from the elements proffered by the respondents in the Delphi-like exercise, when presented with the problem situation. These are all informational, and need to combine in order to generate a winning concept.

The elements identified by the brand Delphi® method constitute the raw material for the conjoint analysis. Table 6.25 shows a collection of promising elements selected from the Delphi-like exercise, and their utility values in the conjoint study. Two conjoint studies were run on the Internet, each with 35 respondents. Half the respondents rated the concepts on the following 1–9 easiness scale: 1 = very hard to eat . . . 9 = very easy to eat. Half the respondents rated the concept on the following 1–9 liking scale: 1 = hate . . . 9 = love.

The conjoint measurement results suggest different dynamics for these elements in the concept, depending upon the mind-set of the consumer. It is clear from Table 6.25 that elements driving the perception of "ease of use" may not be the same as elements that are liked. Indeed, there may be no elements that are very highly liked and very easy to use. The optimal concept, therefore, is some combination of these elements. Depending upon the relative importance of "ease" and "liking," different concepts can be created through optimization.

An idea of what might emerge from this process appears in Table 6.26, which shows three concepts created from the same data set in there, and thus emerge from the development process. The concepts range from 100% utilitarian (maximize "ease of use"), to 100% hedonic (maximize "liking" without any attention to ease of use). The intermediate concept shows one of the many possible combinations of concept elements that constitute a compromise.

DISCUSSION

Although the paradigm here concentrated on the innovation of a new product for a problem situation (eating on the go, and specifically in the car), most interesting aspects are the experiences of the author and his colleagues in developing, explaining, and popularizing the paradigm. There are some aspects worth noting and commenting on.

PARADIGM FOUNDATIONS—SYNCRETISM IS NECESSARY, SYNERGISM IS DESIRABLE

As the technology today becomes increasingly sophisticated we have found that there is no 'one stop shop' that can provide all of the details. Even knowledge of

TABLE 6.25
Utility Values for 24 Elements Selected from Brand Delphi® Inputs, and Tested in an Internet-Based Conjoint Task

EL	Text	Easy	Like
	Additive constant	54	56
C2	Special container designed for use in the car ... no more crumb problems all over the front seat	9	1
D2	It's sealed on all sides ... so you don't mind eating in the car	8	−1
D5	Longer edges make it easy to hold the sandwich	7	−1
B3	Bite-sized sandwiches ... more convenient to eat while driving	7	−3
B1	Bite-size sandwich ... easy to eat on-the-go	7	1
D6	Easy to eat while driving	6	2
D4	All sides covered, so nothing can slide out	5	3
A4	Bread that won't fall apart after the first bite ... no runs, no grease, no mess!	5	0
A1	A new bread that holds everything together ... no more spills	4	−6
A5	A bread that keeps wet ingredients like sauce from shooting out	3	−1
B6	Bite-size ... just like finger sandwiches	2	−4
B4	With a spread already inside ... easy to toast, easy to eat	2	−5
C4	Comes in a double layered wrapper	2	−1
A3	Bread that keeps your sandwich intact ... won't slip and cause a mess	1	−2
D1	A smaller portion ... easily fits in palm of your hand	1	−3
C1	Comes in a package you can open single-handedly!	1	5
C3	Packaged to stay warm and can easily be placed in the car without falling over	0	4
A6	A pita pocket that keeps toppings where they belong ... inside the sandwich	0	−2
C6	An easy to open package, so you don't have to hassle when you're on the road	0	1
D3	No preservatives, low in sugar, no saturated oils...made from whole grain and dried fruit	−1	6
C5	Sweet and soft to get you going in the morning...in a self heating package, just tear off the strip and it warms up	−1	−7
A2	A new pita pocket bread ... so you can fill it up with whatever you like and not worry about making a mess	−5	−4
B2	Egg, bacon, and cheese baked inside the bread	−6	−13
B5	Made with French "pateachou" pastry dough ... stuffed with sausage, country gravy, ham and scrambled egg, bacon or cheddar cheese	−18	−19

Note: The two columns refer to criteria. Ease refers to the utility of concept elements with the concept rated on "hard vs. easy to use." Like refers to the utilities based upon the scale "hate vs. love").

the available tools can only be superficial if an individual is expected to span the range from ethnography to databases to depth interviews to ideation to conjoint. The range is simply too large, the task demands too great. Most researchers today, overwhelmed by concrete business problems to solve, simply cannot stay abreast of the large number of available technologies. Furthermore, even if a person could know the

TABLE 6.26
Three Optimum Combinations for the New Easy to Eat Sandwich

EL		Easy	Like
	Additive constant	54	56
	Maximize "Easy" alone		
A4	Bread that won't fall apart after the first bite . . . no runs, no grease, no mess!	5	0
B3	Bite-sized sandwiches . . . more convenient to eat while driving	7	−3
C2	Special container designed for use in the car . . . no more crumb problems all over the front seat	9	1
D2	It's sealed on all sides . . . so you don't mind eating in the car	8	−1
	Total (additive constant + elements)	82	54
	Maximize both "Easy" and "Like" (compromise optimum)		
A4	Bread that won't fall apart after the first bite . . . no runs, no grease, no mess!	5	0
B1	Bite-size sandwich . . . easy to eat on-the-go	7	1
C1	Comes in a package you can open single-handedly!	1	5
D4	All sides covered, so nothing can slide out	5	3
	Total (additive constant + elements)	72	67
	Maximizes "Like" alone		
A4	Bread that won't fall apart after the first bite . . . no runs, no grease, no mess!	5	0
B1	Bite-size sandwich . . . easy to eat on-the-go	7	1
C1	Comes in a package you can open single-handedly!	1	5
D3	No preservatives, low in sugar, no saturated oils . . . made from whole grain and dried fruit	−1	6
	Total (additive constant + elements)	66	69

Note: Created to maximize perception of "easy," perceptions both of "easy" and "liking" (a compromise), and perceptions of "liking" respectively. These combinations come from the conjoint measurement exercise, and from optimization of utilities to satisfy one or two objectives, respectively.

nature of these techniques, it is almost impossible today to be able to weave together a system by combining the techniques. Different groups, across companies and across disciplines, must work together in a syncretistic mode, to combine their expertise into a coherent whole. The individual parties in this combination do not, however, lose their identity, but rather effectively combine to create a more powerful business organism.

The nature of the paradigm just presented here spans the often incommensurate, incompatible range from observation to discussion, from qualitative to quantitative, and from the so-called "touchy-feely" to database numbers. Individuals expert in one of these areas are unlikely to be working side by side with individuals in another of these areas. It is organizations that must cooperate, no longer simply individuals in a large organization. Perhaps this is why the system represents an ecological chain of different organizations, competing for some businesses, but also cooperating to achieve the objective that any single one of the companies could not achieve on its own.

Tools and Technology Are Necessary But Not Everything ... Yet Neither is the Expert Analyst

One of the 'hall conversations' often heard at professional meetings in product design is the growing popularity of technology tools for creative data acquisition, analysis, and knowledge development (Ciborra and Patriotta, 1998; Pawle and Cooper, 2001). Tools, to some, are the panacea that promises to bring product design to a new level of sophistication. At the other end of the spectrum, are the traditionalists, deeply suspicious of product design technologies, longing for the good old days, and staunchly refusing to abandon the insight of the research to the mindless, heartless, soulless machines that often do the data acquisition and analysis. To be sure, tools and technologies are necessary for the aforementioned new paradigm. The Internet above all provides the means to reach and engage the hundreds or thousands of respondents in a parallel model. The computer, the server, the Internet explorer are tools that allow the researcher to reach the consumer. The specific computer applications replace the interviewer with automated presentation, data acquisition, and structured analysis. The cost, the time, the scope, and the power could not be duplicated without enormous expense in a world lacking machines (Moskowitz and Ewald, 2001).

Yet machines cannot work alone. Without the human being framing the problem, identifying an issue, formulating a mind-set *aufgabe* question, and selecting the correct elements, all that is done in the paradigm is to create a high tech monkey typing Shakespeare. That is, without soul, without knowledge and intuition, without the experience to recognize a business problem, the paradigm will simply lead to combinations of features that have no reality or need (Cooper, 1999).

Perhaps the best that can be said about the paradigm is its ability to engage consumers in the development of innovative ideas in a structured manner. The ingoing assumption is that the respondents need not be particularly innovative, nor articulate (although those characteristics are certainly desirable), but the respondents should have some sense of problems on the one hand, and an ability to intuit whether a solution is meaningful or not. The joint effort of dozens or hundreds of respondents, interacting on the computer in real time, provides the necessary process to create solutions, fine tune, and then optimize them. The existence of a foundation study ahead of time puts things into context, and provides an organizing principle into which these solutions can be comfortably embedded. Finally, ethnographic observation permits the developer to intuit how the solution might fit a problem that has become real through the actual observation of people experiencing it.

So What is Really Necessary to Understand Consumers at the Design Phase?

The traditional approach to product design through understanding the consumer has been to begin with fundamentals, move on to different types of testing, and then finish with procedures such as category appraisal and product optimization. These traditional methods have been buttressed by the scientific underpinnings of psychophysics, statistics, and sensory analysis. Psychophysics has given the traditionalists an understanding of how we process physical information to sensory responses.

Psychophysics has taught us about scaling, about relations between physical stimuli and responses, and about sensory processes in general. Statistics, in turn, has given the product designer a sense of whether two stimuli differ from each other, and to what degree. Statistics has taught us to make models, to optimize, and to map stimuli. Finally, sensory analysis has taught us how to collect data in the most proper fashion, how to analyze these data, present reports to colleagues, and in general maintain a professional approach to the entire process.

Yet, as we enter this new century, we find that the traditional methods are not sufficient. They fail because they do not lead to actionable results. They lead to data, to acceptable and standard methods. The problem is that they lead to procedures and not to products. As this chapter continued to emphasize again and again, there needs to be a link between the information that a researcher provides and the design steps that are taken. Good practice provides good science, but not necessarily good design. There needs to be something more, a process that links knowledge of the consumer to design properties of the product. We must involve the consumer, but with the proper method to go deeply into the consumer mind. Otherwise, we are in danger of splitting off the science of sensory analysis with its psychophysics/statistical underpinnings from the practical needs of the product developer. In the end, therefore, this author believes that there will emerge another science of product design, not so closely allied with sensory analysis and statistical testing, but perhaps more aligned with new methods of concept design, product modeling, and novel ways of understanding consumer behavior such as ethnography and "in context" research. The prospect is exciting. Hopefully, this chapter will have given inspiration for such a new venture.

BIBLIOGRAPHY

Abrams, B. 2000; *The Observational Research Handbook*; New York; NTC Business Books (AMA); pp. 1–62.

Amerine, M.A., Pangborn, R.M., and Roessler, E.B. 1965; *Principles of Sensory Evaluation of Food*; New York; Academic Press.

Ashman, H., Beckley, J., Adams, J., and Mascuch, J. 2002; *Teens Versus Adults: The 2001 Teen Crave It Study*; Institute of Food Technologists; Anaheim.

Beckley, J., and Moskowitz, H.R. 2002; *Databasing the Consumer Mind: The Crave It!, Drink It!, Buy It! & Healthy You! Databases*; Institute of Food Technologists; Anaheim.

Borko, H., and Livingston, C. 1989; Cognition and improvisation: Differences in mathematics instruction by expert and novice teachers; *American Educational Research Journal*; 26; pp. 473–498.

Box, G.E.P., Hunter, J., and Hunter, S. 1978; *Statistics for Experimenters*; John Wiley; New York.

Brody, A.L. 1957; Masticatory Properties of Foods by the Train Gage Denture Tenderometer Ph.D. Thesis, M.I.T.

Cairncross, S.E., and Sjostrom, L.B. 1950; Flavor profiles—a new approach to flavor problems; *Food Technology*; 4; pp. 308–311.

Caul, J.F. 1957; *The Profile Method of Flavor Analysis*; *Advances in Food Research*, New York; Academic Press, pp. 1–40.

Ciborra, C.U., and Patriotta, G. 1998; Groupware and teamwork in R&D: limits to learning and innovation; *R&D Management*; 28; pp. 44–52.

Clapperton, J., Dagliesh, C.E., and Meilgaard, M.C. 1975; Progress towards an international system of beer flavor terminology; *Master Brewers Association Of America Technical Journal*; 12; pp. 273–280.

Cleveland, C.E. (1986); Defining Customer Expectations Using Computer Content Analysis; In: *Handbook on Research: Techniques to Solve Common Marketing Problems*; Bank Marketing Association.

Cleveland, C.E. 2001; Quali-Quant techniques with Socrates and Aristotle; In: *Quester Text Processing*; Chapter 3, Genesis Institute.

Cleveland, C.E. 2002; Research Project to determine what the key ideas were when eating fast food in the car; Quester Research Project 02845.

Cooper, R.G. 1999; The invisible success factors in product innovation; *Journal of Product Innovation Management*; 16; pp. 116–133.

Dravnieks, A. 1974; Personal communication.

Ekman, G., and Akesson, C. 1964; Saltiness, sweetness and preference: A study of quantitative relations in individual subjects; Report 177; Psychological Laboratories, University Of Stockholm, Sweden.

Ericson, K., and Stuff, D. 1998; Doing Team Ethnography—Warnings & Advice; Qualitative Research Methods Series 42; SAGE Publications.

Finke, R.A., World, T.B., & Smith, S.M. (1995); *Creative Cognition Approach*; MIT Press; Cambridge MA.

Flores L. 2002; Making idea generation and innovation available on the decision-maker desktop; Working Paper, Amiens Graduate School of Business.

Flores L, and Briggs R. 2001; Beyond Data Gathering, Implications of CRM Systems to Market Research; Proceedings of the 54th ESOMAR Congress, Rome.

Gleick, J. 1999; *Faster: The Acceleration of Just About Everything*; Pantheon Books; New York.

Goldenberg, J., Lehmann, D.R., and Mazursky, D. 1999; The primacy of the idea itself as a predictor of new product success; Working Paper, Marketing Science Institute, Cambridge, MA.

Gordon, J. 1965; Evaluation of sugar-acid-sweetness relationships in orange juice by a response surface approach; *Journal of Food Science*; 39; pp. 903–907.

Green, P.E. & Krieger, A. 1991; Segmenting markets with conjoint analysis; *Journal of Marketing*; 55; pp. 20–31.

Green, P.E. & Srinivasan, V. (1980); A general approach to product design optimization via conjoint measurement; *Journal of Marketing*; 45; pp. 17–37.

Griffin, A. 1997; PDMA research on new product development practices: Updating trends and benchmarking best practices; *Journal of Product Innovation Management*; 14; pp. 429–458.

Gustaffson, A., Herrmann A., and Huber F. 2001; *Conjoint Measurement: Methods and Applications*; 2nd ed. Springer Verlag; Berlin.

Haeckel S. 1999; *Adaptive Enterprise, Creating and Leading Sense and Respond Organizations*; Harvard Business School Press; Boston.

Hall, J. & Winchester, M. 2000; Focus on your customer through segmentation; *Australian and New Zealand Wine Industry Journal*; 15; pp. 93–97.

Hartigan, J. and Wong, M. 1979; A k-means clustering algorithm; *Applied Statistics*; 28; pp.100–108.

Health United States 2002; *With Chart Book on Trends in the Health of Americans*; Government Printing Office; (December 1, 2002).

Hippel, E.V. 1986: Lead users: Source of novel product concepts; *Management Science*; 32; pp. 791–805.

Hoban, T.J. 1998; Improving the success of new product development; *Food Technology*; 52; pp. 46–49.

Holland, J.H. 1975; *Adaptation in Natural and Artificial Systems: An Introductory Analysis with Applications to Biology, Control, and Artificial Intelligence*; University Of Michigan Press; Ann Arbor, MI.

Hughson, A., Ashman, H., de la Huerga, V., and Moskowitz, H.R. 2004; Mind-sets of the wine consumer; *Journal of Sensory Studies*; 19; pp. 10–15.

Jolson, M. and Rossow, A. 1971; The Delphi Process in marketing decision making; *Journal of Marketing Research*; vol. VIII; pp. 443–448.

Levinthal, Daniel A. 1991; Organizational adaptation and environmental selection—interrelated processes of change; *Organization Science*; 2; pp. 307–333.

Lord, J.B. 1999; New product failure and success. In: Developing new food products for a changing marketplace; Brody, A.L and Lord, J.B. eds.; Technomic Publishing Company; Lancaster, PA.

Meilgaard, M., Civille, G.V., and Carr, B.T. 1987; *Sensory Evaluation Techniques*; CRC Press; Boca Raton, FL; Chapter 8; pp. 119–142.

Meiselman, H.L. (1978); Scales for measuring food preference; In: *Encyclopedia of Food Science*; pp. 675–678. Petersen, M.S and Johnson. A.H., eds.; AVI, Westport, CT.

Meiselman, H.L. 2000; *Dimensions Of The Meal: The Science, Culture, Business and Art Of Eating*; Gaithersburg, MD, Aspen.

Mitchell, A. 1983; *The Nine American Lifestyles*; MacMillan; New York.

Moskowitz, H.R., 1981; Sensory intensity vs. hedonic functions: Classical psychophysical approaches; *Journal Of Food Quality*; 5; pp. 109–138.

Moskowitz, H.R. 1996a; Segmenting consumers world-wide: An application of multiple media conjoint methods; pp. 535–551; Proceedings of The 49th ESOMAR Congress, Istanbul, Turkey.

Moskowitz, H.R. 1996b; Segmenting consumers on the basis of their responses to concept elements: An approach derived from product research; *Canadian Journal of Market Research*; 15; pp. 38–54.

Moskowitz, H.R. 1997; From a process to a transaction: Implications for research and the research community; In: *The 1997 CASRO Journal, Council of American Survey Research Organizations*; Port Jefferson, NY; pp. 81–85.

Moskowitz, H.R. (1998); Designing new products in cyberspace—research driven innovation. CASRO Annual Journal; Port Washington, NY; pp. 69–78.

Moskowitz, H.R. 2000; Sensory driven product development: Optimization approaches and analyses; *Australasian Journal Of Market Research*; 8; pp. 31–52.

Moskowitz, H.R., and Ewald, J. 2001; Always On—Bringing market research down to the development engineer, closer to the customer, and into the vortex of product development; Paper presented at the ESOMAR Congress, Rome.

Moskowitz, H.R., Gofman, A., Katz. R., Itty, B., Manchaiah, M., and Ma, Z. 2001; Rapid, inexpensive, actionable concept generation & optimization—The use and promise of self-authoring conjoint analysis for the foodservice industry; *Foodservice Technology*; 1; pp. 149–168.

Moskowitz, H.R., Kapsalis, J.G., Cardello, A., Fishken, D., Maller, G., and Segars, R. 1979; Determining relationships among objective, expert and consumer measures of texture; *Food Technology*; pp. 84–88.

Moskowitz, H.R., Poretta, S., and Silcher, M. 2004; Concept research in food product design and development; Ames, Iowa Blackwell Publishing.

Moskowitz, H.R. and Rabino, S. 1994; Sensory segmentation: An organizing principle for international product concept generation; *Journal of Global Marketing*; 8; pp. 73–93.
MS Interactive 2004; The perception analyzer. Website; http://www.perceptionanalyzer.com/
Pangborn, R.M. 1970; Individual variations in affective responses to taste stimuli; *Psychonomic Science*; 21; 125–128.
Pawle, J.S, and Cooper, P. 2001; Using web research technology to accelerate innovation. Proceedings of Net Effects, Barcelona, European Society of Market Research, 11–30.
Peryam, D.R., and Pilgrim, F.J. 1957; Hedonic scale method of measuring food preferences; *Food Technology*; 11; pp. 9–14.
Rosenau, M.D.Jr., Griffin, A. Castellion, G.A. and Anschuetz. N.F. 1995; *The PDMA Handbook of New Product Development*; John Wiley; New York, NY.
Schutz, H .G. 1965; A food action rating scale for measuring food acceptance; *Journal Of Food Science*; 30; pp. 365–374.
Schutz, H.G., and Cardello, A. 2001; A labeled affective magnitude (LAM) scale for assessing food liking/disliking; *Journal of Sensory Studies*; 16; pp. 117–159.
Stewart, A. 1998; The Ethnographer's Method. Alex Stewart; Qualitative Research Methods Series; 46; SAGE Publication.
Stinson, W.S. Jr. 1996; Consumer packaged goods (branded food goods); In: *Product; New Product Development*; *Product Development*; pp. 297–312. Rosenau, M.D.Jr., Griffin, A., Castellion, G.A. and Anschuetz. N.F., eds.; John Wiley; New York, NY.
Stevens, S.S., 1956; The direct estimation of sensory magnitudes—loudness; *American Journal of Psychology*; 69; pp. 1–25.
Stone, H., Sidel, J.L., Oliver, S., Woolsey, A., and Singleton, R. 1974; Sensory evaluation by quantitative descriptive analysis; *Food Technology*; 28; pp. 28–34.
Systat. 1997; Systat, the system for statistics, Systat Division of SPSS Users manual; Evanston.
Szczesniak, A.S., Brandt, J.A., and Friedman, H.H. (1963); Development of standard rating scales for mechanical parameters of texture and correlation between the objective and sensory methods of texture evaluation; *Journal Of Food Science*; 28; pp. 39–403.
Wheelwright, S.C., and Clark, K.B. 1992; *Revolutionizing Product Development: Quantum Leaps in Speed, Efficiency and Quality*; Free Press; New York.

7 New Product Organizations: High-Performance Team Management for a Changing Environment

Robert E. Smith and John W. Finley

CONTENTS

Different organizations have different notions of the drivers for their new food products. Increasingly, consumers are recognized as the principal drivers, with marketing and their new food product operatives as the food product development organization's consumer representatives. The food product development process must have organizational support and the enthusiastic cooperation of those responsible for the process. Incorporating all meaningful contributors from everywhere as early in the process as possible appears to be the mechanism to optimize and speed progress. Every organization should be positive toward reaching out to extend resources such as universities, suppliers, and consultants. New food product development departments, matrix, project venture, and cross-functional teams each offers benefits and drawbacks, which should be considered in this perspective of organizational objectives and resources.

INTRODUCTION

As with many other industries, during the 1960s the food industry benefited from the investments that resulted in the scientific and technological expansion of the post-Sputnik era. Some of the rapid expansion in science and technology developed in the universities were adapted and converted into new products by the food industry. Universities expanded their food science and technology departments, merging disciplines such as dairy science, nutrition, microbiology, plant science, animal science, engineering, and vegetable crops, to form food science and technology departments. These departments produced both undergraduate and graduate students who were recruited and welcomed by the rapidly expanding technical departments of the food industries. The industry embraced these individuals and incorporated them into rapidly growing research and development programs.

At the time, many major food processors had comprehensive research programs that spanned the spectrum from discovery research to product development programs. And suppliers of ingredients, flavors, equipment, and packaging had equally vigorous development programs targeting the food industries. Throughout this era universities developed high quality basic science, which found its way to the "basic" research in industry. Companies such as General Foods (now Kraft), Ralston Purina, Proctor & Gamble, and Swift explored and developed new technologies, which in turn led to families of new food products. The results of the technologies were reflected in a rapid expansion of innovative new products ranging from sophisticated frozen foods to multilayered dessert products to fabricated meats to pourable salad dressings, and so forth.

During the late 1970s and into the 1980s, corporate America began to retrench and emphasize much more short-term financial return versus longer-term research and development programs. For the corporations to remain fiscally competitive, food product development evolved largely to development of "now" products. Although product development was cut back relative to sales in the food industry, it survived by applying technologies that were in the pipeline or by developing line extensions. Longer-term developments were minimized or eliminated. The technology pipeline was beginning to run dry, because the food industry was moving away from basic research and many universities were often finding more support in applied as contrasted to basic areas. This change in emphasis was fueled by the fact that U.S. government cutbacks in research funding especially for food products limited the ability to perform longer-term or higher-risk research projects. It became evident that balance needed to be re-established between fundamental research and food product development for the food industry. Such a balance has not been achieved, as the two appear to continue to function in independent orbits. Nevertheless, food product development driven by consumer expansion and diversification, competitive forces, financial resources, distribution channel changes, and technological impacts from peripheral disciplines, continues to expand—although, as cited in the other chapters, only infrequently smoothly. This chapter describes the organization structures employed, past, present, and future, to facilitate food product development.

What is the best organizational model for product development? There probably is no correct answer to this question. In any given situation with proper inputs and resource allocations, any structure will work. All organizations have existing cultures that in part will dictate what type of new food product organization will work. In the food industry are research and development and food product development organizations ranging in size from one-person operations to the few mega organizations such as those incorporated in Nestle, Unilever, or Kraft. Certainly, the smaller companies cannot support major longer-term research and development programs. Essentially, all these organizations need to get the food product out the door, and they do. This does not imply, however, that they lack innovation. Frequently small food manufacturers are willing to take greater risk than more conservative larger organizations. Currently, for example, in the areas of food supplements, "healthy" foods, and "functional" foods many small companies have introduced innovative new products to the market. Usually the supporting data for claims or promotion have been limited. The more traditional companies are taking a lower risk approach, usually moving more slowly making sure they can support the health claims with the new products, although the low carbohydrate fad of the early 2000s may belie that thesis.

THE CHANGING FOCUS OF THE WORK PLACE

In the September 1994 issue of Fortune magazine, William Bridges published an article titled, "*The End of the Job*." This article focuses on the changing environment in today's workplace, resulting in the demise of the traditional job as defined by explicit job descriptions. During the previous age of mass production and huge production lines, well-defined jobs were necessary for production and assembly of goods. Frequently research and development organizations in the food industry evolved to be somewhat like production lines. The product was developed (often without any consumer or marketing input), handed off to a pilot plant, and later to marketing and production. Microbiology and quality assurance inputs came late in the development phase to deter disasters. At the last moment, the product was passed to packaging and then to distribution to deliver to sales, sometimes after the production was initiated. Each contributing person, from product design through production and delivery had a specific job or set of responsibilities. This job definition usually determined where the individual's next opportunity would be found in the organization. Employees were evaluated on how they did their "job," not on what problems were solved, or how the company performed as measured at the end of the year. Many employees did their "jobs" in now defunct companies.

The modern employee is presented with unfamiliar risks, but much greater opportunities to develop his or her skills and ultimately achieve greater independence while helping to develop added value to the company. Today, project or even venture teams are replacing many of middle management jobs in R&D and product development. Through the appropriate use of research and product development teams, middle management roles can be significantly reduced. In traditional organizations, the managers fulfilled the role of communicator. They funneled the research and development goals from management to the technical professional and the results back up through the hierarchy and then back down through another function. With the advent of team research and development, the cross-functional teams provide the cross communications between groups such as research and production. This is also facilitated by the use advanced information systems, which provide cross communications on a real-time basis. Another advantage is that e-mail and related electronic communications allow teams to work together from distant locations. A researcher can "watch" a production run from an R&D location half way around the world and contribute ideas in real time without physical travel.

The fast moving organizations operate on projects rather than the traditional structured environments. The "job" as it was known from the onset of the industrial revolution is probably dead. We are evolving into an environment in which work is defined as a series of tasks that need to be completed, but not in the traditional job environment. Instead, the tasks will be completed by groups of qualified individuals assembled together in teams, which have clearly defined goals. In food research and product development the goal may be a new ingredient, a new food product, or a new process. The team will include members who bring together all of the necessary skills to complete the task and not necessarily all in the company. Depending on the specific goal, it is likely all aspects from packaging and design through engineering through production and distribution, quality assurance will be included. The role of marketing is to define the target. In some cases, marketing will develop sub-teams, which include consumers and health professionals to obtain clear definition of the proposed food product. In functional foods, the product definition and the target consumers must be carefully identified. This will significantly improve the team's ability to provide specifications for the product.

WHAT IS A HIGH-PERFORMANCE TEAM FOR FOOD PRODUCT DEVELOPMENT?

A team is defined as a group of individuals drawn together to work toward a common goal. Although this is a simple definition, it is anything but trivial. First, it should be recognized that every individual on a team is present for a specific reason. Whether a "secondary" participant or a frontline functionary, each contributes in some key manner to the success or lack thereof of the team. Each individual is a member subverting his own personal ambitions to the betterment of the team—and that is a guiding principle. World Cup soccer teams and Super Bowl-winning football teams are all composed of individuals who bring special skills to help make the team function as a whole. The dramatic self-removal of France's best goal scorer from the 2006 World Cup final immediately prior to the finish was a vivid example of placing self ahead a team with a consequent loss. Without a skilled place kicker, it is unlikely that a football team would make it to the Super Bowl let alone win it. When players are paid in millions of dollars per season no team carries any "extra" players. Similarly, with teams developing new products or processes, a multitude of skills are required, but there are not likely to be any extra members. Each person is a professional integer to the whole, regardless of the level.

In food research and product development, the high-performance team is composed of members who contribute to all aspects of the project from research foundation, marketing, packaging, culinology™, engineering, sensory analysis, production, distribution through quality assurance. High-performance food product development teams differ from other teams, because they address all aspects of the project. They communicate regularly on an open and frank basis, and all success and failure is shared. Each member is responsible for his or her own contribution and for that of the other team members. The teams are built around trust, communication, and commitment to the final success of the project. The elimination of layers of middle management frees individuals to contribute their technical knowledge to the project team. Such persons may be truly labeled "professionals."

As stated above, with the disappearance of the "job" concept, the level of the team members within the conventional corporate hierarchy, still very much extant in many food companies, is not an issue. In the conventional hierarchy, the team traditionally would be led by the highest-ranking member. In the new version, the team might be coached or sponsored by a higher-ranking person, but the team leader could come from anywhere in the organization, and not necessarily hold a "rank." The critical factor is that the areas of expertise are present to accomplish the goal expeditiously. As mentioned for a food product development effort, the team might consist of a marketing person, a culinologist™, several technical persons with expertise in all of the operations needed, sensory experts, and so forth. Fundamental scientists with appropriate expertise, in areas such as biochemistry, microbiology, enzymology, or polymer chemistry would work with other technical individuals experienced in marketing, sales, manufacturing, distribution, packaging, analytical services, and quality assurance. Such a team accomplishes "buy in" from every area of the company, from every discipline represented in the company that will be impacted by the new food product.

This vertical integration of resources in the food product development team facilitates "getting it right the first time" or more realistically reduces the errors and failures. It is important that these members communicate regularly and openly throughout the process and establish clearly defined conditions of satisfaction for the final food product. The conditions of satisfaction describe what it will take to make the new food product meet the needs of each phase of development. Vertical integration helps assure that all of the product "needs" are met throughout each phase of the development, and minimizes the probability that food products or concepts will be "thrown over the transom to the next guy to fix."

HIERARCHY VERSUS TEAM MANAGEMENT

Research organizations traditionally have been structured like manufacturing organizations in which each individual had a job in a well-defined hierarchy. The

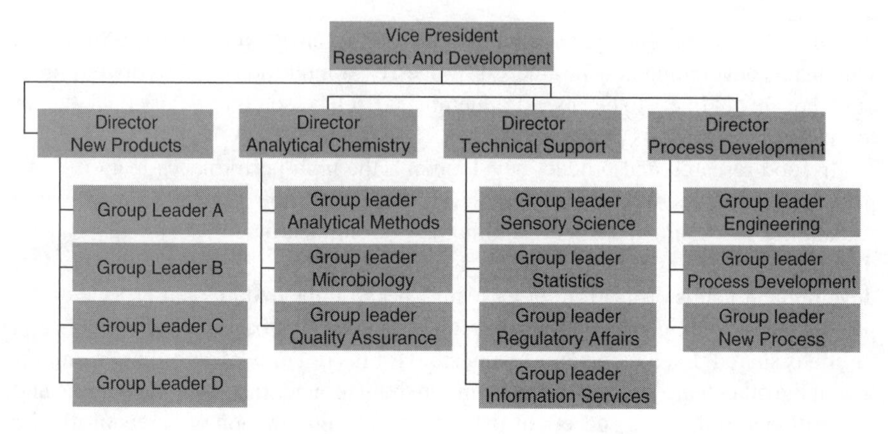

FIGURE 7.1 Generic R & D organization.

top of a generic R&D/product development organization is shown in Figure 7.1. In this traditional-style organization, a number of directors report to a vice president. The directors all have organizations of differing sizes reporting to them. Typically, group leader, scientist, and technician functions will have similar structures below the directors. In this model, the director of food product development is likely to have the largest organization reporting to him or her. Each individual in the organization has a defined "job." This job is carefully spelled out in a job description. In most organizations the group leaders would be in some level of competition to eventually inherit the director's job, and the directors in turn will compete for the for the vice president's job. In an organization of this sort, research and product development programs and assignments typically follow the lines on the chart. In this mode, the assignments generally come down from the vice president to directors, and so forth. A request for a new food product might come from a marketing vice president to the R&D vice president or a Product Development Director who then assigns the technical aspects of the project as appropriate in his organization. No one really has overall responsibility for the total food product development: it is a task: handoff: task.

In Table 7.1, the responsibilities of the manager and the scientist/technologist in the hierarchical organization are summarized. For the past 50 years or so and in many current situations, this is the way research and food product development was performed. In this top-down model, the manager is responsible for connection to the business and the scientist/technologist stays in the laboratory generating new discoveries. The scientist/technologist is judged solely on technical productivity. When we study the success of the industry over the past 50 years, we can hardly say the system does not work. But an 80% new food product failure rate is not a shining success. However, the pressures on contemporary society and resultant changes in the work environment would suggest that another model, such as the team approach, should be considered. In this new paradigm, research and development efforts can be more sharply focused on a specific goal or concept and conducted by *cross-functional* teams.

TABLE 7.1
Manager's and Scientist's Responsibilities in the Hierarchical Organization

- Manager
— Manager knows best
— Plans, leads, organizes, controls, and decides
— Communicates to clients and superiors
— Power to commit
— Evaluated by superiors

- Scientist/technologist
— Mainly performs research and/or development
— New discoveries may often be serendipitous
— Technical excellence validated by peer reviewed publications

The model for team research and development, on the other hand, is to have clearly defined goals and teams that include all of the core competencies required to complete the project successfully. Scientists and technologists in the team-approach must become much more focused on the business and vice versa, marketing, distribution, sales, financial, legal, upper management, and so forth, staff must become more involved in the technical aspects. The team has the responsibility for considering consumer and customer needs (retailer, food services, and so on), business objectives, and technical capabilities. Technical capabilities should be derived from core competencies. When the core competencies are not available to the team, they must be obtained from external sources. As described above, the outside capabilities may come from relationships with suppliers, government laboratories, universities, independent consultants, or even other food companies. It is important to emphasize that this model for product development includes marketing, sales, distribution, legal, financial, and so forth, collaborating simultaneously on the same team with the technical staff and the reverse. In the section that follows on team work, there is greater discussion of the actual team composition and responsibility.

In Table 7.2, the responsibilities of management, scientists, and technologists in this new paradigm are summarized. In the team paradigm, management is essentially provided by the team. A senior management coach, advises, guides, and supports, and so forth, the team to help achieve the common goals. Teams replace the traditional reporting hierarchy in which the individuals on the team report to managers. With the disappearance of middle managers, we see the emergence of leaders who function as coaches, cheerleaders, scouts, friends, prodders, and so forth, to oversee the process *from beginning to end.* These individuals support and nurture employees as senior management has in theory always done. The new employee team member is provided access to the same information that traditionally was in the exclusive domain of senior management. As a result, he or she must accept the responsibility of ownership of the program, project—and even the company. In other words, the team member must

TABLE 7.2
Responsibilities of Management, Scientists, and Technologists in the Team Research Paradigm

- Management
  - — Team members know best
  - — The team sets goals, management coaches for achievement/accomplishment
  - — Communications are facilitated, because all interested parties are team members
  - — Each staff member is empowered to make commitments
  - — Leaders are evaluated by their peers as well as superiors
- Scientists/technologists
  - — Research and development are conducted to support team goals
  - — Search for breakthroughs and incremental movements to facilitate project success
  - — Technology evaluated by commercial success

take responsibility for the task at hand and the future of the organization. *When the team succeeds, everybody succeeds.* The team members participate actively and are held accountable for key decisions related to the project. Ongoing training is essential to support the greater knowledge base of the employee. Research and development team members must understand the needs of marketing, distribution, and be familiar with how decisions that their teams make impact the income statements and balance sheets of the company—and vice versa.

STRUCTURE OF THE HIGH PERFORMANCE TEAM

As implied above, the new food product development organizations may be structured in many different ways. One useful description of organizational structure options can be thought of as segments on a continuum beginning with little or no commitment to the project, that is, obligation to the traditional organization; all the way through to complete team—responsibility only to the team/project (See Table 7.3).

A *functional* structure means the work is performed by the various traditional departments such as chemistry, microbiology, engineering, and so forth, with very little to no project focus. A new food products committee or a food product planning committee may request the work on an ad hoc basis. Such development is usually low risk and probably involves primarily line extensions. The existing functional departmental people know the product, the market, or the business; they meet quickly and readily make the necessary decisions. Functional departmental staffs working on a project, operate as individuals in a group, but not a team. Obviously, despite some past successes, relatively little true innovation occurs in functional structures—one of the many reasons for new food product failures.

Other organizational options include variations on matrix and in the ultimate, the venture team. The *functional matrix* option organizes teams, with people from the

TABLE 7.3
Organizational Structures—Pluses and Minuses

Organization Type	Strengths	Weaknesses
Centralized	Stronger function	Harder to sell ideas
	More technical	Silo focused
	Bigger ideas	Less business focused
	Longer-term	
De-centralized	Strong biz	
	Good collaboration	
	Not silo	
	Easier to sell ideas	
Dotted line	Dots do not mean much	Shorter term
	More productive on the straight line	Less technical
	business responsibility	Hard to grow ideas
Matrix	You have two masters	
	Not easy	
	Less productive	
Venture teams	Good times	Bad times often killed
A broad	Money is there (sign of good	early
B specific	business, good economy)	
Dictator	You have direction	Often not good direction
		Smaller ideas

Net: Any organization will work if you have the right people support from senior management, money, and simple tools.

various departments such as manufacturing, research and development, marketing, sensory, packaging, legal, but the individual still reports to the functional departments. Such food product development is close to current business practice. Functional matrix involves more focus than routing new food product improvements, but the individual departments still manage their internal processes. In a functional matrix, team members are functional specialists, and their departmental managers make most of the decisions—that is, the food product development component is secondary to the primary departmental function.

The middle—the *balanced matrix* option—is for situations in which both functional and project views are critical; neither current food products nor new products are the driver. The compromise of balanced matrix is not regarded favorably for new food products, because the new food product either requires or does not require assertive action. "Halfway measures" would just make for indecision and delay. In some companies, however, department managers might tip too far toward a full-scale venture, or the reverse, toward functional matrix.

The *project matrix* option recognizes the frequent requirement for stronger project push. Team members are project employees first and functional or departmental staff second. A microbiology team member communicates to the microbiology department manager but is part of the project team. Department managers note that their "employees" have subverted their departmental interests to the project and often tilt the project even against the functional department's best interests.

The **venture** option extends the food product development project to its ultimate. Team members are transferred out of the functional departments to work full time on the development project. The venture may be within the regular organization, or it may be placed outside the current organization or company—a *spinout venture*. How far removed, depends on how critical it is that there be no influence on the team from traditional departments.

Implementation issues (such as managing team vs. individual rewards), are especially important to consider when managing dedicated teams—one cannot just assemble individuals and declare them a team. Ventures are not for every project, because the company may not be able to support ventures. Venture teams are risky, hard to establish, difficult to implement, and precarious to evaluate and disassemble.

When persons from different departments or functions within a company are formally assembled to work on a project, issues of priorities are raised. Should the individuals place first priority on the project or on the function from which they are drawn? Obviously, team members want the company to succeed, but they may have different opinions about how any particular new project may contribute to the organizational growth. The packaging development manager may see a new package form as meeting consumer demands and enhancing sales; the production manager may believe his or her costs will increase more than the sales volume; accounting invariably projects to another cost; marketing sees consumers dividing purchases; and so forth. These examples are the reality of new project/venture, approach, not counting the political issues. When food product development project is important and encounters resistances, then the project/venture approach is appropriate, if it can be afforded. If, on the other hand, food product development will entail only minor variations to a standard project or platform, it is possible that functional departments can produce the desire end results.

Role of Senior Management

As the role of the team changes, so does the role of senior management in the food product development organization. In food product development organizations, as increased levels of teamwork take over, less "management" will be required. The workers in the organization will no longer be burdened with the constraints of the hierarchical organization. The elimination of job boundaries will result in a workforce that is much more independent and self-directed than in traditional organizations. Middle management, as we have understood it previously, will erode. Former managers will return to doing substantive work, contributing to teams based on their expertise and experience. There will still be a need for managers who will oversee the food product development process from beginning to beginning, thus assuring that the teams remain focused on the desired end products and that the required resources are made available.

A critical aspect of the remaining leadership role is that of employee coaching. The coaching role includes nurturing and developing high potential employees in the same way senior management does in today's environment. This means that the leader will provide staff with direct access to information that was once only in the domain of the decision makers. The employees will be expected to understand the why and wherefore of the corporation's business strategy to a 'for greater degree' than in conventional environments. The leaders must assure that the vision and strategy of the organization is clearly understood by the team members. Thus, to a greater degree than ever, senior management must become messengers keeping employees informed of the strategic roadmap, and the leaders must become champions of the projects within the organization helping to assure adequate resources to conduct the work in a timely way.

THE SIGNIFICANT ATTRIBUTES OF A HIGH-PERFORMANCE TEAM

Every activity of humankind is based on some sort of communication. High-performance teams are primarily based on precise and continuous communication. Such open and frank communication leads to the second important characteristic of the high-performance team—trust. One of the underlying prerequisites of good communications is critical listening. Each team member needs to feel free to state his or her perspectives and all must listen. Listening to understand and to clarify help build the trust and relationships required for success.

An important feature of the high-performance team is its ability to surface problems in order to accelerate projects. When a problem is identified, the team must come together and help resolve it. Frequently, challenging breakdowns are solved through innovation. Teams typically come up with innovative ways to solve many problems. The team should listen to all approaches to problems and their potential solutions. Breakthroughs should be acknowledged and those who made the breakthroughs recognized as expeditiously as possible, if possible. It is simply human nature that we all like to be acknowledged and recognized for jobs that are well done. By making this part of every team meeting, participants feel reinforced and as a result become much more motivated.

HIGH-PERFORMANCE TEAMS CAN MEET INDIVIDUAL NEEDS

In the future, emphasis will be placed on accomplishing the task, not the location where it must be done. With advanced telecommunications, more employees can work at home or at remote locations and in time frames that they choose. The individual will be able to work within a style that fits his or her personal needs. In the old production line, job geographic proximity was essential. Now technology allows coworkers to live thousands of miles apart and still work "together" in real time. The expanded modern workplace will accommodate much more individuality, with the proviso that the individual is part of a team.

CONCEPT OF THE CHAMPION

The leadership and members within teams also need to "champion" ideas and projects to convince those who control resources that it is in their best interest to come on board. The champion should passionately sell the innovation and its commercial potential but always be cautious not to oversell and create unachievable goals. Typically, the leaders in the organization or the more experienced team members are the *champions*. As such, they speak to those who control resources to communicate to them the technical achievements as well as the commercial ramifications. An advantage of the integrated high-performance team is that members come from various disciplines within the company beyond R&D. For example, when the vice president of R&D meets with the vice president of manufacturing to sell the latest innovation in processing, the vice president of manufacturing is aware that members of his or her staff are already on board with the project. The vice president of R&D then is in the position of championing an idea that should already have some support from manufacturing. Selling the concept or innovation must be conducted at all stages in the development of the project from conception through commercialization. At each stage of food product development, the champion must agree to the targets and hold the team accountable for meeting the targets.

The champion must believe in the innovation and be willing to expend the effort and take the risk. Occasionally, for a breakthrough project, the champion or senior management must be willing to "bet their job" to sell the concept. At the same time, he or she must be willing to share in the responsibility if the project fails. For this reason, no project should be oversold or over promised. The champion must support projects that are good for the company, always keeping in mind the importance of return on investment. Will the project ultimately improve the bottom line of the company? Clear objective evaluation and explanation of status at all phases of the project garnished with enthusiasm and belief in the project will make the role of the champion enjoyable and rewarding for the individual as well as the company.

CLEAR GOAL—UNCLEAR PATH

The high-performance team must have a clearly defined goal, but it is generally safe to say that the path to achieve the goal is anything but clear. When the project is started, the path is usually not clear. For example, in 1961, President John F. Kennedy promised to put a man on the moon by the end of the decade. Nevertheless, in 1961, we had no idea how to accomplish this incredibly complex task. In 1969, we took the "small step for a man," the "giant leap for mankind." The goal was clear; the work to be done was monumental. Teams were established with extraordinary budgets and the task was accomplished. There were, however, a myriad of daily problems that had to be overcome.

On a smaller scale, all of these same things happen with food product development projects. A successful project is the one in which these daily surprises are overcome in a timely and effective way so that the project can keep moving. Planning of the project automatically accounts for the things that do not go wrong. It is the unforeseen adverse occurrences that frequently destroy timelines and projects. In order to overcome this inherent obstacle, we define a project as a series of breakdowns. This accomplishes two things. First, it gives "permission" to identify breakdowns, and second, it surfaces breakdowns before they interfere with project timelines.

The management of breakdowns and problems is one of the most critical factors in the success of a development project. The management team of the project must meet regularly (e.g. weekly) and discuss the progress of the project. These meetings should have an agenda that covers all critical issues and minimizes superfluous discussion. As part of these discussions, it is critical that any breakdowns be brought to the attention of the group. It is every individual's responsibility to identify breakdowns; it is the team's responsibility to "put their heads together" to solve the breakdown. Resolving breakdowns and problems is one of the most important reasons that all team members must try to participate in all meetings. During these meetings, breakthroughs and problems are surfaced, discussions convert issues into solutions, and then the solutions become opportunities.

Breakdowns are inevitable parts of any project, particularly where research and development are involved, because, by definition, research and development are analogous to sailing over uncharted waters. The focus should be on the timely and effective remediation of the issue(s). While the new work environment has fewer rules than the "job-based" environment, there is one cardinal rule on the team: "No finger pointing." When a breakdown occurs, it has happened—nothing can be gained by retribution. The goal is to fix it.

In a real high-performance project, breakdowns and problems should be forced. Milestones should be established on time frames that are aggressive enough to find the weak points. For example, if the breakdown is that the analytical chemistry cannot be done in time for the next step, perhaps a new chemical method is required. Traditional "job-based" organizations would immediately say we need more staff. In the modern American food industry, this is not likely to be the solution. We need to work smarter through better use of new technology. The technology may be developed internally or contracted from an outside vender. Whichever, it must be identified and implemented quickly and effectively without "egos" or "turf" issues getting in the way. "The way we have always done it" no longer applies.

Individual Commitment

The commitment of individuals to the success of the project is the fuel that makes it work. How does this work in the "jobless" environment? The project-driven task environment provides the target and challenge for the individual. Commitment, as always, must come from inside each individual on the team. When individuals are part of a team, there is a mutual support system that encourages enthusiasm and commitment. The successful staff member in the future will rally around the project or the task and not the job. The end goal of getting the new food product into the market is the focus. The committed worker will do what it takes to get it there. The commitment to do what it takes is what makes a breakthrough teamwork. The new environment allows the individual much more free expression such as working at home and a more open and supporting, flexible work environment.

The commitment for the goal of the team needs to be established early and unanimously by the team. The aligned commitment to the project goal is a major first step in the initiation of an effective high-performance team. Again, it is important to acknowledge that the team may not know how it will reach the goal; it is alliance on the goal that is critical.

The individual working in the "jobless" environment must bring a multitude of contemporary skills to the organization. The role or need for the nontechnical manager in the R&D organizations is rapidly going the way of the "traditional job." The individual must come to the team with technical skills and a willingness to expand his or her technical knowledge. In addition to technical skills, they must be adaptable in the way they apply their skills, remaining flexible to make the project work for the goals of the company and not for the individual job. These skills range from marketing through basic science, through manufacturing, engineering, distribution, and sales. Every team that is assembled is unique and is faced with a unique set of problems. The brilliant leadership of the future will know how to organize these teams and then let the teams manage themselves. The leadership should help the teams acquire the needed resources to complete the desired task effectively. *The team must be empowered to make critical decisions and provided with the information and resources to do so.*

THE NEED FOR STRATEGIC PARTNERSHIPS

In modern industry, the costs of research and development are under constant scrutiny. As a result, a food company cannot necessarily maintain in depth experts in all the fields necessary to develop new products. Clearly, companies must have core competencies such as cereal technology at General Mills Inc., chocolate technology at MARS, or carbonated beverage technology at Coca Cola. However, modern food product development requires technologies that frequently extend beyond a company's core competencies. These could be handled by alliances. For example, flavor formulation has become increasingly complex. Low-fat/no-fat products have different physicochemical properties and consequently major modifications must be made in flavor formulations. The integration of the supplier into a food product development team is much more cost effective than establishing an extensive flavor development program within the company. Flavor companies can provide the needed technology and apply the flavor delivery technology over a spectrum of products without necessarily betraying confidentiality of the food companies. In other words, high flavor technology developed for a low fat baked product by a flavor supplier may also have similar application in low fat confection developed by a candy manufacturer. When confidentiality is assured, the strategic partnership between suppliers and manufacturers can lead to new business opportunities and innovative technology breakthroughs. The synergistic relationships help to limit needless duplication of effort within a corporation and encourage the development and support of core competency focused on the strength and strategic position of the company.

Strategic partnerships are emerging between food and pharmaceutical companies. Functional foods or nutraceuticals appear to offer unique opportunities for new healthy food products. Pharmaceutical manufacturers have experience identifying bioactive components in biological systems and either extracting them or producing them synthetically. Food companies have core competencies in the formulation and marketing of food products. The strategic partnership affords both pharmaceutical companies and food companies to develop new markets while building on their own internal strengths. Particularly in an age where few companies can start to develop new core competencies in a completely unrelated field, strategic partnerships are the most effective way to conduct business.

U.S. government laboratories, particularly U.S. Department of Agriculture (USDA) laboratories, sometimes develop innovative technologies, which can be applied by the food industry. Cooperative research is now possible between the scientists in USDA laboratories and industry. When these Cooperative Research and Development Agreement (CRADA) programs are established, research done and partially supported by industry is then available to the industrial supporter who has the first right of refusal on the technology. One product example of USDA research reaching the market place was Oatrim oat derivative. Oatrim was developed at the USDA Northern Regional Research Center. ConAgra worked with scientists at the USDA, and eventually Oatrim was commercialized by them as a low calorie ingredient and fiber source. The USDA gains some benefit from royalties and the corporate sponsor is able to leverage research funds effectively from the expertise and experience of the USDA researchers.

Several universities have developed centers surrounding core competency within the university. For example, the Universities of Georgia and Wisconsin have formed research centers focused on food safety issues. The Center for Advanced Food Technology (CAFT) at Rutgers has built a center based on the physical chemistry of food where there is a strong competency. University of Georgia's Food Product Innovation and Commercialization (FoodPIC) program is an example of assembling all the relevant disciplines for a food product development program into a single site resource. In all cases, industrial members provide support to the centers. These programs provide ready access to the information developed in the centers, a pipeline for new graduates to enter industry, and enhanced university industry communication. Information from these centers can be brought to research and development teams either through corporate monitors learning at the university or inviting university faculty members to actively participate in corporate food product development teams when their specific expertise is required.

Other external resources available are cooperative research with experts in universities or individual consultants. Generally these are established on a one-to-one basis between the researcher and the corporation.

SUMMARY

The organizational nature of the R&D environment is evolving away from the traditional hierarchical organization toward a team approach. Teams allow the reduction in the layers and roles of middle management. These middle managers can then be redeployed in the organization, utilizing their technical skills, experience, and expertise more fully. The key step in this research evolution is to develop high-performance teams for specific projects. These teams include members involved in the entire development and commercialization of the project. Research, marketing, engineering, production, regulatory, and packaging may be involved in the project from the inception. This allows the team to anticipate and solve many problems before they occur, rather than the traditional hand off from one group to another. The high-performance team succeeds, because it is based on continuous open communication. The communication includes careful listening to each other and supporting each other when there are either breakthroughs or breakdowns. The result is higher levels of commitment and trust throughout the project. Teamwork as described above accelerates project activities and provides a more pleasant working environment for all.

BIBLIOGRAPHY

Bridges, William. 1994. The End of the Job. *Fortune*, September.

Cooper, R. G. 1993. *Winning at New Products*, 2nd ed*: Accelerating the Process from Idea to Launch.* Reading, Massachusetts: Perseus Books.

Crawford, C. Merle and Anthony Di Benedetto. 2006. *New Products Management*, 8th ed. Boston, Massachusetts: Irwin/McGraw-Hill.

Wheelright, Steven C. and Kim B. Clark. 1992. *Revolutionizing Product Development*, New York: The Free Press.

8 Food Science, Technology, and Engineering Overview for Food Product Development

Romeo T. Toledo and Aaron L. Brody

CONTENTS

Development of new food products requires an understanding and appreciation for the scientific and technological principles of foods and food processing. Food serves the basic functions of providing nutrition and psychological satisfaction. Almost all foods are complex biochemical and biophysical structures that are inevitably vulnerable to deterioration. Vectors of food deterioration include microbiological, enzymatic, biochemical, and physical. Food science and technology is a discipline, one of whose objectives is to retard food deterioration by applying measured heat, temperature reduction, water removal, blending, and/or chemical additives, individually or in combination. Microbiological food safety demands that all food products be developed and handled so that the probability of an adverse public health incident is minimized. Food engineering is a subdiscipline in which the processes and equipment to effect the thermal, physical, and/or chemical changes are optimized. Food product development must take account of the ability to translate concepts and prototypes into commercially viable entities.

Food is the most primal of human needs. The need for food is manifested by physiological changes in the body, primarily low levels of blood sugar, and a near empty stomach, which signals the brain to make transient messages of hunger. Visual and olfactory signals to the brain may either stimulate or depress appetite. In the most basic sense, food provides the nutrients needed for the body to have the energy to perform muscular movement, sustain growth, or maintain health. The study of nutrient needs to maintain a healthy life is the science of nutrition.

In affluent societies, food also has a social role in addition to nutrition. Eating is a social event, a pleasant activity in an atmosphere where personal or business decisions may be made more readily or where friends and family may strengthen the closeness of relationships. Some social scientists postulate that the eating event is a healthy stimulus that enhances the nutrient acquisition and retards disease.

To a food scientist/technologist, food is more than just a collection of nutrients. Food should provide the consumer with the enjoyment and satisfaction associated with pleasing visual, oral, and olfactory stimuli. Food science and technology in product development also involves manipulation of the chemical constituents and physical properties of food and ingredients to maximize the positive sensory perceptions by consumers of the product.

Unlike freshly prepared foods from fresh ingredients in some restaurants or domestic kitchens, commercial food products are consumed after they have spent time in the distribution system. Although commercially processed foods must attract the consumer by the appearance of the package, repeat purchases depend on how pleasing a sensory experience is perceived when the product is consumed. Thus, food scientists and technologist must not only be concerned with nutrient content and sensory attributes of food immediately after preparation but also with how changes which occur during storage and the method of preparation prior to serving affect product sensory properties.

PRIMARY FOOD COMPONENTS

The largest and most important food component is water, comprising more than 70% of our nutrient intake. Water is the vehicle for nutrient transport and the medium in which most biochemical and physiological reactions take place. The other primary food nutrients arranged in the order of their contribution to the total dietary energy intake are carbohydrates, fat, and protein. These nutrients provide approximately 4, 9, and 4 kcal per gram of food, respectively. Metabolism of these major nutrients produces the chemical energy that powers muscular contraction, allows brain cells to function, and permits the synthesis of compounds the body needs from simpler compounds in the diet. Regulations that govern the labeling of food products as reduced calorie, "light" or low calorie, specify the caloric density of food and the amount of fat present per serving. Generally, the digestible carbohydrates are preferred energy-contributing component of food formulations. Their metabolism consumes oxygen and produces carbon dioxide, water, and other end products, which are readily eliminated through the respiratory system. In contrast, metabolism of fats and proteins involve the liver to convert metabolic intermediates into compounds safely, which are further metabolized or eliminated through the urinary system.

CARBOHYDRATES

Fiber

Carbohydrates constitute the structural and storage organelles of plants and are also present to a much lesser degree in animal tissue. Structural carbohydrates in plants are complex compounds of repeating five carbon sugars called pentosans or repeating six carbon sugars called glucans. The former, called hemicellulose, and the latter, called cellulose, are nondigestible and, along with another complex noncarbohydrate compound called lignin, constitute most of what is considered dietary fiber. Cellulose, hemicellulose, and lignin are fibrous. They do not soften appreciably on hydration and heating, and therefore may impart a rough mouthfeel to food products. Another plant component, pectin, is soluble in hot water but is nondigestible. Pectin, a soluble dietary fiber and a component of cell walls of plants, is responsible for the softening of vegetables on cooking. Changes in the pectin molecular structure cause the softening of ripened fruits. Commercial pectin preparations are used as gelling agents for jams, jellies, and preserves.

Some plant seeds, exudates from the bark or the stem, and leaves of aquatic plants contain soluble complex carbohydrates called gums. Gums dissolve in water to produce viscous solutions and at specific concentrations or in the presence of specific ions or sugar, they would form gels. Gums provide food scientists and technologists and formulators with some of the tools needed to modify body, firmness, and mouthfeel, that is, "texture" of food products.

Sugars

Simple sugars are the immediate products of photosynthesis. They are also the end products of the complete digestion of carbohydrates. Simple sugars cannot be further broken down and still remain sugars. Sugars are sweet water-soluble carbohydrates. The end products of digestion of complex carbohydrates in the alimentary tract, sugars, are rapidly absorbed from the intestines into the blood circulatory system. This characteristic of easy absorption of simple sugars make them a rapid energy source for some individuals, but it could also be detrimental to diabetics who are susceptible to hyper/hypoglycemic swings with intakes of simple sugars. Refined sugars are the common highly purified commercial carbohydrate sweeteners such as corn syrup, high fructose corn syrup, and cane and beet sugar. The refining process removes impurities such as molasses that characterize raw sugar products. Some consumers may tend to consume refined sugars with minimal intake of the other food nutrients, and so when refined sugars are consumed in excess, there might be inadequate intake of other dietary nutrients. One health concern with refined (or other) sugars in the diet is their ability to support the growth of bacteria responsible for tooth decay. These real and perceived health concerns as well as other functionalities must guide a food scientist or technologist in the choice of carbohydrate used in food formulations.

Starches

Starches are the storage carbohydrates in grains and root crops. Starches are complex molecules consisting of long chains of repeating units of the simple sugar, glucose. Since plants also utilize starch as they respire during their resting period in the absence of sunlight, starches are easily digestible by humans although not as digestible as sugars. Starches are the **complex** carbohydrates recommended to provide the majority of dietary calories. In addition to being a source of calories, starches play a significant role in determining the mouth feel of foods.

Native starch occurs as granules in a cellular matrix within grains and storage roots or tubers of plants. Macerating the root or milling the grain releases starch as granules, which can be easily separated from the other components to produce a purified starch. Wheat flour is a mixture of the grain's starch and other components of the grain. Cornstarch is pure starch.

Within the starch granule are two types of polymers, a straight chain amylose and a branched chain amylopectin. These large complex molecules are tightly coiled within the granule. Heating starch in the presence of water permits the granules to absorb water, swell, and eventually release the amylose and amylopectin into solution. When intact starch granules are no longer visible, the hydrated starch molecules in solution are called gelatinized starch. The gelatinization process is manifested by an increase in viscosity. Different starches will have different gelatinization temperatures and different viscosities at specific concentrations. When starch at an appropriate concentration is gelatinized and allowed to cool, a firm gel may form. The gelling properties of the starch determine the ability of that starch to bind water and impart firmness to food products.

Amylose and amylopectin are present in different starches in different proportions. Gelatinized amylose and amylopectin have different characteristics. Gelatinized amylopectin solutions are opaque and do not form firm gels on cooling. Gelatinized amylose solutions, on the other hand, are clear and form firm gels with a tendency to release free water from the gel matrix on cooling. Amylose has a higher gelatinization temperature than amylopectin.

Although most natural starches consist of mixtures of these two polymers, some starches contain much more of one than the other does. For example, waxy maize starch contains almost all amylopectin. Regular cornstarch has more amylose than amylopectin.

Over time, with gelatinized starch, the starch molecules in the solution slowly lose water of hydration and water trapped within the spaces between adjacent molecules is slowly squeezed out resulting in agglomeration of starch molecules. The solution turns turbid and becomes less viscous so that, eventually, starch agglomerates precipitate. This process of conversion of gelatinized starch from a soluble to an insoluble form is called retrogradation. Starch retrogradation is a process that may be reversed by heating in the presence of water. In low-moisture cooked starchy products such as bread, retrogradation is manifested by a stiffening of the structure, or staling, making the product to appear hard and dry. Starch retrogradation rate increases with decreasing moisture content and **reducing** temperature. In frozen foods, however, water is immobilized by freezing, and so there is little water movement from entrapment within the starch molecular network. Thus, retrogradation in frozen foods is minimized if both the freezing and thawing processes are carried out rapidly. Foods containing starch stored under refrigeration above freezing are most susceptible to retrogradation. In frozen foods, the time of holding at low temperature just above the freezing point is the critical time for development of retrograded starch. When starch is used in a food formulation as an emulsifier or a suspension medium for critical ingredients, starch retrogradation can irreversibly alter the desirable product quality attributes.

Because of its straight molecular structure, gelatinized amylose is more susceptible to retrogradation than amylopectin. Starch manufacturers have developed a number of modified starches designed to alter the gelatinization temperature, viscosity enhancing properties, and retrogradation tendency of natural starches. A number of these modified starches are available commercially. Starch manufacturers represent a good resource for food scientists and technologists needing starch with specific functional attributes desirable in a formulation.

Fats and Oils

Fats or oils, also called triglycerides, are compounds that consist of three long chain fatty acids attached to a glycerol molecule. Fats and oils have a similar chemical structure, but different types of fatty acids in the molecule change the physical state of the compound at room temperature. Fat triglycerides are solid at room temperature while oils are liquid at room temperature. When the component fatty acids are highly saturated, each carbon atom in the molecule chain has its full complement of hydrogen atoms, that is, single bonds join the carbon atoms. The molecule then is a triglyceride containing these fatty acids and will be solid at room temperature. Fats composed of saturated fatty acids are stable against reacting with oxygen, and so saturated fats are less subject to oxidative rancidity development.

On the other hand, oils are made up of unsaturated (and some saturated) fatty acids, that is, they do not have their full complement of hydrogen atoms in the molecule; some of the carbon atoms in the chain are joined by double bonds. These triglycerides will easily react with oxygen to produce compounds that impart an oxidized rancid odor or flavor. Catalysts for such reactions include heat, light, salts, and metal ions. Highly unsaturated fatty acids or omega 3 fatty acids are more susceptible to oxidative rancidity.

Physiologically, in animals, fats tend to be present in leaf fat or adipose tissue while unsaturated fats form a component of cellular membranes. The degree of saturation of fats depends upon the source. Fish contains the least saturated of fats in animal foods, followed by poultry. Red meats contain the highest level of saturated fatty acids. On the other hand, plant-derived oils, such as peanut, canola, corn, soybean, and cottonseed are highly unsaturated. Tropical plant oils such as cocoa, coconut, and palm oil contain highly saturated fats. Because of their susceptibility to oxidation, unsaturated oils have often been hydrogenated to saturate the entire fatty acid chain and thus produce a stable solid. Some of the linkages in the process are on opposite sides of the fatty acid chain, or became trans fatty acids, implicated in some human disease issues, and now being removed from commerce by legislative and regulatory fiat.

In the human body, saturated fats cannot be dehydrogenated to transform them to unsaturated compounds. Thus, the need for highly unsaturated fats to form cellular membranes must be met by dietary intake of these compounds. Highly unsaturated fats, also known as polyunsaturated fats are considered essential dietary nutrients. Saturated fats perform no major physiological function and serve only as a source of dietary calories. Fats have nearly double the caloric equivalent of the same weight of carbohydrates and proteins. Thus, high fat products are also high in calories. One of the best ways of reducing caloric content of foods is in formulating them to reduce the fat content.

The melting point of fat used in a food formulation plays a role in how the components blend, how the products hold water and fat as the product undergoes the mechanical and thermal rigors of processing, and eventually affect the texture, appearance, and even flavor of the product. Fat is an excellent carrier for flavors and color. Thus, success of a food formulation may depend on the choice of fat used. Achievement of adequate storage stability against oxidative rancidity development will dictate formulation and packaging options, which are necessarily highly dependent upon the types of fats in the formulation.

PROTEINS

Proteins are the structural component of animal, that is, human, tissue. The muscles that contract and relax to allow the body to move are composed of the proteins actin and myosin, while the connective tissue that separates muscle bundles is collagen. The protein, collagen, also forms the matrix into which calcium and phosphorus are deposited in bones.

Proteins are molecules that consists of a long chain of amino acids or nitrogen-containing organic compounds. There are 22 known amino acids in proteins and the type number and distribution of amino acids in a protein determines the protein type. About half of the amino acids are essential, that is, cannot be synthesized within the body, and so must be supplied from external, that is, food sources. Plant and animal proteins differ in the distribution of different types of amino acids in the molecule. Animal proteins most closely resemble the amino acid profile of human tissue protein, and so these proteins are best for consumption for human growth and maintenance. Plant proteins, however, may contain a low level of some of the essential amino acids, and so more of a variety of plant protein must be consumed in the diet to meet the required protein intake for growth and maintenance of healthy tissue. Some proteins, such as gelatin, lack one or more of the essential amino acids and so have no nutritive value as a protein unless complemented by another protein. Proteins not available as essential building blocks are metabolized, however, for energy and thus will contribute to dietary calories.

During digestion, proteins produce free amino acids. Commercial proteolytic enzymes from microbial sources may also digest proteins *in vitro* to alter their solubility, alter gelled product textural attributes, or to impart characteristic flavors. Proteins are generally bland in flavor. However, as the length of the amino acid chain becomes shorter, the flavor becomes more distinct. Hydrolyzed proteins with high fractions of free amino acids have a very strong flavor and odor. Some may even have a bitter flavor. Hydrolysis of proteins to give the beneficial meaty flavor is a carefully orchestrated process of manipulating enzyme type, temperature, time, and pH to obtain the desired product attributes.

Foods that contain animal muscle as the primary ingredient, and plant protein-based foods must be carefully formulated and processed to take advantage of the ability of proteins to bind water and fat and set into a stable solid matrix. Proteins and carbohydrates have about the same caloric equivalent; therefore, for this purpose, they may be exchanged in food formulae to produce least cost formulations. A number of commercial protein products and protein hydrolyzates are available to food scientists and technologists for product formulations.

Other Food Components

A number of food components are present at low concentrations, and yet they define the characteristic color, flavor, and other sensory attributes of foods.

Acids

After water and the primary nutrients, acids are present in the next higher quantity in foods. Acids are intermediates in plant metabolism; certain plants have the capacity to accumulate specific acids at high enough levels to give a distinctive sour flavor and mouthfeel. The intensity of the sourness is dependent on both the pH and the fraction of undissociated acid. Generally, strong acids give low pH solutions at relatively low acid concentrations and the mouthfeel is harsh on swallowing.

The types of undissociated acid molecules also differ in their interaction with other food components such as carbohydrates to modulate or enhance the sourness perception. In general, sugars modulate sourness, while mineral salts intensify the sensation. Gums such as pectin tend to coat the linings of the throat to minimize the harshness of the acid, but they also prolong the sourness sensation after swallowing. Buffering the acid by addition of one of its salts also tends to reduce the harshness of the sourness sensation. Lactic acid has a slightly bitter aftertaste while tartaric acid leaves a slight scratchy feeling in the throat after the product is swallowed. The most common acidulent, citric acid, is best used in a formulation with a small amount of sugar to leave a smooth nonpersistent sourness sensation after swallowing.

Acids may be naturally present in food products or they may be added ingredients. Citric acid, the primary acid in citrus fruits, is the most commonly used plant acid in commerce. Malic acid is the primary acid in apples, and tartaric acid is the primary acid in grapes. Lactic acid is produced by bacterial fermentation in salt-pickled vegetables such as cucumbers and cabbage and is also the acid in either bacteriologically fermented or spoiled dairy products. Most of these acids are produced commercially by fermentation and are available commercially in pure form for use in food formulations.

Pigments

Color pigments comprise the next lower level food component after the acids. Plants synthesize pigments to produce the various colors and hues of plant products. The most widely distributed of the pigments is chlorophyll, the green pigment in plants. Chlorophyll is water soluble and is unstable with heat, particularly at low pH. Degradation of chlorophyll bleaches the green color or transforms the color to brown. Care must be taken when processing products with green color or else severe color changes such as browning could develop.

The next most widely distributed water-soluble pigments are the anthocyanins whose colors range from pink to deep blue. Anthocyanin color is pH dependent, being red at the low end of the pH scale with the blue intensifying as the pH increases. Anthocyanins also change color to brown when they degrade during storage or during heat treatment of the product. The degraded pigments tend to agglomerate with surrounding degradation products of the molecule to form an unsightly precipitate at the bottom of the container. The presence of oxygen as well as elevated storage temperature tends to accelerate the degradation of anthocyanins, while the presence of sugar tends to slow the rate of degradation. Anthocyanins do not have an attractive color at near neutral pH, and so they are most used as a colorant in high acid frozen sliced fruit and shelf stable juices or beverages.

The natural fat-soluble colors are available in yellow, orange and red. These compounds include carotene, which also is a precursor of Vitamin A and a group of similar compounds, the carotenoids, which could not be transformed into Vitamin A. These fat-soluble colors are heat stable, although slow loss of color can occur during storage if they are stored under conditions that promote oxidation. Natural plant-derived water and fat-soluble color pigments are commercially available for use as food colorants.

Meat pigments normally are not intensified by addition of artificial colorants although on some processed meat products color enhancement by the addition of colored spices such as paprika, turmeric, and annatto is practiced. Meat pigments such as myoglobin vary in concentration in the meat with the age and species of the animal. In raw meats, color changes from dull red with a purplish hue when meat is stored in the absence of oxygen and to a deep red on exposure to oxygen due to oxygenation. The addition of a small quantity ($<0.5\%$) of carbon monoxide (CO) to fresh red meat produces a very stable carbon monoxide myoglobin whose red color is virtually unchangeable, even when the meat is spoiled. The technology was accepted in the United States during the early 2000s but has been the topic of considerable controversy in that it may mask spoilage. Oxidized myoglobin or oxymyoglobin pigments turn to brown metmyoglobin. When cooked the pigment becomes a gray color. Addition of nitrite produces a stable pink color, characteristic of cured meats.

Micronutrients

The micronutrients such as minerals needed for meeting dietary needs are normally not a consideration in food product development, except when formulating analogs using ingredients that do not contain the micronutrients in the product being simulated. In recent years, reduced sodium and enhanced calcium products have been a formulation objective. Removal of sodium affects taste. Incorporation of calcium is a difficult task. Other micronutrients of concern include potassium, iron, and cobalt.

Flavor

Other compounds present in trace quantities contribute to the color and flavor of foods. When volatile, these compounds are responsible for the aroma of food and when nonvolatile they contribute to the taste sensation. Typical nonvolatile compounds are the tannins and high molecular weight alcohols. The volatile compounds have low molecular weight and may be lost during processing. An important component that participates in flavor development during heating is the group of organic compounds called carbonyls. They usually participate in reactions that result in roasted flavors. The flavor of coffee and cocoa, roasted nuts, and roast beef may be attributed to the reactions involving carbonyls. Although these compounds may be naturally present, formulations may be developed where these compounds are added to intensify the desired flavor or color effect. Aroma, taste, and mouth feel together constitute the flavor.

FUNCTIONAL PROPERTIES OF THE PRIMARY FOOD COMPONENTS

Starch, sugars, proteins, and fats interact significantly in food products to establish the mouth feel and flavor. Manipulation of these interactions to produce desirable product attributes tests the creativity of food scientists and technologists and urlinoligists in food product development.

Immobilization of water through hydration of macromolecules or entrapment is a major goal in the design of food formulations. By virtue of their small molecular size, sugars are mobile in solution, and so the sugar molecules may be easily positioned in intermolecular spaces in the product matrix to facilitate immobilization of water and maintain separation of macromolecular species preventing their aggregation. Thus, sugars could be used to minimize starch retrogradation. The gelling of the pectin-sugar system in fruit preserves, jams, and jellies is a good example of the water immobilizing properties of sugars in food systems.

The water immobilization properties of starches associated with gelatinization are a well-known phenomenon, as are the water immobilizing properties of heat setting proteins. Some food ingredients such as fibers and nonheat setting proteins simply imbibe water reducing free water in the macromolecular interstices in the product.

Seldom understood is the role of fat in the immobilization of water. This role is primarily physical. Fat is hydrophobic, and so it can attach to hydrophobic moieties in macromolecules forming a water impermeable membrane that would retard water migration within the product matrix. Examples of protein-fat interactions effective in water immobilization are those that occur in processed meat products.

Because of the hydrophilic nature of starch, starch-fat interactions are not very effective in immobilizing water, but the presence of proteins in starch systems intensifies starch-protein-fat interactions to amplify the water immobilizing effect. Some modified starches have hydrophobic moieties created within the starch molecule making them effective as water immobilizing agents.

Minerals of the ionic or cationic species may enhance or reduce the water immobilizing properties of macromolecules. Mineral ions may crosslink sites in adjacent macromolecules forming interstitial spaces to trap water, or they could promote intermolecular attraction and agglomeration to squeeze water out of the system. Thus, care must be exercised in the selection of ingredients, including tap water used in the formulation to understand fully whether the presence of mineral ions is beneficial, detrimental, or neutral in a system. The hardness of water used in a product is a commonly overlooked factor in food product development. Similarly, the presence of minerals in ingredients, which occurs by virtue of the original source or their addition during manufacturing, could be a factor that affects formulated product properties in a manner that the formulator does not know.

Flavor and color development may also be manipulated by formulation or processing. In particular, roasted flavors develop as a result of reactions between carbonyl compounds and free amino acids in a Maillard reaction. This reaction results in the formation of large complex molecules with a brown color and characteristic roasted flavor. Low molecular weight volatile compounds with a pleasing odor may also be formed. A pleasing roasted flavor is different from a burnt flavor. Flavor development through the Maillard reaction depends on the concentration of carbonyls and amino acids in the food, the moisture content, temperature, and time. Aldose or ketose reducing sugars such as glucose and fructose may provide the carbonyls required by the reaction. Amino acids can come from breakdown of proteins biochemically, by indigenous enzymes, or by added hydrolyzed proteins.

Since reaction time is relatively readily controlled, the most difficult element of ensuring proper extent of the reaction is controlling temperature and moisture content. The reaction appears to be favored by high temperature and low moisture content.

Flavor of Maillard reaction products does not appear to be as good in the presence of excess moisture compared with low moisture conditions. In high moisture products heated at atmospheric pressure, temperature is limited to the boiling point of water, 212°F (100°C). Slow moisture evaporation from the surface and adequate replacement by diffusing water from the product interior results in constant high moisture at the surface, and so Maillard reaction flavor development does not occur until a dry surface crust is formed. (Note that this is one reason for a paucity of desirable Maillard reaction products in microwave cooked food products.) On the other hand, roasting in air at very high temperatures rapidly forms the dry surface crust, because rapid heating vaporizes surface water faster than it is replaced by diffusion from the interior of the material. Thus, for the same time of roasting, the latter results in more intense Maillard reaction flavor development compared to low temperature roasting.

Although fat itself has little or no flavor, its presence in roasted foods intensifies the surface temperature from radiant heating in the oven, and the hydrophobic nature of fat isolates water from the reacting amino acids and carbonyls resulting in more Maillard reaction flavor development compared with that is formed in the same product without fat.

FOOD SPOILAGE

Enzymatic, Biochemical, and Physical Deteriorations

Depending on the product, food spoilage is primarily microbiological in nature, although enzymatic, biochemical, and physical changes also render a product less than acceptable by consumers. Physical changes such as starch retrogradation and loss of turgidity in frozen/thawed fruits and vegetables render products less desirable than their fresh counterparts, but the product is still edible. Biochemical changes in contrast occur over longer times, and so, for a period considered the product shelf life, the product may be still acceptable to consumers. Examples of biochemical changes that occur during processing are nonoxidative browning, unwanted Maillard reactions, and caramelization of sugars. Examples of biochemical changes that occur over a long distribution time are off-flavor development from still active oxidative enzymes in plant products, ascorbic acid degradation in juices, and lipid rancidity development. Although physical changes may be minimized by careful selection of conditions during processing and packaging, retardation of chemical changes may require more severe measures.

Microbiological Spoilage

Microbiological spoilage is the most significant type of spoilage in terms of economic loss to the industry and consumer dissatisfaction with the processed food. Microorganisms that cause spoilage are grouped into the yeasts, molds, and bacteria.

Reproduction of these microorganisms to generate numbers that cause spoilage requires conditions favorable for their growth.

All microorganisms require water to grow. One index used to determine the availability of water to support microbial growth is the water activity. Water activity is generally decreased by reduction of moisture content and lowered by the presence of solutes such as salt and sugar. Water activity is generally independent of temperature, although below the freezing point, crystallization of water removes water from solution as ice and so the concentration of solutes increases, decreasing the water activity. Molds require the lowest water activity to support growth, generally about 0.7. Yeasts are generally inhibited from growing at a water activity below 0.85. Bacteria on the other hand, are generally inhibited from growing at a water activity as low as 0.90. Bacterial spores generally do not germinate at water activity below 0.94.

The pH of food is also a factor inhibiting microbiological growth. Yeasts and molds are not inhibited at low pH levels of most foods. However, the large number of groupings of bacteria and the differences in requirements for growth by the different groups means that only a few species are inhibited at a certain pH while some species may grow at this pH. However, spores of bacteria that are resistant to inactivation by heat or chemical disinfectants, generally do not germinate at pH of 4.6 and below.

Some bacteria called aerobes require oxygen to grow while the anaerobes require the absence of oxygen. Between these two extremes are the facultative anaerobes that can grow under both conditions, and the microaerophiles that will grow even with very little oxygen present. Oxygen content of foods can be manipulated by modified atmosphere or vacuum packaging. pH of foods can be adjusted by acidification. When combined with water activity reduction using solutes in the formulation or removal of water, an effective tool is provided to the food scientist and technologists for altering the medium to prevent microbiological spoilage.

All microorganisms respond to temperature in their rate of growth. The lower the storage temperature, the slower is the rate of growth. The lowest temperature that inhibits growth classifies the microorganisms into the psychrophiles (cold loving), mesophiles (grow at room temperature), and thermophiles (grow most abundantly at high temperatures). Mesophiles are generally the spoilage microorganisms in ambient temperature distributed products. Psychrophiles are spoilage microorganisms in refrigerated products. Since low temperature is effective in slowing microbiological growth, short-term storage is possible without microbiological spoilage when foods are properly refrigerated. Freezing almost always prevents microbiological growth by a combination of low temperature and decrease in water activity as solutes concentrate with ice crystal formation.

If pH, water activity, and oxygen concentration in the packaged food cannot effectively prevent or retard microbial spoilage, the food may be heat treated to inactivate potential spoilage microorganisms. When a food is low acid, that is, the water activity is greater than 0.85 and the pH is greater than 4.6, heat treatments must be performed at temperature above 212°F (100°C) with steam or water under pressure to ensure the inactivation of bacterial spores. When the food is acid, that is, pH below 4.6, heat treatment to stabilize food against spoilage may be adequate in boiling water or even lower temperatures at atmospheric pressure.

While short-term storage of perishable foods may be prolonged by temperature control, reduction of the microbiological population and modified atmosphere packaging, long-term shelf storage is possible only if all microorganisms capable of growing in the food under the conditions that exist in the package, are inactivated by heat treatment or other means.

MICROBIOLOGICAL FOOD SAFETY

Microbiological food safety appears to be the most important problem in food safety in the United States, with up to 5000 deaths and nearly 100 million incidents recorded annually. Food-borne illnesses are overwhelmingly microbiological and almost all from home-kitchen- or restaurant-prepared foods. Very few are from factory-processed foods.

As indicated above, microorganisms use human food as their food and consume the food to produce carbon dioxide and eventually water, but can produce other undesirable end products such as acids and amine and sulfide odors. Usually, the end products are spoilage and not necessarily harmful. Further, spoilage microorganisms often suppress the growth of pathogenic microorganisms. Spoiled foods, however, can be harmful to some persons, and so spoiled food has the potential to be unsafe, and some few food spoilages are always unsafe.

Microorganisms are ubiquitous. Usually, but not necessarily always, the greater the numbers of microorganisms, the greater the problem with respect to spoilage and illness. Food infections are usually related to the quantities of microorganisms, for example, *Salmonella* or *Listeria* infections, which are very hazardous to young, old, and ill persons and pregnant women. In food intoxications, toxins are produced as microorganisms' biological end products. The toxins are very hazardous to everyone. Examples of toxin-producing microorganisms: *Clostridia* and *Staphylococci*. In addition, some "emerging" microorganisms such as *E. coli 0157:H7* can produce adverse effects with only a few microorganisms present.

Microbiological food safety issues may be regarded or obviated by one or more of several mechanisms:

- Heat
    - Pasteurization
    - Sterilization
- Reduced temperature
    - Chilling
    - Freezing
- Reduced moisture
- Chemical additives
- Irradiation

Thermal energy may destroy microorganisms. If mild heat is applied, some heat-resistant microorganisms may survive, and so the product must be refrigerated in distribution. Considerable heat above 212°F (100°C) is required to destroy spores of low-acid anaerobic toxin formers such as *Clostridia*. If sufficient heat is applied to destroy vegetative and spore forms, the product is usually ambient-temperature shelf-stable (provided the product is not recontaminated), although biochemical deterioration will usually occur.

Refrigeration or reducing the temperature above the freezing point slows the rate of microbiological reproduction, but does not necessarily totally arrest the growth of some pathogenic microorganisms. On the other hand, freezing can stop the growth of pathogenic microorganisms. If the product is thawed, however, these microorganisms can grow.

Removal of water is an effective means to retard pathogenic microbiological growth. Included in water removal are drying and water activity adjustment.

Combination or "hurdle" technologies can reduce or, in a few instances, completely stop, pathogenic microbiological growth and toxin production.

The issue here is that food products are at risk to a variety of pathogenic microbiological hazards that must be controlled if the product is to be distributed.

HACCP

The Hazard Analysis Critical Control Point (HACCP) system is a preventive system to attempt to ensure safe production and distribution of foods. It is based on the application of technical and scientific principles to food processing from field to table. The principles of HACCP are applicable to all phases of food production, including agriculture, food preparation and handling, processing, packaging, distribution, and consumer handling and use, although it is mainly used today in food processing. Efforts are underway to extend HACCP to all food processing and into distribution.

The most basic concept underlying HACCP is *prevention* rather than inspection. Processors should have sufficient information concerning this segment of the food system, so that they are able to identify where and how a food safety problem may occur. HACCP deals with control of safety factors affecting the ingredients, product, processing, and packaging. The objective is to make the product safely *and* to be able to prove that the product has been made safely.

The "where" and "how" are the Hazard Analysis (HA) part of HACCP. The proof of control of processes and conditions is the Critical Control Point (CCP) element. Flowing from this concept, HACCP is a methodical and systematic application of the appropriate science and technology to plan, control, and document the safe production of foods.

The HACCP concept covers all elements of potential food safety hazards that will more probably lead to a health risk—biological, physical, and chemical—whether naturally occurring in the food, contributed by the environment, or generated by a deviation in processing. Although chemical hazards appear to be the most feared by consumers, as indicated above, microbiological hazards are the most serious. HACCP systems address all three types of hazards, but the emphasis is on microbiological issues.

Some HACCP Terms

Control: To manage the conditions of an operation to maintain compliance with established criteria. The state in which correct procedures are being followed and criteria met.

Control Point: Any point, step, or procedure at which biological, physical, or chemical factors can be controlled.

Corrective Action: Procedures to be followed when a deviation occurs.

Critical Control Point (CCP): A point, step, or procedure at which control can be applied to a food safety hazard to be prevented, eliminated, or reduced to acceptable levels.

Critical Defect: A deviation at a CCP, which may result in a hazard.

Critical Limit: A criterion that must be met for each preventive measure associated with a CCP.

Deviation: Failure to meet a critical limit.

HACCP Plan: The written document based upon the principles of HACCP, which delineates the procedures to be followed to ensure the control of a specific process or procedure.

HACCP System: The result of the implementation of the HACCP plan.

HACCP Team: Those persons responsible for developing and implementing a HACCP plan.

HACCP Plan Validation: The initial review by the HACCP team to ensure that all elements of the HACCP plan are accurate.

Hazard: A biological, chemical, or physical attribute that may cause a food to be unsafe for consumption.

Monitor: To conduct a planned sequence of observations or measurements to assess whether a CCP is under control and to produce an accurate record for future use in verification.

Preventive Measure: Physical, chemical, or other factors that can be used to control an identified health hazard.

Risk: An estimate of the likely occurrence of a hazard.

Verification: Methods, procedures, or tests in addition to monitoring to determine if the HACCP system is in compliance with the HACCP plan and whether the plan meets modification and revalidation.

Biological Hazard Analyses

A biological hazard is one that if uncontrolled, will result in foodborne illness. As indicated in the preceding section, the primary organisms of concern are pathogenic bacteria, such as *Clostridium botulinum, Listeria monocytogenes, Campylobacter* and *Salmonella* species, and *Staphylococcus aureus* and *E. coli 0157:H7*.

Physical Hazard Analyses

Physical hazards are represented by foreign objects that are capable of injuring the consumer. The HACCP team must identify both physical and chemical hazards associated with the finished product.

Principles of HACCP

HACCP is a systematic approach to food safety consisting of seven principles:

1. Conduct a HA. Prepare a list of operations in the process at which significant hazards could occur and describe preventive measures that might obviate the hazards.
2. Identify the CCPs in the process
3. Establish critical limits for preventive measures associated with each identified CCP.
4. Establish CCP monitoring requirements. Establish procedures for using the results of monitoring to adjust the process and maintain control.
5. Establish corrective actions to be taken when monitoring indicates that a deviation from an established critical limit has occurred.
6. Establish effective record-keeping procedures that document the HACCP system.
7. Establish procedures for verification that the HACCP system is functioning properly.

Principle No. 1: Conduct a HA. Prepare a list of operations in the process at which significant hazards occur and describe the preventive measures.

The steps that precede the development of a HA are:

- Assemble the HACCP team
- Describe the food and its distribution
- Identify intended users of the food
- Develop a flow diagram
- Verify the flow diagram
- Conduct the HA

The HACCP team is responsible to conduct hazard analyses and identify the operations in the process in which hazards of potential significance can occur. For inclusion in the HA list, the hazards must be of a nature that their prevention, elimination, or reduction to acceptable levels is essential. Hazards, which are of low risk and not likely to occur would not require much further consideration when developing the HACCP plan. Low risk hazards, however, should not be dismissed as insignificant and may need to be addressed by other means.

Principle No. 2: Identify the CCPs in the process.

A CCP is a point, step, or procedure at which control can be applied and a food safety hazard can be prevented, eliminated, or reduced to acceptable levels. An ideal CCP has:

- Critical limits that are supported by research and information in the technical literature
- Critical limits that are specific, quantifiable, and provide the basis for a go or no-go decision on acceptability of product
- Technology for controlling the process at a CCP, which is readily available and at reasonable cost
- Adequate monitoring (preferably continuously) and automatic adjustment of the operation to maintain control
- Historical point of control
- A point at which significant hazards can be prevented or eliminated

All significant hazards identified during the HA must be addressed.

Principle No. 3: Establish critical limits for preventive measures associated with each identified CCP.

A critical limit is a criterion that must be met for each deterrent measure associated with a CCP. Each CCP has one or more measures that must be properly controlled to assure prevention, elimination, or reduction of hazards to acceptable levels. Each preventive measure has associated critical limits that serve as boundaries of safety for each CCP. Critical limits may be set for preventive measures such as temperature, time, physical dimensions, humidity, moisture, water activity (a_W), pH, acidity, salt concentration, viscosity, preservatives, and so forth.

Principle No. 4: Establish CCP monitoring requirements. Establish procedures for using the results of monitoring to adjust the process and maintain control.

Monitoring is a planned sequence of observations or measurements to assess whether a CCP is under control and to produce an accurate record for future use in verification. Monitoring serves three main purposes:

- It is essential to food safety management that it tracks the system's operation. If monitoring indicates that there is a trend towards loss of control, that is, exceeding a target level, then action can be taken to bring the process back into control before a deviation occurs.
- Monitoring is used to determine when there is loss of control and a deviation occurs at a CCP, that is, exceeding the critical limit. Corrective action must then be taken.
- It provides written documentation for use in verification of the HACCP plan.

Principle No. 5: Establish corrective action to be taken when monitoring indicates that a deviation from an established critical limit has occurred.

The HACCP system is intended to identify potential health hazards and to establish strategies to prevent their occurrence. However, ideal circumstances do not always prevail and deviations from established processes may occur. If a deviation from established critical limits occurs, corrective action plans must be in place to:

- Determine the disposition of noncompliant product
- Fix or correct the cause of noncompliance to assure that the CCP is brought under control
- Maintain records of the corrective actions that have been taken.

Principle No. 6: Establish effective recordkeeping procedures that document the HACCP system.

The approved HACCP plan and associated records must be on file at the food establishment. Generally, the records utilized in the total HACCP system includes:

- The HACCP plan

Listing of the HACCP team members and assigned responsibilities
Description of the product and its intended use
Flow diagram for the manufacturing process indicating CCPs
Hazards associated with each CCP and preventive measures
Critical limits
Monitoring system
Corrective action plans for deviations from critical limits
Recordkeeping procedures
Procedures for verification of the HACCP system

In addition, other information can be tabulated as in the HACCP Master Sheet:

- Records obtained during the operation of the plan, especially those records of monitoring and verification activities.

Principle No. 7: Establish procedures for verification that the HACCP system is functioning properly.

Four processes are involved in the verification:

- Scientific or technical processes to verify that critical limits at CCPs are satisfactory, that is, validation of the HACCP plan
- Verification that the HACCP plan is functioning effectively
- Documented periodic validations, independent of audits, or other verification procedures
- Government's regulatory responsibility and actions to ensure that the establishment's HACCP system is functioning satisfactorily.

CHEMICAL ADDITIVES

Chemical additives represent another tool for a product development scientist in inducing desirable changes or inhibiting undesirable changes. Chemical additives may be classified as preservatives, processing aids, flavorants, colorants, and bulking agents. When using an additive, it is important that the additive performs a useful function, and that it must be safe under the conditions of effective functionality. The United States Food and Drug Administration (FDA) has a list of additives "Generally Recognized as Safe" (GRAS) under specified conditions of use. When the GRAS compound is used as specified, there is no need for FDA clearance. Other conditions of use may require GRAS affirmation from the FDA. New additives will require extensive testing for safety before FDA will clear them for use in commercially processed foods.

FOOD PROCESSING

Foods are processed to help preserve them, that is, retard microbiological safety issues and quality changes and to enhance their sensory qualities. Among the processes employed are heating, cooling, water removal, concentration, mixing, extrusion and filtration, and combinations.

APPLICATION OF THERMAL ENERGY

Heating of food can usually inactivate microorganisms and enzymes and may produce desirable—or undesirable—chemical or physical changes in foods. Food is highly susceptible to thermal degradation, and so to produce nutritionally sound and microbiologically safe foods, heat treatment should be carefully controlled.

Heat Treatment

Heat treatment may involve application of heat indirectly, as in a heat exchanger, or through direct contact of the heating medium with the food, as in bread baking in a hot-air oven. The main operations involving heat treatment of foods are blanching, preheating, pasteurization, sterilization, cooking, evaporation, and dehydration.

Blanching is a low-temperature treatment of raw foodstuffs generally to inactivate enzymes, expel air, and soften the product. Most vegetables and some fruits are blanched before canning, freezing, or drying. The commonly used types of blanching equipment are rotary drums, screw conveyors, and flume blanchers. Water is the dominant heating medium in drum and flume blanchers; steam is used in the screw-type blancher. Temperature is usually 212°F (100°C) or slightly lower.

Preheating food before canning aids in the production of a vacuum in the sealed container and uniform initial food temperature before further thermal exposure for pasteurization or sterilization. When the product cools, the headspace water vapor condenses, generating a partial vacuum. Liquid or slurry foods are usually preheated in tubular or other heat exchangers. Particulate foods submerged in brine or syrup in an open-top can are passed through a heated chamber, known as an exhaust box, in which steam is the usual heating medium.

Pasteurization is relatively mild heat treatment involving the application of sufficient thermal energy to inactivate the vegetative cells of bacteria, molds, and yeasts and enzymes. Typical heating times and temperatures for pasteurization may be 30 min at 149°F (65°C) or 15 s at 160°F (72°C).

Thermal sterilization is usually accomplished by heating food that is packed and hermetically sealed in containers. During the heating phase, the food is heated for the proper time by applying the heating medium to the exterior of the container. The cooling phase begins immediately after the heating phase. Since the temperature is at its maximum at the end of the heating phase, it should be reduced as quickly as possible to avoid further thermal destruction of the food quality.

Both batch and continuous heat-sterilization processes are used commercially. A still retort permits thermal processing without product agitation. The heating media are normally steam for food in cans or glasses, and high-temperature water with air overriding pressure or steam-air mixture for retort pouches. Continuous processes are less costly in terms of energy, labor, and time than batch processes, but the cost and complexity of equipment for continuous processing are greater.

The temperatures to which food is heated in conventional sterilization processes depend on the pH of the food. A normal temperature range for the heat sterilization of low-acid food (food with pH 4.6 or above and with water activity equal to or above 0.80) is 221–248°F (105–120°C); and a range for high-acid food (pH below 4.6, for example, fruit) is 180–212°F (82–100°C).

Some heat sterilization methods are called high-temperature short-time (HTST) processes. An HTST process usually consists of two separate heat treatments. Liquid food is preheated and then heated to temperatures of up to 280°F (140°C) in less than a few seconds. In another method, low viscosity liquid food is heated by tubular or plate-and-frame heat exchanger. Scraped-surface heat exchangers are used to process highly viscous or particulate-containing liquid foods. The heated food then passes through a holding tube before cooling in a vacuum flash chamber where the excess moisture is flashed off.

Cooling

The rate of spoilage and quality deterioration of fresh food is reduced exponentially as temperature is lowered. Raw meat is generally cooled in still air with high relative humidity to reduce moisture loss. Postharvest cooling methods for fresh fruit and vegetables depend on the type and volume of produce. With hydrocooling, produce is cooled by spraying or immersing in chilled water. Forced-air cooling is accomplished by forced flow of chilled air through the produce. In vacuum cooling, water is evaporated from the surfaces of vegetables or fruits by a vacuum created around the product. This system is especially good for products such as lettuce and spinach, which have large surface areas in relation to volume.

Freezing

Many food products, if properly frozen and handled, maintain acceptable condition for prolonged periods. The shelf life of frozen foods is extended beyond that of the fresh product, because the lower temperature decreases the rate of deterioration, and, as the liquid water is changed to a solid, the solutes are immobilized if the temperature is reduced below the glass transition temperature.

Freezing Systems: Commercial freezing methods include forced air and immersion. In forced air freezing, the heat is removed from the product by cold air. In many freezers of this type, the food is stationary during freezing, but in large installations, the food may move through the freezer or an intermittent or slow-speed conveyor.

Particulate foods such as peas, strawberries, hamburger patties, shrimp, or scallops may be individually quick frozen (IQF) by cold air being forced up through the product. The velocity is sufficiently high to create a fluidized bed, lift the product from the conveyor belt, and keep it agitated so that each particle is individually frozen.

Because the heat transfer rate between air and a solid is relatively low, many freezers are designed to contact cold liquids or solids with the food products. Salt brines and sugar solutions have been used.

Novelty desserts are frozen in molds that are immersed in a refrigerated solution.

In cryogenic freezing, either liquid nitrogen or carbon dioxide snow is sprayed on food products for very rapid freezing without the capital expense of a large mechanical refrigeration system. At atmospheric pressure, liquid nitrogen boils at −320°F (−196°C). Both the vaporization of the liquid nitrogen and the sensible heat required to raise the temperature of the vapor is used to remove heat from the food product.

Concentration

Water-rich liquids are concentrated to remove water and thus:

- Provide storage stability
- Produce saturated solutions
- Produce supersaturated solutions which will form glassy or amorphous solids
- Reduce storage, shipment, and packaging volumes and costs
- Induce flavor and texture changes.

Liquid foods are usually concentrated by thermal evaporation.

Food evaporators usually contain heat-transfer tubes surrounded by a steam- or vapor-filled shell. The liquid to be concentrated flows through the tubes. Part of the water in the liquid is vaporized by the heat provided by the steam or vapor. The water vapor and remaining liquid separate and leave the evaporator through separate lines.

Dehydration

Although some food can be dried beyond its capability of reconstitution, a dehydrated product should be (but not always is) essentially in its original state after adding water. In drying, the water is usually removed below the boiling point as vapor. The basis of dehydration is that drying transfers energy into the product to vaporize the water, and transfers moisture out. The energy supplies the necessary heat, usually the latent heat of vaporization and is responsible for the water migration.

Many foods are dried to a final moisture content of less than 5%, but this content varies with the product and its eventual use. Dehydrated foods lower distribution costs due to the reduction in product volume and weight. Drying typically results in a number of changes including bulk density change, case hardening, and toughening; heat damage, which may include browning; loss of ability to rehydrate; loss of nutritive value; loss of volatiles; and shrinkage. These changes are not necessarily desirable.

The dryer selected depends on food product characteristics, desired quality, costs, and volume throughput. Direct dryers include bed, belt, pneumatic, rotary, sheet, spray, through-circulation, tray, and tunnel. Indirect dryers include agitated-pan, drum, conveyor, steam-tube, vacuum, and vibrating-tray.

Drum dryers apply heat on the inside of a rotating drum surface to dry material contacting the outside surface. Heat transfer is by conduction and provides the necessary latent heat of vaporization. The drum rotates, and a thin layer of wet product is applied. Drying rates depend on film thickness, drum speed, drum temperature, feed temperature, and so forth. The speed of rotation is adjusted so that the desired moisture content of the scraped, dry product is obtained. Quality reduction is minimized by assuring that the product film has uniform thickness and that the dry film is removed completely. An advantage of drum dryers is their capability of handling slurries and pastes of high moisture and viscosity. Products such as mashed potatoes and powdered milk are typical applications.

Spray drying is a method of producing a dried powder out of feed in the form of pastes or slurries, or other liquids containing dissolved solids. Spray drying consists of atomization and moisture removal. Atomization breaks the feed into a spray of liquid droplets. This provides a large surface area for moisture evaporation. The spray is contacted with a hot air, and evaporation and drying produce the desired solid product particles. Spray drying is an important preservation method since it removes moisture quickly and continuously without causing much heat degradation. The disadvantage of spray drying are the high energy needed, the requirement that the product be capable of being atomized, and the potential loss of desirable volatiles as with instant coffee.

Filtration

Filtration processes remove solids from a fluid by using a physical barrier containing openings or pores of the appropriate shape and size. The major force required is pressure to pump the fluid through the filter medium.

Filtration processes cover a wide range of particle sizes. The sieving or screening operations use fairly coarse barriers. The screens may be vibrated or rotated to improve filtration rates.

Equipment for solid-liquid filtration can be classified in a number of ways:

- Whether the retained particles (the cake) or the filtrate (the clarified liquid) is the desired product
- Batch or continuous operation

- Driving force
- Mechanical arrangement of the filter medium

Mixing

Mixing provides more-or-less homogeneous compositions and physical properties for food ingredients and combinations of ingredients. Mixing is also used to facilitate heat and mass transfer, create dispersions, produce emulsions and foams, suspend solids, facilitate reactions, and produce textural and structural changes.

Many different mixing methods are used in order to produce the liquid, solid, dough-like, and foamy mixtures and dispersions in food processing. In general, these methods involve repeatedly subdividing the mixture into discrete domains that are then transported relative to each other so that the domains repeatedly encounter and exchange matter with new neighboring domains.

Liquid mixing is usually performed by rapidly turning screw propellers or radial-flow turbines to circulate liquid in baffled tanks.

SUMMARY

Numerous tools are available to a food product development scientist/technologist for effective formulation of food products for commercial processing and distribution. The scientist must know the chemistry and functional properties of food constituents and ingredients and apply this knowledge to create commercially successful convenient food products for the target market.

BIBLIOGRAPHY

Brody, A. L. and J. Lord 2000. *Developing New Food Products for a Changing Marketplace*. CRC Press, Boca Raton, Florida.

Doyle, M. P., L. R. Beuchat, and T. J. Montville 1997. *Food Microbiology Fundamentals and Frontiers*, ASM Press, Washington, D.C.

Erickson, M. C. and Y. C. Hung 1997. *Quality in Frozen Food*, Chapman and Hall, New York.

Fennema, O., ed., 1985. *Food Chemistry*, 2nd ed. Marcel Dekker, New York, New York.

Glicksman M., ed., 1982. *Food Hydrocolloids*, CRC Press, Boca Raton, Florida.

Lopez A. 1987. *A Complete Course in Canning*, 2nd. ed. vols 1, 2 and 3. CTI Inc. Baltimore, Maryland.

Potter, N. N. and J. H. Hotchkiss 1995. *Food Science*. 5th ed. Chapman and Hall, New York, New York.

Saravacos, G. D. and A. E. Kostarofoulos 2002. *Handbook of Food Processing Equipment*, Kluwer Academic, New York.

Toledo R. 1991 *Fundamentals of Food Process Engineering*, 2nd ed. Can Nostrand, New York.

Whistler, R. L., J. N. BeMiller, and E. F. Paschall eds., 1984. *Starch, its Chemistry and Manufacture*, 2nd ed. Academic Press, New York, New York.

9 Development of Packaging for Food Products

Aaron L. Brody

CONTENTS

Virtually all food products are encased in packaging to protect them from the inwardly hostile natural environment. Packaging represents the product and so is a powerful marketing tool; the development of effective and attractive packaging is critical to the ability to distribute and deliver food that delights consumers. Packaging is not just materials such as paper, metal, glass, or plastic; or structures such as cans, bottles, cartons, or pouches; or a distribution vehicle such as a case. Rather, packaging is an integration of comprehension of product content protection requirements with food process, selection of alternative material/structure combinations, equipment and distribution, with the end user and his/her application, and the ultimate disposition of the spent package.

The role of packaging is to protect the contained food product. Packaging is *always* corollary to the attribute of the food contained.

Packaging is a system whose objective is to protect the contained food product against an always-hostile environment of water, water vapor, air and its oxygen, microorganisms, insects, other intruders, dirt, pilferage, and so on—because a constant competition exists between man and his or her surroundings. Food packaging is engineered to facilitate the movement of a food product from its point of production to its ultimate consumption and beyond.

If there is no food product, there is no need for a food package.

Packaging is arguably the single most important link in the distribution chain that places a product into the hands of the consumer. Realistically, today, food packaging should be regarded as an integral component of the food product contained.

The words *package* or *packaging* have different meanings, intended to convey different images. The *package* is the physical entity that actually contains the product, the can, bottle, carton, pouch, and so on. *Packaging* is the integration of the physical elements through technology to generate the package. *Packaging* is a discipline. The *package* is what the consumer must open to access the food contents.

All definitions of food packaging center about a single concept: the protection of the packaged food product for the purpose of facilitating its journey to the marketplace and use by the consumer. Packaging is that combination of technology materials, machinery, marketing, people, and economics that together provides protection, unification, and communication.

FUNCTIONS OF FOOD PACKAGING

Food packaging's roles depend mostly, but not totally, on the food product contained. The main functions of food packaging are protection, containment, communication, unitization, sanitation, dispensing, food product use, convenience, deterrence of pilfering, and deterrence from other human intrusions such as tampering, and minimizing environmental insult—at a minimum economic cost.

PROTECTION/PRESERVATION

Protection means the establishment of a barrier between the contained food product and the environment that competes with man for the food product.

For example, the food product must be protected to control its moisture content. Most dry products such as instant coffee or sugar candies are susceptible to consequences of exposure to moisture or liquid water; hygroscopic foods such as fried snacks or dry bakery goods absorb water and lose mouth feel. Conversely, most wet products such as lettuce are susceptible to loss of their water content and consequent change in masticatory properties. Even bottled water can lose its water by migration and evaporation and cause package collapse.

Oxygen present in the air reacts with most food products. Only some fresh meats, cheeses, and wines benefit from oxygen. By establishing a barrier between the air and the food product, packaging can help retard the oxidation of vulnerable components in foods—provided that almost all the oxygen in the package is first removed. When appropriately applied, packaging thereby helps retard the oxidation of foods and the related product deterioration. Lipids and flavoring ingredients of foods may be adversely affected by exposure to oxygen.

Carbon dioxide must be contained in liquid products such as beer, champagne, sparkling wine, and carbonated beverages and in the newer carbonated fruits and candies to ensure that the flavor quality of these products is maintained throughout distribution.

Volatile essences and aromatic principles would be lost because of migration and volatilization were it not for flavor-barrier packaging. Simultaneously, the absorption of foreign odors and flavors from the environment is deterred by the presence of low flavor permeability packaging.

In today's complex distribution environment, food products are manufactured, fabricated, grown, and/or transformed in one geographic region, and their consumption is in a geographic area far removed from its origin. Packaging is required to ensure the product's integrity during transit and in storage to extend the time and geographic span required between origination and final use. Time required can be measured in weeks or months, and distances can, of course, be thousands of miles.

Containment

Food packages hold or contain food products. Packaging permits holding or carrying not only what can be grasped in a single person's hands or arms, but also products such as liquids or granular flowable powders that simply cannot be held or transported in industrial- or consumer-sized units if they are not contained food (Figure 9.1). Products such as carbonated beverages and beer could not be consumed at any distance or time from the manufacturing site without packaging (Figure 9.2). Further, the aging of wines and cheeses requires packaging if these processes are to occur without spoilage.

Sanitation

Packaging helps to maintain the sanitary, health, and safety integrity of contained products. Processing and packaging are intended to stabilize food products against degradation during distribution. One purpose of food packaging is to reduce food

FIGURE 9.1 Paperboard basket carrier representative of multipacking of glass bottles.

FIGURE 9.2 Refrigerator paperboard pack of twelve cans—secondary or multipackage engineered for convenience of carrying, placement in refrigerator for inventorying and accessing contents.

spoilage and minimize the environmental losses of nutritional or functional value of the product.

The presence of debris, foreign matter, microorganisms, and the droppings of insects and rodents in food products is intolerable. They are especially undesirable if these contaminants have entered the product after they have been properly processed and stabilized. Packaging acts as a barrier to retard or prevent the entry of environmental contaminants. Packaging also minimizes the contamination of contained products by intentional or casual human contact and its potential for infecting the product.

One of the significant objectives of packaging of moist foods is prevention of reentry of microorganisms after they have been successfully removed from, stabilized, or destroyed within the product. For example, wet food products such as vegetables and meats are thermally processed in hermetically sealed metal cans, glass jars, plastic pouches or trays in order to destroy ubiquitous microorganisms and to ensure against entry of other microorganisms after sterilization. By deterring recontamination, packaging minimizes the risk of disease and infection from food and reduces spoilage that could lead to toxin production, pathogenic microbiological growth, or economic losses.

Food packaging also retards losses of nutrients from biochemical reactions all of which are temperature-accelerated and prevents nutrient losses that could arise from the growth of microorganisms.

Microorganisms grow by consuming the food product and consequently impairing the nutritive value of food products, provided the product is edible after the microbiological activities.

UNITIZATION

Unitization is the assembly or grouping of a number of individual items of products or packages into a single entity that can be more easily distributed, marketed, or purchased as a single unit. For example, a plastic ring or shrink film carrier for six cans of beer, or a corrugated fiberboard (Figure 9.3) shipping case containing 12 bottles of salad dressing is far easier to move than attempting to manually carry 6 individual cans or 12 bottles. Forty-eight individual jars of baby food would be

FIGURE 9.3 Regular slotted corrugated fiberboard case commonly employed for distribution—shipping of primary packages of foods.

impossible for a person to carry. A pallet load of 360 cases is far easier to move as a single unit rather than attempting to load and unload a truck with 360 individual cases.

A paperboard folding carton containing three flexible material pouches of seasoning or soup mix delivers more product to a consumer than does a single pouch. A paperboard carton wrapped around 12 beer bottles provides more desired liquid refreshment for home entertainment than does an attempt to carry individual bottles in one's hands.

Unitization reduces the number of handlings required in physical distribution and thus reduces the potential for damage. Because losses in physical distribution are significantly reduced because of unitization, significant reductions in distribution costs are achieved.

Communication

Packaging is one of our major consumer packaged goods and food communications media. Usually overlooked in the measured media criteria, packaging is a powerful communications link between the consumer or user and the manufacturer/marketer, at both the point of purchase and the point of use.

Self-service retailing that reduces the cost of distributing products from the manufacturer or grower to the consumer virtually could not exist without the graphic communications message on the surface of the food package. The task of communicating identity, brand, can size, price, constituents, nutritional value, source instructions, warnings, warranties, and so on, is the responsibility of the package and/or its label. Considerable information is required by both law and regulation for food products (Chapter 19). This information is designed to assist the consumer, particularly those consumers who are partially cognizant of the need to be fully informed of the contents.

Self-service retailing requires that the package surfaces bear clear, easily seen messages on the identity of the contents. The package on the shelf is the main link between the producer and the purchaser. At this point in its cycle, the *package* and not the *product* is being purchased. The package is the *promise* to the consumer of what is inside that package. Thus, recognizable packaging is extremely important, if not essential. Graphic designs that hide, obscure, or otherwise deceive are self-defeating because consumer frustration is perhaps best expressed by rejection during the purchase-decision process. The absence of a sale is telegraphed to the producer by a reduction in purchase quantity by the retail outlet. On the other hand, probably only about half of all products retailed by today's typical supermarkets receive significant media advertising. Consequently, the package itself is an advertising medium communicating the benefits to be received from the investment of money by the prospective purchaser.

Despite this key role of communications, approximately a quarter of all items are distributed with minimum or no packaging. Fresh red meat, delicatessen items, on-premise-baked products, home meal replacement prepared foods, and fresh produce frequently are distributed with almost no graphics on the package surface.

FIGURE 9.4 Breakfast cereal in unit portion size plastic tub with easy peel flexible material seal for convenience in pouring in milk and consumption from wide mouth "bowl."

FIGURE 9.5 Gable top coated paperboard cartons with and without plastic dispensing spout for chilled juices.

Dispersing and Dispensing

The user or consumer often dispenses food product into readily used quantities. Packaging often facilitates the safe and convenient use of the product (Figure 9.4). Thus, bottles may have push-pull or no-drip tops, cartons may have pouring spouts, salt and pepper shakers and spice containers may have openings through which the product may be shaken or more recently, built-in grinders to foster "fresh" preparation (Figure 9.5).

To facilitate opening, the container, such as a carbonated beverage or beer bottle or can, almost invariably has a "finger-friendly," easy-opening device to expose a pouring orifice. Many packages, such as coffee or snack food cans and syrup bottles, have reclosure devices that permit the user to effectively reseal the package and protect it during reuse.

Pilferage Deterrence

The cost for shoplifting, intentional switching of price markers by consumers, and so on, in self-service retail stores is much too high. Despite increasing vigilance by security people and electronic devices (which increases costs), plus numerous attempts made to deter the problem through packaging and adjuncts such as radio frequency identification (RFID), this staggering amount unfortunately has not been declining. Nevertheless, packaging helps to keep this figure from reaching astronomical heights.

Tampering

The tragic headline events of 1982, affecting McNeil Laboratories' Tylenol analgesic capsule products, and 1986 were this country's most dramatic manifestations of intentional tampering with products. Intentional tampering of fresh fruit and vegetables by consumers has occurred for many years. Intentional opening of packages to taste-test, smell, or examine the contents is not uncommon and is obviously unsanitary. Intentional opening of packages to cause some undesirable event such as the Tylenol poisoning incident is rare, but nevertheless was highlighted by the 1980s incidents. Packaging per se is a deterrent to tampering as has been evidenced by the infinitesimally small number of intentional tamperings with products in the United States.

The laws and regulations promulgated as a result of the Tylenol tragedy have led to overt methods of visibly signaling or further deterring tampering of proprietary drugs. Inspections for evidence of tampering and production of tamper-resistant packages have now become essential functions of over-the-counter drug packaging.

Tamper-evident/resistant packages are not required by law or regulation for foods, but many food packages nevertheless incorporate such features to deter intentional and unintentional tampering. Tamper-evident/resistant packaging is a step in the recently activated programs on food security.

Other Functions of Food Packaging

Other functions of food packaging include apportionment of the product into standard units of weight, measure, or quantity prior to purchase (Figure 9.6). Yet another objective is to facilitate product use by the consumer with devices such as spouts, squeeze bottles, and spray cans. Aerosols not only serve as dispensers, but also prepare the product for use, such as aerating contained whip toppings. Still other forms of packaging are used in further preparation of the product by the consumer, for example, tea bags that are plastic-coated, porous paper pouches, or frozen dinner trays, which were originally aluminum and which are now fabricated from other materials such as crystallized polyester and polyester-coated paperboard. And many packages incorporate reclosure features to foster reuse or continued use.

FIGURE 9.6 Lunch kit with each component individually packaged because of different protection requirements: crackers in one sealed compartment of thermoformed plastic tray, cheese slices in another, ham in another; miniature chocolate candy bar in polypropylene flow wrap; fruit drink in stand-up flexible aluminum foil lamination pouch. All in die cut paperboard carton. Right: package crackers in one compartment, cheese in another; and pudding in thermoformed plastic cup. Display feature is paperboard backing beneath the thermoformed tray.

FOOD PACKAGING: ADVANTAGES

Packaging is a low-cost means of protecting food products, reducing waste, and reducing the cost the consumer pays for food products. Because of the efficiencies in the American food distribution system, its infrastructure, and its packaging systems, losses of agriculture commodities in the United States between the field and the consumer are less than 20%. In other parts of the world, where little or no packaging is used, well over half of the food grown in the field never reaches consumers because of spoilage, infestation, weight losses, and theft.

Packaging permits a larger number of different food products to be available to the consumer. For example, the American food industry alone produces perhaps up to 100,000 different food products. Although some might regard this number as frivolous waste, the food industry, like all other industries in a free-market economy, responds to consumer wants and needs. If there were no market for so many different items, they would not exist, particularly in this period of high recognition of economic value of product inventory. Each year, more than 15,000 new food products are introduced—each in response to some processor/marketer's perception of a consumer desire.

Probably the most significant benefit of food packaging to the American consumer is the safety of the products contained. In contrast to the disease and infection vectors, food-borne infections, intoxications, and adulterations prevalent in food products in other countries of the world, the American food industries produce products that are overwhelmingly safe and beneficial to consumers. Despite the infamous Tylenol tampering or the *Escherichia coli 0157:H7* in bagged spinach incidents, the number of deaths attributed to packaged food products in the United States, due to some error in processing or packaging, stands at fewer than a dozen in a decade. Fewer than a dozen persons have died from toxins produced in American commercially canned foods in the last half-century; that is, the death rate from canned foods is less than six per trillion packages. This is a better safety record than any other entity that we employ in the course of our lifetime.

The use of packaging significantly reduces the garbage, waste, and litter in our streets and solid waste streams. The number of apple cores, orange peels and seeds, bones from meat and fish, feathers from poultry, and outer leaves from lettuce is very markedly reduced by the use of packaged processed meats, not-from-concentrate chilled orange juice, prebreaded frozen chicken, and so on. Thus, the solid waste stream and the unsanitary practices not uncommon in solid waste disposal are reduced markedly by the use of food packaging. Packaging generates less solid waste than the materials it replaces. Further, when in central factory locations, the excess materials such as orange peels, potato peels, bones, and trimmings, can be employed for beneficial and/or profitable results, for example, cattle feed, fertilizer, fermentation media.

In several studies, the U.S. Department of Agriculture has demonstrated that the cost of a food product prepared from a factory-processed convenience food is significantly less than the same food product prepared from scratch raw ingredients purchasable by the consumer. Thus, a cake prepared from a cake mix or chilled fluid batter, an entree reheated in a microwave oven from a home meal replacement tray, or a low-fat ice cream are all significantly less expensive to consumers than the equivalents prepared from basic raw materials. Further, the uniformity of convenience foods is much greater than could be possible with the variability of homemakers and their kitchen facilities, or even individual food service kitchens.

The types of food products available to consumers have multiplied many times in the past four decades. Although some social scientists might not regard this magnification as a measure of progress, consumers apparently feel otherwise. Consumers in the United States today are able to obtain food products not available to their ancestors and certainly not available to wide geographic regions just a generation ago. Thus, shrimp, hearts of palm, guacamole, bagels, enchiladas, sushi, nan—and even ice cream—are products that the consumer has come to expect to be available at his or her retail shop, but were not available in days past. This availability, of course, has in part been made possible by food packaging.

FOOD PACKAGING AND THE QUALITY OF LIFE

Efficient food production processing, coupled with packaging in a distributional infrastructure, has been a key element in the contemporary American industrialized society and standard of living. Until the twentieth century, agricultural commodities were produced and consumed only locally. Foods were dispensed almost exclusively from bulk sources. As the population departed from the fields for the factories in the cities in the early years of the twentieth century, proportionally fewer food products were consumed at the site of production. Thus, food products had to be shipped long distances. The time span between origination and consumption increased significantly and fostered the need for packaging to protect products during their journeys. As the economy developed further, mass production of food products and requisite packaging materials also grew. The numbers of factories packaging branded food products increased and merchandising and marketing became major tools. Because of the efficiencies and economies of mass production, more people and capital could be diverted from agriculture to the manufacture of a wide variety of industrial and consumer products, better housing, and more rapid and efficient transportation and recreational products. As consumer income expanded, more income could be dedicated to leisure, education, travel, and cultural aspects that contribute to our quality of life. The increase in consumer income expanded the requirements for product safety, reliability, and quality. Consequently, this was translated into a need for more and better packaging that could reduce the investment in time and risks, and permit the increase in the selection of products available.

Self-service retailing, an innovation directed at significantly increasing the productivity of retail labor and reducing prices, is a key contributor to the American food marketing system. Packaging is an essential element of our self-service food retailing. Internet and related electronics shopping and subsequent distribution are possible only with packaging.

As our population deviated from an agrarian society, it increased in size and income. The way we were as individuals, families, and communities has changed. The automobile and transportation distances, inter-dependence upon others, entertainment, both active and passive recreation, electronic communication, and the need for self-contained packaged items that could be used without further thought, all are far more important than they ever were in the past (Figure 9.7).

FIGURE 9.7 Sugar shell coated milk chocolate candies in automobile coffee cup holder cup shaped package for "dashboard dining."

FIGURE 9.8 Stand-up retort pouch of chicken breast. Pouch is lamination containing aluminum foil as the main barrier. Product is filled into pouch which is hermetically heat sealed and the subjected to high temperature to sterilize the contents.

The nature of our work force has changed. Women, first the central focus of agrarian society and homemakers in the urban context, are now employees and entrepreneurs. The pressures of sharing homemaking, motherhood, and parenthood have all led to significant changes in the requirements for food products that are safe, convenient, easy to use, and reliable (Figure 9.8).

DISADVANTAGES OF FOOD PACKAGING

Beginning in the late 1960s and persisting through today, counter-packaging elements have been vocal around the world. During the early 1970s—and continuing into today, some individuals (many of whom were elected legislators) saw, in the truncation or abolishment of packaging, a true "yellow brick road" to Utopia. Through their efforts, mountains of legislation and regulation were introduced and argued—with some now the law of the nation, state or locality. As with all similar national debates, most of the points on both sides have been oversimplified.

Nevertheless, it is wise to enumerate some of the often-voiced concerns and criticisms about packaging.

COST

Obviously, because it is comprised of physical materials, equipment, people, and thought, food packaging has an associated cost. At the outset, this cost of packaging is occasionally greater than the cost of the product without packaging, for example, water or carbonated beverages.

Resource Utilization

Because food and other packaging are composed of materials ultimately derived from the Earth, it is a user of resources. Since most packaging is used only once and discarded, it diminishes the Earth's natural resources.

Energy

Because energy is required both to make package materials and to package, packaging per se is regarded as a net user of energy. Further, because almost all plastics are derived from petrochemicals, they are perceived to be more wasteful of energy.

Toxicity

Package materials are composed of chemicals some of which could be hazardous to humans. Since packaging and food product are in proximity with each other, the chemicals of package materials migrate into the food product and can be harmful. Some agenda-oriented persons and organizations have seized upon minute migrants as sources of carcinomas, teratogenic defects, and other diseases. Residuals from materials erroneously or, rarely, inadvertently used in package materials converting operations can be toxic even at minute levels. This last issue has emerged particularly during the recent years of importing converted package materials from Asian countries where standards deviate from those in western countries.

Litter

After use, because they have no further principal use, packages are discarded in greenfields, streets, and waters to become temporary or permanent eyesores on the landscape. Further, the cost of removing package debris, that is, litter from the streets and highways is usually borne by the taxpayer.

Solid Waste

Used packaging that has not become litter fills our landfills, dumps, and land depressions—or even ocean depths in which it is sometimes placed. Further, the cost of transporting this used packaging from the home, office, factory, store, or restaurant to the solid waste disposal site eventually must be borne by the consumer. In the past, these locations were unsanitary havens for insects, scavengers, rodents, and other undesirables.

Waste

To retard theft, packages such as plastic or paperboard hanging on shelves at the checkout counters are many times larger than the products contained and, therefore, represent a major waste of packaging resources. The constant use and discarding of package materials is an anathema to a sustainable source of planetary raw materials.

Since 2000, the concept of sustaining all resources has moved to the forefront of consciousness, with package materials as a signal target.

Deception

Marketers and packagers employ food packaging primarily to hide defects or to deceive consumers into believing the advertising claims on the package surface. Further, packaging is designed to convey the impression that the quantity of contents contained is significantly greater than really present.

Convenience

Since the food packagers are accused of caring little about their customers, they are allegedly careless about their structural design and sometimes have failed to incorporate means to open, close, dispense, apportion, and so on.

Enclosure

Because packaging encloses the product, the consumer cannot directly touch, test, feel, smell, or taste the contained food product. Thus, the consumer cannot obtain sufficient direct information on the nature of the product to make an intelligent purchasing decision.

Multiples

Because 6 or 12 cans or bottles or two pouches are linked together in a single (secondary) package, the consumer is compelled to buy more units than are desired in any single purchase decision.

Sanitation

The package allegedly introduces microorganisms to contaminate the contained product, or at least fosters the entry of microorganisms, to recontaminate the food product.

Quality

The package extracts natural attributes of the contained product and reduces its quality. By definition, a processed and packaged food product is perceived inferior to that original harvested from the ground, plucked from the trees, removed from the water or fashioned by human hand.

The word "plastic" frequently appears in contemporary English conversation and literature to convey the notion of synthetic, unvarying, not real; "packaged goods" are often under the same accusation umbrella: they are counter to fresh, natural, organic, and sustainable.

CONCLUSION

The preceding list is but a sampling of views of those who would resolve major national and humanistic issues by significantly reducing or putting an end to packaging and especially food packaging. Most of the arguments against food packaging have a shred of basis in fact, but many are wholly imaginary. In the face of these apparent disadvantages, the overwhelming majority of consumers continue to use packaged foods. To paraphrase Sir Winston Churchill, "Packaging is the worst of solutions, except for all the others."

Food package developers must be cognizant that some consumers, journalists, regulators, legislators, educators, and other opinion influences believe all, many, or some of the negatives about packaging. Thus, package structure and design, especially for the increasing volume of natural, organic, nutraceutical, and so on, foods, must take account of the perceived value of packaging.

PRESERVATION REQUIREMENTS OF COMMON FOOD CATEGORIES

It has become increasingly important to integrate packaging into the total food product and distribution system if the objectives of delivering safe and high quality food are to be achieved. The upcoming section addresses various food product categories and their preservation needs, many of which may be achieved in part through packaging. The potential beneficial effects together with the issues of packaging are introduced. Package materials and structures are described in the perspectives of their food preservation characteristics. Finally, the development steps are enumerated in the context of new food product development.

MEATS

Fresh Meat

Most meat is offered to consumers either fresh or recently cut, with little further processing to suppress the normal microbiological flora present from the contamination received during the killing and breaking operations required to reduce carcass meat to edible cuts. Fresh meat is vulnerable to microbiological deterioration (Figure 9.9). Microorganisms can be as benign as slime formers; can adversely affect quality such as stink producers; or may be pathogens such as *E. coli O157:H7*. The major mechanisms to retard fresh meat spoilage are initial sanitation and temperature reduction, which is often coupled with reduced oxygen during distribution to retard normal spoilage microbial growth. Reduced oxygen also leads to fresh meat color being the purple of the myoglobin pigment. Exposure to air converts the natural meat pigment to bright cherry red oxymyoglobin characteristic of most fresh meat offered to and accepted by most retail consumers. Reduced oxygen packaging is achieved through mechanical removal of air from the interiors of gas barrier multilayer flexible material pouches closed by heat sealing the end after filling. A recent, and controversial, development has been the incorporation of carbon monoxide into the

FIGURE 9.9 Barrier bag of primal cut of beef for distribution to retail back room for further cutting and packaging or food service operation for cutting and cooking.

meat to produce a permanent red color that persists throughout distribution regardless of environmental variables.

Ground Meat

Because of the consumer desire and the producer need to effectively use raw materials, about 40% of fresh beef is necessarily ground beef to enable the preparation of hamburger sandwiches, sauces, loaves, casseroles, and related foods (Figure 9.10). Ground beef was originally a by-product, that is, the trimmings from reducing muscle to edible portion size. The demand for ground beef is so great that some muscle cuts and carcasses are specifically ground to meet the demand. Grinding the beef always further distributes microbiological flora from and beneath the surface throughout the mass and thus provides a rich substrate for microbiological growth regardless of temperatures. Relatively little pork is reduced to ground fresh form, but increasingly during and since the 1990s, significant quantities of poultry meat are comminuted and offered fresh to consumers, both on their own and as a relatively low cost substitute for ground beef. About half of ground beef is coarsely ground at abattoir level and packaged under reduced oxygen for distribution under refrigerated temperature to retail grocery backrooms to help retard microbiological growth. The most common distribution packaging technique is pressure stuffing into chubs which are tubes of flexible gas barrier materials closed at each end by metal clips. At retail level the coarsely ground beef is finely ground to restore the desirable red color and to provide the consumer with the desired product. Intact retail cuts and portions are placed in expanded polystyrene (EPS) trays which are overwrapped with plasticized polyvinyl chloride (PVC) film. The tray materials are fat and moisture resistant only to the extent that many trays are internally lined with absorbent pads to absorb the purge from the meat as it ages and protein tightens to squeeze out serum and/or the meat deteriorates in the retail packages. The PVC materials are not sealed but rather tacked so that the modest water vapor barrier structure does retards loss of moisture during distribution. Being a poor gas barrier, PVC film permits access of air, and being only

FIGURE 9.10 Case-ready ground beef in high oxygen/high carbon dioxide modified atmosphere package: solid plastic barrier tray with heat sealed barrier flexible film lidding.

tack sealed, the oxymyoglobin red color is retained for the short duration—usually less than 10 days—of chilled retail distribution. The brief distribution period is due largely to the inherent relatively short shelf life of the product.

Case-Ready Meat

For many years, attempts have been made to shift the retail cutting of the beef and pork away from the retailer's back room and into centralized factories. This movement has been stronger in Europe than in the United States but has increased in the United States, especially as part of the highly efficient discount retailers such as Wal-Mart, Target, and so on, as part of their super-center systems (Figure 9.11).

Case-ready or centralized retail meat packaging involves the cutting and packaging at central factory locations under hygienic conditions to reduce the probability of microbiological contamination. The package is usually in a gas barrier structure, typically gas/moisture barrier EPS or ethylene vinyl alcohol (EVOH) laminated solid polypropylene trays heat sealed with polyester/EVOH/sealant gas barrier closure film. The internal gas usually is altered to a high oxygen /high carbon dioxide internal atmosphere. The high oxygen concentration fosters the retention of the consumer desired oxymyoglobin red color while the elevated carbon dioxide suppresses the growth of most spoilage microorganisms but not many pathogenic microorganisms. Using these or analogous technologies, refrigerated microbiological shelf lives of intact retail cuts may be extended from a few days to as much as a few weeks or for up to a week for ground beef permitting long distance distribution, for example, from a central factory to a multiplicity of retail establishments. One thesis with central packaging of ground beef is that the probability of the presence of the *E. coli 0157*:*H*7pathogen occasionally present from unsanitary practices is reduced.

FIGURE 9.11 Case-ready fresh red meat intact cut in high oxygen/high carbon dioxide modified atmosphere expanded polystyrene barrier tray with barrier film closure.

Alternative packaging systems for case-ready beef and pork include the master or mother bag system used widely for many years for cut poultry in which retail cuts are placed in polyolefin film overwrapped EPS trays and the trays are multipacked in gas barrier pouches whose internal atmospheres are carbon dioxide to retard the growth of aerobic spoilage microorganisms. Another popular system involves the use of gas barrier multilayer plastic trays with heat seal closure by means of flexible gas and moisture barrier. Conventional nongas barrier trays such as EPS may be overwrapped with gas/moisture barrier flexible films subsequently shrunk tightly around the tray to impart an attractive appearance. Other systems, all of which involve removal of oxygen, include vacuum skin packaging in which a film is heated and draped over the meat on a gas/moisture barrier tray. The film clings to the meat so that no headspace remains meaning that the color of the contained meat is purple myoglobin. In one such system applied during the 1990s but since largely abandoned, the drape film is a multilayer whose outer gas barrier layer may be removed by the retailer exposing a gas permeable film that permits the entry of air that reblooms the pigment and restores the desired color. Variations on this double film system include those in which the upper film is not multilayer but rather independent flexible materials one of which is gas/moisture barrier and the other of which is gas permeable to permit air entry to restore the red color.

Carbon dioxide and reduced oxygen in gas barrier packages retards microbiological growth (CO_2) under chilled distribution while a permanent real color is produced by the small quantities of carbon monoxide. This "solution" to color retention has been the subject of major controversy since the red color is retained regardless of microbiological quality. Advocates suggest that expiration dates, ballooning packages, spoilage odors, and visible slime are alternative signals of deterioration.

Processed or Cured Meat

Longer-term preservation of meats is achieved by curing using agents such as salt, sodium nitrite, sugar, seasonings, spices, and smoke and processing methods such as cooking and drying to alter the water activity, add natural antimicrobials, provide a more stable (nitrosomyoglobin) red color, and generally enhance the flavor and mouth feel of the cured meats (Figure 9.12). Cured meats such as frankfurters and ham are often offered in tubular or sausage form which means that the shape is dictated by the traditional process and consumer dictates. Because of the preservatives, refrigerated shelf life is generally several times longer than for fresh meats. Because the cured meats are not nearly as sensitive to oxygen variations as are fresh meats, the use of reduced oxygen to enhance the refrigerated shelf life is quite common (Figure 9.13).

Packaging for reduced oxygen packaging of cured meats is drawn from a multiplicity of gas and moisture barrier package materials and structures depending on the protection and the marketing needs: frankfurters are generally in twin web vacuum packages in which the base tray is an in-line thermoformed nylon/PVDC or EVOH web and the closure is a heat sealed polyester polyethylene terephthalate polyester (PET)/PVDC or EVOH flexible material. Sliced luncheon meats and their analogs are in thermoformed unplasticized PVC/PVDC or polyacrylonitrile (PAN) trays heat seal closed with PET/PVDC or PET/EVOH flexible lidding. Sliced bacon packaging employs one of several variations of barrier film skin packaging (in contact with the surface of the product) to achieve the oxygen barrier. Ham may be fresh cured or cooked, with the cooking often performed in the package. The oxygen barrier materials employed are usually a variation of nylon/PVDC or EVOH film in pouch form. More recently, reclosure has become of paramount importance with a variety of flexible barrier plastic pouches in reusable nonbarrier plastic bowls, easy open

FIGURE 9.12 Cured meat in inert gas in surface printed barrier flexible pouch.

FIGURE 9.13 Cured meat in inert gas within barrier flexible film pouch in paperboard secondary carton for in-store rack display.

FIGURE 9.14 Baked beans in multilayer barrier plastic "bucket" can with full panel easy open closure.

nonbarrier plastic bowls, and so on, being employed to satisfy consumer desires for convenience (Figure 9.14).

Poultry

Poultry is largely chicken, but since the early 1990s turkey has become a much more important category of animal protein. Further, chicken has increasingly penetrated the cured meat market as a meat that is less expensive and nutritionally and functionally similar to beef and pork or as an ingredient. Since the mid-1970s, poultry processing has shifted into large size highly mechanized killing and dressing operations. In such facilities the dressed birds are chilled in water close to their freezing points after which they are usually cut into retail parts and packaged in case-ready form: EPS trays overwrapped with printed PVC, polyethylene or coextruded barrier film. Some European and a few U.S. poultry companies apply air chilling. The package is intended to appear as if it has been prepared in the retailer's back room, but in reality, may be only a moisture and microbial barrier. Individual retail packages, however, may be multipacked in gas barrier flexible package materials to permit gas flush packaging and thus extending the refrigerated shelf life of the fresh poultry products.

All meat products may be microbiologically preserved by thermal sterilization in metal cans. Product is filled and the container is hermetically sealed usually by double seam metal end closure. After sealing, the cans are retorted (cooked under high pressure) to destroy virtually all microorganisms present, and cooled to arrest further cooking. The sealed metal container serves as a gas, moisture, microbiological, and so on, barrier to ensure indefinite microbiological preservation. Even when evacuated, cans or jars do not, however, ensure against further biochemical deterioration of the contents, a common occurrence.

SEAFOOD

Fish is among the most difficult of all foods to preserve in their fresh state because of their inherent microbiological populations many of which are psychrophilic, that is, capable of growth at refrigerated temperatures. Live fish and shellfish grow in relatively low temperature ocean and fresh waters, and so contain biological systems evolved for the natural chilled temperature. Some seafoods may harbor a nonproteolytic anaerobic microbiological pathogen, *Clostridium botulinum type E* capable of toxin production at chilled temperatures without signaling spoilage through generation of odorous proteolytic breakdown compounds.

Packaging for fresh seafood is generally moisture resistant but not necessarily resistant to microbial recontamination. Simple polyethylene film is employed often as a liner in corrugated fiberboard or foamed polystyrene (insulating) cases. The polyethylene film serves not only to retain product moisture but also protects the structural case against internal moisture.

Seafood may be frozen, in which case the packaging is usually a form of moisture resistant material plus structure such as polyethylene pouches or polyethylene coated paperboard cartons.

Canning of seafood is much like that for meats since all seafoods are low-acid and so require high pressure cooking or retorting to achieve sterility in metal cans,

One variation unique to seafood is thermal *pasteurization* in which the product is, under reasonably clean conditions achievable in contemporary commercial seafood factories, packed into plastic cans or analogs. The filled and hermetically sealed cans are heated to temperatures of up to 80°C to effect pasteurization to permit several weeks of refrigerated shelf life.

Dairy Products

Milk

Milk and its analogs and derivatives are generally excellent microbiological growth substrates and therefore potential sources for pathogenic microbiological growth. For these reasons, almost all milk is thermally pasteurized or heated shy of sterility as an integral element of processing. Refrigerated distribution is generally dictated for all dairy products that are pasteurized to minimize the probability of spoilage.

Milk is generally thermally pasteurized and packaged in relatively simple polyethylene coated paperboard gable top cartons (Figure 9.15) or extrusion blow-molded high-density polyethylene bottles for refrigerated short-term (two plus weeks) distribution. Such packages offer little beyond containment and avoidance of recontamination as protection benefits. Obviously, such packages retard the loss of moisture and resist fat intrusion. In recent years, fluid milk packaging has been upgraded to incorporate reclosure, a feature that has been generally missing from gable top polyethylene-coated paperboard cartons. Further, in recent years, environmental conditions for packaging the product have been upgraded microbiologically to enhance

FIGURE 9.15 Gable top paperboard cartons with plastic dispensing spouts have been commonly used for containment of milk and juice products.

refrigerated shelf life in a process called extended shelf life (ESL). The product is aseptically processed, (i.e., sterilized) and filled under *clean* conditions for refrigerated distribution to permit up to 90 days of shelf life. This technology for flavored and branded milks has fostered a totally new and successful product category.

An alternative packaging system, popular in Canada and a few other countries, employs polyethylene pouches. This variant has been enhanced by re-engineering into aseptic format, a system that has not been widely accepted in the United States for fluid milk. Flexible film pouch systems are generally less expensive than paperboard and semi-rigid bottles, but may be less convenient for consumers.

In other geographic regions, aseptic or sterile packaging is employed to deliver microbiologically ambient-temperature shelf-stable fluid dairy products. The most common aseptic processing technology is ultra high temperature short time (UHT) thermal treatment to sterilize the product followed by aseptic transfer into the packaging equipment. Among the UHT processes are plate heat exchangers and direct steam injection or infusion. Three general types of aseptic packaging equipment are employed commercially: vertical form/fill/seal in which a paperboard/polyethylene/aluminum foil composite package material is sterilized by hydrogen peroxide prior to formation; erected preformed paperboard composite cartons which are sterilized by hydrogen peroxide spray; and bag-in-box in which the plastic pouch is presterilized by ionizing radiation. The former two are generally employed for consumer size bricks/blocks while the last is applied to hotel/restaurant/institutional sizes, largely for ice cream mixes. Fluid milk is sometimes pasteurized, cooled, and filled into bag-in-box pouches for refrigerated distribution.

Cheese

Fresh full moisture cheeses such as cottage cheese manufactured from pasteurized milk are generally packaged in thermoformed polystyrene or injection molded polypropylene tubs or larger size polyethylene pouches for refrigerated distribution. These package forms do not afford significant protection beyond barrier against recontamination, that is, they are little more than rudimentary moisture loss and dust protectors because the refrigerated distribution time is a relatively short as several weeks.

Fermented Milks

Fermented milks such as yogurts fall into the category of fresh cheeses from a packaging perspective, that is, they are packaged in either thermoformed polystyrene or injected molded polypropylene cups or tubs, flexible material tubes or extrusion blow-molded high-density polyethylene (HDPE) bottles, to contain and to protect against moisture loss and microbial recontamination. Because the refrigerated shelf life is relatively short, however, few measures are taken from a packaging standpoint to lengthen the shelf life. Aseptic or ESL packaging of such desserts is performed to achieve extended ambient or refrigerated temperature shelf life especially of squeezable tubes. The basic systems employed for cups include one with preformed cups and the other, thermoform/fill/seal. Squeeze tubes from flexible materials are formed on vertical form/fill/seal machines which may be clean or aseptic.

Recently, aseptic packaging of dairy products has been complemented by ultra clean packaging on preformed cup deposit/fill/seal thermoform/fill/seal systems and vertical form/fill/seal tube packaging. In these systems intended to offer extended refrigerated shelf life for low-acid dairy products, the microbicidal treatment may be hot water or peracetic acid to clean the package material surfaces. The same package sterilization systems may be employed to achieve ambient-temperature shelf stability for high-acid products such as juices and related beverages.

Cured cheeses such as cheddar, Swiss, or process are subject to surface mold spoilage as well as to further fermentation by the naturally present microflora. These microbiological growths may be retarded by packaging under reduced oxygen and/or elevated carbon dioxide. To retain the internal environmental condition, gas barrier package materials are commonly used on a commercial basis. Generally, flexible barrier materials such as nylon plus PVDC or polyester/PVDC or EVOH, are employed.

In recent years, shredded cheeses (and smaller unit portion sizes) have been popularized largely as a result of packaging technology to reduce mold growth and permit reclosure (Figure 9.16). Shredded cheeses have increased surface areas which thereby increase the probability of microbiological and especially visible mold growth. Gas packaging under carbon dioxide in gas barrier material pouches is mandatory to achieve any finite chilled shelf life. One feature of all shredded domestic cheese packages today is the zipper or slide reclosure (Figure 9.17).

Ice Cream

Ice cream and related frozen desserts obviously are distributed under frozen conditions. The fluid product must be pasteurized prior to freezing and packaging. The

FIGURE 9.16 Flexible cheese pouch with easy open and reclosure using slide.

FIGURE 9.17 Shredded cheese in inert gas flushed flexible barrier pouch with zipper reclosure features.

FIGURE 9.18 Ice cream cartons; square round paperboard bodies with insert injection molded plastic paperboard reclosure lids.

packaging is basically moisture resistant because of the presence of liquid water prior to freezing and sometimes during removal for consumption. Water resistant paperboard, polyethylene coated paperboard and polyethylene structures are usually sufficient for containment of frozen desserts (Figure 9.18). In recent years, the application of round corner rectangular convolute wound paperboard tubs closed with insert injection molded snap on tops has generated an easier to open, scoop, and reclose package, thus increasing sales in the premium and even super-premium category. Super-premium ice cream such as Ben and Jerry's are often packaged in cylindrical spiral wound paperboard canisters closed with paperboard snap on hoods—a "throwback" package to communicate nostalgic quality.

FRUIT AND VEGETABLES

In the commercial context, fruits are generally high acid (pH < 4.6) and vegetables are generally low acid (pH < 4.6). Major exceptions are tomatoes, which commercially are regarded as high-acid "vegetables" although they are really fruit, and melons and avocados which are low-acid fruit.

The most popular produce forms are fresh, fresh-cut, and minimally processed. Fresh fruit and vegetables are living, "breathing" entities exhibiting the physiological consumption of oxygen and production of carbon dioxide and water vapor. From a spoilage standpoint, fresh produce is more subject to physiological than to microbiological spoilage, and measures to extend the shelf life are designed to retard such respiratory reactions and water loss (Figure 9.19).

The most fundamental means of retarding fresh produce deterioration is temperature reduction ideally to near the freezing point but more commonly in

FIGURE 9.19 Modified atmosphere flexible film pouch of fresh-cut vegetables with special characteristics of high gas permeability to permit controlled entry of air to obviate respiratory anaerobiosis.

commerce to the 4–7°C range. Temperature reduction also reduces the rate of microbiological growth which is usually secondary to physiological deterioration and follows it. Some fruit, such as tomatoes, and vegetables such as potatoes, are sensitive to near-freezing temperatures and so must be held at temperatures not below 10°C for fear of biochemical deterioration.

Since the 1960s, alteration of the atmospheric environment in the form of modified or controlled atmosphere preservation and packaging have been commercial to extend the refrigerated shelf life of fresh produce items such as apples, pears, strawberries, lettuce, spinach, and now fresh-cut vegetables and fruit. Controlled atmosphere has been largely confined to warehouse and transportation vehicles such as trucks and seaboard containers. In controlled atmosphere preservation, the oxygen, carbon dioxide, ethylene and water vapor levels are under constant control to optimize refrigerated shelf life but this is usually not achieved through packaging. For each class of produce a separate set of environmental conditions is required for optimum preservation effect. In some modified atmosphere (MA) packaging (MAP), the produce is placed in a package structure and an initial atmosphere is introduced. Normal produce respiration (consumption of O_2 and production of CO_2 and H_2O) plus the permeation of gas and water vapor through the package material and structure, drive the interior environment towards an equilibrium gas environment that extends the produce quality retention under refrigeration. Other systems inject mixed gases of appropriate concentrations so that, in effect, equilibrium is established immediately.

The target internal atmosphere is to retard respiration rate, but reduced oxygen and elevated carbon dioxide independently or together retard the usual microbiological growth on fruit and vegetable surfaces. One major problem is that fresh produce may enter into respiratory anaerobiosis if the oxygen concentration is reduced to near extinction or zero. In respiratory anaerobiosis, the metabolic pathways produce fermentation and hence undesirable flavor compounds. To minimize the production of these undesirable end products, elaborate packaging systems have been and continue to be developed. Most of these involve mechanisms to permit small quantities of air into the package to compensate for the oxygen consumed by the respiring produce. High gas permeability plastic films, micro-perforated plastic films, plastic films disrupted with mineral fill, and films fabricated from temperature-sensitive polymers have all been proposed and/or used commercially.

The imperative for reduced temperature is emphasized in MAP because the dissolution rate of carbon dioxide in water is greater at lower temperatures than at higher temperatures.

Since the late 1980s, fresh-cut vegetables, especially lettuce, cabbage, and carrots have been major products in both the retail and the food service market. Cleaning, trimming, and size reduction lead to greater surface to volume of the produce and to the expression of fluids from the interior to increase the respiration and microbiological growth rate. On the other hand, commercial fresh-cutting operations generally are far superior to mainstream fresh produce handling in cleanliness, speed through the operations, temperature reduction and judicious application of microbicides such as chlorine. Although some would argue, on the basis of microbiological counts found in fresh-cut produce in distribution channels that fresh-cut produce is less safe than uncut produce, the paucity of cleaning undergone by whole produce coupled with the relative absence of reported adverse public health incidents leads to the opposite conclusion that fresh-cut produce is significantly safer microbiologically. Such arguments surfaced in 2006 with fresh-cut organically grown spinach when some persons claimed that packaging fostered *E. coli 0157:H7*, an unproven and illogical allegation.

Packaging of uncut produce is really a multitude of materials, structures and forms that range from old and traditional, such as wire bound wood crates, to inexpensive, such as injection molded polypropylene baskets, to polyethylene film liners within waxed corrugated fiberboard cases. Much of the packaging is designed to help retard moisture loss from the fresh produce or to resist the moisture evaporating or dripping from the produce (or, occasionally, its associated cooling ice) to ensure the maintenance of the structure throughout distribution. Some packaging recognizes the issue of anaerobic respiration and incorporates intentional openings to ensure passage of air into the package, as, for example, perforated polyethylene pouches or netting for grapes, apples or potatoes.

For freezing, vegetables are cleaned, trimmed, cut and blanched prior to freezing and then packaging, or prior to packaging and then freezing. Blanching (a mild heating process to destroy enzymes and drive out air) and other preprocessing operations reduce the numbers of microorganisms. Fruit may be treated with sugar to help retard enzymatic browning and other undesirable oxidations. Produce may be individually quick frozen (IQF) using cold air or cryogenic liquids prior to packaging or frozen after packaging as in folding paperboard cartons. Most frozen food packages are relatively simple monolayer polyethylene film pouches or polyethylene coated solid bleached sulfate paperboard to retard moisture loss.

Canning of vegetables to achieve long-time ambient-temperature microbiological stability is conventional for low-acid foods, with blanching prior to placement in steel cans, today all welded side seam tin-free steel in the United States, with two-piece rapidly replacing the traditional three-piece, followed by hermetic sealing using double-seaming and retorting and cooling. Canned fruit is generally into lined two or three-piece steel cans using hot filling sometimes coupled with postfill thermal treatment. Increasingly one end is easy-open for consumer convenience. Tetra Pak's highly publicized Tetra Recart™ retortable composite barrier paperboard cartons have been employed for vegetables. The decline is consumption of canned foods, often attributed to relatively poor quality from retort overcooking is being combated by reduced total thermal processes, shaped cans, easy open cans and plastic-film lined cans (Figure 9.20).

In recent years, fruit has been filled hot into plastic gas/moisture multilayer barrier tubs and cups prior to heat sealing with barrier flexible plastic materials to achieve ambient-temperature microbiological shelf stability or extended refrigerated temperature shelf life. These plastic packages are intended to provide greater convenience for the consumer as well as to communicate that the contained product is not the "overprocessed canned food."

Tomato Products

The highly popular tomato-based sauces, pizza toppings, and pasta sauces, and so on, must be treated as if they were low-acid if they contain meat as so many do.

FIGURE 9.20 Easy open metal top on can for ease of access to the contents.

FIGURE 9.21 Glass jars for low-acid foods such as chili offer product visibility and shelf differentiation from mainstream versions in metal cans.

For marketing purposes, tomato-based products for retail sale are commonly in glass jars (Figure 9.21) with reclosable metal closures, although in large sizes with low surface to volume ratios, multilayer barrier plastic jars are being used for added consumer convenience (Figure 9.22). The glass jars are often retorted after filling and hermetic sealing. Some such products are in steel cans, and occasionally, some have been packaged aseptically in composite paperboard barrier bricks/blocks.

FIGURE 9.22 Pasta sauce in glass jar with hermetic steel closure. Retorted to sterilize contents. Full body panel shrink film label depicts contents without showing them.

Juices and Juice Drinks

Juices and analogous fruit beverages may be hot filled or aseptically packaged. Traditional packaging has been hot filling into steel cans and glass bottles and jars. Hot filling is sterilizing the product outside of the package, filling sterile hot fluid product into the package so that the product sterilizes the package interior, hermetically sealing and cooking after closure. Aseptic packaging, described above for paperboard composite cartons, is being applied to polyester bottles using various chemical sterilants, usually hydrogen peroxide to effect the sterility of the package and closure interiors. Most fruit beverages today are hot filled into heat set polyester bottles capable of resisting temperatures of up to 85°C without distortion, some counter pressured with nitrogen. Hermetic sealing of the bottles provides microbiological barriers but the polyester is a modest oxygen barrier and so the ambient-temperature shelf life from a biochemical perspective is relatively limited. Polyester bottles have been a major driver for increase fruit beverage consumption due to light weight and convenience of use (Figure 9.23).

For more than twenty years high-acid fluid foods such as tomato pastes and non-meat containing sauces have been hot filled into flexible pouches for food service applications. The hot filling generates an internal vacuum within the pouch after cooling by virtue of steam condensation so that the contents are generally ambient-temperature shelf stable from a microbiological perspective. Package materials used are usually laminations of polyester and aluminum foil or metallized polyester with linear low density polyethylene (LLDPE) internal sealant to achieve an hermetic heat seal. Some efforts have been made to employ transparent gas/water vapor barrier films in the structures: polyester/EVOH laminations with LLDPE sealant. Transparent flexible pouches offer the opportunity for the retail consumer to see the contents

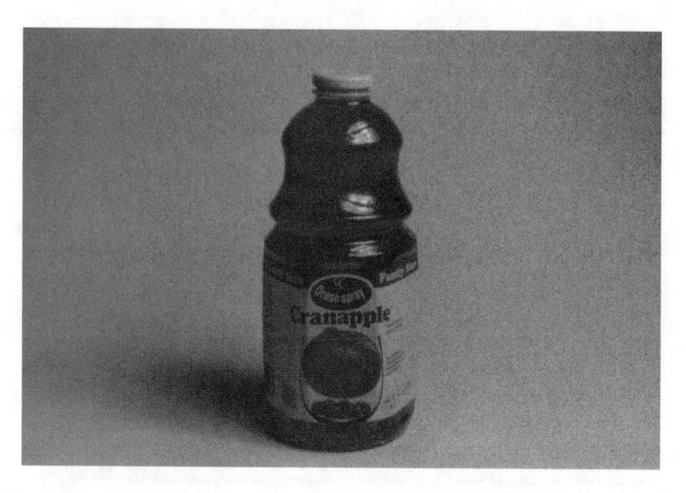

FIGURE 9.23 Juice drink in polyester bottle with wraparound label hiding collapsing panels engineered to maintain structure after hot filling and cooling.

or the food service worker to identify the contents even without being able to read the label.

Other Products

A variety of wet food products that do not fall clearly into the meat, dairy, fruit, or vegetable categories may be described as prepared foods or ready meals, a rapidly increasing segment of the industrialized society food market. Prepared foods are those that combine several different ingredient components into dishes that are ready to eat or nearly ready to (heat and) eat. If within a can, the thermal process must be for the slowest heating component requiring a thermal differential meaning that much of the product is overcooked to ensure microbiological stability. If frozen, the components are separate but the process of freezing reduces the eating quality. The preferred preservation technology from a quality retention or consumer preference perspective is refrigeration, which, of course, offers the least shelf life preservation effect.

Incorporation of multiple ingredients from a variety of sources correctly implies many different sources for potential microbiological contamination. With refrigeration as the major deterioration barrier, problems are obviated mostly by reducing the time between preparation and consumption to less than 1 day, under refrigeration above freezing, plus cleanliness during preparation. As commercial operations have attempted to prolong the quality retention periods beyond same or next day consumption, enhanced preservation or "hurdles," for example, pH reduction, oxygen removal, and so on, have been introduced.

Packaging for air-packaged prepared dish products is often thermoformed oriented polystyrene or polyester trays with oriented polystyrene dome closures snap locked into position, or some variation, that is, no gas, moisture or microbiological barriers of consequence. Refrigerated shelf life of air-packaged prepared foods is measured in hours or at best, days. When the product is intended to be heated for consumption, the base tray packaging may be thermoformed polypropylene or crystallized polyester with no special barrier closure. For reduced oxygen or MAP the tray material may be a thermoformed coextruded polypropylene/EVOH/polyethylene with a flexible gas/moisture barrier lamination closure heat sealed to the tray flanges. Refrigerated shelf life for such products may be measured in weeks—with 2–3 weeks being typical and biochemical changes being the principal deteriorative vector.

For several years, concepts of pasteurizing the contents, vacuum packaging and distribution under refrigeration have been both debated and commercial in both the United States and Europe. Sous vide is the most publicized (and controversial) package of this type. In the sous vide process, the product is packaged under vacuum and heat sealed in an appropriate gas/water vapor barrier package structure such as flexible aluminum foil lamination. The packaged product is thermally processed at less than 100°C to destroy spoilage microorganisms and then chilled for distribution under refrigerated or, sometimes in the United States, frozen conditions. The United States' option is to ensure against the growth of pathogenic anaerobic microorganisms. A very similar technology is cook/chill in which the pumpable products such as chili, chicken a la king, or cheese sauce are hot filled at 80+°C into nylon/polyethylene pouches which are immediately chilled (in circulating cold water) to 2°C and then distributed at temperatures of 1°C in food service channels. The hot filling generates a partial vacuum within the package to virtually eliminate the growth of any spoilage microorganisms that might be present. Of course, any heat resistant pathogenic microbiological spores present may survive, grow, and, in extreme circumstances, produce toxins, an adverse reaction retarded (but not obviated) by refrigeration throughout distribution.

Dry Foods

Removing water from food products markedly reduces water activity and its subsequent biochemical activity, and thus also significantly reduces the potential for microbiological growth since microorganisms do not grow at water activities below 0.8.

Dry products include those directly dried from liquid form such as instant coffee, instant tea, dry milk, and so on. The liquid is spray-, drum-, or air-dried, or even freeze-dried, to remove water.

Moisture entering dry foods can change physical and biological properties. Engineered dry products include beverage mixes such as blends of dry sugars, citric acid, color, flavor, and so on; and soup mixes which include dehydrated meat stock plus noodles, vegetables, meats, and so on, which become a particulate-containing liquid on rehydration with hot water. Such products must be packaged in moisture resistant structures to ensure against water vapor entry which can damage the contents by increasing the propensity to biochemically react.

Products containing relatively high fat such as baked goods or some soup mixes also must be packaged so that the fat does not interact with the packaging materials. Flavoring mixes that contain seasonings and volatile flavoring components can unfavorably interact with interior polyolefin packaging materials which can scalp or remove flavor from the product if improperly packaged or even disrupt (e.g., delaminate) the package material. Packages for dry products must be well sealed; for example, provide a total barrier against access by water vapor, and, for products susceptible to oxidation, also exclude oxygen after removal of air from the interior of the package.

Fats and Oils

Fats and oils may be classified as those with and those without water. Cooking oils such as corn or canola oil and hydrogenated vegetable shortenings contain no water and so are relatively stable at ambient temperatures.

Unsaturated lipids, that is, those containing fatty acids with double bonds, are susceptible to oxidative rancidity. Because of their unsaturated fatty acids, oils are more subject to oxidative rancidity than fats which are solid at room temperature, but both are usually sparged with and packaged under nitrogen to reduce oxygen. Hydrogenated vegetable shortenings, although declining because of their trans fat content, generally are packaged under nitrogen that is, without oxygen in spiral wound composite paperboard cans to minimize oxidative rancidity. (For food service, polyethylene lined cartons are more common.) Edible liquid oils are packaged in injection blow-molded polyester bottles usually under nitrogen.

Margarine and butter and analogous bread spread products consist of fat plus water and water-soluble ingredients which contribute flavor and color to the product. Often, these products are distributed at refrigerated temperatures to assist in quality retention. Fat-resistant packaging such as polyethylene-coated paperboard, aluminum foil/paper laminations and parchment paper wraps, and polypropylene tubs are used to package butter, margarine, and similar fat-based bread spreads.

Cereal Products

Dry breakfast cereals generally are so low in water content ($<5\%$) to be susceptible to water vapor absorption and so require good moisture and well as fat-barrier packaging. Further, packaging should retain the product flavors. Breakfast cereals are usually packaged in coextruded polyolefin films fabricated into pouches or bags inserted into or contained within printed paperboard carton outer shells. Sweetened cereals may be packaged in aluminum foil lamination, metallized plastic film, or gas barrier plastic films or flexible laminations to retard water vapor and flavor transmission.

Soft bakery goods such as breads, cakes, and muffins are highly aerated starch-based structures subject to dehydration and staling. In moist environments, baked goods are often vulnerable to microbiological deterioration as a result of surface growth of mold and other microorganisms. To retard water loss, good moisture barriers such as coextruded polyethylene film bags or polyethylene extrusion-coated paperboard cartons are used for packaging, with better barrier achieved when the package is fully closed.

Hard baked goods such as cookies and crackers generally have low water and often high fat contents. Water can be absorbed, however, so that the products can lose their desirable mouth feel properties and become subject to oxidative or even hydrolytic rancidity. Package structures for cookies and crackers include fat- and

FIGURE 9.24 Saltine crackers stack packed in slug format in coextruded polyolefin flow wrap with four slugs in paperboard carton secondary package.

moisture-resistant coextruded polyolefin film pouches within paperboard carton shells and thermoformed polystyrene trays overwrapped with polyethylene or oriented polypropylene film (Figure 9.24). Soft chewy cookies are packaged in high moisture barrier flexible laminations often containing metallized film to enhance the package moisture barrier and appearance simultaneously.

Salty Snacks

Dry salty snacks include dry cereal or potato products such as potato and corn and tortilla chips, and pretzels, and include roasted nuts, all of which have low water and high fat contents (Figure 9.25). Snack food packaging problems are often compounded by the presence of flavorings such as salt, a catalyst for fat oxidation. Snacks are usually packaged in flexible pouches made from multilayer oriented polypropylene or metallized oriented polypropylene (which may also enhance appearance) to provide low moisture and gas transmission. Salty snack food producers depend on rapid and controlled product distribution to minimize fat oxidation. Many salty snacks are packaged under nitrogen both in flexible pouches and in rigid containers such as paperboard composite canisters, thermoformed plastic, or extrusion blow-molded bottles to extend shelf life and widen the distribution range (Figure 9.26). Often, semirigid canisters may be treated with inert gas to remove oxygen. Generally, light, which catalyzes fat oxidation, harms snack products, and so opaque packaging is often employed.

Candy

Chocolate (Figure 9.27), a mixture of fat (cocoa butter, chocolate liquor, milk) and nonfat components such as sugar, usually has low water activity and so is not

FIGURE 9.25 Potato chip pouch made from metallized oriented polypropylene to minimize moisture gain and loss of internal inert gas to prolong quality retention period.

FIGURE 9.26 Gusseted side wall stand-up flexible laminated pouch for dry baked goods.

vulnerable to microbiological growth, but is subject to slow flavor change due to inherent biochemistry. Ingredients such as nuts and caramel are susceptible to water content variation and subsequent fat oxidation (Figure 9.28). Chocolates, which are generally shelf stable at ambient temperatures, are packaged in fat-resistant papers or plastic films and moisture/fat barrier such as oriented or oriented pearlized polypropylene film—often sealed with cohesive or pressure sensitive adhesive to avoid the heat of heat sealing and consequent thermal damage (Figure 9.29).

Hard sugar candies (Figure 9.30) are flavored amorphous sugars which are very hygroscopic because of their extremely low moisture contents down to 1%. Sugar

FIGURE 9.27 Solid chocolate bar in classical aluminum foil/paper inner wrap and printed paper sleeve.

FIGURE 9.28 Chocolate coated confections and sugar confections are protected against moisture gain or loss.

candies are packaged in low-moisture-transmission packaging such as unmounted aluminum foil, oriented polypropylene film, or metallized oriented polypropylene film, or cast polypropylene film.

Obviously, this listing is only a sampling of the many alternative packaging forms offered and employed commercially for foods subject to immediate microbiological or biochemical deterioration. An entire encyclopedia is required to enumerate all of the known options available to the food packaging technologist with the advantages and issues associated with each.

FIGURE 9.29 Sugar shell coated chocolate peanut candies in unit portion size in flexible film pouch.

FIGURE 9.30 Hard sugar candies in a variety of moisture and flavor barrier packages that still convey the fun aspects of consuming the product-oriented polypropylene film flow wrap for individual pieces multipacked in printed oriented polypropylene pouch, In the foreground is a tube of candies in classical aluminum foil paper combination with printed paper sleeve.

PACKAGE MATERIALS AND STRUCTURES

Package Materials

In describing package materials, different dimension conventions are employed depending on the materials and their origins, and the systems commonly used in that industry segment.

Paper and Paperboard

The most widely used package material in the world is paper and paperboard derived from natural cellulose sources such as trees, bushes, or shoots. Paper is the less used in packaging because its protective/barrier properties are almost nonexistent and its usefulness is almost solely as decoration and dust cover. Paper is cellulose fiber mat in gauges of less than 0.010 in. When the gauge or caliper in industry parlance is greater than 0.010 to perhaps as much as 0.040 in., the material is paperboard, which, in various forms, can be an effective structural material to protect contents against impact, compression, vibration, and so on. Only when coated and laminated does a paper or paperboard structure such as a carton or case become a good moisture and gas barrier. For this reason, despite their long histories as a package material, paper and paperboard are only infrequently used as protective packaging against moisture, gas, odors, or microorganisms.

Paper and paperboard may be manufactured from either trees or recycled paper and paperboard which are present in large volumes in commerce. Virgin paper and paperboard, derived from trees, has greater strength than recycled materials whose fibers have been reduced in length due to the multiple processing. Therefore, increased gauges or calipers of recycled paper or paperboard are required to achieve the same structural properties. On the other hand, because of the short fiber lengths, the printing and coating surfaces are smoother. Paper and paperboard are moisture sensitive, changing their properties significantly and thus often requiring internal and external treatments to compensate for moisture change and thus ensure performance.

Metals

Two metals are commonly employed for package materials: steel and aluminum (Figure 9.31). Steel is traditional for cans and glass bottle closures, but is subject to corrosion in the presence of air and moisture and so is almost always protected by coatings or other materials such as plastic film. Until the 1980s, the most widely used steel protection was tin plating which also acted as a base for lead soldering of the side seams of "tin" cans. When lead was finally declared toxic and removed from cans during the 1980s in the United States, tin was also found to be superfluous and declined as a steel can liner. Replacing tin in "tin-free" cans was chrome and chrome oxide.

In almost every instance the coated steel is further protected by organic coatings such as vinyls and epoxies which are really the primary protection.

Steel is rigid, an ideal microbial, gas, and water vapor barrier, and resistant to every temperature to which a food may be reasonably subjected. Because steel to steel or steel to glass interfaces are not necessarily perfect, the metal is complemented by resilient plastic to compensate for the minute irregularities in closures.

Aluminum is lighter weight than steel and easier to fabricate. Therefore aluminum has become the metal of choice for beverage packaging in the United States and favored in other countries. As with steel, the aluminum must be coated to protect

FIGURE 9.31 Steel soup can with full panel easy open top.

it from corrosion. Although it is the most used material for can making in the United States, wall thickness is very thin for economic reasons. Aluminum cans therefore must be internally pressurized from carbon dioxide or nitrogen to maintain structure. Thus, aluminum is not widely used for food can applications in which internal vacuums and pressures change as a result of retorting.

Aluminum may be rolled to very thin gauges to produce foil, a flexible material with excellent microbial, gas, and water vapor barrier properties when it is protected by plastic film. Aluminum foil is generally regarded as the only "perfect" barrier flexible package material. Its deficiencies include a tendency to pin holing especially in thinner gauges and to cracking when flexed.

In recent years, some applications of aluminum foil have been replaced by vacuum–aluminum metallization of plastic films such as polyester or polypropylene, and by barrier plastics such as ethylene vinyl/alcohol.

Glass

The oldest and least expensive per se package material is glass, derived from sand. By itself, glass is renowned as a perfect barrier material against gas, water vapor, microorganisms, odors, and so on. Closure however, may, be less than perfect. Further, the transparency of glass is often regarded by marketers and consumers as a desirable property. On the other hand, food packaging technologists may view the transparency as less than desirable because visible and ultraviolet radiation accelerate biochemical and particularly oxidative reactions.

Glass is energy intensive to produce, heavy, and, of course, vulnerable to impact and vibration even though it has excellent vertical compressive strength. For these reasons, glass is being gradually displaced by plastic materials in industrialized societies.

Plastics

Plastics is a term describing a number of families of petrochemically derived polymeric materials each with different properties. The overwhelming majority of polymers are not suitable as package materials because they are too expensive or toxic in contact with food or do not possess properties desired in packaging applications, or cannot be readily fabricated into packages. The most commonly used plastic package materials are polyethylene, polypropylene, polyester, polystyrene, and nylon, each is quite different in properties. Plastics are often being combined with each other and with other materials to deliver desired packaging properties.

Polyethylene

Polyethylene is the most used plastic in the world for both packaging applications. Polyethylene is manufactured in a variety of densities ranging from 0.89 grams/cc or very low density (LDPE), to 0.96 or high density (HDPE). It is light weight, inexpensive, impact resistant, relatively easily fabricated, and forgiving. Polyethylene is not a good gas barrier and is generally not transparent, but rather translucent. Polyethylene may be extruded into film with excellent water vapor and aqueous liquid containment properties. Low density polyethylene film is more commonly used as a flexible package material (Figure 9.32). Low density polyethylene is also extrusion coated onto other substrates such as paper, paperboard, plastic or even metal to impart water and water vapor resistance or heat sealability. A variety of polyethylene is offered for different applications.

Although used for flexible packaging, high density polyethylene is more often seen in the form of extrusion blow-molded (squeezable) bottles with impact and drop resistance, good water, and water vapor barrier, but poor gas barrier properties.

Polypropylene

Polypropylene is a polyolefin with better water vapor barrier properties and greater transparency and stiffness than polyethylene. Although somewhat difficult to fabricate, polypropylene may be extruded and oriented into films that are widely used for making pouches particularly on vertical form/fill/seal machines. In cast (unoriented) film form, polypropylene is a good twistable candy wrapper.

Polypropylene's heat resistance up to about 133°C permits it to be employed for microwave-only heating trays and for plastic or plastic/aluminum foil retort packaging (Figure 9.33).

FIGURE 9.32 Printed stand-up flexible laminated pouches of dry foods function as moisture barriers while displaying well on retail shelves.

FIGURE 9.33 Retort pouch of rice in glass-coated flexible lamination to permit microwave reheating.

Polyester

A relatively difficult-to-fabricate polymer, PET is increasingly the plastic of choice as a glass replacement in making food and beverage bottles (Figure 9.34). Polyester plastic is a fairly good gas and moisture barrier and is difficult to fabricate, but in bottle, tray or film form is dimensionally quite stable and strong. Heat resistance in partially crystallized form is sufficient (up to about 85°C) to permit its use in hot fillable bottles such as for juices and juice drinks. When increasingly crystallized the heat resistance rises to the level of being able to resist conventional oven heating temperatures. For this reason crystallized polyester is employed to manufacture dual

FIGURE 9.34 Flavored milk in unit portion size polyester bottles with printed full panel shrink film labels.

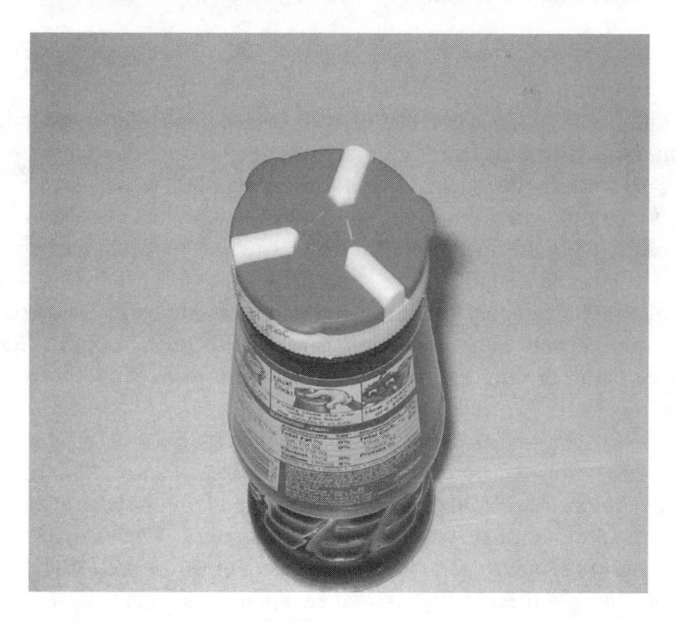

FIGURE 9.35 Polyester ketchup bottle with multiple dispensing closure targeted at children to permit ease of use.

ovenable trays for heat-and-eat prepared foods. Dual ovenable means that the plastic is capable of being heated in either conventional or microwave ovens.

The transparency of polyester makes it highly desirable from a marketing standpoint after taking into account the issues of light sensitivity of contained foods (Figure 9.35).

Nylon

Nylons are family of polymers noted for their very good gas barrier properties. Moisture barrier tends to be lower than those in polyolefin polymers. Nylons are somewhat hygroscopic meaning that gas barrier may be reduced in the presence of moisture. Gas and water vapor barriers are enhanced by the multilayering with polyolefins and high gas barrier polymers. Nylons are thermoformable and both soft and tough, and so are often used for thermoformed processed meat and cheese package structures in which the oxygen within the package is reduced to extend the refrigerated shelf life.

Polystyrene

Polystyrene is a thermoplastic with little moisture or gas barrier. It is, however, very machineable and, in pristine or oriented form, usually highly transparent. Structural strength is not good unless the plastic is oriented or admixed with a rubber modifier which reduces the transparency. Polystyrene is often used as the easy and inexpensive tray material for prepared refrigerated foods. Expanded or foamed polystyrene is a commonly used opaque thermoformable tray material—with few barrier properties. Major uses are for meat and poultry trays and short term take-out food trays.

Polyvinyl chloride

Polyvinyl chloride (PVC) is a polymer capable of being modified by chemical additives into plastics with a wide range of properties. The final materials may be soft films with high gas permeabilities such as used for overwrapping fresh meat in retail stores; stiff films with only modest gas barrier properties, readily blow moldable semirigid bottles or easily thermoformed sheet for trays. Gas and moisture barriers are fairly good but must be enhanced to achieve "barrier" status.

PVC falls into a category of chlorine-containing polymers which are regarded by some environmentalists as less than desirable. For this reason, in Europe and, to a lesser extent in the United States, PVC has been resisted as a food packaging material.

Polyvinylidene chloride

Polyvinylidene chloride (PVDC) is a high gas, moisture, fat, and flavor barrier plastic that is so difficult to fabricate on its own that it is almost always used as a coating or film on other substrates to gain the advantages of the properties. PVDC is another chlorine-containing molecule that is looked upon unfavorably by environmental interests.

Ethylene vinyl alcohol

Ethylene/vinyl alcohol (EVOH) is an outstanding gas and flavor barrier polymer which is highly moisture sensitive and so must be combined with polyolefin to render it an effective package material. Often EVOH is sandwiched between layers of polypropylene which act as water vapor barriers and thus protect the EVOH from moisture.

Edible and Biodegradable Plastics

By admixing hydrocarbon-based plastics with photo activators or in organics such as minerals, sensitivity to natural environmental conditions is enhanced so that the plastic may break physically into smaller pieces and "disappear."

Edible packaging such as zein or casein proteins, pectins, lipids, and so on, have been intensively studied for decades at academic levels. With plasticizers, some can achieve packaging properties approaching those of hydrocarbon plastics, but economics and the imperative to protect the "edibles" with plastic have retarded their commercialization.

Since the turn of the twenty-first century, a number of newer "biodegradable" or "compostable" plastics have been commercialized and promoted as part of the "sustainable" packaging movement. The most widely publicized is polylactic acid (PLA) from NatureWorks LLC, a part of Cargill. PLA is the result of polymerization of lactic acid produced by fermentation of dextrose obtained by hydrolysis of corn starch. Thermoformable and transparent, PLA has a maximum use temperature of about 40°C, has poor moisture and gas permeation properties and is hygroscopic. Although the suppliers compare PLA with polyester, many food packaging technologists regard it as an expensive analogue of polystyrene. PLA is claimed to be compostable, but only under specially controlled conditions.

Package Structures

Currently, rigid, semi–rigid, and flexible forms are the most common commercial structures to contain foods. Paperboard is most common in the form of corrugated fiberboard cases engineered for distribution packaging. In corrugated fiberboard, two webs of paperboard liner are adhered to a central or fluted medium section imparting the major impact and compression resistance to the structure. Folding cartons constitute the second most significant structure fabricated from single wall paperboard. Folding cartons are generally rectangular solid shape, although in recent years, odd shapes have entered and often are lined with flexible films to impart the desired barrier.

Metal cans have traditionally been cylindrical shaped probably because of the need to minimize problems with heat transfer into the contents during retorting. Recently, metal and particularly aluminum has been fabricated into tray, tub, cup, and undulating shapes, and embossed for greater consumer appeal with consequential problems with measuring and computing the thermal inputs to achieve sterilization. During the 1990s shaped cylinders entered the market again in efforts to increase market share. A few have been applied for cans requiring thermal sterilization, but barrel and distorted body cans are not rare in Europe for retorted low-acid foods. Analogous shaped cans are being used for hot filling of high-acid beverages.

Noted for its formability, glass has traditionally been offered in a very wide range of shapes and sizes including narrow neck bottles and wide mouth jars. Each represents its own singular problems in terms of fabrication, closure, and, when applicable, thermal sterilization.

FIGURE 9.36 Polyester condiment bottles featuring wide mouth closures with large dispensing openings, hinged snap reclosure, inverting to foster rapid dispensing and full panel shrink fully printed film labels.

FIGURE 9.37 Fruit hot filled into barrier plastic cups and multipacked in paperboard wraparound sleeve.

Plastics are noteworthy for their ability to be relatively easily formed into the widest variety of shapes. Thin films can be extruded for fabrication into flexible package materials (Figure 9.36). These flexible materials may then be employed as pouch or bag stock or as overwraps on cartons or other structures, or as inner protective liners in cartons, drums, cases, and so on. Thicker films, (i.e., greater than 0.010″ or 10 mils), designated sheet, may be thermoformed into cups, tubs, and trays for containment (Figure 9.37). Plastic resins may be injection or extrusion molded into bottles or jars by melting the thermoplastic material and extruding or forcing it, under pressure, into cooled molds that constitute the shape of the hollow object, that is, the bottle or jar.

PACKAGING DEVELOPMENT

The primary packaging development path is the development of the total system from concept to marketplace, and the secondary path is the development of the package itself as an integral part of the whole system. Many food marketing organizations call on package materials suppliers, converters, consultants, graphic designers, and so on, to assist in the development of packaging suitable for their food product.

THE TOTAL SYSTEM PATH

The total system path involves management, marketing, consumer research, food science, and technology sales, engineering, manufacturing, and the packaging development departments or staffs in coordinated efforts, and includes the following steps:

- Definition of the goal (initial reason for the development, identification of their consumers, and their uses for the packaged product forecasting of market potential, projection of acceptable developmental cost and final package cost);
- Package development path;
- Marketing testing (planning, execution, analysis of results feedback, and remediation);
- Decision whether to proceed, modify and retest, or to drop further effort;
- Full production (planning, execution of scale-up to full production);
- Product launch.

THE PACKAGING DEVELOPMENT PATH

The packaging development path involves packaging management and in-house, supplier, consultants, and so on, staffs. It comprises the following steps:

- Definition of the food product properties as they relate to package technical requirements;
- Description of consumer and consumer use;
- Definition of package technical and functional requirements;
- Definition of package marketing and hence structural and graphic design requirements;
- Identification of legal and regulatory requirements;
- Creation of alternatives and selection of potential package designs and materials;
- Involvement of production and engineering for equipment;

- Estimate of the probable time, resource requirements, and cost of development;
- Decision to proceed;
- Sample package preparation and testing for technical performance, consumer opinion, and economic feasibility;
- Evaluation of package materials and structures for the assurance of absence of inherent defects that can adversely affect the food contents;
- Purchase specifications;
- Quality assurance specifications;
- Suggest added caution when considering offshore naturals and structures because of reliability, quality, and toxicity issues;
- Shelf life testing;
- Consumer testing;
- Market testing, when applicable;
- Market launch;
- Feedback and remediation.

The initiation for the development may originate from any source. A new food product or technical development may produce a more efficient or less-costly process, a new source of supply may develop for a cheaper or higher-performance material, a competitor introduces a new package that leads to steal, marketing may want to create a "new image," or, as is common today, consumers signal their desire for a new feature or package.

Marketing research should enter the picture at this point to establish basic criteria. What is the predicted market potential for the proposed product? Does the projected selling price include a sufficient margin to recover the developmental cost and invariably incremental costs, in a reasonable time? Will the consumer pay a premium if acquired? What special features might the package need to have to command such a premium? Sketches or models of proposed packages can be included in the marketing research element of the investigation. If careful marketing research is not performed at this point but is deferred until a test market stage if performed, considerable development money may be spent only to find that the package meets all requirements except economics.

Definition of Food Properties Affecting Packaging Technical Properties

Package requirements are dictated by contained food product requirements. The general physical form of the product influences the type of package to be used. It is therefore necessary to know, for example, whether the food unit is large or small; a solid or liquid or a combination, such as a form or an emulsion; massive, chunky, granular, or powdery, if a solid, or watery and thin, or thick and viscous, if a liquid; and soft and light or hard and dense.

It is necessary to comprehend special properties of the product that will require special features in the package. Is the product sensitive to temperature? Must it be protected against extreme heat or cold? Will it be marketed frozen? Will the entrance of moisture or evaporation of the product make it unsaleable?

Definition of Package Technical and Functional Requirements

The functional requirements for a new package must be defined precisely, accurately, and completely if the development of the package is to be accomplished with economy and dispatch. All too often a package development nears completion and then the food packaging technologist is informed by marketing, management, or product development, "This is not what we want! You did not understand that another or different attribute is needed!" Therefore, it is vital to the success of food packaging development to be complete in fact gathering so as to define the target accurately. A checklist is helpful to keep track of and properly organize the process.

From the data supplied by markets and the information gathered about properties of the product itself, the food packaging technologist enumerates the properties the package must have and at the same time begins to select from among the vast array of materials and package structures for consideration. In addition, unless a radical departure from the mainstream is desired, experience can be drawn on for guidance. For this reason, it is extremely desirable to know what packages have been or are being used for similar products. A study of the advantages or disadvantages of these packages will help in making a selection. Conventional packages will reveal the current market unit sizes or counts, existing prices, whether product visibility is expected, and so on, and will offer some evidence on the shelf life that may be anticipated.

In considering the types of primary and secondary packaging required for a product, it is necessary to determine the nature of the handling, storage, and distribution cycle from the point of manufacture to the point of consumption. Packaged product should not be introduced into commercial distribution without thorough shelf life testing. As many expert opinions, predictive models and analogues as have been suggested, the only reasonably reliable method to determine how a packaged product will perform in the marketplace is actual real-time/real-condition testing. Even here, however, testing is sometimes insufficient to determine what will happen in real-life commercial situations. Thus, shelf life testing should be followed up with field observations to confirm the test results or to modify the product, packaging, or distribution conditions.

The effect of climate in a given market area must be considered. High altitudes (or air freight) can cause packages to expand, stress seals, or even explode. Thus the environment within the transportation vehicles and within warehouses is equal in importance to the type of packaging chosen.

A number of package limitations are imposed on the packaging technology by the very nature of the package manufacturing process. For example, some methods of forming aluminum cans or containers will not permit the depth-to-diameter ratio to exceed specified limits. Other limitations may be imposed by the production line for the product itself. Filling speeds may require a wider mouth. Production line speeds may dictate what type of heat sealing compound or labeling adhesive must be used. Thus the food packaging technologist should be cognizant of all phases of packaging and product manufacturing that can affect the package's functional requirements.

Definition of Package Design/Marketing Requirements

Marketing staff should study the market and propose the design requirements for the package. The packaging technologist must be aware of these proposals from the outset, as they may place boundaries on the technical and functional properties of the package.

Existing conventional commercial packaging will reveal what convenience features may be expected by the consumer in the marketplace. These include such items as easy-opening devices, dispensers, reclosure features, and pilferage protection.

Identification of Legal and Regulatory Requirements

The food packaging technologist must be fully aware of all legal and other regulatory restrictions that may influence the choice of packaging material or design.

The packaging technologist must also be aware of the nonstatutory regulations that may affect his choice of package. Industry-accepted standards may not include a dimension or size desired. Some religions (particularly the Jewish and Muslim faiths) have strict rules applying to packages that are used for foods.

Selection of Potential Package Designs and Materials

Working together with graphic design professionals, the food packaging technologist should prepare an array of optional package designs considered to be technically feasible, including specifications on materials, methods of manufacture, potential sources, and estimated costs. This list will be subjected to a preliminary screening, and the preferred few possibilities will then be rendered into artists' computer drawings or actual package mockups so that some idea of consumer preference can be obtained by marketing research.

Estimation of the Probable Cost of Development

The food packaging technologist should have a grasp of the magnitude of the cost. For each proposed package design he/she should be able to estimate the developmental cost needed to bring it to the marketplace and the probability of success. He/she should also be able to point out whether new or modified packaging equipment might be required. Engineering must be involved in the process.

Testing for Technical Performance, Marketing, and Economic Feasibility

The food packaging technologist should obtain a number of sample packages for each design concept so that evaluation tests can be conducted. He/she may purchase the packages or make them by hand, by pilot machinery, or on a regular manufacturing line. He/she may choose several alternative materials or fabrication methods for each concept. During this procedure, he/she eliminates only those items which are probably extremely difficult to manufacture or too costly. When the packages are in hand, he/she subjects them to product compatibility tests, design fulfillment tests, and distribution evaluation.

Product compatibility tests determine whether the package in any way tends to adversely affect the product and, almost as important, whether the product tends to adversely affect the package. Design fulfillment tests determine whether the package meets the preestablished design and performance criteria. Will it hold the desired quantity, will it dispense, does it provide critical protection needs, does it provide minimum shelf life requirements? Distribution tests determine whether the package will survive the normal handling to be expected in passing through distribution channels. Abuse testing determines the margin of safety built into the package.

Finally, it may be desirable to submit the best concepts to consumer testing. This helps gauge the consumer's reaction to the design and the probability of its being a commercial success. Unexpected consumer dislikes may eliminate some concepts or require some design modifications.

The food packaging technologist must be familiar with the costs of materials and of the final package forms so that in selecting a package structure and design he/she will not choose one that is too expensive for the product or market need. He/she walks a fine line between inadequate packaging, which may lead to complaints or loss of sales, and overpackaging, which gives unnecessary protection at too high a cost. In the beginning it is best to err towards overpackaging, as process refinements or new technological developments frequently permit the package costs to be reduced without loss of performance in the marketplace.

It is also difficult to judge how much the consumer will pay for the added value provided by a new product-package combination.

Decision to Proceed to Marketing Test

Management must make intelligent, informed decisions. One or two package concepts have been tested. Limited production runs have indicated that the package can be made and used. Distribution tests and shelf-life tests have indicated that it is functional. Estimated costs are within budgetary guidelines. All that remains is to determine whether the consumer will buy the product and at what price. Management must decide whether the cost of a marketing test is justified.

The decision is one of the most critical management decisions in the development of a food product, and the package design is a major factor in whether the test market will succeed or fail. Up to this point, only development costs have been risked. Beyond this point not only are larger expenses for materials and services risks, but the entire company image is exposed to the adverse reactions that may occur in the marketplace should the package fail to perform. A successful package development may or may not help a product attain marketing success, but a package failure almost certainly will damage the chances of a good product attaining success.

Market Testing, if Used

The planning, execution, and analysis of results of marketing tests are primarily the responsibility of marketing personnel. The marketing test, if used, will indicate whether the consumer will buy the product and at what price, in what sales units, and so on. It can also signal whether sales are sustained, indicating repurchase, or whether they taper off, indicating one-time buying, signaling product deficiency. It will reveal shortcomings in the package design, if any, and manufacturing problems.

In recent years, test markets are less used in order to accelerate time to market and to minimize altering the competition. After examining the results of marketing and consumer tests, management must decide whether to terminate the program, make modifications and retest, or go into full-scale production.

BIBLIOGRAPHY

Brody, Aaron L. 1989. *Controlled/Modified Atmosphere/Vacuum Packaging of Foods.* Trumbull, CT: Food & Nutrition Press.

Brody, Aaron L. 1994. *Modified Atmosphere Food Packaging*. Herndon, VA: Institute of Packaging Professionals.

Brody, Aaron L. and Kenneth S. Marsh. 1997. *The Wiley Encyclopedia of Packaging Technology* 2nd ed. New York: John Wiley & Sons.

David, Jairus, Ralph Graves, and V. R. Carlson. 1995. *Aseptic Packaging of Food.* Boca Raton, FL: CRC Press.

Meyers, Herbert M. and Murray J. Lubliner. 1998. *The Marketer's Guide to Successful Package Design.* Chicago: NTC.

Paine, Frank A. and Heather Y. Paine. 1983. *A Handbook of Food Packaging.* London, United Kingdom: Blackie & Son Ltd.

Robertson, Gordon L. 2006. *Food Packaging* 2nd ed. Boca Raton, FL: CRC Press.

Soroka, W. 2003. *Fundamentals of Packaging Technology* 3rd ed. Naperville, IL: Institute of Packaging Professionals.

Wiley, Robert C. 1994. *Minimally Processed Refrigerated Fruits and Vegetables.* New York: Chapman & Hall.

10 New Food Products: Technical Development

Stanley Segall

CONTENTS

Translating concepts into prototype products that can be viewed, sniffed, tasted, and savored is a choreographed series of events. Beginning with the concept and its definition and proceeding into a product innovation charter and a product protocol which ultimately describes the physicochemical entity to be generated, food product development requires intimate blending of research findings, science, technology, imagination, experience, skill, and not a little bit of art. Development is just that, a process of initiation and advance, error, iteration, adaptation and reiteration, directed towards an elusive goal of a nearly perfect manifestation of the product concept.

THE NEW PRODUCT

What exactly is a "new" food product? New products can come in many forms depending on just who is classifying them. For example, although referring to only the first three types as "truly new," Hoban (1998) identifies what he refers to as the following six types of new products, calling them:

- "Classically Innovative" (e.g., the first frozen juice concentrate, squeezable yogurt);
- "Equity Transfer Products" (e.g., a restaurant deciding to market its distinctive salad dressing through supermarket outlets using its well-known restaurant name);
- "Line Extensions" (e.g., deciding to add a green color to an existing beer to exploit specialty marketing);
- "Clones" (imitations, "copies") (e.g., producing and marketing your own, but nonunique, version of a lemon and lime carbonated beverage to compete with a well-known competitor's product);
- "Temporary" (e.g., special bunny-shaped chocolates for Easter and sold only at that time); and
- "Conversion" (e.g., replacing one size container in your line with another size of the same product).

There could be many other categories by which this list may be expanded. For example, to these there may be added a seventh category designated adaptations such as altering a product concept to "improve" its uniformly sized and shaped formulated potato chips that can nest in each other and thereby stack, or an eighth, "Private Labels" (e.g., having another company or contractor process package their item in your packaging but marketing it through your own distribution system—you retain brand ownership).

Regardless of how many categories or how they might be classified, in almost every case a new food product means a product not previously marketed or produced by the organization for which it is developed or made available. Some products are really news that is, classically innovative in the most obvious meaning of the word, some are modifications of existing products, others are imitations or copies of competing or already existing products, and still others may be minor modifications in shape, size, color, or packaging. In some cases, this can even include food products already available in the marketplace and obtained fully market-ready from sources outside one company and introduced without modification or with only simple packaging or other minor modification, into another company's marketing and distribution network.

Types of New Products

Novel: Intellectual Property

Patents

The most obvious "new product" would be one that is novel, unique, and distinctly untried, unfamiliar, or even previously nonexistent such as microwavable popcorn or coffee-enhanced carbonated beverages. Often, but not always, products that meet this description fall into the category of invention, especially if they are innovations in technology, and as intellectual property may be eligible for protection by patent. A patent more effectively locks out competitors and may be looked on as a contract between the inventor or patent assignee and the government (representing the public) in which the inventor is guaranteed a limited term of exclusive use in exchange for providing a full written description and disclosure of the invention. In the United States this exclusivity period is normally 17 years from the time of granting of the patent. For certain items subject to lengthy delays by the U.S. FDA before obtaining permission to market, such as food additives, color additives, (or drugs) and so forth, additional time can be granted for the period of patent protection.

What can be patented? Exactly what can be patented is defined in the U.S. patent laws found in Title 35 of the United States Code.

The language of the patent law is subject to wide interpretation and is perhaps best stated by Burton Amernick in describing the scope of patentability, to "include anything under the sun that is made by man" (*Diamond v. Chakrabarty*, 1980) a conclusion engendered by a Supreme Court case involving the patenting of a living organism, a genetically engineered bacterium developed for breaking down crude oil and thus useful in dealing with oil spills (Amernick, 1991). Despite this seeming universality, there are limits to patentability. Items such as natural laws, printed materials (except for copyright, trademark or trade dress), naturally present materials, natural phenomena, and abstract ideas, and so forth, are not usually patentable.

On the other hand, foods that might include elements such as functional compounds, material compositions, manufacturing methods, methods of altering or modifying natural product characteristics, and so forth, incidentally or collectively could be considered potentially patentable. Protection of intellectual property by patenting is impacted by numerous variables, for example, publication before application for patenting, different provisions of foreign governments, and so forth.

At the outset of this chapter on new food product technical development, it is important to recognize that, oddly enough, it is not always obvious whether or not the result of a new product development project, or any step or part of the project, represents a protectable, unique, and novel food product. Later review of the final and complete technical report on the project by the company's legal staff is crucial to determining if there are protectable element to the product, process or other aspects of intellectual property.

For this reason, as well as simply being sure to be following one of the basic tenets of good laboratory practices, all work and discussions regarding any part of a development project must be faithfully and regularly (e.g., daily) documented, in writing, using a fixed page laboratory-style notebook or computer input. Scratch pads, scrap paper, backs of envelopes, laboratory paper towels, and so forth, make poor recording documents, but if they exist they should be saved. A loose-leaf notebook or a computer recording document might be useful as a working process practice but, unless backed up with a hard document or mechanism not easily subject to alteration, might later prove problematic since they are subject to after-the-fact alteration. The written document must be carefully dated and witnessed by a knowledgeable colleague.

For a more exhaustive treatment of patentability the reader should consult more specific literature on patent law, for example Amernick's *Patent Law for the Nonlawyer* (1986), or a patent attorney, or attorney dealing with protection of intellectual property.

Trade secrets

Sometimes there are decisions within a company to forgo the protection afforded by patentability to avoid the concurrent necessity of public disclosure. In these cases it is still possible to retain the advantage of exclusivity, perhaps for even longer than that, which might be gained from patenting. Companies can attempt to retain commercial advantage by use of trade secret protection. There is a great deal of variability in the legal protection of trade secrets and is dependent on prior case law and the state involved. Nearly anything, a formula, a process, a piece of equipment, a set of data, or anything else, which allows one company to gain and maintain an advantage over competitors who do not have knowledge of the particular item or practice, can be viewed as a trade secret.

On the other hand, to be protected as a trade secret, the information must be so secret that it would not be easy to obtain by usual, proper, or legal means without consent of the owning party.

Information relating to trade secrets is protected against use of illegal means of disclosure. For example, bribing an employee of a company in a competing company to reveal trade secrets, or deliberately hiring away an employee from a competitor with the intention of obtaining trade secret from that employee, or, even without prior intent, utilizing information on a trade secret in the possession of a current employee who had obtained the information as a result of previous employment by a competing company, all these would be illegal in most jurisdictions.

To be fully protected, trade secret information should be really new but need not meet the extent of requirements for a patent. Very often, manufacturing methods and practices are best protected by the trade secret method. Of course, utilizing the trade secret method exposes a company to the risk that their exclusivity could be ended by legitimate independent discovery of the secret by a diligent competitor.

Trademarks

One final item in this list of "really new" products may not actually be a product, though it often is associated with one, and that is the provision of a trademark or service mark. For a genuinely new product, even if it is patented, it is wise to protect the name or appearance of the product by use of a specific word, symbol, or combination of the two to identify the product in such a way as to distinguish it from that of the competition, particularly if the products are similar. For example, Aspartame (Nutrasweet) bears a distinctive trademark and thus distinguishes itself from similar nonnutritive sweeteners such as saccharin (Sweet 'n Low). These marks can be registered in the U.S. Patent and Trademark Office and once registered this status can be so indicated on the package.

A tradename is not the same as a trademark or service mark and is not ordinarily registerable unless it is used as a tradename while at the same time registered as a trademark. A tradename is usually used only to identify a company, unless also registered as a trademark or service mark.

Competitive Matching or Cloning

One of the most common types of "new" product development projects involves the need to have a product, which directly meets a competitor's challenge. If your competitor is marketing a specialty pizza that has an unusual character, for example a spicy chicken topping or a strong basil flavoring, you may be called upon to provide your company with exactly the same product (even if the name and packaging is different). This would be a direct duplication. In many instances all that is required is a matching product.

In other cases, this may really mean that what is required is essentially a duplicate product though very often it may be simply something very much similar rather than an exact duplication. Slight but noticeable differences may exist simply due to the difficulty involved in exact duplication. Differences may be deliberately built in. In either event, this difference could be advantageous. Thus your pizza topping may indeed be a spicy chicken topping but you could emphasize one spice more or less than the competitor to provide your product with a signature flavor.

In many cases there is actually an advantage to having a slight, but distinct, difference. For example, if your company decides that a competitor's chocolate ice cream is pulling down your sales of chocolate ice cream, it may be wise to not just duplicate your competitor's chocolate flavor, but to go one better by providing a signature difference, perhaps a tangy note from the incorporation of just a touch of another flavor to the chocolate, perhaps a hint of orange, brandy, or coffee or other distinctive, or signature, flavor note.

In some cases, a company may be marketing a limited line of items, which logically requires expansion to a series of items. For example, there have been carbonated beverage manufacturers that traditionally marketed only one or two product items (e.g., ginger ale and tonic water or just cola, etc.) for which they were well known and which sold quite well but in limited volume compared to the overall demand for carbonated beverages of other flavors. Such companies discovered that they were permitting competitors to gain a foothold with consumers who were seeking additional flavors in other venues.

Sometimes, just to protect your company's market position for its flagship products, and to compete successfully with competitors, which might otherwise command a larger share of shelf space in the retailer's outlets, it may be necessary for a food manufacturer to develop an entire line of products even if in the aggregate they may or may not be highly profitable in and of themselves. Of course, your company may discover that in producing, packaging and marketing the more extensive line of products required that not only does this technique make it more difficult for a competitor to command shelf space for his product, but the added sales of your new beverage products may also improve the revenue stream of your company. Further, economics of scale begin to come into play as improvements in utilization efficiency of the marketing, manufacturing, and distribution systems add significantly to the bottom line.

The "new and improved" product may only be a new container size or shape, or a new flavor addition to change the aroma of the product, or even the same product in a container, which has improved characteristics or looks similar to a popular competing product, that is, it has been adapted but it is still a "new" product. It has to be developed, tested, and added to the existing product line. By definition and in practice, if it has not already existed in the company product line, it is a new product.

In other cases, there may be need for considerable technical study. For example, when it is decided that your food product is now going to incorporate wheat bran because someone states that there is evidence that this item reduces incidence of a colon cancer in rats, or oat bran because another researcher states that this reduces circulating cholesterol levels, or ascorbic acid because it prevents product discoloration or supplies a needed nutrient, much more has to be done with product formulation than simply incorporating the new ingredient. Stability, compatibility, mouth feel, sensory acceptance, all these things and more must be extensively tested, and legal restrictions regarding health claims in a food product as well as implications for label requirements must be looked into in order to comply with various regulations.

Reinventing the Wheel

Besides the truly novel new product, the simply duplicative new product, or the adaptive new food product types mentioned thus far, a high probability exists that the new food product your company tells you is now required may in fact already exist. *One of the less obvious but truly critical responsibilities of new food product development professionals is to know what is already available in the marketplace, which might be readily adapted to fill the needs of your company.*

Regardless of whether or not your company would reject or consider such an approach, you must be aware of what already exists in the market and be sure it is taken into account during any discussions regarding the nature of a new food product. It is often wise to take advantage of prior developments and existing manufacturing capabilities of another company to provide your organization with a ready-made "new" product. It is always recommended to look at the existing market place to determine if a supplier could quickly and economically provide your company with a product rather than incurring all of the risks, costs, and delays involved in beginning the development from scratch. If you can locate an already existing source of a product, what is often called a "private label" or "contract" manufacturer, the lead time from concept to availability can be reduced from a sometimes substantial delay of months or longer, to just weeks. In addition, this can be a way to really test the market for the new item without the need for risking possible additional capital investment for development and manufacture. A major caveat is the danger of loss of product quality control or dilution of brand identity, but there are many ways for this to be overcome.

It is possible to request only the most minor alterations in color or size or shape or packaging to provide for integrating the "new" food product into your existing line.

Variations on this approach have included purchasing and supplying the packaging independently; using a distinctive color in your company's food product; or developing an exclusive die or other piece of manufacturing equipment for use only with your product to provide exclusive brand identity; or simply buying the potential supplier (sometimes called a merger) or working out a supply exclusivity agreement with the manufacturer; all commonplace practices.

In the domestic appliance industry, it has not been unusual for an equipment manufacturer to market an item under its own brand name while making the identical or near identical item under private label for a different and even competing marketing company or for another appliance marketer or as a "store brand" item. This practice is obviously well known in the food product field where supermarkets both manufacture their own products and contract with brand owners to provide a "store brand." Of course, under these circumstances it is still necessary for an entire set of technical control procedures to be imposed. While that is a subject unto itself, it often falls to the food product development group to determine the parameters and specifications for the product and to develop, test, and implement the appropriate technical quality control procedures just as it would if the entire product were developed in-house in their own laboratory. In fact, this latter aspect is almost more important for a purchased new product than it would be if the entire product were manufactured and controlled in-house.

The lesson here requires that before launching elaborate development plans, do not "reinvent the wheel" until you are sure that company management has been made aware that a desired new product already exists and only then consciously makes a decision and gives the order to proceed with a new in-house development.

In summary then, new food products take many forms, from the truly unique and novel to the mundane and blatantly imitative. They can even be some other company's product adapted into your company in the thinnest of disguises. The important concept is that they represent an addition to your company's product line, which is usually designed to improve or protect the competitive position of your company and/or to improve the overall bottom line.

THE CONCEPT

Where does the idea for a new food product originate? As indicated in Chapters 5 and 6, companies themselves consider very important sources of new ideas to be marketing research, technical research and development, customers, and other consumer product companies. Sources like university research, scientific journals, suppliers, consultants, and trade events were rated as less important but still of value (Hoban, 1998).

On the other hand, even some of the more unlikely or lesser-rated idea areas can still be quite important. For example, the significance of nutrients as new product ingredients has been well documented.

Where do the ideas come from? A simple answer would be to assume that new product ideas can spring from anywhere. While you might be tempted to think this a logical answer, experience shows that, with rare exception, a successful process is ordinarily much less fortuitous and much more deliberate. As the cliché' asserts, new products are conceived from 0.5% inspiration and 99.5% determination and perspiration.

There must be a corporate climate that encourages the mind-set needed to see the possibilities that might lead to a new product suggestion. The company has to provide clear incentives, which encourage and reward employees to look critically and carefully at current products, at changing market conditions, at competitor's products, at customer needs, at production techniques, at new ingredient materials, at new technologies, at changing demographics, at legal, safety, and environmental factors, in fact at anything going on around them, and relate all this to what might then result in an improved or new product.

Ideas can indeed come from just about anyone (see Chapters 5 and 6). They might originate with your customers, employees professional colleagues, management, consultants, trade or professional meetings, technical literature, often the popular literature, and surprisingly enough, even competitors, or perhaps we should be saying, especially competitors. The impact of competitive activity is often the principal driving force behind those "hurry-up, we need the item yesterday" cries familiar to all technical and product development personnel. Remember, every one of your competitors is also charged with new food product development and therefore the industry collectively represents a huge expansion of effort, ideas, and incentive.

Competition is very often one of the strongest forces impacting on new food product development activity. Remember, the competition is also your most severe and marketing and technically competent critic. They will find any flaws in your products and may point them out to you by their actions even before you see them.

Monitor their actions and activity; it may be crucial to maintaining your own new or improved product activity. Every product marketed by a competitor has to be examined and evaluated to determine what, if any, response will be made by your company.

While competition is an important source of pressure for new product activity, care must be exercised to avoid unthinking reaction to every competitive move. Ill thought-out or panic reaction to competitive actions or pressures can detract from important and perhaps more significant development already in process in your own organization. While response to competitive action must be maintained, care must be taken not to be so negatively reactive that your own projects suffer or die. All R&D programs operate with a limited set of resources and they must be used rationally, wisely, and systematically. *Panic or knee-jerk reactions must be avoided.* Decisions about new product development must be made with a clear idea of whether they are likely to fit with the company's objectives and its corporate culture.

Ideas by themselves, no matter what their source, are not always going to get you well thought-out possibilities for new products. There has to be a mechanism for gathering in suggestions and evaluating them on an organized basis. This can be as simple as the old-fashioned "suggestion box" or as institutionalized as a standing committee, which seeks out ideas, reviews and evaluates them in a systematic fashion and, reports regularly to an appropriate decision maker, or just about anything in between. Even concepts, the translation of ideas into tangible words or prototypes, are not necessarily valid until thoroughly evaluated and developed into working prototypes.

Research and development and marketing groups must maintain constant surveillance on the activity and signals coming out of the market.

It is the responsibility of the R&D and/or product development teams to translate good ideas and company-approved ideas, into marketable products; however, R&D cannot operate as an independent agent in the development of new products. A well-run company has a series of decision points built into its organizational structure. Input must be sought from all the players in a company at every step along the way in the development of a new food product. The role of the R&D product development team is to build what marketing and management decides is needed and not necessarily or always what may appear to them to be interesting in the short run. The R&D product development team must be careful to remember that they are a support organization and not the reason the company exists.

EVALUATION, INITIATION, AND REVIEW

Regardless of the sources of new food product ideas and concepts, a formal mechanism for their systematic evaluation must be present. In small organizations, particularly those headed by a strong founding or entrepreneurial figure, it is often simply the decision of one person. This may be very successful and is certainly very time efficient, but this method tends to lead to highly subjective decisions concerning which ideas to pursue or reject and is prone to a high risk of failure. A more rational approach involves review of needs and ideas by more than one person, preferably by a standing committee, which meets on a regularly scheduled basis but also can be called on immediately if an emergency, for example, competitive activity, should arise.

Ideally, the composition of the group should be relatively small and include representation from the key areas of the company, in particular, marketing, sales, R&D, production, packaging, and so forth. In some cases, especially where interim commitment of supplementary development funds is necessary, representation from finance is included. The committee/group can meet as often as weekly or as infrequently as monthly or quarterly to go over suggestions from various sources and to review the company's perceived needs in the new food products area. The committee can also be called on at any time to deal with changes in the market, for example, the impact of a competitor's new or improved product; or a new food such as nutraceuticals or low carbohydrate or no trans fat, or the sudden action of a regulatory agency on a product or component of a product, or the unanticipated negative performance of an existing product.

Even unexpected changes in the availability or pricing of a product component might require a rapid new product response. The authority of the committee/group can vary from simply being advisory to an appropriate decision-making operating officer of the company, such as the marketing head or the Chief Operating Officer. If the group is headed by a decision maker, he/she can authorize the initiation of the new product process itself, usually assignment of responsibility, initial assessment of technical feasibility and preliminary estimation of a development budget. In some cases, this same committee hears progress reports from R&D/product development concerning ongoing development projects.

Ideas for product improvements related to the manufacturing process often best come from the employees actually engaged in day-to-day manufacture. Many times these suggestions, no matter how important, are merely production improvements but sometimes they can result in truly new products. An unusual example of this technique, some years ago, was the suggestion by technicians originally working with the lyophilization of arterial sections for medical materials that, using the same equipment for lyophilization of lunch meats and fruit pieces, produced a rather tasty and well-preserved and easily rehydratable dried food item. The name of the process was changed to freeze-drying, and of course, "the rest is history."

In a more conventional example, production line personnel in a breakfast cereal manufacturing plant producing large single serving-sized briquettes found during meal breaks that these made a nice snack item when broken into bite-sized pieces. This eventually led to a suggestion resulting in simply manufacturing the original briquette as a party snack of small size to begin with, and this eventually opened an entire new market to what had previously been only a breakfast food producer. In both these instances, it took someone to observe and listen to the workers, someone who could see the value of their suggestion and also "sell" the idea as one worth pursuing as a new food product.

THE POLITICS OF THE NEW FOOD PRODUCT DEVELOPMENT DECISION

New food product ideas can and often do originate with the technical personnel in a company, but this is not always the best source. This is because, with this as a source, technical feasibility and production compatibility sometimes overshadow financial and marketing considerations. It is perfectly reasonable for ideas to come from the laboratory, or from production, but unless the marketing department and consumers have reviewed the suggestion, even if technical development is eventually highly successful, the result may be a wonderful product, which will not or cannot be willingly or profitably marketed by the company.

In a well-organized company, it is generally the marketing department that plays the key role of characterizing and describing ideas that eventually translate into practical descriptions of desired new products, and this role is logical. Since they are the representatives of the consumers and have to market and perhaps sell the product, it is the marketing department that should have the first and final words on whether to initiate or continue development of a new food product.

It is not at all uncommon for budgeting to be set up in such a way that the marketing department operates as a profit center and actually provides all or part of the food product development funds and recovers its expenditure when the new product becomes an income-producing part of the company line. In these cases, the performance of the marketing head is directly impacted by the fate of new products, and so these administrators have a strong voice in the critical decisions concerning initiation of a new product development project and without their input, nothing happens.

In almost every situation, it is sales and marketing that is on the firing line where buyers (customers, potential customers, and consumers) show why they use certain products, describe modifications they would find desirable, point out undesirable features of company and competitive products, and express their opinions concerning what they would like to have as an ideal product.

It is an unwise and foolhardy R&D or product development manager or director who today attempts to build a new product without first getting the marketing people on board. Production can often come up with great ideas for product and process improvement, but R&D by its very nature is constantly critical of its own products and is, and should, always be looking for ways to "improve" products. While R&D can present really clever and innovative variations in products, it is marketing that has to market the product, and it is marketing that has its fingers on the pulse of the marketplace—it is the surrogate for the consumer.

It is wise to bear in mind that the best and most unique products in the world are of no value if there is a lack of enthusiasm and support from marketing. Companies are in business to generate sales volume and profits, not to simply make and market products. A wise and effective R&D/product development leader, particularly one specifically responsible for building new food products for the company, will keep this thought ever in mind and develop clear, open, strong lines of communication with marketing and never commit to major budget expenditures for new or modified food product development without consulting closely with their company's marketing arm.

Others too have a role in the decision making and so there are additional important company departments that must be considered. Engineering and production will eventually have to produce the item, and it is therefore critical to keep them informed as to progress and to solicit their input in order to smooth the way for an untroubled scale-up of the new product into the pilot or existing production facilities and schedules. Finance will certainly have a lot to say about whether a product can be profitable or whether one direction of development might be more financially advantageous than another. Purchasing must be involved since they will be responsible for seeing to the most cost-effective purchase system for component ingredients and materials. Legal and regulatory individuals must also be consulted and kept informed. In some cases, unless they are under the control of marketing, it may be necessary to call in the Public Relations and advertising departments for their inputs.

Different companies may have still other areas, which must be considered for input and information. Balancing all of the various constituencies and interests within the company, while keeping a careful eye on the competitive climate in the marketplace, calls for an R&D/product development manager/director who can orchestrate all of the disparate elements necessary to assure success for a new food product's development. The leader is not merely a technical whiz in a white coat or a marketing MBA with sleeves rolled-up, but is a politically savvy individual able to exercise all the skills of a trained diplomat while simultaneously working out the technical and marketing details involved in putting together all the components of a new product.

ELEMENTS OF PRODUCT DEVELOPMENT

Once a decision has been made to proceed with the development of a particular new food product, a series of factors must be taken into consideration within the several elements of the project. These factors include: the nature of the product; the types of expertise required; the composition of the development team; and the time-priority of this project in relation to other development projects already in progress. The elements themselves can be listed as follows.

THE NATURE OF THE NEW PRODUCT

Usually, one of the first limitations on any new food product is selection of an upper limit for cost because of the universal impact this has on every other aspect of a new product: cost of materials, cost of packaging, cost of new equipment, cost of distribution, and cost of marketing. The degree of difficulty likely to be encountered in the development of any particular new product as well as the length of time the project is likely to take will be heavily impacted by the extent to which the product is, or is not, a departure from items already in the company pipeline.

If the project involves modification of an existing food product, then the characteristics and behavior of the product are probably well known to the development team and correspondingly less time need be allocated for the learning curve than if the project involves something totally novel to the company or especially if it is truly unique to the industry. The same might be true if the team will be working with familiar ingredients and raw materials rather than materials new to the company. If it is the latter, are the new materials readily available and compatible with available storage and handling facilities? Will new control and purchasing parameters need to be set? Decisions must be made as to whether there is a need to keep as close to existing materials and production and distribution methods as possible or to what extent there can be some or great departure from familiar or compatible ground.

If the new food product is likely to be perishable but all the other company products are ambient temperature shelf-stable, this will require new distribution methods compared to existing products unless analogous items already exist in the product line. It is necessary to consider or specify parameters such as item size, portion size, multiple or single serving, nutrient content, shape, moisture, lack of need for or extent of end preparation, microwavability, reconstitutability, the need to meet special restrictions (and whose) such as religious, environmental soundness, "organic" or "natural" designation, "meatless," fat and/or sugar free, and so forth.

Must the product fit into currently available production, packaging, storage, transportation, marketing, and distribution systems or should new systems be considered? Must the new product fit into the existing company regulatory compliance systems or will new compliance requirements be tolerated? Will the new package label present special problems in terms of design and compatibility with the company's other items? Finally, there must be consultation with the appropriate legal and regulatory authorities in the company regarding consideration of the degree of intellectual property protection desired or inherent in the new product, or for that matter, the extent to which the nature of the new product is restricted or its development direction dictated by the need to avoid infringing on the intellectual property rights of others.

The Development System

In order to determine whether a given proposed new food product development project should be carried out, the various elements that might be applied to any project must be considered. The product development plan may be prepared and tracked on a computer program such as Microsoft Project. A systematic approach whether to carry out a specific new food product development project requires consideration of all or some of the following development steps:
Step 1. Evaluation

(a) As previously noted, ideas must be gathered, examined, evaluated and prescreened

(b) Those ideas deemed worth looking at further must be assessed for technical, safety, legal, and regulatory feasibility, the availability of appropriate technical facilities, and the likelihood of fitting in with the company mission (see Chapter 2)

(c) Ideas must be converted into concepts understood by target consumers for evaluation and modification

Step 2. Parameter Setting: (sometimes called "Product Protocol")

(a) A definition or description of what the end product must be, or exactly what will constitute successful completion of the project or product, must be agreed to

(b) Upper and lower quality and shelf-life tolerances for all product characteristics must be set

Step 3. Cross-Functional Technical Operations: consideration of one or more aspects of a series of technical functions (see Chapter 7):

(a) Ingredient selection

(b) Formulating system

(c) Manufacturing unit operation selection

(d) Initial cost analysis

(e) Physical and microbiological characteristics evaluation

(f) Quality evaluation procedures (including testing protocols)

(g) Safety evaluation (including initiation of Hazard Analysis Critical Control or HACCP development)

(h) Functionality testing

(i) Sensory testing

(j) Stability and shelf-life determination

(k) Packaging and labeling requirements

(l) Plot production

(m) Full legal and regulatory assessment

(n) Semiproduction run and cost analysis

(o) Consumer testing

(p) Preliminary data evaluation

(q) Full production run and cost analysis

(r) Final selection of control and HACCP procedures

(s) Test marketing

(t) Full data evaluation

Step 4. Budgeting

Every potential project must proceed through Evaluation (step 1) and should be permitted to go further *only* if the decision is positive. Not every project will require passing through all of the development steps noted above. The project can be rejected after consideration of step 1, or even step 2. There can even be a decision to reject based on consideration of any of the technical operations units (step 3) if it is determined that there is reason to expect that one of these operations cannot be successfully carried out or concluded satisfactorily. On the basis of a careful assessment of which these elements, particularly the technical operations (step 3), will be applied as part of the new food product development, an estimated budget figure and a timeline can be attached to the project. By applying this information, the qualifications and number of team personnel can be assigned and development costs further refined to complete the setting of the initial budget (step 4) for the project. At this point a rational decision can be made concerning final approval and initiation of a particular project.

BUDGET

The budget is one of the key variables required to decide whether or not to move from the evaluation step to further consideration of a development project. Besides its importance in determining the nature of the team required to conduct a development project, budget is also a critical factor influencing both the time required to develop a new food product and the extent to which the new food product is premarket tested. For all these reasons, great care must be exercised in setting these figures. It may be true that a high budget can lead to more complete development of the new food product and fewer challenges in the initial market introduction, but these advantages come at the cost of time delay in getting the product to market.

A delicate compromise must be reached between the desire of technical personnel to come up with a nearly perfect product and the anxiety of the marketing staff to get the product to the consumer as quickly as possible. Obviously, a proper balance must be struck once the decision has been made that the initial idea should be pursued further. Clearly, "right sizing" in determining the budget for a new food product development project is the key to a successful management decision.

As important as this decision is, budgets developed before actual initiation should not be viewed as carved in stone. Prior to final review and formal approval of a technical development plan for the project, the budget should be considered as preliminary. A final budget requires that a detailed plan of technical development operation have been drawn up by the R&D/product development management and subjected to final review.

The Development Team

Development teams can vary in size from a single technical person to a series of very elaborate specialty groups, or anything in between (see Chapter 7). Regardless of the number of functions to be performed, the development team must operate as a cohesive unit and not simply a series of disconnected technical groups. Once the decision has been made as to just what technical units need to be applied, the development team will consist of all of the members from each discipline. It is critical that each member of the team understand and participate in the gradual evolution of the project from market need to product launch.

The project must not be treated as a series of disconnected operations, with responsibility for each group ceasing when it perceives its particular portion of the task to be completed. What Takeuchi and Nonaka (1997) called "the old sequential, 'relay-like' approach in which the product is handed from one functional department to another will not meet the demands of today." They believe that product development is a "holistic" process, one in which, while the project is handled by each individual in the team, it is always treated as a team effort, a system they termed the "cross-functional team concept." Cross-functional teams favor cross-pollination of ideas during the project. Team composition is not necessarily fixed for the entire term of the project but can change as more or less expertise is needed for various tasks.

In some cases, all the elements of the development team exist in-house while, in other situations, the use of outside specialty groups may be quite appropriate and cost-effective. A flavor house might wish to develop a special flavor mixture for exclusive use by a potential brand owner, which lacks its own flavor compounding expertise. If the form of the flavor product is simply a mixture of essential oils and various aldehydes, ketones, esters, and alcohols, it may well assign the project to a single flavor compounder. In this case the expertise needed resides entirely in one person who, working alone, at least initially, compounds and formulates one or more products for evaluation by the food product development team. This is a team of one. On the other more complex and also requires formulation in an emulsified water dispersible base and perhaps homogenization and/or perhaps further formulation into flavored and food-colored syrups for soft drink use, then more than one type of expertise might be required. In this latter case, there may be need for specialists in emulsification technology and color stability to be added to the team. In either case, there may or may not be need for sensory evaluation, for shelf-life study, for microbial safety, and so forth. If the expertise exists in house then they would be added to the team. If they are not available in house but are deemed necessary, then they can be added on a consulting basis or by turning to outside specialty or testing organizations.

In recent years, depending on the nature of the substances used in compounding and formulating, it has been found there could also be a need to review legal, regulatory, and label status. Any or all of these could be done in house or partially by use of consulting specialists depending on how the company has organized management of governmental and legal affairs.

If there is a need to determine consumer acceptance to a degree beyond the usual relatively small in-house sensory evaluation capability of many companies, then the rather extensive sensory team would have to be part of the development protocol.

Together with concerns imposed by the nature of the product, any technical considerations that must be taken into account, and the availability of the particular expertise required, the size and make-up of the food product development team is also impacted by costs. Part of the decision-making process in considering the set up of a particular new food product project must necessarily involve the size of budget assigned the project. Budgeting figures are justified by the nature of the product, the number of variables requiring evaluation, the difficulty of its technical operation and the priority level given to the project plus an assessment of the potential financial return likely to result from a successful project. Obviously, the higher the budget, the greater the number of elements that can be assigned to the development team.

In some cases, the decision to utilize a smaller or more limited team may be made for various reasons (cost, availability of personnel or facilities, etc.). This can still lead to a successful result in developing the new product, it may just take more time or some elements of evaluation and testing may not be extensively carried out. For example, in smaller food companies it is not unusual that a choice may be made to conduct sensory testing in house rather than undertaking the more time-consuming and more expensive full-scale consumer testing, or there may be a decision to bypass some of the pilot plant runs. Often, the nature of the innovation in an existing product to develop a new product is so minor that a large number of the technical operations steps can be bypassed or truncated. Simply changing the shape or color of an existing product may obviate the necessity for an extensive development team or elaborate project protocol.

Outside Help

Outside expertise can be obtained from a number of sources. A good pool of highly qualified "no cost" assistance can endurably be found from ingredient and equipment suppliers. The limitations of this approach include a lack of confidentiality; a tendency for data favorable to the supplier's products to be emphasized; the presentation of only partial data; lack of access to data not necessarily favorable to the supplier's products but important for disclosing information on functional behavior, even if negative; a lack of experimental control; and an implied obligation to the supplier.

Despite these limitations, there are times when this supplier sourcing approach can be used for preliminary or exploratory data, for a limited or very narrow item of information, and for exploration of noncritical or nonproprietary parameters.

Another source of outside technical assistance would be consideration of government agencies like the National Center for Food Safety and Technology, a research facility of the U.S. FDA, which will partner with industry as part of a research consortium at the Illinois Institute of Technology's Center for Food Safety and Technology and can provide expertise and technical facilities in areas such as food safety, quality assurance, biotechnology, food processing, and packaging technologies. The Eastern Regional Research Center of U.S. Department of Agriculture in Philadelphia can enter into R&D agreements with companies to utilize its expertise and facilities for commercialization of technologies. The National Center for Agricultural Utilization Research in Peoria, Illinois specializes in developing new uses for agricultural commodities to make new added value products. Intellectual properties can be protected by patents and patent licensing and cooperative research agreements provide for commercialization (Giese, 1997).

Other sources of technical assistance can be found in universities. Many universities that house Departments of Food Science and Technology also have established technical centers, research institutes or new food product development groups, which form partnership arrangements with industry and offer a variety of technical facilities and technical expertise.

An example is the Food Product Innovation and Commercialization (FoodPIC) group of the University of Georgia, housed in Griffin, Georgia. Commercial research and development consulting firms are an excellent source of high quality assistance at all levels, including technical, analytical, sensory, marketing, legal, and regulatory. While there may be some level of insecurity regarding confidentiality, in general, legal and contractual considerations do assure a high degree of proprietary information protection. In many instances, particularly with the private consulting firms, you have a choice of selecting as much or as little technical assistance as deemed necessary. In some cases, it may be useful to consider "farming out" the entire food product development project. Going this far is justified only if you do not have the in-house expertise or facility to carry the project or your in-house food product development capacity has been exceeded. The major drawback to the external consultant approach is usually expense, but such an approach could still be cost effective if the alternative were no product and the anticipated direct or indirect revenue from adding the new product to your company's line would quickly permit cost recovery on a timely basis and subsequently generate profits. Often, because of outsourcing efficiencies, the net cost of employing external consultants can be much less than in-house development when all costs are computed.

THE TECHNICAL DEVELOPMENT PLAN

After subjecting the new food product concept to careful scrutiny within the parameters laid down by the Development System approach, a formal plan for the initiation and implementation of the technical development must be prepared. The plan should be a document, which lays out the rationale justifying the development of the product, any background history or significant factors bearing on the "go-ahead decision", any special aspects noted by various elements of the evaluation group, a clear statement of the benefit the product will bring to the company, and commits company management to the success of the project.

This plan will select the most appropriate technical steps likely to result in successful achievement of the development objective and include a detailed budget, account for resources needed, suggest tasks for the specific teams required, lay out significant review points or milestones in the project, and present a tentative timetable for intermediate and final goals. The technical development plan should be reviewed by appropriate technical, production, marketing, and financial groups for final review and initiating decision. While it is expected that operation of the project will be guided by this plan, there must remain sufficient flexibility and openness of mind to alter any step or timetable as the project develops if the intermediate findings indicate that changes are required. This is not to say that the plan has little or no importance and can be easily changed, particularly when other areas of the company must be able to reliably plan for key steps like pilot runs for sample production and setting up any required consumer testing, and so forth.

The plan should be a document drawn up prior to actual experience in the development and operating world, and thus is simply a "best estimate" of the most likely path to success. If that real world experience dictates a need for accommodation and change, then, the cost in terms of time and money of any proposed change must also be taken into account. A plan is at best an intellectual exercise, and while management hopes that their technical judgment is up to the task of providing an accurate and correct predictor of success, it is still only a plan. It is the goal that is likely to remain fixed.

THE DEVELOPMENT PROCESS

INFORM

New food product development projects seldom come to the product development teams as a total surprise. R&D/product development management often apprises their personnel of items under consideration, in part as a technique for giving everyone a sense that they are part of the overall development team and in part as a mechanism providing informal feedback to the team for use in the evaluation process. This practice is often the best way to avoid speculation and rumor. Even where there are no regular information sessions, the "grapevine" often passes information along. For these or other reasons, there is usually no great surprise when management officially informs technical/product development personnel about the initiation of a new food product development project. The team should be identified and then presented with a full briefing from the R&D/product development management on the nature of the new project, its significance for the company as a whole and for them in particular, the reasons for its selection, the nature of any known or potential competition, the feasibility of the project, any particular technical challenges and particular attributes, and initial projections for major target dates (start date, initial "bench" prototype, laboratory or bench production, pilot production, pilot testing, shelf-life studies, semi-production run, safety studies, quality control procedures, consumer evaluations, initial production run, and so forth, and the desired final turn-over or completion date). The technical team should examine the technical development plan, be encouraged to offer their comments, and otherwise be encouraged to buy into the project.

Initiation

The first phase of initiation usually involves a discussion session at which all team members are expected to discuss, criticize, suggest and generally "brain storm" the project to identify where to begin, even if it is obvious, and agree to or modify the technical steps needed, and review the target dates for their feasibility. Hopefully, they should come out of this initial session in substantial agreement so that they "buy-in" to the project and the development process.

The second phase of initiation can vary depending on the nature of the project. Often it involves the assembling of ingredients, hardware, and equipment. Sometimes it may require information gathering before this can even begin or this aspect can be carried out simultaneously. Information gathering can be as simple as reading the specification sheet developed by the evaluation group or it can be as complicated as a comprehensive literature search, require identification of sources of information, internet search, trips to a library or a nearby technical center or university, or wherever and to whomever the information can be located, consultation with supplier, review of competitive products, and so forth.

Based on what can be determined from the various information sources and deliberations by the team members, additional information requirements should be identified, any additional equipment or ingredients shall be ordered or obtained, determination should be made concerning the possible need for any modifications in the original technical development plan, need for any analytical or other preliminary work (modification of equipment or ingredients etc.) should be identified, intermediate goals should be set, and, finally, initial specific action assignments made.

ITERATION

Trial runs are made starting with the initial formulation (see Chapter 12). Rapid evaluation is made on the resulting trial results using appropriate testing methods. Modifications are made in the formulation or the ingredients or the process and the process continued through as many iterations as needed to reach a point where the resulting bench product is deemed ready for its first general assessment. Based on these preliminary evaluations, the iterative modification process continues until the result is judged to be a reasonable prototype. Methods of judgment at this stage might be simply "in-lab," but there should be progression to at least small in-house formal sensory testing using a sensory panel evaluation methodology (see Chapter 13).

It is not unusual to employ focus groups or small consumer panels to obtain information suggesting useful changes or confirming the direction of the development. Progress assessment must include continuous marketing input so that agreement can be reached as to when the bench product has met the goal originally set for meeting the desired product definition. At this point preparations may proceed for moving to the pilot or contract production stage and beginnings made towards defining the technical label declaration and the packaging need characteristics.

PILOT OPERATIONS

Although some products move directly from the bench to the production plant, this is a risky approach,and moves to "short-cut" the development process and bring the product to market more quickly can prove costly in terms of lost material and technical time costs if (usually when) something goes wrong. More importantly, while any technical mistakes can still be corrected, a major risk is loss of confidence in the new food product by consumers, marketing, and production. Nearly undetectable minor deviations at the bench level can become major defects at the production level. Many potential production problems can be avoided by using an intermediate stage before moving to a large scale-up. This can occur in a pilot plant, a small scale facility equipped with small capacity commercial production equipment where production runs of the product can be made on a "simulated" production run basis.

In some cases, particularly where no separate pilot production facility is available, food product is manufactured in the actual production plant but on a reduced or limited volume basis, in semi-production run. The purpose of pilot or semi-production runs is to gain experience in larger than bench scale production of product with no or minimal interruption of the regular production of current products, and to do so on a modest scale. This practice permits the team to optimize cost, translate scientific or laboratory terminology into production terms, determine any special equipment or production technique needs, fully test any new or modified equipment involved, provide for further modifications in formulation designed to improve the production process, more fully test quality control procedures, and accumulate larger quantities of product for further testing and evaluation. A more accurate assessment of costs can also be worked out. Pilot plant operations may be performed in a contract processing/packaging company.

At this stage plans can be formulated for the quantities of product needed for shelf-life studies, microbiological studies, and consumer sensory and acceptance testing. The packaging requirements can be determined at this stage as well, although in most cases this should have been carried out as part earlier stages. The pilot stage is a good place, if it has not been done earlier, to bring production staff into the team on a full scale. Besides familiarizing them with the product, their feedback in terms of assessing the likely behavior of the product in the full scale production process, especially predicting where production potholes are likely to become problems, and offering suggestions for any product changes to address this area can be crucial to the eventual production of the final product on a full scale basis. R&D/product development must always remember that the production department will eventually take full responsibility for the product so the earlier they "buy in" to the product, the better.

PROJECT COMPLETION DETAILS

Having completed what R&D/product development frequently considers the "important" phase of the project, the formulation of the new product, there is a temptation on the part of food product development specialists to gloss over the remaining development processes, considering anything more merely "minor details." The usual anxiety on the part of marketing to scoop the competition frequently adds to these pressures. This is analogous to the busy chief of a surgical team, having performed what he/she considered the crucial part of an operation, dashing out of the operating theater while tossing a casual instruction back to the surgical assistants that the real work having been done, they can now finish up. Remember that old aphorism attributed to Lawrence Peter Berra, "it ain't over 'till it's over!" A project is not "over" until everyone designated as having responsibility for the project has had a chance to review the results and formally sign off. Even after everyone is satisfied with the item finally produced, there are still important details critical to the commercial success of the product that have to be attended to before the book can be completely closed on the development. For example, even though not strictly part of the "technical development" phase, R&D/product development personnel should expect to monitor initial or start-up production for a "shake-down" period. In addition, the following development considerations should be considered or satisfied, though not necessarily in the sequential order presented here. In point of fact, many of these and other prior development areas could be carried out concurrently. These issues are usually matters of budget level, personnel available, time pressures, and corporate culture.

PACKAGING SELECTION

The exact nature of the packaging to be used for a product is dependent on a large number of factors. This topic is covered extensively elsewhere in the text in its own sections (see Chapters 9 and 18) so that no great detail will be given in this chapter, but, packaging is a crucial element in any product and the brevity of its mention here should not be interpreted as a measure of its significance. It should be patently obvious that in the twenty-first century foods are not marketed and distributed in bulk at the consumer level. For the past century and a half or more, foods have been contained and distributed in some type of closed package. Suffice to say that the packaging system selected must meet the protection and technical needs of the product, the demands of the particular manufacturing process, the marketing objectives set for it, the intermediate and retail distribution system into which the product goes to market, and the handling, storage, and use conditions imposed by the final user, the target consumer. Conditions such as compatibility or noninteraction with the product, appropriate gas transfer control, chemical and physical protective qualities, shelf life, microbiological and chemical safety, environmental compatibility, the expectations of the market (competitive or customary practices), filling and handling equipment compatibility, fitting into some over-all corporate practice, cost, visual design, and so forth. All these factors, and more, impact on the selection of an appropriate packaging system.

Packaging requirements must *not* be an afterthought left to a separate unit within the company. The need for packaging expertise must be considered as part of the team requirement process from the very beginning of planning for new product development.

Shelf-Life and Consumer Testing (Including Failure Limit Testing)

Samples of finished product are usually produced in the pilot or semi-production facility, packaged in the container analogous to that intended for commercial use, and stored for preset times under selected environmental conditions. In any shelf-life study, it is necessary to establish the limits of acceptability and failure for a product, otherwise known as tolerance limits. Once this has been identified, suitable testing methodology can be selected or developed. It is necessary to know how long and under what environmental conditions (e.g., temperature, humidity, light, etc.) a product can be held an still be acceptable to the target consumers. It is also necessary to know under what conditions the product is likely to fail, sometimes called failure limit testing. The definition of just what is acceptable and what levels of acceptability are desired should have been part of the initial agreed-upon description (Development System, Step 2, Parameter Setting) of what constituted an acceptable product. This information is necessary to be able to set parameters of market life (when to "pull" the product), and to determine how well your product performs compared to those of competitors. It is necessary to know whether problems of significant microbiological or chemical safety importance occur under expected marketing, distribution ,and consumer storage conditions (see Chapter 16).

Ordinarily, sensory (color, taste, odor, mouth feel) characteristics determine acceptability; however, in some products additional or other factors may be the determining attributes.

For nutraceuticals or for foods with required or stated constituent content, for example the nutrient value of an infant formula or a product claiming specific quantitative content of n-3-fatty acids, acceptability may also be determined by ability to meet stated label contents or regulatory requirements.

If crispness is the required characteristic, as in potato chips, then judgment could be either or both analytical and sensory, or both.

Where quantitative content of a specific ingredient or particular physical characteristics are the determining factors of acceptability or quality, judgment of acceptability can be made by analytical laboratory methods. In cases where acceptability or quality is determined by organoleptic or sensory qualities, human sensory perception methods must be employed (see Chapter 13). These can vary from small-scale in-house panels with suitably trained and qualified taste experts, usually employing a limited number of judges (10–50), to large scale out-of-house consumer panels made up of large numbers (100–1000 or more) of untrained consumers. Consumer panels work best when only limited information is requested (yes or no, acceptable or unacceptable, etc.) and are time consuming and expensive.

When properly conducted, consumer tests yield very valuable information.

Very often, the consumer panel is used to determine initial acceptance or rejection of new products while in-house panels mainly function for quality control after products reach the production stage. Of course, highly trained in-house panels are also used to assist in food product development, but they are used primarily in an analytical manner. to get detailed information on specific or detailed aspects of the product. They do not function well to predict over-all consumer acceptability. However, when trained to detect sensory characteristics previously determined to be associated with consumer acceptability or technical limits selected to judge product quality, expert panels can be very effectively used to inexpensively evaluate product shelf life.

Sensory analysis and evaluation is a valuable and complex area with unique technical technique. These topics are covered in Chapter 13.

Regulatory Compliance, Quality Control, and Label Requirements

All materials, ingredients, and procedures that become part of or come into contact with the product or its ingredients during manufacture must be continuously monitored to assure that all meet the test of being permitted for use in or on foods by the appropriate regulatory body, for example the U.S. FDA, or U.S. Department of Agriculture (see Chapter 19). If the product is designed to meet FDA Standard of Identity for the particular product, the exact nature of permitted ingredients is very narrowly proscribed. The vast majority of processed food products require detailed label declaration of all ingredients. If the product is designed for marketing to particular consumer groups expecting that it meet special standards (e.g., Kosher, vegetarian, organic, natural, etc.), this becomes part of product requirements. This extends to all ingredients, to the contact surfaces of processing equipment, to all package materials and includes all water. R&D or equivalent technical group must set up and monitor all production procedures to assure adherence to Good Manufacturing Practice (GMP) standards and HACCP, including the training of production personnel and maintenance of accurate production records. In addition, appropriate procedures required for compliance with environmental regulations affected by the production of the new product must be developed and monitored. In an increasing number of cases, adherence to water use and limitation regulations must be monitored. This includes proper handling and disposal of solid and semi-solid waste materials and processing water. It may even include fuel, power, and utility requirements.

It is incumbent on a competent technical group to develop appropriate procedures for analytical, microbiological, safety, environmental contaminant, and sensory monitoring of product quality. Producing a physiologically safe food product is both good business sense and an absolute regulatory requirement. The means that analytical procedures must be developed concurrently with food product formulation. Providing and utilizing such procedures in a systematic manner is part of a GMPs requirement. As such, quality control can be viewed as fulfilling a regulatory requirement, maintaining cost control, and assuring a continuing level of acceptance in the market. An R&D/product development new product development project cannot be considered complete until these procedures have been appropriately developed, thoroughly tested, and are in place at the time production is authorized.

Quality control or quality assurance is usually considered part of the production monitoring process, but it must be considered by R&D/product development during the development of new food products. Since no product can be any better than the components from which it is prepared, all of the requirements and procedures noted for the finished food product must also apply to ingredients. Specifications for ingredients must be developed concurrent with formulation development. All ingredients must have an appropriate specification standard prepared by a technical or culinology group so that the purchasing department can have parameters against which to judge and compare the cost and quality of materials supplied by vendors. These variables would include safety, physical, chemical, microbiological, and even sensory standards and the methodology to properly monitor these qualities. R&D must bear in mind that quality control is a dynamic process. The process does not end at the point of product hand off to production. The quality control process is ongoing quality improvement, that is,: process improvement, new equipment evaluation, cost monitoring, field monitoring, alternative and substitute ingredient evaluation, and even reformulation.

By or before the project has reached the pilot or semi-production stage, R&D/product development must assemble the information required for the label. For nearly all processed food products, this means a list of all ingredients arranged in descending order of quantitative predominance (by weight or percentage) starting with the quantitatively largest item and called by the appropriate or permitted name. In addition, a nutrient declaration that meets the requirements of the 1990 Nutrition Labeling and Education (NLEA) and its amendments must also be prepared.

If there is to be any health claim included as part of the label, special care must be exercised to assure compliance with regulatory guidelines (Porter, 1996) (see Chapter 19). If there is a desire to meet the specific requirements of special designations, such as Kosher or organic or other indicators of particular quality, these must be considered.

Even items like the size of print and placement of items on the label may be subject to regulation. This can be carried out in-house if that particular expertise is available either in R&D or in the responsible legal and regulatory departments of the company. Very often, this specialized requirement is best met by use of experienced professional outside experts who are well versed in the complex and somewhat arcane requirements of the act. This information must be supplied to the graphics personnel responsible for preparation of the printed label, but it is the responsibility of both the technical personnel and the regulatory and legal personnel to review all proposed labels before final approval and use. Obviously, the marketing department has its requirements from a sales point of view and may require particular messages and items like the Universal Product Code (UPC) (Bar Code) symbol for price scanning.

DOCUMENTATION

Early in this chapter we referred to the need to carefully document every step of the way in the development of new food products. When the dust has at least partially settled and the new product looks like it is on its way, it is incumbent on R&D/product development to review what transpired during the project, distill out the essence of what has been accomplished, and carefully write up a detailed report on all aspects of the project. At its very least, this is a history; at its best it documents the value of R&D/product development to company progress. In part, this is necessary to explain just what the company has received for its budgetary expenditure (or investment) and who has contributed, and how. It is also a technology transfer document. It is necessary to record the experience gained so that future projects can benefit from what might have been learned along the way; thus the report is a means of assuring technology transfer. Every new food product development project is different but every project is also the same. Lessons learned in one case are often transferable to another. This valuable asset should not be lost to the ephemeral memory of sometimes transient personnel. In addition, the project must be reviewed by the appropriate parties within or authorized by the company for any intellectual property "gold," which can be mined out of this rich source.

We place the documentation reminder both early and late in this chapter on new food product development in the hope that the reader will have learned what we term the central dogma of new product development: as important as they are, new product ideas are not products until a process of exploiting that idea has been developed and carried out, and that a new product development process is not truly complete until the product is on the market and the final report has been written.

APPENDIX

Planning and Scheduling of New Product Development Projects

New food product projects are complex, time-consuming and inherently risky. Projects involve many different activities, performed by different people across multiple functions in the organization. There is necessary time sequencing to certain sets of activities while other activities can be performed independently. Because it is vitally important to conserve both time and organizational resources, new food product projects must be carefully planned and orchestrated. New food product success in most cases is dependent on getting to market as quickly as possible but not at the expense of skipping or short-circuiting vital steps in the development process.

Once a new food product development project is approved, project planning must take place. The project team must determine the full range of tasks or activities to be performed, the order in which the activities must be performed, and the expected time to complete each activity. One of the most important steps to be completed in setting up the project is the determination of activities, the completion of which if is delayed, will close the entire project to be held up. These are termed critical activities. In addition, the personnel, financial, and capital resources required to complete each task must be determined, along with the administrative support needed to keep the project on-time, within-budget, and providing the deliverables required at each stage of the project. Responsibility for completing each task involved in the project must be determined and communicated. Finally, program budgets are drawn up and agreed on.

Project planning network diagrams can be used to visually display the project. The essential elements of such networks are those discussed in the preceding paragraph. We can use computer project management programs and/or network diagrams to both plan and control the project. Prior to commencing the project, the planning network can help project administrators plan an efficient schedule. Alternative planning networks can be evaluated to assess the impact on completion time and cost. Activities that can be performed in parallel can be identified, as can critical activities. By performing activities that are not time dependent at the same time, overall project completion time can be shortened. After commencing the project, administrators can compare actual time to complete activities versus projected time, and actual budgets to projected budgets, and identify variances. These variances indicate potential problem areas in both budgeting and project completion timetables. One of the uses of critical path analysis is the development of contingency plans to redeploy resources from "slack" activities (which are those activities that if we fail to complete on schedule will not hold up the entire project) to critical activities.

The development and launch of a new food product can be a relatively simple process, an extremely complex process, or can represent an intermediate level of complexity. The simplest processes are those involving simple line extensions such as new flavors for a line of isotonic beverages or chilled tea. The most complex processes are those for which new food and/or packaging technologies are being employed, and severe technical challenges must be met; plus, there is a need for significant development of and investment in new plant and equipment. While both of these processes follow the same basic logic of a concept phase, a development phase and implementation phase, the number of activities at each phase is greater by orders of magnitude for the latter process. Figure 10.1 and Table 10.1 demonstrate the basic logic, sequence and range of activities involved in a new food product development process.

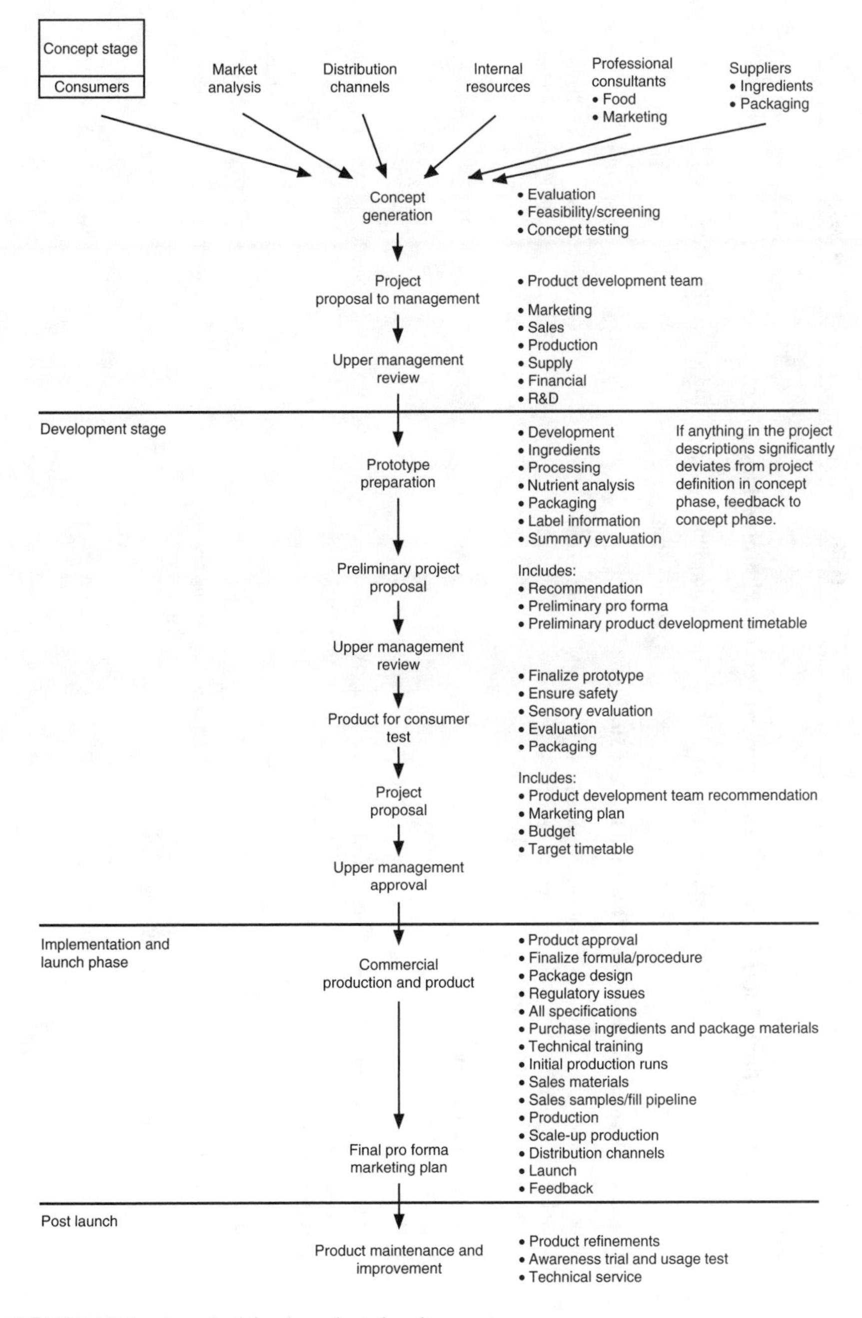

FIGURE 10.1 A typical food product development sequence.

TABLE 10.1
Representative New Food Product Development Project Planning Document

Tasks	Resources Required	Responsibility	Budget	Time Required	Contingency
		Product Strategy			
Objective: Develop and Implement New Food Product Strategy					
Marketing research inputs • Product requirements • Pricing • Characteristics of consumers, users, etc. • Size of markets • Growth rates	• Market, consumer, product and industry experience information source(s)	• Marketing manager	$ X	X months	• Develop new products without marketing information and subsequently determine markets
Develop basic marketing strategy • Identification of target markets	• Team of marketing and product technology	• Marketing manager • Product development management	$ X	X weeks	• Develop new products without marketing information and subsequently determine markets
New product • Focus groups/research from consumers	• Marketing research • Focus group facilities • Willing consumers	• Marketing research manager	$ X	X months	• Develop new products without marketing information and subsequently determine markets

• Selection of food product mix based on marketing research inputs	• Team of marketing and product technology	• Marketing manager • Product development management	$ X	X month	
• Determination of new product requirements from consumers/users	• Marketing research				
• Selection of new products based on both marketing and product requirements • Prices/costs • Properties					
• New product development strategy	• Technical • Product development	• Product management • Marketing manager • Technical manager	$ X	X month	• Develop new food products without strategy or plan • Modify existing products
• Imitate existing product	• Marketing				
• Improve on relevant existing product properties	• Product development				
• Duplicate commercial product to price on more economic basis					

Continued

TABLE 10.1
Continued

Tasks	Resources Required	Responsibility	Budget	Time Required	Contingency
Product Strategy (continued)					
• Innovate totally new products • Extend current product line • Internal development • External development					
• Use existing company production equipment					
• Product development planning • Objectives and tasks • Times and schedules • Establish mileposts (stages/gates) • Pro forma costs and budgets • Resources required • Human • Physical • Consumables	• Technical staff • Project planners	• Project planner	$ X	X weeks	• Proceed with minimal or no planning

• Physical facilities • Financial	• Technical management				
• Presentations to and approvals from management					
• New product development		• New product management • Product development • Technical manager			• Performed by external agency
• Implementation of plan • Laboratory/kitchen • Culinology™ input • Scale up • Pilot plant			$ X	X years	
• Establishment of test protocols	• Technical staff	• Technical manager	$ X	X months	• Use FDA protocols
• Evaluation of chef's samples versus • Existing products • Alternative iterations	• Internal laboratory	• Technical manager	$ X	Ongoing over X years	• Independent external laboratory

Continued

TABLE 10.1
Continued

Tasks	Resources Required	Responsibility	Budget	Time Required	Contingency
		Product Strategy (continued)			
• Evaluation by consumers	• Users/finished products	• New product management • Marketing manager • Sales manager	$ X	X months	• No alternative
Production procedures • Scale up • Equipment • Materials • Operating parameters • Measurements and controls	• Technical management	• Production manager	$ X	X months	• Add-on to company's existing line
Production specifications • For all new products	• Technical quality management	• Quality assurance	$ X	X month	• Use Technical Staff
Quality assurance protocols	• Technical management	• Quality assurance	$ X	X month	• Use Technical Staff
Cost development	• Financial Analysts • Cost inputs • Capital requirements	• New product management	$ X	X week	
Establish procedures to monitor product use in field and to feedback information	• Field technical staff	• Quality assurance	$ X	X week	• Use Technical Staff

Pricing

Objective: Establish Pricing for New Food Product

Establish product costs for new product • Materials • Inventory • Labor • Utilities • Projected variables • Projected indirects	• Production • Purchasing • Technical • Accounting/financial prior cost development data and information	• New product management	$ X	X months	• Set price based on competitive product pricing • Set arbitrary low introductory costs
Capital investment required	• Engineering • Production • Financial	• New product management	$ X	X month	• No return on investment (ROI) computation required, with associated risk
Operating capital required	• Accounting	• New product management	$ X	X week	• Guess
Develop total costing	• Accounting	• New product management	$ X	X week	• Set price based on competitive product pricing
Determine prices	• Marketing research	• New product management	$ X	X month	• Point-in-time pricing
Determine lower and upper limits of pricing	• New product management	• New product management	$ X	X week	• Price based on competitive product pricing only

Continued

TABLE 10.1
Continued

Tasks	Resources Required	Responsibility	Budget	Time Required	Contingency
		Pricing (continued)			
Determine margins to generate appropriate returns for business venture	• New product management • Accounting	• New product management	$ X	X week	• Let market pricing dictate margins
Ensure that pricing generates cash flow to drive the new product	• Accounting	• New product management	$ X	X week	• Ignore, with attendant risks
		Business Plans—Synthetic Confetti Pricing			
Establish pricing strategy • Launch • Steady state	• New product management	• New product management	$ X	X week	Price based on competitive product pricing only
		Communications			
Objective: Communication Strategy for New Food Product					
Identify target markets for the new product category	• From marketing research and strategy	• New product management • Marketing manager	$ X	X week	• No communications program
Characterize target markets by needs and perceived needs	• Marketing research	• New product management	$ X	X months	• Marketing management decision
Determine marketing strategy and objectives	• Marketing data	• New product management	$ X		• Go directly to tactics

Develop communications strategy based on marketing strategy
• Objective
  • Launch
  • Steady state
• By product
• By market
	• Marketing strategy
	• Inputs on targets from research
	• Communications manager
		• New product management
		• Marketing management
			$ X	X month	• No communications strategy
					• Tactical only

Message to each target market	• Writer(s)	• New product management	$ X	1 month	• Single message to all

Communications category
• Media conference
• Hard copy pieces
• Media releases
• Trade promotion
• Trade show stand
• Trade show presentations
• Advertising
• Schedules
• Art
• Budget
  • Launch
  • Steady state
	• Communications manager
	• Advertising agency
	• Public relations agency
	
	• Trade show management
	• Trade show management
	• Advertising agency
	• Communications manager
	• Art
	• Financial budget person
		• Marketing manager
		• Communications manager
			$ X	X months	• General—no targeting
					• Internal resources

Continued

TABLE 10.1
Continued

Tasks	Resources Required	Responsibility	Budget	Time Required	Contingency
		Communication (continued)			
• Measurement of results • Modifications of plan	• Marketing research				
Implementation	• Communications manager • PR agency • Advertising agency • Trade show management agency	• Communications manager within group	$ X	• Launch—X months • Reminder—ongoing	
		Distribution			
Objective: Develop Distribution Channels to Move Product from Production to Retailer/Consumer					
Select location(s) of production	• Production	• New product management • Production management	$ X	X months elapsed	• Contract with copacker
Determine products to be made at each site, if more than one production site is selected	• Production		$ X	X months elapsed	• Produce to order; still requires distribution
Identify users to be served • By type of product • By volume • Frequency of delivery • By location	• Marketing • Marketing research	• New product management	$ X	X months	• Respond to user orders

Develop logistical plan to balance
• Production
• Inventory
  • Production
  • Distribution
• Delivery
	• Distribution manager	• New product management	$ X	X month	• No plan, not acceptable

Determine distribution policy • Direct delivery • Brokers with control of distribution • Costs for each • Delivery to users for each • Geographic dispersion • Construct alternative strategy for each • Select one or more	• Distribution manager	• Distribution manager • New product management • Marketing management	$ X	X month	• No policy, not acceptable • Go directly to distributor • Use company's current distribution
Identify distributors, if distributor system is selected: brokers; distribution warehouses; etc., depending on strategy selected	• Distribution manager • Marketing research	• New product management • Distribution manager	$ X	X months	• Use predetermined list • Use company's distribution system

Continued

TABLE 10.1
Continued

Tasks	Resources Required	Responsibility	Budget	Time Required	Contingency
		Distribution (continued)			
Communicate with distribution channels	• Purchasing • Distribution manager	• Distribution manager	$ X	X months	
Negotiate with distribution channels	• Purchasing	• Distribution manager	$ X	X months	
Establish interactive electronic data interchange between distribution channel and production	Information technology (IT) Manager	• New product management • Distribution manager	$ X	X months	• Paper transactions • Telephone transactions
Select method for product movement, e.g., truck • Use external distribution • LTL • Full truckload • Independent • Contracted • Own fleet	• Distribution manager	• Distribution manager	$ X	X week	

Negotiate and contract with carrier	• Distribution manager • Purchasing manager	• Distribution manager • Purchasing manager	$ X	X month		
Implementation	• Distribution manager	• New product management • Distribution manager	$ X	Ongoing		
Sales						
Objective: Develop Sales Strategy for Product to Buyer						
Marketing strategy • Target market products • Pull strategy with users • Distribution intermediaries	• Marketing plan • Marketing manager	• New product management • Marketing manager	$ X	X month		• No marketing strategy—not acceptable
Sales strategy • Employ company's sales • Employ separate sales force • Independent sales • Broker sales force • Distributor sales	• Sales manager • Marketing manager	• New product management • Marketing manager	$ X	X week		• Go directly to one predetermined sales strategy

Continued

TABLE 10.1
Continued

Tasks	Resources Required	Responsibility	Budget	Time Required	Contingency
		Sales (continued)			
• Evaluate alternatives • Determine costs for each • Project probable sales outcome for each		• New product management • Sales manager			
• Select best alternative(s)	• Sales manager • Research • Financial analyst	• Sales manager	$ X	X month	
Identify target markets to be contacted by Sales • Purchasing agents • Retail • Hotel, restaurant Institutional (HRI) • Distributors • Others • By name, location, telephone, fax, e-mail, etc.	• Research • From communications plan	• Sales manager	$ X	X months	• Respond to general inquiries
Determine message to be employed for each target market category	• Marketing manager	• Marketing manager	$ X	X weeks	

Determine type of sales person desired for each target category	• Human resources • Sales manager	• New product management	$ X	X week	
Determine training required for sales persons for each category • Product • Industry • Company's policies/culture • Message	• Marketing manager • Sales training • Sales manager		$ X		• No formal training program
Develop sales training program • Product benefits • Marketing support • User needs • Competitor counters	• Sales training • Sales manager	• Sales manager • Marketing manager	$ X	X months	• Direct field training by sales and/or marketing manager
Determine number and type sales person for each target category • Centralized • Geographic dispersion • Inside • Field	• Sales manager	• Sales manager	$ X	X week	• Inside sales only
Collateral materials • Brochures • Videos • Internet home page • Etc.	• Communications manager • Advertising agency • PR agency	• Sales manager	$ X	X month	• Use existing collateral materials

Continued

TABLE 10.1
Continued

Tasks	Resources Required	Responsibility	Budget	Time Required	Contingency
		Sales (continued)			
Develop budget for sales	• Sales manager • Financial analyst	• Sales manager • New product management	$ X		• No budget—not acceptable
Hire required sales staff	• Human resources • Sales manager	• Sales manager	$ X	X months	• Transfer from inside company • Transfer from Technical • Use brokers • Inside sales only
Sales training	• Sales training • Physical facility • Sales staff • Technical management • Marketing Management	• Sales trainer • Sales manager	$ X	X week	• Field training only
Sales staff dispersion • Target market • Geography	• Sales manager	• Sales manager	$ X	X week	• Broker sales force • Distributor sales force • Inside sales
Inside sales and customer service • Procedures • Responses • Staffing implement	• Sales manager	• New product management • Sales manager • Product management	$ X $ X	X month Ongoing	• Directly to production scheduling

BIBLIOGRAPHY

Amernick, B. A. 1991. Protection of Intellectual Property, In *Food Product Development From Concept to Marketplace,* Graf, E. and I. S. Saguy, eds., New York: Van Nostrand Reinhold, pp. 365–378.

Brody Aaron, L, and Lord John, B. 2000, *Developing New Food Products for a Changing Marketplace*, Boco Raton, CRC Press.

Crawford, Merle and Anthony DiBenedetto, 2006, *New Products Management*, Irwin: McGraw Hill.

Giese, J. 1997. "Technical Centers Facilitate Food Product Development," *Food Technology.* June, pp. 50–54.

Hoban, Thomas J. 1998. Improving The Success of New Product Development, *Food Technology.* January, pp. 46–49.

Pessemier, E. 1982. *Product Management Strategy and Organization.* 2nd ed. New York: John Wiley & Sons. pp. 20–23.

Peterson, Kenneth, R.B. Handifield and G. Ragatz, A Model of Supplier Integration into New Product Development," *J. Product Innovation Management*, 20 (4), 2003, pp. 284–299.

Porter, D. V. 1996. Health Claims on Food Products: NLEA, *Nutrition Today*, 31:35–38.

Pszczola, D. E. 1998. The ABCs of Nutraceutical Ingredients, *Food Technology*, March, pp. 30–37.

Resurreccion, A, *Consumer Sensory Testing for Product Development*, Aspen.

Takeuchi, H. and I. Nonaka. 1997. The New Product Development Game, Harvard Business Review (January–February 1986); In *Managing Teams in the Food Industry*, P. Hollingsworth, ed., *Food Technology.* November, pp. 75–79.

Wheelwright, Steven C. and Kim B. Clark, 1992. *Revolutionizing New Product Development*, Free Press, New York.

11 Innovative New Food Products: Technical Development in the Laboratory

Alvan W. Pyne

CONTENTS

Long considered the dual fountains of new food products, kitchens and laboratories, are essential elements but not the totality. Blending superb comprehension of scientific principles marries with creativity and daring—the synergy of culinology and food science. A food product prototype is only one component that, when acceptable, must be subsequently translated into a food product that can be commercially produced, packaged, and distributed to consumers who would receive a true reflection of the original promise of the concept. Implementation in a food factory with specifications, quality assurance (QA) tools and packaging represents challenges for those team members who have been involved from the beginning but are now responsible.

INTRODUCTION

Numerous books and articles have been written on the subject of new product and food product development, and many courses, seminars and workshops offered to help marketing and technical managers understand this "nebulous" area of business and how to succeed at it. Food product development is a demanding, fast-paced, high-risk/high-benefit part of a business that requires dedicated inputs from multiple sources, for example, marketing, food scientists/technologists, culinologists when available, engineers, plant operations, external suppliers, and financial and business management, all working closely together to achieve tightly defined objectives. Such a food product development team usually must work under difficult time constraints and pressures following disciplined schedules all designed to achieve a timely competitive advantage in the marketplace. This is certainly the case for innovative new product development. It is acknowledged, however, that most food product development activities are carried out to deliver line extensions and brand expansion of existing product lines, a process requiring far less risk and development effort than that expended for development of innovative new products. Line extensions generally are based on development of new flavors, new packaging regimes, package sizings, that is, basically variations on a basic and often successful theme.

In this chapter, we focus on truly innovative food product development as differentiated from the more common line extension and brand expansion activities, although the basic steps required to take products into the marketplace are similar (Figure 11.1).

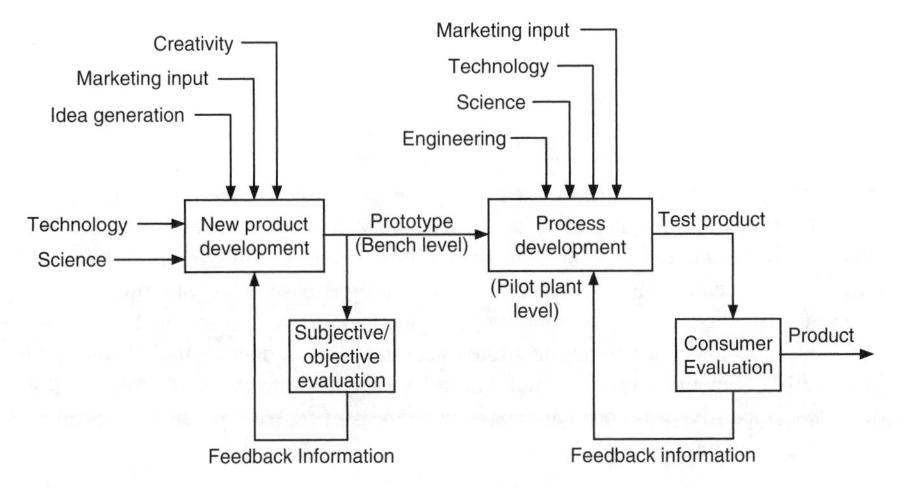

FIGURE 11.1 Interaction of key elements in new product development.

The three fundamental truths to understand about this game of new product development are as follows:

- The probability of success is not high.
- You never *really* know whether you have a success or not until you have taken the plunge to find out if it is consumers are willing to purchase.
- The rewards, if successful, can be very great.

New product development is difficult work. If you are going to play the game of new product development, you have to have the constitution for it and a willingness to roll the dice, back your hand, analyze the game and know when to quit. Participation in the game requires the willingness of you and management to risk considerable funds to accomplish committed and agreed-upon objectives.

The process of new food product development is at best a delicate and tortuous path. New product development in any organization must have a bonafide champion, must be nurtured and encouraged and must be supported by top management. In general, marketing leads and the technical function follows, working as a team to develop and commercialize new food products.

In addition to requiring the assembly of a tightly knit multidisciplinary team to develop and deliver the product, there must be the appropriate and continuing consumer input, which requires essentially introducing the identified target consumer into the food product development picture at the very early stages of the project and maintaining that input throughout. Consumer involvement is achieved by performing relevant studies and panel testing protocepts with a variety of target consumers to identify the target consumers' needs or wants, or to assemble a rational basis to persuade the consumer that they really desire the proposed new food product, not nearly as easy to accomplish.

Generally, the food product concept has been defined based on the inputs of marketing, technical and now often, culinary personnel. The technical development department is challenged with the task of translating the desired consumer food product concept and attributes into technical terms and identifying and employing approaches to ensure that the types of desired attributes are delivered within the food product according to the defined cost parameters; shelf-life considerations; and desired flavor, texture, appearance, attributes, and so forth. It is imperative to perform such activities, keeping in mind that the main objective is to deliver a food product that will be acceptable to the target consumer and possess at least one fully defined point of difference from the competition. In order to ensure that this is the case, the team must be well aware of the substitute or competitive products in the market or the potentially competitive products that could be introduced to the marketplace. If no directly competitive food products exist, awareness of real and potential substitutes must be paramount.

SETTING THE STAGE

PROJECT/PRODUCT PLANNING

New food product development activity should be integrally tied into the business planning process whatever be the business unit or corporate level of the processing, marketing, distribution, organization, and so forth. *A solid commitment at the highest level of management/ownership to support the activity with the required resources is indispensable.* Provision should be made for timely review of progress reports and previously identified milestones to keep the product planning and implementation process on schedule.

MARKETING RESEARCH BRIEF

In general, the marketing component provides briefs to the culinary and technical development groups based on the information provided in Table 11.1. For a new food product, the anticipated points of distinction from competition should clearly be defined. The projected method of preservation and distribution ,for example, shelf stable, dehydrated, heat processing, refrigeration, freezing, and so forth, should be established. These parameters will impact food product safety, shelf life, packaging and distribution considerations.

THE TECHNICAL FEASIBILITY STUDY

In response to a marketing research brief, the technical development group should perform technical feasibility studies providing the information delineated in Table 11.2. The probability of technical success and any technical challenge areas should be identified and defined. As much as possible, the recommended processing and types of equipment should be identified as early as enumerated. In general, the technical feasibility studies should provide enough information to reach a decision to move on to the next step of in-depth food product development.

TABLE 11.1
New Product Development

Phase 1: Technical Feasibility Study—Information Requested from Marketing

A. General Product Description
   1. Concept
   2. Anticipated consumer need and type of use for product
   3. Closest competition on market today
   4. Anticipated points of difference from existing competition
B. Marketing Positioning of Food Product
C. Anticipated Distribution Channel—refrigerated, frozen, ambient temperature shelf-stable
D. Targeted selling price for given unit size
E. Targeted plant cost per unit*
   1. Anticipated number of units to be produced for
      a. Test market, if used
      b. Launch
      c. Commercial production
   2. Labor, utilities
   3. Ingredients
   4. Package materials
F. Anticipated packaging
G. Anticipated shelf-life requirements
H. Anticipated production facilities required
I. Anticipated date of production introduction
   1. Test market
   2. Roll out
J. Extent of anticipated line extension possibilities
   1. Test market, if used
   2. Roll out
J. Extent of anticipate line extension possibilities

* Critical piece of data

SELECTING AND ORGANIZING THE TEAM

SELECTION OF TEAM MEMBERS

The degree of success of the new food product development activity very much depends on the selection of appropriate team members. Not only should the members be competent and highly skilled in their particular discipline, be it technical or non-technical, but they should also work harmoniously together supporting one another in pursuing the defined objectives. Generally, the team leader is selected from the marketing department, but as the progress of the new food product development unfolds, this leadership may be shared to a greater degree with the technical management as the product development activity moves into process development, process engineering, pilot plant, scale-up, launch, and eventual commercialization.

TABLE 11.2
New Product Development

Phase 1: Technical Feasibility Study—Expected Results from Technical Feasibility Study

A. Review of prior art/patent literature—should be an indispensable starting platform
B. Overall assessment of chances of technical success within specified framework
   1. Potential technical red flag areas (potholes)
   2. Estimated cost/time requirement to circumvent red flag areas
C. Provide a range of prototype products based on ingredient cost/quality relationship
D. Breakout of cost formula
E. Ingredient specifications
F. Estimated and desired shelf life
G. Technical recommendations on types of packaging alternatives
H. Preliminary estimate of production requirements
   1. Processing considerations
   2. Types of equipment required
I. Comparison of alternative production facility candidates based on expected production requirements
J. Estimated cost, time, and extent of further development work required based on preliminary assumptions

DEFINING THE BUDGET

The new product plan is generally prepared outlining the objectives or rationale for the new product development, the expected impact on the business, and the probability of marketing and technical success. This living plan provides a basis for identifying the required budget and the resources for carrying forward the new food product development activity. The product plan also identifies agreed-upon schedules, time sequence charts, Gantt charts, the mutually agreed-upon milestones, and essentially provides the road map to move the product forward to full commercialization, all techniques now widely available by computer project management methods.

It is imperative that realistic budgets be formulated and adhered to as closely as possible as the food product development moves from concept to launch and commercialization. In smaller and entrepreneurial companies, new food product development schedules may be compressed. The levels of bureaucracy are not as nearly cumbersome, and a sense of urgency can be more firmly established. The tendency to use external sources to supplement in-house resources may be easier to establish within smaller companies. In many cases the innovative spirit required to bring the products forward at a fast pace may in fact be more resident in small, more flexible and nimble companies than in the larger multidivisional/multibranch companies. In fact, in order to increase "speed to market," larger companies are increasingly trying to emulate the smaller companies in developing the desired flexibility and reduced bureaucratic levels in introducing new food products.

PHASE 1: PROTOTYPE DEVELOPMENT

Following a brief and objectives from marketing, a research and development or equivalent technical group generally prepares prototypes using the resources of research chefs or culinologists, test kitchens, and food scientists/technologists working at bench level to develop preliminary initial formulae or recipes. The objective is to develop prototype food products that can be presented to marketing management to initially determine if they are in line with marketing's expectations. The final jury however is not composed of marketing or business or technical people, but rather of consumers. It is essential that the input of the target consumer be factored into the initial preparation of these prototypes and that repeated testing of formulae be performed using consumer inputs to generate the feedback from the consumer as to the acceptability, desirability and/or changes required in these prototypes (see Chapter 12). Based on feedback from this testing, adjustments then must be made to optimize the sensory properties of the food product, that is, the flavor levels, sweetness level, the amount of seasonings, texture, thickness, color, and so forth.

SYNERGY OF CULINOLOGY™ (REGISTERED TRADEMARK OF RCA, RESEARCH CHEFS ASSOCIATION) AND FOOD SCIENCE AND TECHNOLOGY

An approach used by some companies is to employ chefs in a kitchen or the laboratory to prepare or formulate the desired early prototype food products utilizing ingredients and recipes to deliver product prototypes deemed acceptable by marketing management and confirmed via consumer testing (see Chapter 12). Working in conjunction with food scientists/technologists, these chefs/culinologists modify the recipe/formulae and eventually together all concur with food formulations based on the numerous chef preparations and the practical considerations of the food scientists/technologist. The objective is development of a product at a reasonable cost that will perform within the desired constraints. Regardless of the approach, one underlying variable must be kept in mind constantly, and that is *the product must taste good to the target consumer*. Thus the flavor and flavor delivery systems are of paramount importance in development of the prototypes and eventually the finished product to ensure the primary desired product attributes.

INITIAL FORMULATION

Perhaps the early formulation efforts may best be envisioned by some illustrative examples. Let us assume that the defined task is to deliver a superior low-fat salad dressing product that will possess superior mouth feel, texture, flavor definition, and balanced flavor compared to the existing competitive products on the market. Initially, a comparative evaluation of the current competitive products in the marketplace is performed identifying the strengths and weaknesses of such products. The technical group then conducts a search to identify any emerging technologies that could provide the superior product attributes that are desired. Technologies could come from internal development capitalizing an ongoing long term research programs designed to identify and develop low-fat texturizing systems and/or utilization of external technologies identified as potential candidates to deliver the desired product attributes. Increasingly, research and development organizations are probing the potential for introducing and incorporating the external technologies to augment their internal development programs or to capitalize on an identified technology to deliver a proprietary positioning. Regardless of the source of the technology, the important challenge is to harness it as quickly as possible, fit it into the matrix, and apply it to deliver the proposed and desired product attributes.

Such technologies can come from a variety of food and/or nonfood industries, applying the concept of lateral thinking to utilize a particular technology in a new and innovative approach. For example, spray dispensers can direct small quantities of dressing to precise location on the salad. Certain drug delivery systems can be utilized and defined to provide flavor delivery systems. Microencapsulation technologies can be utilized to provide the desired point of difference in releasing a flavor on demand at the point of preparation or consumption.

If possible, the technology employed should be proprietary, on the basis of acquisition, licensing, achieving a patentable position, or adapting/enhancing piggybacking on the other technologies.

In many cases, the food product development team requires development staff to work closely with food ingredient suppliers. In some cases the suppliers are recruited as members of the team, testing a number of ingredients components that were developed specifically to achieve required product attributes.

SHELF-LIFE CONSIDERATIONS

By this time the decision will be made as to how the product is to be marketed, that is, ambient temperature shelf-stable, frozen, refrigerated, or dehydrated. It is advisable to present the product to the consumer in this format taking into consideration the end use, the method of preparation. For example, if the group is designing a dry soup mix it is very important at the outset to establish the convenience aspects of the preparation of the resulting quality of the final prepared soup. Based on the refinements of the prototypes intensive sample preparation at the bench and/or kitchen level is undertaken to develop a product as close as possible to the objectives and the product attributes. At this point, decisions have been made with regard to the specific ingredients to be used, the cost constraints, and the parameters for the product preparation at the consumption venue. Once the modes of preservation and marketing are identified, shelf-life testing should be initiated to determine sensory, biological and biochemical issues that might be encountered under such conditions. If feasible, accelerated shelf-life testing can be conducted in which the temperature of the product is raised to determine, for example, if there could be any substantial problems with regard to chemical interactions, autoxidation of oils, and any potential microbiological problems. Once the formula has been agreed upon by technical and marketing, with consumer inputs, then extensive shelf-life studies should be carried out prior to test marketing of the product. (See Chapter 16.)

PRODUCT SAFETY

The format for distribution of the product to the consumer must be carefully defined. Should the product be ambient temperature shelf-stable? Is it to be frozen, refrigerated, dehydrated, or heat processed in order to present the desired product quality aligned with the desired product safety? Above all, the product must be safe for consumption under the defined processing parameters, product preparation, and product consumption parameters.

EVALUATION OF THE PROTOTYPES

Once initial prototypes have been prepared and tested in the laboratory, it is necessary to return to the target consumer with these prototypes to determine if, in fact, these prototypes match the concepts that were originally developed with target consumer inputs. An approximate costing of the prototype formula are calculated prior to this testing to ensure that the delivered product will be within the guidelines of the cost parameters initially identified for the product. Consumer testing must be used to obtain feedback concerning confirmation of the concept, overall product acceptance, the level of enthusiasm expressed for the product, some indication of intent to purchase and repurchase, and the flavor intensity levels and mouth-feel and texture parameters (Chapter 13). This information is used to refine and fine-tune the formula. Generally several sensory panel tests are required for optimization.

PHASE II: BENCH LEVEL FORMULATION

SELECTION OF INGREDIENTS

A delicate balance must be maintained between the cost of selected ingredients and the desired quality of the product. The projected cost of the end product is defined in conjunction with the anticipated margin and selling price of the product. The formula must be fine-tuned to ensure that the anticipated ingredients from the suppliers will perform and that sufficient quantities of the ingredients are available within the constraints of the product cost structure. For example, it is extremely risky to formulate a product with an ingredient from a flavor house or food supplier that is currently only available at bench level or pilot plant quantities. The danger is that the ingredient supplier stream may not be adequate to keep up with the test market and eventual nationwide or worldwide roll-out.

PANEL TESTING

Once in-depth product development is initiated it is necessary to go back to the consumer with these samples to determine if, in fact, they match the concepts and prototypes that were originally discussed and presented to the consumer groups.

Consumer testing is used to obtain feedback concerning confirmation of the concept (Chapter13). Read out is obtained regarding the overall product acceptance, the level of enthusiasm expressed for the product, some indication of intent to purchase and repurchase, the flavor intensity levels and mouth-feel and texture parameters. This information is used to refine the formulation. Generally several sensory panel tests are required to optimize the formula.

PHASE III: PROCESS DEVELOPMENT

Once bench level samples and formulae have been agreed upon by the product development team and proper consumer input has been factored into such development, the focus shifts from the bench level to the pilot plant to scale up the formulation and carry out test runs based on the early selection of the equipment and unit operations required to produce the product. Generally a decision is made as to whether the method of production will be batch, semi-batch, or continuous, and the selection of the equipment will be in line with that decision. The objective is to obtain the best quality food product with the most economical throughput through the pilot plant operations. At this point, the food engineer should cooperate very closely with the food scientist/technologist to ensure that the desired quality is preserved and that the resulting product from the pilot plant line is as close as possible to the standard samples. Inevitably, with the scale-up problems involved, the product from the pilot plant line will have to be fine-tuned to obtain as close a match as possible to the bench samples. Many times, it is not possible to come up with an exact match and so compromises are made in order to produce a reasonable product as close as possible to the originally defined criteria.

If test marketing is employed, food processing/marketing companies may use pilot plant produced product for test marketing prior to making a decision to invest capital to modify or build a plant to produce the new product. An alternative is to use toll processors (i.e., contract processor/packagers) to produce test market quantities in order to generate enough data to justify further capital investments. Generally, further consumer testing is required to further fine-tune the initial finished product off the processing line.

The product development team computes an estimated Free On Board (FOB) plant cost based on the selected ingredient formulation throughput and fixed and variable costs required to produce the product. Specifications are firmed. A Quality Assurance (QA) manual including a Hazard Analysis Critical Control Points (HACCP) plan is drafted. If the product is to be produced in an existing plant, the product development team takes up residence at the plant working with the factory staff to ensure a smooth transition to pass the responsibility of producing the product to the plant personnel. The food product development team leader works closely with the plant personnel to ensure that he/she is fully satisfied that information concerning the product and the knowledge required to produce the product has been adequately transferred to the plant level.

PROCESS OPTIONS

Once the bench prototypes have been prepared and approved, translation from bench preparation to pilot plant production involves identification and selection of required processing equipment to produce the product. If the new product is potentially patentable, a preferred approach would be to apply for and try to obtain a patent involving both the product and process if indeed novelty is involved in the processing of the new product. Primary equipment required, such as various mixing vessels and pumps, are generally available at the pilot plant or at the commercial production site. Novel processing equipment can be custom designed by having technical and internal or external process engineers work closely with the equipment suppliers. In some cases, the required processing equipment may actually be built from scratch by a process/engineering department or outside engineering firm. Throughput designs, flow rates, ease of cleaning the equipment, computer-controlled operation of the equipment, and so forth, are all key considerations in designing and laying out the required processing line.

PRODUCT SPECIFICATIONS

Once the product formula and the method of product have been agreed upon, a commonly accepted practice is to prepare a product "write out," which will detail all of the specifications for the product, the formula of preparation, the processing conditions, and the QA, and/or Quality Control (QC) specifications and HACCP program for ensuring the expected quality of product. The Research and Development (R&D) Department, working in close conjunction with marketing management and the process engineers and the plant management, generally prepares the product "write out." It constitutes essentially the "Bible" for the products with no deviation from the product write out permitted without express approval and sign-off of marketing, R&D, process engineers, and plant management. R&D people travel to the plant to assist the plant personnel in producing the product and will be in charge for the first test market production runs after which time the responsibility will be turned over to the plant management with R&D assisting in a consulting capacity. The QA Department, generally at the Division level or at Corporate level, is responsible for preparing all QA delineation for the new product introduction, and the QC Department of the plant is normally responsible for the day-to-day monitoring and testing of the daily production of the product at the plant.

PHASE V: TEST MARKET PRODUCTION

If employed test market production of a new product may be carried out using contract processor/packagers or using pilot plant facilities at the companies' process development facilities or utilizing existing processing lines in the companies' plants. In recent years, test marketing has been less employed to truncate the development process and to minimize rapid competitive counter actions.

CONTRACT PROCESSING/PACKAGING

If this approach is selected, it is necessary to diligently screen the potential contract processor/packagers to determine the suitability of the processing/packaging facility; to ensure that good manufacturing practices are in place, that the facility is acceptable as a food processing/packaging site, and that adequate controls are in place to ensure the confidentiality of the production. A contract must be negotiated that will provide the desired insurance of consistent product quality and safety,and which requires microbiological monitoring and testing of the products as specified by the R&D Department and its HACCP plan. Moreover, the R&D Department will be on the site working closely with the contract processor/packager to ensure that the proper product quality is delivered. In negotiating the contract it is necessary to ensure that the cost parameters are clearly understood, responsibilities are delineated, liabilities are properly established and that both management of the food marketing brand owning company and the contracting firm are committed in writing to provide the required input to ensure proper product production and packaging.

Pilot Plant Production

If internal pilot plant production is selected, generally the R&D staff works in line with the pilot plant staff to ensure that the established processing line is operating correctly, that all controls are in place, and that consistent product can be produced from this site. Monitoring of the test market production is provided by the R&D Department utilizing in-house analytical and microbiological facilities supplemented by outsourcing, if necessary, to perform some of the required test monitoring.

Test Market Production in an Existing Plant

If this approach is selected, R&D management must work closely with process engineering and plant personnel to either produce the food product on an existing or modified processing line or on a newly engineered and constructed processing line. In either case the product write out is followed very precisely. Any changes that are made must be agreed upon by all parties involved, and the personnel involved are committed to be on-site as long as necessary to ensure a smooth roll-out of the initial test market production.

COMMERCIALIZATION/LIMITED ROLL OUT

Once the test market production, if applied, has been successfully completed and the test market or other tests to predict consumer purchase response has resulted in positive response, the product development team begins to transfer responsibility for commercialization and launch of the food product to plant operations. The hand off is facilitated by the product development team ensuring an orderly transfer of all the product information in the form of a product write out. The product write out is essentially the "Bible" for the product, and contains specific details with regard to the ingredient specifications, product formula, processing conditions, QC and QA procedures required, that is, everything to ensure that the food product is correct and will have the desired product quality. Sometimes the commercialization is initially limited to a regional launch, followed by a national roll out, and in some cases the company will move ahead with an initial national roll out. The plant manager assumes responsibility for continuous production of the new food product.

TECHNOLOGY SKILLS AND PRODUCT TRANSFER

Given that the product is a success, the company may see fit to launch the new product in other markets and even set up the new product in various countries to exploit the technology more fully to obtain increased sales. Once the required information is furnished to the new location, the product development team may be required to work with staff and management at these locations. The food product usually must be fine-tuned to the desires of the particular consumer market involved. Ingredients may have to be substituted, but, in general, the intent is to ensure the adequate transfer of the product knowledge, the skills, and the technology to capitalize on the investment made by the company.

PRODUCT AUDIT/FOLLOW UP

Management is extremely interested in the potential success of the launch of the new product. Comparisons are made between the projected return on investment, the sales volume, and, most importantly, repeat sales. All of this is rigorously tracked to determine if the company has a successful food product, or not. Lessons learned from such an audit can be valuable providing a basis for honing skills in pursuing the next new product and to carrying it forward to commercialization.

In some cases the product development team will follow the product and actually provide the resources for a new business unit or a newly formed division based on how successful and broad based the product commercialization is, and, in other cases, the product development team will regroup or be partially reorganized to tackle the daunting challenge of doing it over again and bringing out another new product that hopefully will be a success for the company.

Costing of the prototype formulation is prepared prior to this testing to ensure that the delivered product will be within the guideline of the cost parameter identified for the product.

One of the most difficult tasks in setting up a new product development effort is to achieve a satisfactory interaction between people of very diverse backgrounds, training and personalities. Experience shows that many times the technical individual starts his/her career in industrial new product development with the perception that new product development really centers around the technical activities involved in the process. Perhaps this is because in the technical training the individual has not had that much exposure in general to business training and therefore does not have an understanding of the *total* process involved to develop and market new products. At any rate, it is not too long before a person realizes that in order to be effective in corporate structure; he/she must provide meaningful inputs to the management responsible for directing the overall effort.

By now it is obvious that new product development encompasses a multidisciplinary approach where such teams must work together and communicate frequently to accomplish agreed-upon objectives. There must be a continuous exchange of information and activities and there must be established a mutual trust and respect for the individual members of the team.

The team effort referred to at the beginning of this chapter is much more important in carrying out food product development to develop such truly innovative products that will have a defined distinctiveness in the marketplace.

TIME CONSTRAINTS

The food product development process is indeed an intense, complicated activity and operation to manage. Selection of the product of the product development team is crucial not only in terms of identifying the multifaceted competency of each member, but also their compatibility and their ability to interact and work together under very tightly controlled time constraints. Since the product development team works in an environment that is really a microcosm of a small business and functions best with a certain degree of entrepreneurship infused into the group.

SHELF-LIFE TESTING

As indicated above, the decision must be made early on how the product is going to be distributed. It is important to translate the prototypes into the desired mode of preservation and distribution as soon as possible to determine what effect the mode of processing and preservation will have on the finished food product quality.

MARKET

Test Market Or Launch Production In A Newly Designed And Built Facility

Infrequently, this approach is used to establish a greenfields plant to produce the new product. In this case, all of the foregoing must be carefully integrated to ensure that the production facility is acceptable and that consistent quality product will be rolled off of the processing line.

PACKAGING

An integral part of the product is in fact its package. The package engineering design for the product is of paramount importance to ensure the compatibility of the package with the product characteristics providing for attractive marketing, safety, and QA of the packaged product. The Packaging R&D Department, if present, is closely involved in the product development and there input is a very important input to the whole procedure of product development and eventual test market roll out. Alternatively, external resources such as suppliers and/or consultants should be employed. Details of this interaction and input are covered elsewhere in this publication (Chapter 9).

PROJECT PLANNING

It is imperative at the outset to establish the ground rules for how the food product development process will proceed. This requires determining that the project leader of product development will be rotated to the marketing, R&D and process engineering, and production groups as the project moves from concept to commercialization or whether one project leader will be in charge of all phases of this development (Chapter 7). Whatever be the decision, it is important to establish who the key players are going to be and what their roles will be and the established ground rules as to how the project will product development will be carried out. In many situations the senior leader for food product development is selected from the division or corporate marketing staff and the leadership role will be shared at least initially with the key management in the technical department. The product development team must be clearly identified, must be adequately communicated by corporate management, and be firmly supported by the management at all levels. All members of the food product development team must feel that it is their responsibility, in fact, so that the team operates in the spirit of entrepreneurship ideally under a corporate umbrella and functions as if they were bringing on a new business for the company, particularly if the product is considered to be truly a innovative. A very rigorous approach generally is not required to develop line extensions or brand expansions but cases where a major innovative product is desired and the commitment is present to commercialize such a product, the endeavor must be fully recognized and supported by management as a major initiative during the planning cycle of the company. Generally some kind of a time sequence chart is prepared using the required inputs of the team,for example, a Gantt chart that will identify which particular elements of the product development schedule will be executed and by whom and completion dates firmly identified and agreed on by the members of the team. The project coordinator may be appointed to ensure that all inputs are dovetailed and project is kept on schedule. Progress will be reviewed when agreed-upon milestone dates, and interim reports will be prepared and circulated to members of the team and to the supporting managements. Senior management will be involved from the point of granting approval and lending support to the team effort and will be apprised on the basis of timely reports and presentations as the development process unfolds.

BUDGETING

But if the new product involves the input of substantial emerging technologies and innovative thinking, these resources for such must be harnessed very early in the process and can be available to management as needed. Product development is a very demanding process. Extreme and tight planning are required to establish what the expected lead-time would be in terms of introducing the new product. In fact how is the new product going to be protected by filed patents or trade secrets? Such a decision is really a function of the company's business objectives and the timing involved. In any case it should be expected that once the new product is launched the competition will try to come out with me too products either by trying to get around the patent if the patents have in fact been issued, or to come out with their rapidly developed versions of a similar product. Part of the new product may involve the licensing of technologies. From other sources, the lead-time be clearly stated as well as commitment and the expiration date of such underlying patent must be clearly stated. If the new product involves the development of a new ingredient based on certain "mother" patents, it is desirable to carry out the work to ensure that the patent portfolio can be enlarged and extended by coming up with new applications and new innovations building on the so-called mother patent.

ONGOING DEVELOPMENT

Once the product is launched, depending on the degree of initial success, it may be necessary to reformulate the product to provide for a fine-tuning of consumer preferences, pricing strategies, and/or line extensions of the new product.

DISPOSITION OF THE PRODUCT DEVELOPMENT TEAM

The product development team that has been assembled on an ad hoc basis in order to deliver the product to the market place may follow the product into commercialization depending on the degree of innovation of the food product. In some cases a new division or new business unit may be organized to support the commercial product. In that case members of the product development team may follow the course of success of the product and be employed as an integral part of that emerging business unit. Otherwise they may revert back to their respective organizations and begin the arduous task of developing another new product, which will require similar disciplines that were adhered to in bringing out the previous product.

REGULATIONS

If the product in fact includes a new ingredient, it may be necessary to travel along the tortuous route to getting government approval, that is, Food and Drug Administration (FDA) approval, for utilizing the ingredient or to assume the risk of utilizing the new ingredient in the new product. At any rate, substantial documentation of the potential risk involved using the ingredient, required toxicology studies, and so forth, are all required. All of this must be carefully prepared and be available to the regulatory agency for the company seeking approval for the ingredient or for defending its position in launching the new product. Primary concern, obviously, is that of product safety, whether it is from a perceived microbiological threat, chemical interaction, preservation, and so forth. The company launching the new product assumes liability for the product. However, all facets of the product development team are involved in sharing such liability, including the toll processor, ingredient suppliers, and so forth. It can be seen that new food product development can indeed be a risky game, but the rewards can be very high. For those who play it well, it is indeed a satisfying experience to reach the marketplace and achieve a truly competitive advantage as compared to line extensions and brand expansions.

12 Improving the Success Rate of New Food Product Introductions

Mark Thomas

CONTENTS

The addition of a Research Chef Culinologist™ to the new food product development team can increase a food or food service company's success rate by producing products more in line with consumer needs and desires. This chapter covers the areas most impacted by the Research Chef Culinologist™ from organization through ideation and the development and protection of The Gold Standard.

BACKGROUND

"The history of the table of a nation is a reflection of the civilization of that nation." So starts the preface, penned by Auguste Escoffier, to the first edition of *Larousse Gastronomique* (Escoffier, 1938). This landmark tome became the quintessential reference book on the art of cooking. Cooking has been appreciated as an art form throughout the industrialized world and has been treated as such due to the skills of its practitioners. Chefs through the ages have traditionally been allowed the "creative temperament." Those of us who have worked in the food and food service industry have experienced both the creativity and the temper of those that bear the title chef. Perhaps it was the behaviors of some of the more temperamental practitioners that seem to have permanently locked the role of chef into the role of artist. "Becoming a chef is a career-long process. Cooking is a dynamic profession—one that provides some of the greatest challenges as well as some of the greatest rewards." So states the introduction to *The Professional Chef*, the primary textbook produced by The Culinary Institute of America (CIA, 1996).

The decade of the eighties saw the beginning of a cultural revolution concerning the profession of chef. Celebrity chefs were just beginning to edge into the periphery of the consumer spotlight—Wolfgang Puck and Paul Prudhomme being perhaps the first to become recognized personalities. Today there are television networks dedicated to nothing but food, chefs, and creative personalities.

It was also during this time that another ground-breaking book made its way onto the scene. Ground breaking, because in 1984, Harold McGee had the temerity to combine the disciplines of cooking and science in *On Food and Cooking, the Science and Lore of the Kitchen*. In his introduction, McGee stated that "Most writers on food either ignore the scientific principles, high or low, that underlie cooking, or else disparage the value of such information on the grounds that art cannot be reduced to the test tube." (McGee, 1984) It was the realization that food science and technology was about FOOD; and the scientist, technologist, and chef all had his/her hands in on the development of a product designed to meet a consumer's need, which created a new professional niche.

The Research Chef evolved on two fronts: chain restaurants and food manufacturers, each with a fairly unique skill-set need. The objective at chain restaurants was to develop meals and dining experiences that would meet a guest's needs, while the food manufacturer was more product specific, based on their business model. What became increasingly clear was that whether the Research Chef was working for a food processor or a restaurant, his/her job was to produce food to please the consumer. In this, the skill set is the same.

Recognizing that this group was truly a hybrid, whose needs were not being met by either the Institute of Food Technologists or the American Culinary Federation, a new professional organization was founded. Formed in 1996, by a group of food professionals with a common interest in the challenges facing the profession, the Research Chefs Association (RCA) has rapidly grown to approximately 2000 members. RCA has become the premier source of culinary and technical information for the food industry.

The RCA has as its vision that Culinology™ will become the universally recognized integration of culinary arts with food science and technology, and its practitioners will define the future of food.

As food companies rushed to fill culinary positions in Research and Development (R&D), they encountered many challenges in assimilating this creative person into their organizations.

THE ROLE OF THE RESEARCH CHEF

Once an organization accepts the need for a chef to join their new food product development (NFPD) team, their role must be defined clearly for all members of the team. In a general sense, the Research Chef is charged with keeping the culinary efforts on strategy through objective creative culinary development. Their skill set places them in a position to interpret consumer culinary expectations and needs with a focus on consumer or customer culinary satisfaction while maintaining the authenticity of the product. Throughout the NFPD process, the Research Chef becomes the guardian of the Gold Standard.

By understanding the multidisciplinary team approach, as well as the iterative process, the Research Chef becomes an integral part of the process from ideation to roll out and after action reporting.

ORGANIZATIONAL STRUCTURE AND PROCESS

Much has been written about organizing the new product development team for success. This portion focuses on productively integrating the Research Chef into the management team. Many manufacturers learned early on that just having a chef on staff did not automatically engender success. Bringing a creative mind into a scientific structure was often a recipe for disaster. The best analogy would be an Apple® Mac® trying to operate in a PC network.

Finding a chef with the right personality may seem like the key, but it starts out well before that step. It begins with the corporate culture and strategy, because an effective Research Chef is truly a key element in your management team in the NFPD arena. Can you point to a clearly defined NFPD strategy in your company's business plan? If you cannot, you are missing the foundation for all your efforts.

Finding the right match also involves the basic human resource functions: a well-written and concise job description, clear reporting structure, and clearly defined roles, responsibilities, and authority. It is also highly advisable to utilize personality screening tools. No one needs an "old-school chef," one who is dictatorial and abusive, one who is prone to throw the pots and pans around the kitchen when things do not go their way.

Most food companies have left the old-time structure where the NFPD was handed off from one department to another for a multidisciplinary approach. This approach allows team members to float in and out of the project as needed, while the core team moves ahead. As can be seen in Figure 12.1, an illustration of the "old way," the chef is only involved in the development of the product and has no vested interest in the remainder of the process. The NFPD effort was often driven more by "What can we make?" than "What will they buy? What need can we solve?"

The multidisciplinary approach involves more people at appropriate times during the process and is very customer-needs driven. As can be seen in Figure 12.2, the chef is involved throughout the process and all team members have a vested interest in the outcome. In this case, they all look good together, or look bad together.

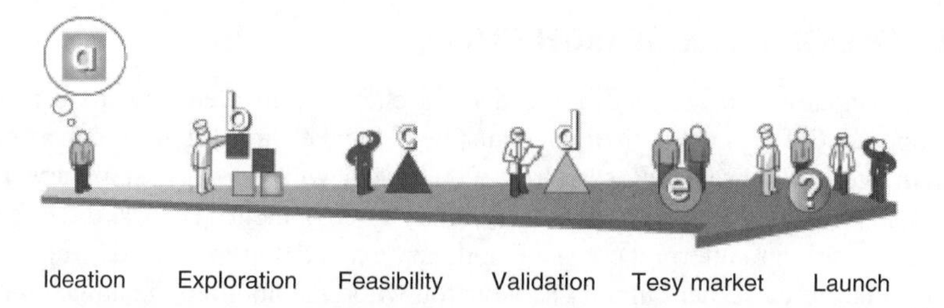

FIGURE 12.1 The NFPD process – the old time approach. (From The Culinary Institute of America, Copyright 2006. Used with permission.)

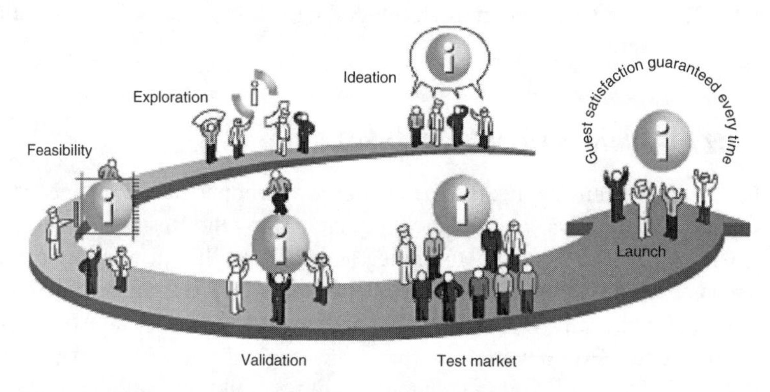

FIGURE 12.2 The NFPD process – the multidisciplinary approach. (From The Culinary Institute of America, Copyright 2006. Used with permission.)

The next element in the organization is an appropriate budget to fund the NFPD effort and the Research Chef. Funds should be allocated relative to the strategic role new products are expected to play in the company's revenue generation. R&D costs vary within the food industry and among the distribution channels, but routinely fall in the 0.2–2% of sales range. As a comparative example, the integrated circuits industry routinely spends on the high-end of the spectrum, at 25% of revenue on R&D (Maxim Integrated Products, 2006). Another example of new product investment is from the drug industry. Tufts University estimates that it costs more than $800 million to introduce a new drug, and that cost includes costs associated with failures (Tufts University, 2001). Smart companies plan for the obvious reality and costs of new product development—both successes and failures.

When developing a budget for new product development, there are three basic areas to consider as costs (shown in Table 12.1).

The obvious are just that, the costs involved with any employee. The hard costs are those expenses directly attributable to the effort. R&D is a clinical discipline that requires the ability to replicate results. That fact alone should substantiate the need for a facility to conduct the new product development. Food manufacturers, for the most part, usually have a separate facility or area in the plant dedicated to the NFPD effort. Restaurant operators are not always in the same position. Multiunit chains most often have a facility that mirrors the restaurant kitchen, where development

TABLE 12.1
Budgeting R&D Cost Categories

Obvious	Hard Cost	Soft Cost
Wages	Facility	Competitive evaluation
Taxes	Equipment	Creative exploration
Benefits	Supplies	
Expenses	Testing	
General and administrative	Support	

work can be performed. Unfortunately, this is not true in all cases. Many small and mid-scale chains still conduct their development work in an operating restaurant kitchen. Nothing increases the risk of error in the developmental process than to attempt to conduct your daily research work around the hustle and bustle of the kitchen staff. Of course, the reality exists that R&D occurs in the small independent restaurant operator's kitchen each time they develop a new menu "special of the day." This assertion does not mean to imply that successful R&D cannot occur in an operating kitchen, but rather that risks increase dramatically. Those risks might appear manageable and acceptable to a 5- or 15-unit restaurant chain, but when the numbers of restaurants exceed 25, they probably are not.

Other hard costs that are often overlooked are testing and outside support. There are a wide range of analytical and market or product research tests that can be used to validate the NFPD process. These tests can be used, in part, to build the Gold Standard specifications.

Lastly, the most overlooked budget areas are the soft costs associated with keeping the chefs, their taste buds, and their creativity at the top of their game. Funds should be available for competitive evaluation samples and visits, whether by a packaged foods manufacturer or restaurant. The restaurant operator's product is an experience made up of food, service, and décor. The culinary creativity of a Research Chef also has to be nurtured, whether through professional development courses like the Worlds of Flavor program offered by The Culinary Institute of America's Continuing Education Department, or independently by doing city or regional restaurant tours. Both approaches work to fuel the creativity required for ideation and development of new products.

STRATEGIC DIRECTION

Discussed above is the need for clear direction as part of the company business plan. Dr. Lord's chapter (see Chapter 3) on *New Product Failure and Success* entitled "Product Development Activities Must Start with and Flow from Business Unit Strategy" clearly lays out the parameters of a strategy. These four paragraphs are incredibly strong and are applicable to packaged foods manufacturers, large restaurant chains, small chains and independent foodservice operators, as well as industrial food ingredient suppliers alike.

Using restaurants as examples, there are many reasons why consumers use a given brand for a given meal occasion. There are also many reasons why they do not use a brand. It is all related to the perceived core competency of the brand. Is it reasonable to expect that Starbucks Coffee would offer excellent pizza, or Domino's would offer excellent soft-serve ice cream? It is the management team's responsibility to understand their core competency and most importantly, the consumer's perception of it.

Top management in the company must fully embrace and be involved in the new product development process by setting the direction and getting people and resources aligned to support it. This is more than just establishing and communicating the corporate vision, it is making certain that the resources—people and finances—are in place so that the work gets done correctly and efficiently.

CHANNEL MARKETS

The Research Chef has a unique position on the NFPD team as it is his/her job to view the culinary landscape and with that view, interpret the needs of the customer. But who is that customer and how do their needs differ from other customers? The Research Chef must always keep in mind who the ultimate customer is—the consumer. The consumer is responsible for the success or failure of all new products, no matter what the channel of distribution. If the trend is to more healthful eating, and Mediterranean cuisine is seen as beginning to surge in popularity, and more and more Spanish restaurants are opening in the bigger cities, perhaps the challenge is to create a new consumer product to ride this coming wave.

Where did that wave start and how knowledgeable is today's consumer about food? All a professional or a consumer needs to do is turn on the television at any time of the day, or night, and you are certain to happen upon several shows about food. Viewer demographics and psychographics cross all imaginable boundaries—age, sex, race, education, and earnings. What unites the group is their insatiable curiosity about food, the more exotic the better. Do not be bothered by the fact that fewer than 10% of them would even attempt to replicate what they witness on the shows. What is happening is the viewer is becoming more educated about cooking styles, methods, and foods. Their knowledge base—and their expectations—have grown exponentially since the 1990s. They have learned about braised lamb shanks and Osso Bucco. What does that mean for developers of new food products? They had better deliver authenticity, and that is one of the primary responsibilities of the Research Chef.

Each of the three principal channels—retail, food service, and industrial—have nuances that must be considered by the Research Chef. Retail packaged foods require attractive, informative packaging that is designed to catch the consumer's eye and interest while protecting the quality and safety of the food for a predetermined amount of time. Food service's packaging needs are predominantly to protect the food through the distribution process until it reaches the restaurant for further processing. Industrial accounts are similar in packaging needs to foodservice. Differences aside, the similarities in the NFPD process are numerous. Once the distribution channel is identified, the process can begin.

IDEATION

The earlier in the process that the Research Chef becomes involved the more effective he or she can be. Ideation is the genesis of the creative process, often beginning with exercises designed to eliminate barriers. This period is not constrained with filters or judgments on the merits of the ideas; it is a time to capture the essence of the idea for further review.

In most companies, the formal process of ideation is managed by the marketing function. The important element to remember is that ideas come from all sources. Examine the following list and imagine how different the ideas for new restaurant menu items could be from each group. The focus of the ideas will be directly related to their stake in the outcome. For example, ideas from corporate operations may be skewed toward cutting labor and maintaining consistency, while field operations may be influenced by their need to keep it simple and to eliminate orphan ingredients.

- Formal, managed ideation sessions
  - Internal personnel
  - External personnel
    - Consumer/customer
    - Professional
  - New food product development team
    - Operations
    - Corporate
    - Field
  - Finance
  - R&D
  - Purchasing
  - Marketing
    - Corporate
    - Field
- Competitors
  - Direct
  - Indirect
- Customers
  - Current users
- Heavy
- Moderate
- Light
  - Tryers/rejectors
  - Nonusers
- Employees
- Vendors
  - Product
  - Technology
- Other restaurant segments (that you do not compete in)

- Retail
- Trade shows
- International

International Farmers Markets can be found in most major metropolitan areas and are excellent places to begin the ideation process. Walk the aisles of an international market and you will see that a majority of the products do not have any English on the packaging (other than the stick-on label, to allow importation into the United States). How many markets sell live eels, sheep heads, cow blood, and every imaginable part of a pig? More than 100 varieties and brands of soy sauce from seven different countries were available in a market located in Atlanta. Research Chefs do not often make the statement "What is that?" or "I've never seen produce like that before." International Farmers Markets are a treasure trove of ideas that pique the interest of the Research Chef. Consider beginning your next ideation session after walking your NFPD team through one of these markets.

CONCEPT DEVELOPMENT

Once the ideas have been generated, they must be transformed into concept statements and some level of filters applied, to begin to narrow the number down into a manageable set. The Research Chef assists at this point by working with marketing to ensure that the descriptive terms are realistic and can be produced to a degree of authenticity.

The Concept Statement contains four elements:

- The statement starts out as a product definition and builds, promising the consumer or guest a certain experience.
- Performance characteristics are added in an attempt to increase the level of interest.
- Sensory characteristics (flavor, spice, texture) are also added until the consumer or guest has a pretty good picture of what to expect.
- Price is added at the end; so the consumer (or guest) can decide if he or she would be interested at that selected level.

A proliferation of new food products has been seen at the retail level that are convenience oriented and time savers for the home consumer. For example, there are many offerings in the poultry category of cooked, sliced, or diced chicken breast pictured in Figure 12.3. One only needs to look at the packaging to deduce what the concept statement is.

We use only our finest white meat chicken and turkey in PERDUE® SHORT CUTS® products. Free of preservatives and artificial colors, seasoned and roasted or grilled to perfection, PERDUE® SHORT CUTS® items are perfect to serve on the side or as the star of any meal GET A HEAD START ON A FRESH MEAL™.

The Research Chef has input into several key areas of this statement, which should be considered before the concept is shared with the consumer.

FIGURE 12.3 Perdue foods honey-roasted chicken (From Perdue Foods website November, Perdue and Short Cuts are registered trademarks of Perdue, Inc, Salisbury, Maryland, 2006.)

- Free of preservatives and artificial colors—has implication on appearance and shelf life. The packaging department also has to support this position.
- Seasoned—what is the appropriate flavor profile? In this example it is honey roasted.
- Roasted or grilled to perfection—will the product deliver on those characteristics that define perfectly grilled or roasted foods? Too often words are used as marketing modifiers without regard to authenticity. The Research Chef would challenge many products in this regard.

Whenever descriptors that attribute certain characteristics to the food are used, the Research Chef must make sure that the product delivers on the claim. In the above-mentioned case, the product claims to be roasted. What are the quality indicators of a roasted product? According to *Cooking Methods, Dry Heat Methods* (The Culinary Institute of America, 2003) from the DVD series produced by The Culinary Institute of America, a properly roasted product generally comes from a tender product to start with, which has been seared to caramelize the color of the skin. The product will have a crispy crust and a moist interior. Following that description, does the product have those attributes? Each one of the 12 cooking methods has specific quality indicators that should be recognized in the development of concept statements and the Gold Standard product. The Research Chef is accountable to produce authentic food products that have the quality indicators that the concept statement promises to the consumer.

Once the Research Chef has had the opportunity to have input into the concept, it is ready for consumer screening and ranking on purchase intent.

Now is the beginning of the bench work for the development of the protocept, a physical embodiment of the written concept. This is often shared with consumers in focus groups to help them understand the product better.

The Research Chef is also tasked with developing procedures, evaluating ingredients, refining recipes, optimizing operations, profit improvement, food safety, and quality assurance while beginning the development of the prototype.

To accomplish all this successfully, the Research Chef must focus on consistent customer satisfaction. Inherent in this statement is the implication that he or she knows the customer or target consumer. A major risk to successful new product or menu item introduction is the failure of the chef to truly understand the target customer. One example might be a chef developing new sandwiches for an elder care facility. The chef loves artisanal breads and has found a local baker to produce an incredible seeded crusty roll, delivered fresh daily. Unfortunately, 99% of the residents have dentures. First, the roll is too crusty for the residents to bite through and chew. Second, the seeds get caught under the dentures causing lacerations and bruises.

Another example is the Mexican restaurant chain that opened a unit in Dearborn, Michigan. They were very proud of the fact that all their food was prepared fresh, even the refried beans that they made every day and served with every platter. Business was going strong until word got out that the beans contained lard. Lard, as we all know, is rendered pork fat. Dearborn is home to the largest Muslim population in the United States, whose religion forbids the consumption of any pork product. These are just examples of what can happen when the uniqueness of customer base or target consumer is not recognized.

Once feedback has been received from consumers on the new concepts, it is time to start working on the product, beginning with the concept statement and ingredient selection. Some of them will be easy—chicken—for the roasted chicken. Some will be functional—gums, starches, salts—to support the product. Others will be discretionary, supportive of the concept statement.

Simple questions start with: Are the ingredients available and obtainable? This is not only a reference to the distribution reality, but to seasonality and a secure, stable supply. Other questions about the supply are most often handled by the procurement function, but not always. That is why the Research Chef has to methodically quiz any potential vendor or supplier before falling hopelessly in love with the next exotic ingredient. It is not always the exotic ingredients that can trip you up.

The rising popularity of many ingredients can impact the markets and food product developers. The original fajitas introduced so many years ago utilized outside skirt meat, considered the best quality. Two things happened to turn the supply situation upside down. First, skirt meat is not considered muscle meat by the Japanese government and therefore could be imported into Japan without a negative impact on their imported beef quota.

Then the popularity of fajitas in the United States grew at an astronomical rate and restaurants soon found themselves using inside skirt as well as outside. Today, cap, lifter, blade, and other cuts are all used to support America's appetite for fajitas, both in the restaurants and at retail.

So, how are ingredients and products selected? Answering a logical progression of questions leads the Research Chef to the appropriate vendors.

- ✓ What products are currently handled in distribution?
- ✓ Do any of our current suppliers make the product?
- ✓ Who makes the product locally?
- ✓ Who makes the product nationally?
- ✓ Who imports the product from overseas?

For each product you look at, there is another checklist.

- ✓ Does this product have the characteristics that are in the concept statement?
- ✓ Flavor/Taste
- ✓ Texture
- ✓ Aroma
- ✓ Appearance

Is this product within the price range required by the finished product?

- ✓ Reverse-cost engineering
- ✓ Pack size
- ✓ Is the manufacturer reliable?
- ✓ Approval/Inspection
- ✓ Contracts
- ✓ Past history

Research Chefs rarely work in a vacuum; they receive input from many sources during the development process. However, during the actual bench work phase of food product development, the taste buds and judgment of the chef creating the food product are relied upon entirely. It is up to that culinary professional to begin filtering out those samples that do not make the grade.

Additional input during the iterative process generally comes from three sources:

- *Internal Informal*

The Research Chef can walk down the hall and recruit members of the NFPD team or other employees to try the products. The benefit is that it is fast, cheap, and easy. The risks are that the results are prone to influence (there might be one or two very strong opinions and egos that sway others), and they may not reflect the taste opinions of your target consumer base.

- *Internal Formal*

A flavor panel can be scheduled with an in-house group that has been screened and trained on how to conduct a proper evaluation. Again, there are cost and time benefits to this approach, but there is a risk of missing the target consumer.

- *External Consumer*

Consumer panel and run preference and rating tests can be scheduled and conducted. This will not only tell us which of the products they like but also give us

diagnostics to adjust various characteristics like flavor and texture. This is expensive and time consuming but yields the best and most actionable data.

Prototype development is an iterative process, a repetition that evolves with the learning received from our testing process. Each bit of learning is incorporated into the prototype until we have met the consumer expectation. This can be validated by testing for purchase intent.

Once the ingredients have been selected and the protocept or prototype has been developed and screened with consumers to determine purchase intent, specifications need to be written for all the ingredients. A specification is made up of a list of the product attributes that can be used by purchasing to negotiate competitive contracts and for operations to verify that the products being delivered are acceptable for use.

The following information should be listed on any specification. Some companies use

- Name or description of the product
- The specification, a physical description of what the product is, looks like, size, weight, grade, chemical components, or ingredients.
- Brand name if appropriate
- Packaging
- Receiving
- Storage

For example:

Potatoes

Russet Burbank variety from Idaho. US #1 or US #2. Minimum average weight 12 oz. each. US #2 generally acceptable January–April. US #1 generally acceptable year round. Grade specified should be the condition of the potato at the time of delivery.

French, Kingston, or Green Giant Brand.

50 lb corrugated fiberboard case.

Store in a cool dry place. Do not refrigerate.

Many developers make the mistake of specifying a brand, which is not necessarily a specification. A specification contains measurable, observable characteristics that are important to the quality of the end product. Elements such as solid content, ingredients, size, weight, and federal grading are all parts of a specification. Brands become part of the specification, not the specification itself.

THE GOLD STANDARD

At this point in the process, the prototype has been identified and validated by consumer testing. The gold standard exists. Whether the Research Chef works for a food manufacturer or a restaurant company, the next steps are the same.

It is time to tear the recipes apart and examine all of the components and processes again. Now you have to define everything involved in the new product in order to develop quality assurance and quality control standards to protect the gold standard.

- *Physical Specifications*

Imagine you are the manager on duty in the restaurant and the new menu items have just been rolled out. You are observing the plates of food as they come out of the kitchen, heading for the guest's table. How do you know the product meets specifications?

Appearance is the first issue to be addressed. Is the item plated correctly? Are all the components present and in the proper amount? Even as a fall back, hot food hot, cold food cold is a good first step.

A description or word picture of the menu item can be written, presenting all the characteristics that can be observed or measured by the manager, without damaging the menu item. This word picture then becomes a quality assurance specification and a tool for use in the restaurant.

Steak is evenly colored with visible grill marks, no excessive charring, seasoning lightly visible, and cooked to guest's degree of doneness. Vegetables are bright and steaming, lightly coated with butter, visible on the plate. Potato is hot (170°F minimum) and topped with guest's selection. Butter and cheese, if present, are melting.

Does that description provide the manager on duty with a quick mental checklist to make sure that the guest gets what he/she is expecting? This is the first or last step (depending on your perspective) in the Gold Standard protection process.

- *Process Specifications*

Drilling down into the menu item presents the next layer of Gold Standard protection, specifications that control the components of the plated menu item. Most often these are items that are prepared in the restaurant, like the vegetables, or are purchased and held for use in assembling the menu item, like butter, cheese, and sour cream for topping potatoes.

The restaurant kitchen can be compared to any manufacturing process, whether you are manufacturing frozen beef stew or automobiles, the basics are the same. You are assembling components and adding value. Every nut, bolt, bumper, and brake that goes into an automobile is controlled by a process specification. In the car it might be how many foot pounds of pressure should be applied to hold the brake pads on. In the kitchen it is how long to fry chicken tenders in 350°F oil until the internal temperature reaches 165°F and the exterior is a crispy golden brown. These instructions are all part of the recipe.

It also equates into measurable and observable characteristics for each component that can be reviewed prior to assembly to ensure the finished menu item meets the Gold Standard. Each component in the manufacturing process should have a short list of characteristics that can be observed and measured. They also should be presented in a way to allow for one way to answer: yes or no. It either is, or is not. Standards, by definition are nonnegotiable, and a simple yes or no indicates whether the product meets standards.

- ✓ Appearance
- ✓ Aroma
- ✓ Taste
- ✓ Temperature
- ✓ Texture

An examination of the quality indicators that might be found for a salad mix used for many items on the restaurant menu.

Lettuce leaves are bright in color, no visible rust, discoloration, or rot. Clean smell, fresh taste. Cold (must not exceed 40°F maximum), crisp, and moist.

If each one does his/her job, the Gold Standard is protected and the guest's expectations are met. Every item on the cook line should have defined quality indicators. This allows the management to conduct a line check, verifying that every ingredient to be used in that shift meets the quality standard.

Restaurants that consistently enforce their standards and conduct line checks by shift have lower food costs and higher quality ratings than their competitions. Of course, they understand what it means to protect the Gold Standard.

- *Ingredient Specifications*

There is another level to the component specification, one touched on above, the ingredient specification. This is the actionable description of the characteristics of the ingredient that Purchasing uses in the exercise of their responsibility to obtain the best product at the best price.

- *Analytical Specifications*

Many tests can be run to support the Gold Standard by further defining the chemical and textural properties. Often, these properties become part of the ingredient specification.

- ✓ Microbiological for food spoilage
- ✓ Pathogenic microorganisms for food safety
- ✓ Chemical components
- ✓ Solids
- ✓ Sugar
- ✓ Salts
- ✓ Textural qualities

One last area to mention, which is used predominantly by food processors and occasionally in the restaurant industry, is the use of analytical specifications to define the Gold Standard. This is used for products that are the identity of the brand or restaurant chain themselves. If concern exists that any changes to this menu item might lead to negative consumer reaction (the risk is extremely high), then you may consider this approach.

An excellent example of this can be found in Chapter 13 (page 293, Figure 12.2 in the first edition, 2000) This is a graphical representation of the comparison of the characteristics of two products. By comparing them in this manner, they are not viewed as the same. As a matter of fact, one has a distinct burnt flavor. This is another way to protect the Gold Standard.

The Research Chef, along with the NFPD team, has the responsibility to make sure that the standards that are set for the Gold Standard are clear, concise, actionable, and nonnegotiable. If the team does not define the Gold Standard, it can not be protected.

SCALE-UP

The new products that have been developed by the Research Chef in the pristine environment of a test kitchen have been approved and are ready for test production or in-restaurant testing. Before that can happen, the formulations must be scaled up to the appropriate size for commercialization. See Chapter 14 for a more detailed discussion of scale-up. There are two areas of consideration that the Research Chef has to address first: batch size for ingredient accuracy and process validation.

- *Ingredient Accuracy*

The bench sample of the Gold Standard needs to be converted into a recipe size that is appropriate for production or restaurant usage. Sales, labor, storage, and supply (deliveries), all impact the decision as to how much of a product to prepare. A recipe for guacamole is used to illustrate this. The recipe made in the test kitchen called for 5 Haas avocadoes and made 2 pounds of finished guacamole. It is estimated that the restaurant will normally use 10 pounds of guacamole per day.

The avocadoes in the test kitchen yield about 1⅓ pounds of pulp, and so 8 pounds of avocado pulp is required to make a 10 pound batch of guacamole. Should you write the recipe calling for 22.8 Haas avocadoes, or should it be 8 pounds of Haas avocado pulp (approximately 23 avocadoes 40 count)?

That small increase in additional pulp should not make a difference in the finished product.

- *Unit of Measure*

The next ingredient is fresh garlic. The Gold Standard called for 15 g, minced. One tablespoon is 15 g and the batch size is 5 times the size of the Gold Standard batch. The initial impulse to use five tablespoons would not be correct.

The first statement about one tablespoon equaling 15 g is true directionally, not absolutely. One teaspoon of water, yes. One teaspoon of salt, or dried basil leaves or saffron, no.

The Research Chef knows the formula or recipe calls for 75 g, which is just over 2.5 ounces. The appropriate step is to mince 75 g of garlic and find the appropriate measuring device that is in use in the restaurant kitchen or production facility that holds that amount. Always try and keep it to a single measure. It is much more accurate to use a 1 gallon or metric measure to add water to a soup as opposed to 16 cups.

- *Consistency of Measure*

Once the Research Chef has calculated the changes in as simple a format as possible for the recipe, the next step is to determine the range of tolerance for critical items in the recipe. Critical items are best described as those that have the most impact: flavorings, seasonings, spices, bases, and thickening agents. Next, consider those items that may be susceptible to measuring inconsistencies like flour, eggs, and protein products.

The guacamole formula has identified garlic, salt, lime juice, and cilantro as ingredients that can have a high impact on the finished product. Experimentation showed that a one half-cup measure, moderately packed, of garlic is what the recipe requires.

The Research Chef must then validate the recipe by making several batches of the guacamole for comparative purposes.

1. A batch of the Gold Standard
2. A production size batch of the new Gold Standard
3. A production size batch, but pack the garlic into the half-cup measure
4. A production size batch, but lightly fill the half-cup measure.

The samples are then tasted to determine the answers to the following questions regarding the production tolerance level for garlic in the guacamole recipe.

✓ Is there a discernable difference among the four products?
✓ Are any of the products objectionable?
✓ Do you have to make any adjustments to the recipe measurement amounts prior to rolling into the operational test or scale-up?

Should consumer input be required at this point? Many factors influence the decision, the most important—how big is the risk? Is the difference that great? If it is, the Research Chef may have to reformulate to get back to the original Gold Standard.

The salt, lime juice, and cilantro have a synergistic and cumulative impact on each other, and must be evaluated following the same protocol as the garlic, before this recipe is ready.

- *Process Validation*

The process or recipe procedures for the new product or menu item must undergo the same scrutiny that the ingredients underwent. *The production floor or kitchen is a completely different environment than the test kitchen.*

- *Handling*

Changing the batch size from 1¾ pounds of pulp to 8 pounds also has an impact on the mixing of the product. It was very easy for a skilled Research Chef to gently combine the ingredients in a stainless steel bowl with a spatula, leaving a chunky texture, but ensuring an even blend of seasonings.

By increasing the amount of pulp, it became more difficult to "make sure ingredients are evenly blended," of course to some, evenly blended means smooth.

The problem is solved by revising the recipe. Now the lime juice, garlic, and other spices are mixed in a small bowl with 2 pounds of the pulp, until the mix is smooth. This is added to the remaining pulp, tomatoes, and onions. Because of this alteration, the quality indicator must be changed to reflect that guacamole is chunky with visible pieces of tomato and onion, and has a rounded fresh avocado flavor accented with garlic, lime, and cilantro.

- *Cooking*

The changes that occur when you move from the bench top to production are amplified for cooked items like a soup or chili prototype. Let's cite an example. Chili, prepared in 200-gallon steam-jacketed kettles will be different in many ways from the prototype that was approved. Everyone loved the original chili, swearing that it was the best they had ever had.

The chili was prepared in 1-gallon batches on the range top in an eight quart stock pot to simulate the kettle. First a little oil, then onions, peppers, garlic, and then beef was browned. Next tomatoes, seasonings, and some masa harina were added to complete the product.

The cooking procedure that was employed was a two-step process. When the meat was first added to the stock pot on the range, it immediately came into contact with the surface and began the dry heat browning process. This operation set the protein into an initial texture. The meat then began to caramelize and develop a flavor typical of a grilled meat, as the moisture evaporated. Once this process had occurred, the remaining ingredients were added, changing the cooking process to moist heat cooking, where a stewing process occurred.

The steam-jacketed kettle had significant differences that were not recognized. First, 4 pounds of meat in the test kitchen and 800 pounds in the production operation reacted differently. More meat was in contact with the cooking surface in the test kitchen, which had an increased rate of evaporation and a quicker development of the caramelized protein flavors. The meat in the kettle never had the opportunity to develop the flavor and texture, because not enough was in contact with the cooking surface. Second, the rate of evaporation was much slower because of the kettle configuration and the amount of product involved. The end result was that the product in the kettle was cooked under moist heat conditions and developed a softer, mushier texture and a different flavor than the test-kitchen product.

Again, the iterative process continues with adjustments being made along the way. Each time a change is made, the resulting product should be compared with the original Gold Standard.

ROLL OUT

The new products have passed the operations test and the market testing, and after a few minor revisions are ready for the roll out. It is time for the moment of truth. Do the new items get rolled out or not?

At this point, the NFPD meets to review the data and to make a decision. Each team member reports on their area of concern, covering the positives and negatives. Adjustments are made where necessary and next steps are set.

The Research Chef now slips into the quality assurance advisor role, monitoring the product for any signs of weakness and supporting the rest of the team. This time period can be very beneficial because it is here that valuable learning about the process can take place. Being able to stand back, observe, and learn will make the next product introduction even more efficient.

POST-ACTION REPORT

Following the introduction of a new item, it is time to reconvene the NFPD team to discuss the product and the process. There exists an enormous knowledge base among the team that, if given the opportunity, can be shared to make the team much more effective, like the lean, mean new product machine it can be. All members have to check their egos at the door and come to the meeting prepared to answer these three questions:

- ✓ What went right?
- ✓ What went wrong?
- ✓ What could we have done better?

On the basis of those answers: Revise the process and celebrate your success!

CONCLUSION

The NFPD process is an iterative process, best managed by a multidisciplinary team. The team derives its charter from the company strategy, supported by top management with appropriate funding. The Research Chef is a key team member, providing creative culinary support through an understanding of the customers' needs. The interpretation of those needs and the development of food or meal-based solutions is the primary function of the Research Chef. The authenticity of the Gold Standard and its protection throughout the process is the responsibility of the Research Chef.

BIBLIOGRAPHY

Culinary Institute of America 2002. *The Professional Chef*, 7th ed. John Wiley & Sons, New York.

Culinary Institute of America 2003. *The Cooking Methods Series, Dry Heat Methods*, Volume 1.

The Culinary Institute of America 2006. Used with permission.

Maxim Integrated Products 2006. *Third Quarter Earnings Report*, November 2, 2006.

McGee, H. 1984. *On Food and Cooking*. Charles Scribner's Sons, New York.

Montagne, P. 1938. *Larousse Gastronomique*. Crown Publishers, Inc., New York.

Perdue Foods website November 2006. Perdue and Short Cuts are registered trademarks of Perdue, Inc., Salisbury, Maryland.
Press release 11/30/2001. Tufts University.
Research Chefs Association 2005. Atlanta, GA.
Trademark property of Apple Computer, Inc.

13 Consumer Sensory Testing for Food Product Development

Anna V. A. Resurreccion

CONTENTS

Consumer sensory testing is necessary in food product evaluation for product development guidance, product improvement and optimization, and maintenance. The testing methods employ untrained individuals and larger sample sizes than required in sensory analytical test methods. Consumer affective tests appear to be relatively simple to plan and conduct. Unfortunately, the seemingly easy task is often invalidated by improper data-collection methodology, which invariably leads to faulty interpretation of results. The validity and reliability of the consumer testing methodologies are extremely important if test results are to be used as a basis for business decisions.

The primary purpose of consumer sensory tests is to assess the personal response by current and potential customers of a product, product ideas, or specific product characteristics. Consumer evaluation concerns itself with testing certain products using untrained people who are or will become the ultimate users of the product. These products are evaluated on the basis of appearance, aroma, taste, smell, touch, and hearing. Validity and reliability of the consumer testing methodologies are extremely important. Consumer sensory testing is necessary throughout the various stages in the product cycle. These stages include the development of the product itself, product maintenance, product improvement and optimization, and assessment of market potential.

Sensory acceptance tests are conducted during product development for product development guidance, to screen products, and to identify those products that are significantly disliked and those that match or exceed a specified target product for acceptance. Sensory acceptance tests indicate the acceptance of a product without the package, label, price, and so on. The implicit goal behind any and all sensory evaluation efforts in the food industry is to enhance quality to improve appearance, flavor, and texture as perceived by consumers in order to influence their food choices (translated into purchase) at the point of sale.

Food products created today are designed to satisfy the needs and wants. Product changes may arise from a variety of sources. One of the recommended strategies to assess a concept early in the development stage is by using a focus group.

In a focus group, consumer responses are used to qualitatively assess the concept or to identify critical attributes of a food product or concept. Beyond the concept development, prototype products that possess the critical attributes that are valued by the consumers are developed. At this stage, consumer sensory tests are useful tools for product guidance. The objective of the research is to assess performance potential and product guidance for further development, with the goal of maximizing acceptance.

Alternate product prototypes are narrowed down to a manageable number of final prototypes. Acceptance tests give an estimate of product acceptance. Although acceptance tests are an essential part of the decision-making process in new product development, a successful acceptance test does not guarantee success of a product in the marketplace. Sensory acceptance tests are not a substitute for comprehensive consumer testing.

INTRODUCTION

Today's consumer is confronted with choices from thousands of new product introductions and line extensions yearly. Manufacturers compete for valuable retail shelf space. This competition results in a multitude of new food product introductions that fall short of the product development team's expectations, visions, and hopes (Lord, 2000). No less than 80–90% of new food products introduced fail within 2 years of introduction, resulting in enormous loss to the U.S. food industry (Morris, 1993).

Sensory analysis provides the product development team with information on the sensory properties of products, ingredients, or other related information and services, and how these relate to consumer perception of quality characteristics and to consumer liking. This chapter discusses how sensory evaluation activities are used to arrive at outcomes for each milestone in the product development process.

SENSORY SCIENTISTS ARE AN INTEGRAL PART OF THE PRODUCT DEVELOPMENT TEAM

Establishment of cross-functional new product development teams facilitates assembly, coordination, and use of necessary resources from the many functions required to successfully launch a new product. The cross-functional teams may include product managers, culinologists™, product developers, packaging technologists, engineers, marketers, and sensory practitioners who work in various stages in the food product development process, from the very early stages to postlaunch. Communication between cross-functional team members is enhanced, resulting in synergism, increased speed to market, and potential for success of new product development efforts.

Sensory evaluation has proven to be a necessary and powerful tool to guide food product development (Moskowitz, 2000). The primary role of sensory analysis is to provide information on how the sensory characteristics of products, ingredients, or other related information and services relate to consumer liking and the quality characteristics as perceived by the consumer. The sensory scientist in the food product development team provides assistance in implementing the studies to collect data from consumers as well as trained panelists. Finally, the sensory practitioners provide invaluable guidance in interpreting results (Moskowitz et al., 2003).

MEASUREMENT OF PRODUCT QUALITY

Much of the market success or failure of a new food product results from consumer perception of its sensory quality. Therefore, in new food product development, the very first step is the identification of the food product followed by identification of critical quality attributes of the product, by regular consumers of a product. When the quality attributes important to consumers have been determined, appropriate designs for systematic product optimization can be developed.

New product research can be classified into two test categories: (1) qualitative and (2) quantitative. Both types of tests have a role in the different stages of new product development.

Qualitative research is used to identify *critical* quality attributes. One of the most frequently used qualitative research methods is the focus group. The focus group is particularly valuable in language generation and very early stage product development (Sokolow, 1988). Focus groups are composed of 8–10 consumers recruited to fit specific demographic, attitudinal, and usage characteristics. Focus group testing should be conducted by an experienced moderator who helps to stimulate and direct the discussion (ASTM, 1979; Sokolow, 1988). Focus groups are effective in identifying those critical product attributes that are considered important by consumers, and should be included and maximized in the product (Resurreccion, 1998). Focus groups are likewise important to identify characteristics that are undesirable to consumers, and should be minimized or eliminated from the product. After the critical quality attributes have been identified, the team should use quantitative tests. Focus group results can be used to assist in questionnaire design for succeeding quantitative tests.

Quantitative research includes tests conducted to measure overall acceptance or consumer affective responses to the critical attributes, defined as those attributes that determine a product's preference or acceptance. These critical attributes can be quantified by using descriptive sensory analysis or physicochemical measurements, and the intensities or magnitude of specific attributes can be related to consumer responses. Consumer testing of new product prototypes or market candidates can provide valuable information to product developers (Lawless and Heymann, 1998).

Among the cross-functional team members, the sensory scientist should be best qualified to organize and conduct consumer sensory testing. Substantial literature on consumer marketing studies exists, but only a relatively few references on consumer sensory techniques have been published. Academic courses on consumer testing methods and the details of consumer sensory testing of food products are not commonly offered. Lawless and Heymann (1998) indicate that novices are frequently trained through "shadowing" an experienced sensory researcher in industrial practice. There are a few general guides to consumer sensory testing, including the American Society for Testing and Materials (ASTM, 1979) and a reference by Resurreccion (1998) on consumer sensory testing for product development.

Consumer Input Is Critical

Commercial success for any product requires consumer input throughout each step in the food product development process. A product needs to satisfy a consumer want or need to be successful. *No one can guide the food product development team better than the consumer can.* Moskowitz (2000) listed three major reasons for obtaining consumer input as (1) product acceptance—food manufacturers realize that consumer acceptance, first and foremost, determines product success, (2) speed to market—consumer input shortens development time, and (3) the cost of product testing is minimized.

Sensory Research in the Product Development Process

Consumer and sensory descriptive analysis tests play a role in every stage in the food product development process. Four major stages in the food product development process and the appropriate consumer or sensory tests conducted during each stage are listed as follows:

Stage I Product strategy and definition—Consumer focus groups
Stage II Implementation and marketing—Consumer surveys, consumer sensory tests, and descriptive analysis tests

Stage III Product commercialization—Consumer sensory tests
Stage IV Product launch and evaluation—Consumer sensory tests

These consumer and sensory tests are described as follows:

1. *Consumer surveys*: provide information on product characteristics that consumers seek in a product, which can lead to an assessment of preferences and acceptances, negatives and positives, and norms. Macfie and Thomson (1994) and Meiselman (1996) provide technical information to guide the development of a consumer survey questionnaire.
2. *Descriptive sensory analyses*: provide information on the intensity of the sensory attributes in a product, and are used to monitor changes and document perceived sensory attributes of existing and new products. The descriptive test is usually conducted by the sensory group. Applications of descriptive sensory analysis are found in Gacula (1997) and in the ASTM publication by Hootman (1992).
3. *Consumer tests*: provide information on the degree of liking and consumer perception of intensities of product attributes, for product improvement, new findings, preferences, and other acceptability measurements. For details on consumer testing for product development, consult Resurreccion (1998).

Consumer sensory tests and descriptive sensory analysis are techniques. Selection of all sensory tests should be based on the objectives for the study. The question that must be asked is "What is the test intended to measure?" If the objective is to quantify consumer acceptance, then a consumer affective test is appropriate; if the objective is to describe the characteristics of a product, then the test should include a descriptive sensory analysis of product attributes.

Consumer Acceptance Tests Are Not a Substitute for Marketing Research

Consumer sensory acceptance tests are neither substitutes for nor a competitive alternate to the standard large-scale market research test. Therefore, consumer sensory acceptance tests should not be used in place of large-scale market research tests, when the latter are required. The consumer sensory acceptance test is carried out by the sensory scientist whereas market research is conducted by the company's marketing function. The consumer sensory test with 50–100 participants is a relatively small test compared to the large market research test that involves more than 100 respondents. The consumer sensory acceptance test quantifies consumer acceptance of the product or acceptance of its sensory properties during the early stages of food product development; market research focuses on characteristics of the consumer and identifies what appeals to the consumer, and is conducted in later stages. Both types of research complement each other, and rely on different testing procedures. Both are vital in the development of new food products.

Acceptance and Preference Tests

Consumer sensory affective tests are conducted throughout the different phases in the product development process. Two approaches to consumer sensory acceptance testing are the measurement of preference and the measurement of acceptance (Jellinek, 1964). Acceptance tests measure consumer acceptance or liking of a product. Consumer acceptance of a food may be defined as (1) an experience, or feature of experience, characterized by a positive attitude toward the food and/or (2) actual utilization (such as purchasing or eating) of food by consumers. In contrast, preference tests measure liking of one product, in a set, over another (Stone and Sidel, 1993). The methods most frequently used by sensory practitioners to determine preference and quantify acceptance are the paired preference tests and tests employing the 9-point hedonic scale, respectively. Acceptance measurements can be made on a single or any number of products and do not require comparison to another product. The questions asked during acceptance tests are, "How much do you like the product?" (Stone and Sidel, 1993) or "How acceptable is the product?" (Meilgaard et al., 1991; Stone and Sidel, 1993). In preference tests, the questions asked are: "Which product do you prefer?" or "Which product do you like better?"

Preference methods are used to determine differences in preference, but not differences of products; discrimination tests should be used for this purpose. A preference test is occasionally used to determine whether the company "gold standard" or target food product is preferred over products from the different plants, or made with ingredients from different suppliers. However, if the objective of the test is to determine whether the "gold standard" or target food product is different from the product from different plants or suppliers, the discrimination test is the more appropriate test.

Acceptance scales. The 9-point hedonic scale is the most frequently used scale for acceptance and has been validated in the scientific literature (Stone and Sidel, 1993). Nine-point scales, however, may not be optimal for some scaling purposes, such as when working with children. Adaptations of the hedonic scale work better with younger consumers, such as the 9-point facial scale for children 8 years or older (Kimmel et al., 1994; Kroll, 1990). The 3-, 5-, and 7-point facial hedonic scales were found to be more useful when working with 3, 4, and 5 years old children, respectively (Chen et al., 1996).

Ratio scales were proposed by Moskowitz (1974) as a method to quantify acceptance and preference, and the magnitude estimation method had been validated by experimental psychologists for many years, and is a scaling method of choice in many research laboratories. However, it is not as frequently used by sensory practitioners as the 9-point scale.

Methods Used in Acceptance and Preference Testing

The three types of tests used in consumer acceptance testing are (1) paired preference, (2) ranking, and (3) rating tests. However, two of the most frequently used tests to measure consumer preference and acceptance are the paired preference test, and ratings of acceptance using one or another form of the hedonic scale.

Paired preference tests. To execute this test, the researcher presents a panelist with two samples, typically simultaneously but occasionally sequentially. The preference test is simple and straightforward to execute, and works well even when the consumer panelists have few reading or comprehension skills. The panelist is asked the question, "Which sample does the panelist prefer?" The instructions may or may not force the panelist to make a decision. If a forced choice is imposed, the panelist may not give a "no-preference" response (ASTM, 1996) and must indicate a preference for one sample over another. The paired comparison method to elicit preferences may likewise be used for multiple-paired preferences within a series of samples, such as one standard product against each of several experimental samples. One or more sample pairs of products may be tested during a session. The upper limit for the number of sample pairs that can be tested is determined by physiological and psychological constraints (Resurreccion, 1998).

Paired preference test—advantages. Paired preference tests are easy to organize and implement. There are only two orders of presentation possible, A–B and B–A. However, each order of presentation must be used an equal number of times during the test, or position bias will likely occur. In many cases, a panelist usually evaluates only one pair of products in a test, and no replications are conducted without a replication. If the "no-preference" choice is used, the number of consumers answering a preference for either sample may be decreased because some of the panelist will choose the "no-preference" option. Thus, when the "no-preference" option is introduced, a larger number of consumers, such as 100 or more, is needed. Furthermore, if the researcher chooses to include the option of "dislike both equally" and/or "no-preference," the analysis should be performed only on those responses that show a definite preference for one of the two samples. Stone and Sidel (1993) suggested that if at least 5% of participants select the "dislike both equally," it is necessary to seriously question either the appropriateness of the products being tested or the sample of consumer panelists recruited to participate in the product evaluations.

Paired preference tested—disadvantages. The paired preference test is less efficient than a rating test because only one response per product pair is obtained as opposed to one response per product, when a rating test is used. Furthermore, when the "no-preference" category is not used, it is virtually impossible to determine whether or not both products were disliked. Because preference testing asks for an overall evaluation, the testing may be especially susceptible to unintentional biases that can arise, such as the effect of slight differences in placement on the serving tray, serving temperatures, sample volume, and variability of the sample from the manufacturing process, among others (ASTM, 1996). Executional issues also emerge in preference tests, especially when the panelist evaluates more than two products in a multiproduct test. This situation may result in considerable interaction due to flavor carry over from one product sample to the next. If a sequential presentation is selected for the paired preference test, memory can be a confounding variable.

Paired-preference test—data analyses. In the analysis of the paired preference test, the probability of randomly selecting one of the two products is 50%, or one out of two. The null hypothesis states that the consumers will pick each product an equal number of times; the probability of the null hypothesis is $p = .5$. When analyzing the results of the paired preference test a two-tailed test is used, because there is no prior expectation regarding which of the two products will be preferred by the consumer population. When there is an expectation that the new product will be preferred, the appropriate test is then a one-tailed test (Stone and Sidel, 1993). Analysis of the paired preference tests is accomplished using any of the following statistical methods: binomial, chi-square, or normal distributions, or use of tables. Basic sensory evaluation texts (Lawless and Heymann, 1998; O'Mahony, 1986; Resurreccion, 1998; Stone and Sidel, 1993) detail all of the analyses methods used for paired preference data.

Binomial distribution and tables. Published binomial probability calculations are manageable for small panels, but as the number of observations becomes large, the binomial distribution begins to resemble the normal distribution (Lawless and Heymann, 1998). Before calculators and computers, these calculations were quite cumbersome. However, the task was made simpler after Roessler et al. (1978) published tables (Table 13.1) that use the binomial expansion to calculate the number of correct judgments and their probability of occurrence. Use of the tables greatly simplifies the procedure of determining whether or not a preference for one sample over another is statistically significant in paired preference tests.

The Chi-Square (χ^2) *test.* Typically, researchers use the binomial test to analyze preference data. However, researchers may often use the chi-square test to confirm hypotheses about frequency of occurrence. The chi-square test may be used to obtain the same type of information as the binomial test, but it has the advantage that it can be used to test hypotheses when the response falls into one of several categories, not just two categories. Calculation of the chi-square statistic can be obtained from standard texts such as O'Mahony (1986).

Paired preference test—misuse. The paired preference test is often combined with other tests, resulting in biased results. To be more specific, the paired preference test should not be combined with discrimination tasks. Preference questions should not follow a difference question or vice versa. Consumers recruited to participate in a preference test must be naive users of the food product and will not qualify as panelists in discrimination tests. Furthermore, they should not be asked to focus on differences first, and then to another type of task, to test for preference. Third, unlike in consumer tests, panelists used in a discrimination test are not recruited to represent the target population of a food product sample. Finally, a discrimination test is an analytical test whereas the paired preference tests are preferred for a sample.

RATING TESTS

Rating tests—Hedonic measurement. The 9-point hedonic scale (Peryam and Pilgrim, 1957), developed over 50 years ago, is widely used to measure food acceptance. It is a simple, unambiguous rating scale, used for many years to measure the acceptance of a food and to provide a benchmark number with which to compare products, to compare batches, and to assess the level of acceptance of products in a competitive category. The consumer's task is easy: record the degree of liking, using the scale. There are four presumably equally spaced categories for liking, a neutral point, and then a corresponding four presumably equally spaced categories for disliking. The

TABLE 13.1
Minimum Numbers of Agreeing Judgments Necessary to Establish Significance at Various Levels for the Paired Preference Test (Two-Tailed, $p = .05$)[a]

Number of Trials (n)	Probability Levels 0.5	0.01	Number of Trials (n)	Probability Levels 0.5	0.01
7	7		32	23	24
8	8	8	33	23	25
9	8	9	34	24	25
10	9	10	35	24	26
11	10	11	36	25	27
12	10	11	37	25	27
13	11	12	38	26	28
14	12	13	39	27	28
15	12	13	40	27	29
16	13	14	41	28	30
17	13	15	42	28	30
18	14	15	43	29	31
19	15	16	44	29	31
20	15	17	45	30	32
21	16	17	46	31	33
22	17	18	47	31	33
23	17	19	48	32	34
24	18	19	49	32	34
25	18	20	50	33	35
26	19	20	60	39	41
27	20	21	70	44	47
28	20	22	80	50	52
29	21	22	90	55	58
30	21	23	100	61	64
31	22	24			

[a] Value (x) not appearing in table may be derived from $x = Z[(n + n + 1)/2]$, where n = number of trials, x = minimum number of correct judgments, if $x = 1.96$ at a probability (α) = 5%, and Z = 2.58 at probability (α) = 1%.

Source: Adapted with permission from Roessler, E.B., Pangborn, R.M., Sidel, J.L., and Stone, H. 1978. *Journal of Food Science*, Institute of Food Technologists. From Anna V.A. Resurreccion, 1998, *Consumer Sensory Testing for Product Development*, p 242, Table E.1. Copyright (1998). Aspen Publications. With kind permission of Springer Science and Business Media.

hedonic scale categories are:

- Like extremely (=9)
- Like very much (=8)
- Like moderately (=7)
- Like slightly (=6)
- Neither like nor dislike (=5)
- Dislike slightly (=4)

- Dislike moderately (=3)
- Dislike very much (=2)
- Dislike extremely (=1)

A considerable amount of thought and research was performed to structure the scale, in terms of the number of categories and anchors in each category. Modifications to the 9-point scale continue to be suggested and used, such as eliminating the neutral point (neither like nor dislike), or simplifying the scale by eliminating options such as the "like moderately" and "dislike moderately" points on the scale, or truncating the endpoints by eliminating the "like extremely" and "dislike extremely" points; however, these suggested modifications were either unsuccessful or had no practical value. These modifications would introduce biases such as the problem of end-use avoidance, that is, the hesitation of panelists to use the end categories on the scale. A concern has been that the scale is bipolar scale, but the analysis of the results is unidirectional. However, there is no evidence that consumers have difficulty with the scale and that the statistical analysis presents a problem (Stone and Sidel, 1993). Another major concern is the number of available scales versus the use of the scales by panelists. Truncating a 9-point scale to a 7-point scale may leave the consumer panelist with only a 5-point scale. It is, therefore, best to avoid the tendency to truncate scales (Lawless and Heymann, 1998). In using shorter scales, such as a 3-point category scale, the researcher loses a great amount of information about the preference among products. Continuous, unstructured line scales with three word anchors, one at each end and the midpoint, do not raise this objection.

In some instances, an adaptation of the 9-point hedonic scale in the form of a 9-point facial scale proves useful with children. The 3-, 5-, and 7-point facial hedonic scales were found to be appropriate for 3, 4, and 5 year old children, respectively (Chen et al., 1996). Figure 13.1 shows examples of the facial hedonic scales.

Rating tests—food action rating scale (*FACT*). The food action rating scale (FACT) was devised by Schutz (1965) to measure acceptance of a product by a population, and at the same time combine the attitude toward a food with a measure of expected action that the consumer might take to consume (or not consume) the food. The FACT scale has nine categories, as shown in Table 13.2. The researcher presents the samples sequentially, generally in a randomized order, balanced across the panelists. The panelist selects the appropriate statement and ratings are converted to numerical scores to facilitate statistical analysis of data (Resurreccion, 1998).

Ranking Methods

Ranking is an extension of the paired preference test with greater than two samples. Thus, many of the advantages of the paired preference test apply to ranking. These include simplicity of instructions to participants, requiring a minimum amount of effort to conduct, requiring relatively uncomplicated data handling, and making minimal assumptions about level of measurement because data are treated as ordinal, meaning that only the rank order conveys information and nothing else (Lawless and Heymann, 1998). Three or more coded samples are presented simultaneously; samples presented are sufficient in quantity so that the panelist can retaste the product.

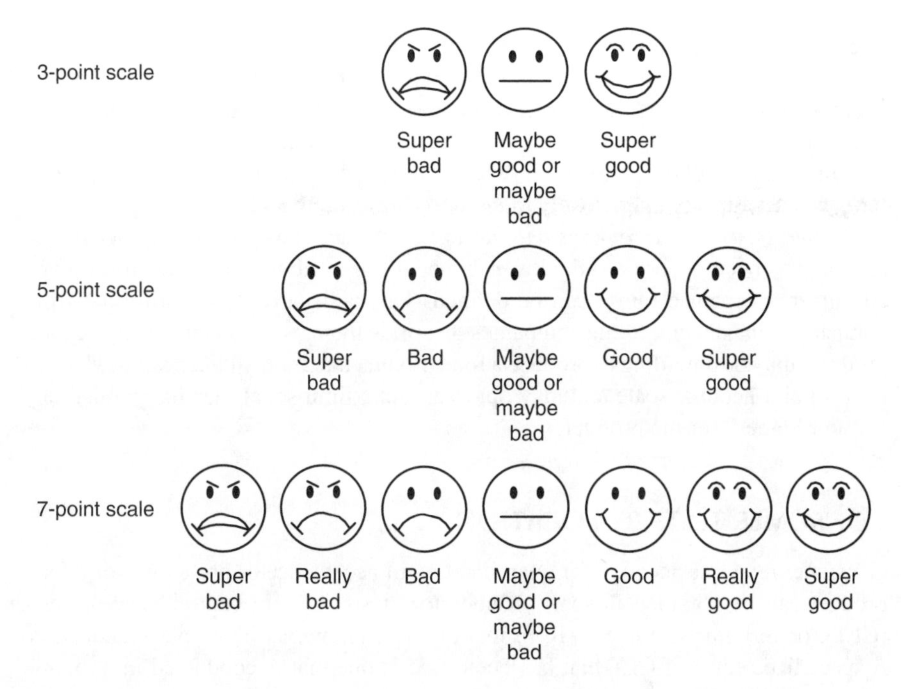

FIGURE 13.1 Facial hedonic scales used to determine preschool children's preference (From Chen et al., 1996. *Journal of Sensory Studies* by Food and Nutrition Press, Inc. Reprinted with permission from Blackwell Publishing. From Anna V.A. Resurreccion, 1998, *Consumer Sensory Testing for Product Development*, p 22, Figure 2.2. Copyright (1998). Aspen Publications. With kind permission of Springer Science and Business Media.)

TABLE 13.2
Descriptors Used in the Food Action Rating Scale by Schutz (1965)

I would eat this food every opportunity I had.
I would eat this very often.
I would frequently eat this.
I like this and would eat it now and then.
I would eat this if available but would not go out of my way.
I do not like it but would eat it on an occassion.
I would hardly ever eat this.
I would eat this only if there were no other food choices.
I would eat this only if I were forced to.

Source: From *Consumer Sensory Testing for Product Development.* 1998, p 22 by Anna V.A. Resurreccion, Table 2.2 Copyright 1998 by Aspen Publications. With kind permission of Springer Science and Business Media.

The number of samples tested depends upon the panelist's span of attention and memory as well as on physiological considerations. With untrained or naive panelists, no more than four to six samples should be included in a ranking test (ASTM, 1996). The panelist assigns an order to the samples according to his or her preference. The disadvantage of the paired preference method is likewise found in ranking in that rank order evaluates samples only in relation to one another.

Scaled-based norms for consumer acceptance. Once consumer acceptance for the product has been quantified, the lower boundary of attribute intensities for a given consumer acceptance rating can be defined. This is usually done with input from company management. Some companies that pride themselves on the quality of their products may be unwilling to produce a food product that is only "liked slightly" (=6) on a 9-point hedonic scale and may opt to accept a limit set at the "like moderately (=7) or higher" for the product.

CONSUMER TESTS BY LOCATION

Consumer responses needed for quantification of acceptance or preference of a food product can be obtained in sensory laboratory tests (LT), in central location tests (CLT), or in home use tests (HUT) that are also known as home placement tests. A specialized form of CLT that has been used is one that is conducted in a mobile laboratory (Resurreccion, 1998).

SENSORY LABORATORY TESTS

Sensory laboratory tests, as the name implies, are conducted in the sensory laboratory of a food manufacturing company, consulting firm, research organization, or university. The laboratory test is the most frequently used by R&D, when the objective is to measure consumer acceptance. The panel is generally composed of consumers who are recruited from a list obtained from various sources or a consumer database. The consumer database, assembled and maintained by the sensory practitioner, is composed of prerecruited consumers screened for eligibility to participate in the tests, and who have agreed to participate. Usually 25–50 consumers participate in a test; however, 50–100 responses are considered adequate (IFT/SED, 1981). Thus the sensory laboratory test may run over several sessions to accommodate the required number of participants depending on available facilities. The recommended number of products per sitting is two to five.

Sensory laboratory tests—advantages. The major advantage of using a sensory laboratory to conduct a consumer affective test is its location and the convenience of the location of the testing facility to the researchers. The laboratory test is particularly appealing to manufacturing companies because of the accessibility of the laboratory to a large number of employee participants, wherein any number of employees can be recruited to participate in the tests on short notice. Although the sensory laboratory provides a convenient location for the research team and is accessible to employees, it is less convenient than a central location for local residents or other consumer panelists who will participate. When the consumer panelists used in the laboratory test are employees of the company who have participated previously in consumer tests, the researcher often can dispense with the orientation, thus saving time.

Among the other consumer tests, the laboratory test allows for the most controlled conditions. Control over sample preparation and testing conditions, including lighting and environmental conditions, is maximized in a sensory laboratory. Control of the product evaluation environment includes control of the testing environment such as odors, lighting, noise, and other distractions. Further, the laboratory test is conducted in individual partitioned booths that isolate panelists from each other. Lighting can mask color differences and other appearance factors so that panelists can focus on either other sensory attributes or on acceptance.

Most sensory laboratories are located adjacent to a fully equipped kitchen. This siting facilitates control over many factors such as sample preparation, the duration and temperature of cooking, holding and reheating, and serving. Often, these well-controlled laboratory tests are designed for a limited number of panelists—for example, a total of 25–50.

Another major advantage of conducting a consumer test in the sensory laboratory is the rapid turn-around time for data to be analyzed and results obtained, because of the proximity to the data processing facilities. When a computerized sensory data entry and analysis system is employed, feedback of results is almost instantaneous and available shortly after the last consumer completes the test.

Sensory laboratories testing—disadvantages. The major disadvantages of conducting consumer tests in the sensory laboratory generally pertain to the nature of the panel that is used. By having the laboratory in the corporate complex, there is great temptation to use corporate employees as panelists, simply on the basis of costs. It presumably costs "less" to work with an employee than to recruit an outside consumer panelist, although a true accounting of the costs would probably reveal that the corporate employee actually costs more—it just that the costs do not appear as line items and "extras." The sensory practitioner needs to remember the nature of the risks associated with using employees in product maintenance tests; employees should not be used in food product development, improvement, or optimization tests. If a decision to use employees in food product maintenance is made, care must be taken to recruit only those employees who are not familiar with the production, testing, or marketing of the product. In addition, employees' rating patterns must be compared to that of a representative panel.

If consumers are screened from a prerecruited consumer database, a majority of consumers who would be most willing to participate are those whose homes or place of employment are located close to the sensory laboratory. In addition, local residents who participate in laboratory tests may be biased due to the belief that all the food products being evaluated are originating from a specific company or plant. If care is not exercised to recruit an appropriate panel, then the demographic characteristics of the panel may be skewed on characteristics such as income, education, race, and any other factors that may have an influence on usage patterns for the food.

Another disadvantage is that a sensory testing booth is very different from a real eating environment and the realism of the laboratory test can be questioned (Hersleth et al., 2005). In some studies, laboratory measurements of food preference were shown to be poor predictors of consumption (Cardello et al., 2000; Kozlowska et al., 2003). An example of the limited information that may be obtained from a laboratory test is the distortion of normal consumption such as sipping from a sample cup versus drinking from a full glass. These limited procedures may influence detection of positive or negative attributes (Meilgaard et al., 1991; Stone and Sidel, 1993). Further, when the food is prepared or tested in the laboratory, product performance may be different from what the researcher would observe in the more natural home-use situations. A final disadvantage of the sensory laboratory test is the limited amount of time that the consumer is exposed to the product, compared to a HUT.

Sensory laboratory tests—panel size. The consumer laboratory test panel is composed of consumers who are prerecruited and screened for their eligibility to participate in the tests. Otherwise, if the panelists are not prerecruited consumers from adjacent neighborhoods (nonemployees), they usually are company employees. Good practice usually works with 25–50 responses per product, at least 40 are recommended by Stone and Sidel (1993) and 50–100 responses recommended by IFT/SED (1981). Consumers differ greatly from each other, and so it is important to have a sufficient number of ratings. In a consumer test consisting of only 24 panelists, it is difficult to establish a statistically significant difference in a test with the small number of panelists. However, even with small panel sizes, it is possible for the sensory practitioner to identify trends in the responses and to provide direction to the food product development team charged with developing the new food product. With 50 panelists, statistical significance of differences has a far greater chance to emerge. Stone and Sidel (1993) warn that different products have their own requirements. Smaller panels can generate results that are highly significant when the intrinsic variability within a sample is small, so that sample-to-sample differences can easily appear. In contrast, when the intrinsic variability within a single food product is high from sample to sample, then larger panels are needed.

Sensory laboratory tests—panelist recruitment. Regardless of the test location, preference or acceptance tests require different selection criteria for panelists compared to those used in discrimination or descriptive tests. The composition of the preference and acceptance panels must be defined by the marketing and sensory functions of the company. In acceptance tests, the characteristics of the panelists should match those of the target market for the product being tested.

One approach to successful recruitment of panelists involves the development of a database of consumer households with members who may be available for testing. In most cases, the demographic characteristics of the individuals in the database are entered upon initial recruitment and updated on a regular basis. Recruitment of consumer participants may be done through a market research agency or by designated recruiters in the sensory practitioner's team who recruit using telephone calls, referrals, or personal contacts. Other methods often used to recruit panelists are random selection from a telephone directory, random digit dialing, posters in retail stores, mailing lists from organizations, assorted consumer databases, and intercepts at malls, shopping areas, or restaurants. A combination of these methods is generally necessary and thus frequently used when developing the initial database.

Employees versus nonemployees. Employees should not be used in affective tests unless a sufficient amount of testing has been conducted to ensure that the employees used are not knowledgeable about the products tested, and exhibit food consumption and preference patterns similar to that of the target market for the product.

Using local residents. One popular and efficient approach to recruitment brings local residents from surrounding neighborhoods into the laboratory. Using local residents as panelists is convenient for the sensory team, making it easier for them to meet recruitment needs and schedule tests. The availability of motivated local panelists allows tests to be completed more rapidly, reduces work interruptions on employees who participate, and reduces hidden costs associated with using employees as panelists. However, the method of recruiting local residents involves prerecruitment, database management, scheduling and budgeting activities, and accessibility and security issues. A system must be established to contact and schedule local residents for a test and remind them, as needed. A budget and means for providing incentives or honoraria to panelists is required; these are not necessary in tests involving employees.

When using local residents, access of the laboratory to panelists should be limited so that they do not wander in restricted areas. All panelists should be escorted at all times for security reasons. The advantages of using local residents are that participants are usually highly motivated; the majority will show up promptly for each test and are willing to provide considerable product information.

Other options. Another option for a company is to develop and maintain a satellite or off-premises laboratory test facility. The off-premises solution might well eliminate many problems associated with bringing local residents to the company premises. This option will be more costly than bringing people to the sensory laboratory. Another alternative will be to contract with a sensory evaluation company, assuming they have the necessary sensory test resources.

In each case, the product development team will need to assess its current acceptance test resources and anticipated workload before determining which, if any, of these options is feasible. There is no question that consumer acceptance tests need to be conducted in new product development. The question is where the testing will be conducted; who the test subjects will be, and what will be the relative costs of the different alternatives will be (Stone and Sidel, 1993).

Sensory laboratory tests—screening. For a given test, consumer panelists will need to qualify according to predetermined criteria that describe the target consumer for the product. A recruitment "screener" should be developed by the sensory practitioner in collaboration with the "client," that is, the person who requests the test. The recruitment screener should be designed to ensure validity of the screening process and allow for intermediate and final tallies of participants, to ensure that quota requirements are being met.

The selection of prospective participants should be conducted as rigorously as possible. Consumer panelist recruitment should be monitored, whether the job of recruitment is done by the sensory function or contracted to a firm that handles consumer recruitment. The undesirable, but often utilized, practice of using those consumer panelists who can be most conveniently contacted, such as those people who live close to the facility, should be discouraged. Similarly, the use of friends and relatives of project staff should be avoided. If used, these individuals may skew the demographic characteristics of the panel and bias the results. Demographic criteria such as age, gender, frequency of product use, availability during the test date and other criteria such as employment with the sponsoring company or similar business concerns, and other security screening criteria are often used. In addition, ethnic or cultural background, occupation, education, family income, experience, and the last date of participation in a consumer affective test may be used. Frequency of participation is an important characteristic to be aware of. With declining response rates, inexperienced recruiters often opt to work with a limited set of individuals who are willing participants in tests, to minimize recruitment time and effort, and, as a consequence, one group participates more frequently than is needed for the test to be valid.

Besides product usage or demographics, it is important to screen potential participants for food allergies. Finally, it is generally important to work with panelists who do not have an in-depth knowledge of a food product. All persons who have in-depth knowledge of the product, or those who have specific knowledge of the samples and the variables being tested, should not be included in the test unless the test is designed specifically to include them.

Sensory laboratory tests—attendance and no shows. An increasing problem, is "no-show" rate at the test session, where a panelist agrees to participate, but something comes up at the last minute and the panelist either phones in or, just as frequently, fails to appear. The problem of "no-shows" can be reduced by a number of steps, such as overbooking, according to a historical "no-show" rate, obtained from previous tests. A "no-show" rate of 20% may be considered high for a panel of consumers recruited from a database, but a much higher rate—possibly close to 50% or higher would be expected when the panel is prerecruited through store intercepts or telephone directories and membership lists, and invited to participate in laboratory tests for the first time. Thus, to ensure participation, select participants who live within a 30-min traveling radius to the test facility, give participants clear directions and a map to help them find the location, and plan test dates that do not conflict with major community or school events. Reminder letters, phone calls, or mail-outs of brightly colored postcards to post in a visible location—such as the home refrigerator; adequate honoraria, rewards, or incentives; and overbooking for the test increase the attendance rate. The most important practice is to have participants commit to the process, and make participants understand the importance of their attendance, promptness, and the value of their participation.

Sensory laboratory tests—incentives for participation. The amount of incentive paid to participants may differ according to many factors including length of the test, location of the test and associated travel expenses, and incidence rates of qualified participants. Payment increases the motivation of panelists, and allows the panelist to set aside a considerably greater amount of time to participate.

Payment is provided after the session is completed. Certain organizations may have restrictions on the type of incentive that may be awarded to panelists, and so incentives vary considerably. Incentives for participation may be provided in the form of cash honoraria, product samples or coupons, selection from a gift catalog, gift certificates and tickets to special functions such as sports events or concerts, or donation to charity or a nonprofit organization.

Sensory laboratory tests—location and design of the test facility. The location of the sensory laboratory is important because location determines the laboratory's accessibility to panelists. The laboratory should be located at a place that is convenient for the majority of the test respondents to access. Laboratories located in places inconvenient to panelists will not only reduce the number of consumers that will want to participate, but also limit the type of panelists that can be recruited for the tests. When the sensory laboratory is located in a company facility, it should preferably not be situated near a noisy hallway, lobby, or cafeteria, because of the possibility of disturbance during the test (ASTM, 1996). The panelists should not be able to hear the telephones nor hear or see other offices or laboratories, food production or laboratory equipment, company employees, visitors walking or conversing, and so forth. Mobile phones should be prohibited or at worst, muted. If the sensory laboratory is located in such areas to increase accessibility to panelists, the laboratory should be equipped with special sound-proofing features. If the sensory laboratory is located in a remote area within the company, the remoteness of the facility may be a negative factor.

The sensory laboratory should be planned for efficient physical operation. Further, the laboratory should ensure for a minimum amount of distraction of panelists caused by laboratory equipment and personnel, and often emerging from the interactions between panelists themselves (ASTM, 1996). The laboratory should feature separate food preparation and testing areas. These areas must be adequately separated to minimize interference during testing due to food preparation and serving operations. The sample preparation area should not be visible to the panel. Individual partitioned booths are essential to avoid distraction between panelists; these should be designed, however, so that they do not elicit the feeling of isolation from the rest of the panel.

The physical environment must be pleasant. It is important to have a reception area, separate from the testing and food preparation areas where panelists can register, fill out demographic, honorarium, and consent forms, and be oriented on testing procedures before and after a test, without disturbing panelists who are doing the test. This area, away from the test itself, encourages social interaction, and allows for the administrative tasks to be done in a comfortable environment, both of which lead to a more pleasant experience for the panelists.

The test area must be maintained as free from extraneous odors as possible. Adequate illumination is required in the evaluation areas for reading, writing, and examination and evaluation of food samples. Special light effects may used either to emphasize or hide irrelevant differences in color and other aspects of appearance. To emphasize color differences, different techniques are used. Examples are spotlights, changes in the spectral illumination by changing the source of light from incandescent to fluorescent, changing the types of fluorescent bulbs used, or changing the position of distance of the light source. To de-emphasize or hide differences, the researcher may use a very low level of illumination, although this can make the evaluation experience less than pleasant. Special lighting such as sodium lights, colored bulbs, or using color filters over standard lights may likewise be used.

Sensory laboratory tests—number of products per sitting. For tests involving employees who cannot spend too much time away from their work assignment, the recommended number of food products per sitting is 2–6. Nonemployees are available for longer periods of time and can evaluate more products, especially if they are paid for their participation. The number of food products that can be evaluated depends on the panelist's motivation, which will be obvious through complaints, such as "inability to test more products," or overt displays of boredom. More samples can be presented to the panelists if a controlled time interval between products is allowed and only acceptance is measured. Kamen et al. (1969) observed that under such conditions, consumers can evaluate as many as 12 products.

The use of complete block experimental designs for consumer testing designs is recommended even when panelists have to return to complete the entire set of test samples. The complete block design requires that each panelist acts as his/her own control. If a complete block design is not feasible, incomplete block designs may be used, but it is recommended (Stone and Sidel, 1993) that incomplete cells per panelist be kept to a minimum of one-third or less of the total number of samples being evaluated by the panelist.

The Central Location Test

Central location tests (CLT) are frequently used for consumer research by both sensory and market researchers. CLTs include a variety of techniques and procedures. The tests are conducted in one or, more frequently, several locations away from the sensory laboratory, in a public location accessible to a large number of consumers who can be recruited to participate in the tests. These tests are usually conducted in a shopping mall or grocery store, school, church or hotel, food service establishment including cafeterias, or similar type of location that is accessible to large numbers of potential consumer panelists (Resurreccion, 1998).

Central location tests—advantages. The major advantage of the CLT is the capability to recruit a large number of naive consumer participants and thus obtain a large number of responses, which allow for stronger statistical analyses. In addition, only consumers or potential consumers of the food products may be recruited to participate in the tests, and no company employees are used. The well conducted CLT, properly recruited and executed, will have considerable impact and validity because actual consumers of the product or product category are used. CLTs enable the collection of information from groups of consumers under conditions of reasonably good control (ASTM, 1979) compared to a HUT, but are less controlled than the laboratory test. Further, several food products may be tested, compared to that recommended in a less well-supervised HUT.

Central location tests—disadvantages. The major disadvantage of the CLT is the distance of the test site from the company. An example is a mall location, often rented by the researcher for the purposes of a single test. The company sensory practitioner, sensory staff, and company employees from other functional groups must travel to the off-site central location. Often, the central location will have limited facilities, equipment, and resources necessary for food preparation and conduction of the test. Examples of food preparation and testing facilities that may be lacking in a central location are space for a registration area, suitable sample preparation areas, and individually partitioned booths for the sensory test. However, suitable equipment and testing space can be obtained, often at additional cost, and a small number of trained personnel may be used so that preparation of samples and serving of products can be better controlled. If the test product samples require extensive sample preparation and product handling, then the benefits of the lower cost of mall intercepts maybe offset and reduced by the lack of food preparation facilities, the increased time, and cost of testing, associated with the preparation of a large amount of sample or its transport in the prepared state to the test site.

Sensory research often requires special facilities to accomplish the test objectives. The inadequacy of facilities limits the types of tasks that can be performed by the consumer panelists, and, of course, limits the ability of the consumers to evaluate the product under the highly controlled experimental conditions of a laboratory test or under the actual use conditions provided in a HUT. In a CLT, the potential for distraction is high. The problem of "walkouts" should be taken into account when planning the test. The large number of panelists that can be recruited for this type of test is a disadvantage just as it is an advantage. Although the location is ideal for recruitment of a large panel, problems include the time and staffing requirements needed to conduct the test. Data are usually collected by trained interviewers rather than by self-administered questionnaire, adding to the number of personnel needed for the test, and increasing the time required to collect data from a given number of respondents. Logistics and choreography of the test session play an increasingly important role as the number of panelists in a test session increases.

In the CLT, products are evaluated under conditions that are far more artificial than the conditions encountered at one's home or restaurant, where the product is typically consumed. The conditions for testing limits the nature of information and, as in most short tests, the researcher can only ask a limited number of questions.

When using mall or store intercepts, panelists generally are not paid for their participation. By not paying the panelists, the sensory practitioner obtains possibly less than optimal data. For example, intercepting shoppers in a mall or shopping area will generate a panel of people who are busy with their shopping activities, and who are not willing to spend more than a few minutes on the interview. Supermarket shoppers have been observed to have a greater time constraint, because, more than retail general store shoppers, they need to return home to store perishables. This short time of testing limits the participants' exposure to the food samples being tested, therefore limiting the information that can be obtained. In contrast, a HUT allows considerably longer exposure to the product, thus permitting more data to be gathered about the consumer's feelings regarding the product (Stone and Sidel, 1993). Similarly, those laboratory tests, wherein the consumers are paid for several hours of testing, generate more data. The consumers are paid for their participation, are expected to participate in a longer test, and can do more complex tasks. When the CLT is used primarily in tests to screen food products and/or to define what type of consumer accepts the product or prefers it to that of a competitor, then the longer exposure time provided by the HUT is often unnecessary. When testing a product that may change over the normal recommended home-storage period, however, the CLT would not provide information on consumer acceptance during actual home-use conditions. In this case, a HUT would provide more realistic acceptance data.

Central laboratory tests—panel. The CLT consumer panel is composed of individuals recruited from a database or consisting of prerecruited consumers (so-called prerecruited panelists). More frequently, consumers are intercepted at the central location to participate in the tests, screened for eligibility and, if qualified, are immediately recruited to participate in the tests. Several central locations may be used in the evaluation of a product to determine regional or demographic (socioeconomic status) effects. The recommended number of food products to be evaluated per sitting in a CLT is one to four.

In a CLT, usually 100 (Stone and Sidel, 1993) or more consumers (responses per product) are obtained, but the number of responses may range from 50 to 300 (Meilgaard et al., 1991), especially when consumer segmentation is anticipated (Stone and Sidel, 1993). The increase in the number of consumers in a CLT compared to the laboratory test is necessary to counterbalance the expected increase in variability due to the inexperience of the consumer participants and the "novelty of the situation" (Stone and Sidel, 1993). Tests that use "real" consumers have considerable face validity and credibility.

Shopping mall intercepts. At a store or shopping mall, consumer intercepts would be the best method of recruitment. Interviewers use visual screening for gender, approximate age, and race to select likely looking prospects from the traffic flow. They approach prospective panelists to ask the necessary screening questions. When a person qualifies as a panelist, the test is explained, and the person is invited to participate. Usually, a small reward for the panelist's participation in the form of money, coupons, or other types of incentives is offered. This method of recruiting by intercepts has become more popular and has been used more often during recent years.

Panelists from civic, social, or religious organizations. Recruitment of panelists may be done through civic, social, or religious organizations such as clubs of different types, church groups, chambers of commerce, and school associations. For recruiting their members to participate in the tests, these groups receive a single award. These groups may provide homogeneous groups of participants, which can help the recruitment process. Often the group allows the researcher to conduct the interviews in facilities provided by the organization, such as a church or school hall, meeting room, and so forth. The organization may agree to provide a number of panelists of a certain demographic characteristic, such as elderly consumers, homemakers with young children, or teens, at intervals to fit the testing schedule. In other cases, the organization provides the project personnel with lists of members who may be called, screened, and qualified ahead of time and scheduled. One of the disadvantages of using such groups is that they represent a narrow segment of the population (Stone and Sidel, 1993), and more often than not possess demographic and food consumption patterns that are similar. In fact, it is likely that members of these groups may know each other or may be related to one another. Their demographic responses as well as their food preference patterns may therefore be skewed.

Other sources of consumer panelists. Often the sensory practitioner sets up a booth or temporary facility at a convention, fair, industrial show, or similar event, where crowds of people are likely to congregate. Booth visitors or interested onlookers are invited to participate. In this case, the test is usually limited to a brief evaluation such as a "taste test" between two samples (ASTM, 1979). Additional sources of panelists that have been used for CLTs are newspaper, radio, and TV advertisements; flyers at community centers, grocery stores, and other business establishments; referrals from current panelists; letters to local businesses requesting their employees to become panelists; purchased mailing lists of consumers in a geographic location or telephone directories; and using random digit dialing to recruit consumers by telephone.

Consumer location tests—screening for consumer panelist. Careful screening for consumer panelists helps to prevent bias. The screening criteria for acceptance and preference tests are very important to establish. A thorough understanding of study objectives is needed to identify who should participate. When screening the test participants it is important to ask the questions in an unambiguous fashion, to be sure that the consumers can properly identify themselves as the relevant target. However, the actual screening criteria should be undisclosed to participants, so that the consumers will answer with actual information about themselves, and not what recruiters want to hear.

There are a number of issues to keep in mind when briefing the staff who will do the recruiting, especially when the staff recruits the consumer panelists by intercepts at a shopping mall. The central location should be selected on the basis of matching the demographic profile of shoppers to the target consumer of the test food product. Recruitment should be scheduled during the peak traffic hours for the desired participant. For example, if full-time employees are required, recruitment must be scheduled before or after regular working hours or during the lunch break. It is not advisable to recruit panelists who are active panelists for another company. Recruiters need to include a "security-check" question to ensure that the panelist does not work for a competitive company, an advertising agency, a market research company, and so forth. Individuals or immediate household members of individuals whose occupation is related to the manufacture or sales of the test food product or in market research should be disqualified from participating. Professional panelists, or panelists who claim to frequently participate in sensory tests should be avoided because they no longer perform as naive panelists, and so the frequency of participation of a panelist in tests within a product category may also be one of the screening criteria. Frequency of use of the product may be an important consideration.

Often, income level and other demographic characteristics such as age, marital status, gender, and education may be used as screening criteria for potential panelists. In some cases, the number of persons in a household, whether these persons are adults, adolescents, or children may be important. Whether the individual is the primary purchaser or preparer of food in the household may likewise be important. Availability and interest of the panelist will determine whether or not they can be scheduled. Potential panelists who are allergic to the food product should immediately be disqualified. Other consumer characteristics may be important, depending upon the specific study, the nature of the food products, and the characteristics of the target market for the food product.

Central location tests—orienting panelists about test procedures. The orientation of consumer panelists should be limited to providing only information that is absolutely required. Generally, this information describes the logistics and procedure of the test. The orientation may deal with explanations regarding the features of the booth area, such as the sample pass-through door, signal lights, and so forth. Other parts of the orientation may deal with the use of data acquisition devices, such as computerized ballots, instructions on using a light pen, touch screen, mouse, or keyboard.

The orientation must be carefully planned to avoid any opportunity for altering the panelists' attitudes toward any of the food samples to be evaluated.

Central location tests—preparations and serving considerations. As in a laboratory test, the sensory practitioner will need to plan for preparation and presentation of samples, the duration and temperature of holding after cooking until serving, portion size of serving, and the method of serving. Considerations include the use of a carrier, dishes on which food is served and whether or not to include bread, crackers, or water for rinsing, and whether or not to include supplies for expectoration. The control of the testing environment should likewise be considered, and every attempt to minimize distractions and bias should be avoided.

Central location tests—number of panelists in a single test session. The number of consumers to be handled at one time depends on a number of factors that include the food product type, the capacity of the testing facility, and the number of technical support staff available to conduct the test. One or two panelists at a time may be recruited to participate, or a larger group of panelists maybe handled at one time. Too many panelists in one area will encourage inattention or interference, and if the panelist to technician ratio is too high, panelists' mistakes or questions that arise during the course of the evaluation may go unnoticed.

Central location tests—number of products to be evaluated. The sensory practitioner should consider carefully the number of samples that can be ideally presented at a mall location and in any test that employs the intercept method of recruiting panelists. The maximum number of food products to be evaluated by a panelist at each session at a CLT should be four, with fewer samples being better, at least from the standpoint of the quality of data. Panelists are more likely to walk out in the middle of the test if the interview is lengthy, complicated, unpleasant or boring.

Central location tests—what test method should be used. Any of the standard affective test methods may be used in CLTs. The decision on what test to use depends on the objectives, the food product, and the panelists. Usually, in a CLT, the information is collected through a self-administered questionnaire. When a self-administered questionnaire is used, it should be easy to understand, not take too long to answer, or be a source of boredom. The attention span of unpaid panelists must be seriously considered. For example, if there are one or two samples, then it is permissible to ask several questions about each. However, as the sample number increases, the number of questions that must be answered should be decreased accordingly.

Unpaid panelists intercepted at a central location will often not agree to a test that will take longer than 10–15 min. Most panelists' interest and cooperation may be maintained for this length of time. Longer tests should be avoided unless there are special circumstances, such as giving the panelists a substantial cash incentive. Prerecruited panelists who are asked to come to a central location may be told beforehand how long the test will last. For example, in such cases, the panelist may be told that the test would take an hour.

On the test dates, panelists are assembled singly or in small groups in the test area where trained sensory personnel conduct the test. The number of sensory personnel assisting in the test will vary with the size of the group being handled, the stimulus control requirements in any given case, and the test procedure. In many cases, one-on-one interviewing is required and in other cases, it may be sufficient for one person to handle 4–5 or as many as 12 people responding to a self-administered questionnaire at one time.

The panelists are given a brief orientation in the form, or written or oral instructions to assure adequate understanding of the test, including the number and type of test product samples to be tested, presentation of samples, waiting periods between samples, and other details of the test. The panelists are informed about the type of information to be collected by a brief review of the questionnaire.

Central location tests—proper handling of test samples. The planning for handling of test samples for a CLT is more involved than a laboratory test because the samples must be transported from the laboratory to the test site. Sufficient amounts of sample for the test should be provided—the test personnel have no access to additional amounts of sample. Sample containers need to be clearly identified and prelabeled with appropriate sample identifying codes and then inventoried. Sample preparation instructions need to be written in detail, as clearly as possible, and should include holding time and temperature conditions. Timing of sample presentation must be planned so that sample preparation coincides with serving times. CLT panelists are bound to walk out if they perceive that their time is wasted by the slow presentation of samples. A dry run of the entire procedure is often necessary to ensure that the schedules are correctly timed. If preparation is going to be conducted in the test site, then all necessary equipment for food preparation should be available. Temperature holding devices such as food service warming lamps or steam tables should be used if needed. The training of personnel in sample preparation and equipment use is absolutely necessary if these procedures affect the quality of the product being tested. Sample serving specifications should be clearly outlined. Finally, appropriate serving utensils such as plates, plastic cups with or without lids, scoops, knives, or forks to be used in serving and testing should be specified.

Central location tests—test facilities. Facilities available at malls or retail establishments for a CLT vary widely. In general, the testing should be conducted at a central location that will be convenient and accessible for the participants. Ample parking space or access to public transportation must be available. Much CLT testing occurs in settings that are less than optimal. Agreed upon requirements for CLT facilities are that the facilities should allow adequate space and equipment for preparation and for presentation of the product in a controlled environment. The facilities should provide proper control of the "physiological and psychological test environment," including adequate lighting, temperature, and humidity control for the general comfort of panelists; freedom from distractions such as odors and noise; and elimination of interference from outsiders and between subjects (ASTM, 1979).

Central location testing is often contracted out to a marketing research company that operates its own testing facility. A CLT facility may be purchased, constructed, or leased for this purpose. In such cases, proper control can be maintained in the test area by constructing a sample preparation laboratory, panel booths, and adequate ventilation systems.

CLTs using groups may be conducted in public or private buildings such as churches, schools, firehouses, and so forth, familiar and convenient to members of the group. This test offers the least amount of environmental control as testing in these sites is usually done without booths, controlled lighting, or ventilation. Noise and odors may pose a severe problem. Sample preparation must be done in advance, at company headquarters; test samples, and serving and sampling equipment need to be transported to the central location. Clean up and waste disposal often involves hiring of a custodian, when the test is conducted in these facilities.

Ideally, panelists should test the products in individual partitioned booths to isolate panelists from distractions from other panelists or test personnel. When booths are not available, panelists should be separated physically from sitting as close to each other as possible, and definitely not facing each other, to minimize distractions and influence of other panelists. Seating should be comfortable and at an appropriate height for the table or counter where testing will take place.

When using a test site for the first time, it should be visited at least 1 week prior to the test date to make sure it will deliver appropriate controls. If the test site has been used previously, a visit to the site one or two days before the test date may be sufficient to implement minor changes. Final details may be arranged during this visit. Central location food preparation equipment should be tested to ensure they are in good working order. For example, ovens should be calibrated prior to the test day. Food samples, ingredients, and other supplies previously examined for quality should be set up for the test. Storage areas and containers for samples and supplies that were requested in advance should be examined. A separate dry run of the sample preparation procedure, using the available equipment should be made as far ahead as possible, to ensure a smooth implementation of the CLT.

As in laboratory tests, samples should be prepared out of the sight of panelists and served on uniformly labeled plates, sample cups, or glasses. The serving area for the CLT should be convenient to both the test personnel and the sample preparation area. The serving areas should have the serving scheme in large legible numbers, posted for test personnel to markoff as they serve samples to each panelist.

Large trash containers need to be positioned in the sample preparation area and the serving areas. Arrangements for waste disposal from the test site during and after completion of the study need to be made. Disposal or storage of samples that are not used should be handled according to the client's directions. These instructions regarding the handling of the product should be agreed upon beforehand.

Panelists need a comfortable place to wait when they arrive for the test. The reception area should provide sufficient space for panelist registration, orientation and waiting for the test, and payment of panelists, and a place for snacks and social interaction after a test. In infrequent cases, an area for childcare is designated to allow panelists with children to take turns at babysitting, for parents involved in food product testing. It is important to establish rules regarding children and to post these rules so that fewer problems will arise.

It is essential that the project leader reviews the complete test protocol, in a briefing session before the actual test, to determine whether the test personnel are familiar with their assigned tasks, the procedures, and the serving instructions. It is a common practice for an agency to conduct a briefing with their personnel prior to the test. The briefings are an important opportunity to review the instructions for the study and explain any special requirements. If a script will be used by sensory personnel, a dry run of the reading of the script is conducted to identify problems, if any. The importance of not deviating from a script is emphasized. The serving scheme for test samples is explained during this time.

Central location tests—mobile laboratories. Over the past two decades, a new form of central location facility has been developed, which can be best described as being a mobile test laboratory. A fully equipped mobile laboratory with complete facilities for food preparation and sensory testing is driven to and parked at a central location. Tests are conducted within the mobile laboratory. In this type of test, the panel consists of consumers who are intercepted at the test site in order to participate in the tests. There are usually no prerecruitment criteria except for age, and visual screening is used for demographic characteristics. Usually 75–100 or more responses can be obtained easily in one location. Often the test is conducted in one location and the mobile laboratory is then driven to other locations where additional days of testing are conducted. Data are collected by trained interviewers rather than by self-administered questionnaires, adding to the number of personnel needed for the test and the time required to collect data from a given number of respondents. The recommended number of products to be evaluated in a mobile laboratory is 2–4.

The major advantage of using the mobile laboratory for a consumer affective test is the ability to recruit a large number of "real" consumers for the test, in test facilities with experimentally controlled and environmental conditions conducive for food preparation and testing, similar to that in a laboratory. The disadvantages of the mobile laboratory test are the expense of maintaining the mobile laboratory and logistical arrangements that have to be made prior to the test for parking and power supply.

Home-Use Tests

Home-use tests are also referred to as home placement or in-home placements tests. This test requires that the test be conducted in the participants' own homes. The HUT is used to assess product attributes, acceptance/preference, and performance under actual home-use conditions, in participant's homes. The HUT thus provides additional and valuable information regarding the product, which may not be obtained in any other type of consumer affective test. From the food product developer's point of view, HUTs provide information about the sensory characteristics of a product under uncontrolled conditions of preparation, serving, and consumption.

HUTs are valuable in obtaining measurements about products that are not possible to obtain in a CLT or a laboratory setting, which are limited by the short period of panelist's exposure to the product. HUTs are designed to gain information and preferences, acceptance, and performance. The HUT questionnaire can also obtain information about product preparation and attitudes, and ask other relevant questions specific to the product, including attitudes about the product category or about the different brands and manufacturers in the product category.

Home-use tests—advantages. The major advantage of the HUT is that the products are tested in the actual residential environment under actual normal home-use conditions. Thus, testing in the consumer's home is considered to be closer to optimal compared to laboratory and CLTs. Another advantage of conducting HUTs is that more information is available from this test method, because one may obtain the responses of the entire household on each household member's usage of the product. Responses may be obtained not only from the participant, who is usually the major shopper and purchaser of food in the household, but also from the other members of the entire household. In many instances, other pieces of information such as the of competitive products in the home during the test, usage patterns of all household members, and other data that would be useful in marketing the product may be obtained.

The HUT method can be used early in the product formulation phase to test a product for acceptance or preference as well as for product performance (Hashim et al., 1996). Information regarding repeat purchase intent obtained from the consumer participating in this test would be more useful than the corresponding information one obtains from a sensory laboratory test or a consumer CLT.

Participants in a HUT should be selected to reflect the target population. If the participants in a HUT are prerecruited from an existing database and screened, the participants are aware of their role and the importance of the data collected. Their awareness and ongoing motivation will likely generate a high response rate. Finally, HUTs lend themselves to product delivery by mail. When using the mail to conduct a HUT, the researcher can achieve speed, economy, and broad coverage, factors that improve the efficiency of the test (Resurreccion and Heaton, 1987).

Home-use tests—disadvantages. The main disadvantages of the HUT are that it requires a considerable amount of time to implement, to distribute samples to participants, and to collect participant's responses. A minimum of 1–4 weeks is needed to complete a HUT (Meilgaard et al., 1991). Lack of control is another disadvantage; little can be done to exert any control over home-use testing conditions. Once the product is in the home of the respondent, the potential for misuse is high (Stone and Sidel, 1992). This lack of control may result in large variability of consumer responses. Consumers' homes vary greatly when it comes to food preparation, time of day when the sample is consumed, and variation in the products consumed along with the test sample.

The test design for a HUT must be as simple and unambiguous as possible—it is best to conduct a HUT on only one or two samples. Otherwise the test situation would be too complex for most respondents. Therefore, the HUT is not appropriate for evaluation of several different samples at one time.

The HUT is the most expensive test in terms of actual food product cost. The required sample sizes of test products evaluated by participants to they make their judgments are generally bigger than the amount of test product served to panelists in a laboratory or CLT. Further, the food product must be packaged in such a way that the package does not bias the results. Others might argue that the larger amounts of test product provided allow the food to be tested under actual home conditions, such as drinking rather than sipping a beverage, which would be the case in a laboratory, central location, or mobile laboratory test. Distribution costs may become a major expense associated with this test. There is also the cost of producing the samples for the test. Depending upon the nature of the samples and the number of test participants, the test samples may be produced "on the bench," or in some larger cases may require more expensive "scale-up" production facilities, which runs into far more money.

If participants have not been prerecruited and screened from an existing consumer database, the HUT participants will likely be less aware of the importance of their role and the importance of the data being collected. In such cases, response rates will be lower than expected.

Home-use tests—product placement. The cost of the food products placed in participants' homes adds considerably to the cost of the test. On the other hand, if the HUT consumer panel size used is reduced because of these costs, then the information one would obtain from the HUT is correspondingly limited. Furthermore, when placement of products is by mail, perishable and nonmailable products usually cannot be tested. More expensive courier services are required.

If HUT participants have not been prerecruited and screened from an existing database, the participants will likely be less aware of the importance of their role in this test and the importance of the data being collected. In such cases, response rates will be lower than desired unless respondents are monitored and reminded, by telephone, about their testing tasks. HUT questionnaires must be self-explanatory, clear, and concise, because there is no opportunity for personnel to explain and elaborate as one would be able to do in a personal interview. Visual aids, if needed, can only be limited to graphics on a questionnaire. Finally, the researcher needs to realize that the consumer respondent will be able to read all the items on the questionnaire beforehand, thereby limiting the impact of sequencing of questions.

Home-use tests—panel size. The sample size for the HUT should be large enough but not unduly large, so that the researcher can get the necessary number of respondents within a reasonable length of time. Most research agencies will not guarantee a response rate. Panelists' cooperation and nonresponse may influence the results of the respondent samplings (ASTM, 1979). The assumptions are held that nonresponders react similarly to those who respond—fortunately, there are instances where mail or courier panel results closely approximate probability sampling results. Most of the time, participants of HUTs conducted by mail or courier are represented as being quota samples that match U.S. census distribution. People who volunteer for HUTs and sustain their interest in these tests may be different from the normal user of the product type. Also, it is often alleged, though seldom proven (ASTM, 1979), that panel members who are used repeatedly respond in a manner that is not typical of consumers.

Home-use tests—recruitment. In HUTs, food product samples are tested in the consumers' own homes rather than in a laboratory or central location. There are various methods to locate and recruit participants, and to deliver the test product to their home, including by mail, as discussed previously. One HUT recruitment method is to prerecruit participants from a consumer database and screen for panelists' eligibility to participate in the tests. However, a common practice is to involve those employees of a company who have little or no responsibility for production, testing, or marketing of the food product. There is some risk associated with employee panels in food product maintenance tests. Employees should not be used in product development, improvement, and optimization studies.

Due to the uncontrolled conditions of home-use testing, a larger sample than that required for a laboratory test is recommended for a HUT. Usually more than 50–100 responses are obtained per product. The number of responses needed in the HUT varies with the type of food product being tested and with the experience of the respondent in participating in the HUT. With "test-wise" panelists, Stone and Sidel

(1993) recommend that a reduction in panel size could be made. Because these panelists know how the test is conducted and feel more comfortable in the test situation, the testing will be less susceptible to error and the result of psychological variables associated with being a subject. In multicity tests, 75–300 responses are obtained per city in three or four cities (Meilgaard et al., 1991). The number of food products to be tested should be limited to one or two.

One recruitment approach is to use the telephone for preliminary questioning to establish qualifications and solicit cooperation. Telephone recruitment may be random within a given area or from lists of members provided by cooperating organizations. This recruitment approach has an advantage in that the recruiting is less expensive when qualified respondents are expected to occur at low frequency. It is probably harder to gain cooperation over the telephone. The sensory practitioner is faced with the problem of delivering food products and test instructions. Furthermore, more effort may be required to place the food products after the initial contact, because participants may be more geographically dispersed.

Another recruitment approach is to conduct a door-to-door household survey in areas selected on the basis of having a high probability of producing desired subjects. An interviewer asks questions to establish qualification and to obtain background information that might be useful in analyzing the results. When a qualified family is found, the product is immediately placed in that household, instructions given about the interview, and then the interviewer makes arrangements for a recontact, or tells the participant how to return the ballot or respond by telephone or e-mail. An advantage of this recruitment approach is that it permits distributing the sample as desired in a given territory and eliminates the problem of food product delivery. Finally, the participants are easier to contact for the final interview. However, the method may be more expensive because much effort is wasted in contacting unqualified people.

Mall intercept or in-store recruitment has been relied on to an increasing extent for many types of market research studies. Mall intercepts, the mainstay of CLT tests, are also useful for HUT recruitment. Interviewers recruit prospective participants in the mall or store, and quickly screen them to determine whether or not they meet requirements. Consumers who agree to participate in the HUT can be given the products virtually “on the spot,” with little delay. Sometimes, bulky or perishable products must be delivered to the respondents’ homes at a later date, after the recruitment was arranged at the mall.

The advantage of mall recruitment for HUTs is that it permits screening a large number of prospects relatively quickly and cheaply, and can be particularly helpful when trying to locate users of a low-incidence food product type. A serious disadvantage of mall recruitment for HUTs is that it may be time consuming and expensive to recontact the respondents for personal interviews, since the participants may live in a region far from the mall, or their residences may be dispersed across a wide area. Further, some target consumers do not frequent malls.

If HUT panelists are to be reached by mail or e-mail, then a necessary step is to develop databases of mailing lists, keep these databases current by rotation, and replace drop-out participants. Developing a database involves initial location of the respondents, contact, gaining cooperation, acquiring necessary information, and maintaining sufficient interest for sustained participation.

Contacts can be made through the telephone through random digit dialing, newspaper advertisements, recruiting through organizations, and so forth, or by e-mail. Prospective panel members provide information about themselves, which is placed on file and retained for a certain period of time. The frequency with which a panel member may be contacted varies widely. Usually, HUT participants are given some incentive for their participation, which keeps them interested in remaining on the database.

The key to obtaining a valid sample of HUT respondents is the initial selection and gathering of participant data, which allows for a "smart" selection of participants. The usual approach is to seek protection in large numbers, with lists containing thousands of names and representing more or less the national census distribution of the population. Census information is obtained by means of questionnaires. Demographic data such as location, the number and identity of family members, family income, race, education, and occupation are nearly always available. Beyond this first step, it is a matter of product usage and proper panelist motivation. The information necessary to find and recruit panelists is generally stored in a searchable database. Drawing the actual sample for a given study is a mechanical matter of identifying families with the desired characteristics, then randomly selecting the desired number of participants to contact, of which some percent will agree to participate.

Home-use tests—facilities and procedures. HUTs are conducted in the respondents' own homes. No controls of the environment or the testing conditions can be made. The researcher has no control over the test, other than providing detailed instructions and hoping that the panelists comply. In HUTs, the food products are tested under actual-use conditions. Responses from one or several members of the family are usually obtained, either at the time of preparation/consumption or later by telephone "callback," mail return of a completed questionnaire, or e-mail. Given the lack of control that can be exerted by the sensory practitioner over the test, it is best to keep matters simple. The HUT design should be as easy to perform and uncomplicated as possible. The focus should be on consumer overall acceptance and acceptance of specific attributes.

Additional considerations for HUTs. HUTs require additional considerations and preparatory steps, which occur before the test. The relatively simple considerations can produce problems if not done correctly. Issues to resolve include the nature of packages and labels, and product preparation instructions that are clear and complete. Product can be either mailed or delivered to consumers' homes, and the specific self-administered HUT questionnaire must be appropriate to the task.

Home-use tests—product samples. The packages used in HUTs are often those in which the product will be marketed, but should be plain and have no graphics. Product packages should only be labeled with the sample code number, preparation instructions, contents, and a telephone number to call in case questions arise.

Preparation instructions must be complete, unambiguous, and must clearly describe exactly how the products are to be prepared, to avoid confusion. Instructions on how to complete the questionnaire should be included. Pretesting the instructions and the questionnaire is important to the success of a HUT, and is always necessary.

The length of time for testing each product dictates the number of products to include in the HUT. The number of products to be tested should be limited to two, primarily because of the length of time needed to evaluate each product. Often HUTs are designed to get information about food products in actual use, which may be once over several days or daily over 4–7 days. The goal of the test is to capture all information surrounding "normal consumer use." The participant generally completes a separate, parallel questionnaire for each product. When more than two products are included in the test, the HUT will require longer time. Further, the risk of nonresponse will be greater due to one or more reasons such as loss of the questionnaire, trips out-of-town, or loss of panelist interest.

There are three methods that can be used to place the samples in the HUT participant's homes. These methods are to: (1) mail or courier the product, (2) deliver the product, or (3) have consumer panelists come to a central location to obtain the product. Providing the participant with all the food products during the beginning of the test, at a central location, minimizes the direct cost of the test, because the logistics demand only one appearance by the participant and one appearance by the test coordination group. Lowered direct costs can be achieved by mailing products to the consumers who have been prerecruited by telephone or mail to participate in the test. Mailing products to subjects has considerable risk and should be avoided except in rare instances such as when the products are microbiologically shelf-stable, and are of a shape and size that lends well to mailing. Examples of shelf-stable products that could be mailed to participants are pecan halves (Resurreccion and Heaton, 1987) and small products that can resist the impact of temperature changes and other rough conditions encountered during the mail out. The problems with using the mail to place products with participants is the time required, the choice of mailers, microbiological safety, and the not-too-infrequent happening that the package may be received by someone other than for whom it was intended, so the participant never tests that particular product. More sensitive products may be shipped by courier.

Providing two or more products simultaneously, rather than sequentially, is likewise not recommended because it allows the participants to make direct comparisons, and increases the probability of recording responses for the wrong product on the questionnaire (Meilgaard et al., 1991), invalidating the results. For these reasons, Stone and Sidel (1993) recommend that cost containment not be enforced in a HUT, if the precision of the results are critical. The researcher may end up wasting a lot more money due to cutting critical steps to minimize costs.

The practice of having consumers recruited for the HUT receiving products at a central location is efficient and attractive to the sensory practitioner planning the test (Hashim et al., 1995). If there is a second sample, consumers will return to the central location to submit the completed questionnaire and the now-empty product package, and then receive the second batch of product and a new questionnaire. This method distributes the travel across the participants. When local residents are used, this HUT strategy for product distribution is more effective and cost-effective than delivering the product to consumer's homes. In certain cases, when the method of sampling for a HUT is to intercept and qualify prospective participants, qualified participants may be given the first of the samples immediately after qualifying. The participants then return to the same location to pick up the next sample and questionnaire, and return the empty sample container and questionnaire.

Home-use tests—measuring product acceptance. The primary objective of the HUT is to measure overall product acceptance. In order to maintain uniformity and continuity (Stone and Sidel, 1993) between different methods of testing such as the laboratory test, CLT, and HUT, the sensory practitioner should use the identical acceptance scale for each test. The HUT questionnaire can include other measures and tasks besides rating scales for acceptance. These are attribute diagnostic, product performance, and paired preference questions.

In addition to overall acceptance, consumer acceptance of specific product attributes such as appearance, aroma, taste, and texture may be obtained, but the results should not be used as one would use results from a sensory analytical test. Stone and Sidel (1993) raised objections to using diagnostic questions, and recommended that diagnostic questions be excluded from the test. Their rationale is that researchers cannot directly validate these attribute ratings and that overall acceptance requires the evaluation of the product as a whole, without direction to what sensory aspects are considered important. They believe that by calling the participant's attention to a specific set of attributes, there is always the possibility that the participants will be biased.

Often, more than overall acceptance is needed from a HUT. Questions pertaining to product performance, "How easy is it to pour milk from this stand-up pouch?" or purchase intent, "If available in the market, how often would you purchase this product?" can be easily obtained.

Some researchers include a paired preference question on products that were tested during two different time periods. Although the products were evaluated sequentially, this final question, if added, relies on the consumers' memories to remember the product that was tested the week before and to compare it with the product currently being tested. This use of the paired preference question is inappropriate and should not be used. The 9-point hedonic scale for each product yields more useful information regarding each product from which paired preference response can be extrapolated.

Home-use tests—data collection. There are various ways of obtaining critical postuse information about the respondent preferences, attitudes, and opinions. They are: personal interview, use of a self-administered questionnaire that is mailed back, telephone interview, and electronically by e-mail through the Internet. One or more questionnaires to be answered during and/or at the conclusion of the testing may be posted. Space may be provided for comments. Obviously, if necessary, telephone interviews may be conducted to amplify written or electronic responses.

Interviews. The most effective approach among the data collection methods is the personal interview. The personal interview allows the researcher to review procedures with the consumer participant and determine whether or not the product was evaluated properly. The interview method allows the interviewer to answer any questions regarding test procedures or food product-use conditions. Respondents are often more motivated and involved when a personal interview occurs in the beginning of the evaluations. The interview method offers possibilities not offered with a self-administered questionnaire. For example, the interviewer can control the sequence of questions by using visual displays such as concept cards, pictures, or representations of rating scales. Probing for clarification of answers by the interviewer or asking for more detailed information is also possible. The major disadvantage of the interview is its cost. If personal interviews are selected as the data collection method, then it is extremely important that all interviewers undergo rigorous training prior to the interviews, to ensure lack of bias. Dry runs of the interview process should be conducted using actual consumers who are not going to participate in the HUT.

Self-administered questionnaire. In many cases, the food product samples are either mailed or handed personally to respondents with instructions and a self-administered questionnaire to be completed, while and after the products are used. Alternatively, the consumer may be contacted by e-mail. The questionnaire is returned by mail to the researcher after it is completed. The self-administered questionnaire is designed in such a way that instructions and questions are easily understood without need for assistance. It is advisable in such cases for a self-administered questionnaire to list the telephone number or e-mail address and name of a contact person who can answer any questions regarding the test. This method is less labor intensive than telephone interviews and, if conducted properly, eliminates interviewer bias and reduces the cost of the interview. The disadvantages of using a self-administered questionnaire are that cooperation may be poorer and the response rate is lower if the instructions are not well understood or the respondent has inadequate reading skills. The impact of the sequence of questioning cannot be controlled, as the entire questionnaire can be read by the participant prior to starting the test. Furthermore, there is no opportunity to correct respondent errors or probe for more complete information unless the respondent telephones the contact person.

Telephone interviews. With proper planning, data may be collected by telephone interviews after the participant has received the product and evaluated it. Telephone interviews have the advantages of lower costs and greater speed of data collection; therefore, less time is involved and thus lower costs are incurred. Furthermore, telephone interviewing makes it easy to use procedures appropriate for random sampling. In addition, telephone interviews make it easy to work with participants who are geographically dispersed. The interviewer can conduct the interview in the evening, and thus lower the nonresponse rate. Compared to mail interviews, the telephone interview's advantages are: greater speed of data collection, ease of maintaining random samples, lower nonresponse rates, and ease of getting "hard-to-reach" respondents such as employed men and teens.

Several limitations of telephone interviewing exist. Compared to personal interviews, telephone interviews need to be relatively short or the participant may just terminate the telephone call. Thus, rapport with the panelist is not as good in telephone interviews compared to personal interviews. Telephone questionnaires have several constraints such as the difficulty to get complete open-ended responses, typically possible from personal interviews, and the impossibility of using visual displays. Furthermore, only rudimentary scaling can be used in a telephone interview. Information that can only be gathered through observation, such as the race of the respondent or the type of housing they live in, is not possible to obtain in a telephone interview. Telephone interviews are more expensive to conduct than mail questionnaires. The telephone interview is often conducted to limit costs compared to personal interviews, particularly in those cases where the respondents may be difficult to contact due to geographic constraints or limited availability during specific time periods.

Telephone interviews may result in poorer cooperation, with lower response rate that researchers observed in personal interviews. Further, the range of questioning is often limited because visual displays are not possible to use, and probing, if used, may not be as effective because verbal descriptions lack the power of visual displays. As with all questionnaires, the telephone interview questionnaire should be pilot-tested with a small sample of respondents, with special emphasis on those questions that require the use of a scale. It is extremely important to ensure that both supervisors and telephone interviewers undergo extensive training on how to ask the questions and how to administer the interview process. Quality control measures that may be employed are audits of the interviewing process.

Telephone interviews may originate from a central location, where a bank of telephones is installed, or from the interviewer's home. The central location permits immediate supervision of interviewers to handle problems as they arise. The telephone linkages should permit interviews with participants to be conducted in any location in the United States.

It is evident that with the high distribution of home computers and electronic linkages, Internet and e-mail interviewing is not only feasible, but may also be more desirable than telephone. Although most such communications would be asynchronous as of this writing, direct communications combining two-way audio linkages and one-way visuals should be expected in the near future. Computer-savvy consumers are required, but as the proportion of the population knowledgeable with computers increases, this aspect will diminish as a screening variable.

All things considered, the HUT is generally reserved for the later stages of sensory evaluation testing with consumers. HUTs have a high degree of face validity. HUTs provide the researcher with the tools to measure how other members of the family react to the product, under the normal conditions of use, at home. The most important considerations are that in most cases the HUT may be more costly than other sensory tests, requires more time to complete, and lacks environmental controls (Stone and Sidel, 1993).

The test location itself may influence the ratings. One might believe that a carefully executed test should generate bias-free information, no matter which testing location has been chosen to conduct the test. Some research shows that just finding the testing location may not influence the rating. For example, Hersleth et al. (2005) showed that changing the environment and the degree of social interaction in the consumer tests exerted no significant effect on hedonic ratings for unbranded popular cheese samples. The authors attribute the high degree of consistency to the absence of a meal context during testing, similar expectations during the testing, and a high familiarity with the samples tested. Similarly, Pound and Duizer (2000) measured responses for overall liking of unbranded, commercial chocolate in four venues: in a teaching laboratory, a formal sensory laboratory, and a central location and in a HUT, respectively. They found similar results from testing in the various locations.

However, other investigators found the opposite—that test venue plays an important role. King et al. (2004) found that introduction of context effects in a CLT can improve the ability to predict actual liking scores in a real-life environment. They found that the relation between context effect and consumer acceptance may not be consistent within and across meal components; thus meal context had the strongest effect on tea, social context had a strong negative effect on pizza, environment had a weak positive effect on pizza and tea and a negative effect on salad, and choice had a positive effect on salad. These results led the authors to conclude that context variables affect product acceptance, but the relation between context effect and consumer acceptance may not be consistent within and across meal components. Consistent with these findings, Hersleth et al. (2003) found that hedonic ratings for Chardonnay wines were higher in a reception room, where some socializing occurred, compared to acceptance measured in partitioned booths in a laboratory. The effects of context factors may be significant when testing certain foods, and should be considered in planning an acceptance test.

CHARACTERIZING THE FOOD PRODUCT

In the product development process, product characterization is carried out by using descriptive analysis testing and/or physicochemical measurements. These tests are used to define the limits or range of product properties or attributes that correspond to acceptable products.

PHYSICOCHEMICAL MEASURES OF PRODUCT QUALITY

Product characterization may be conducted using various instrumental or physicochemical methods. Only when no instrumental or physicochemical test can be used to validly and reliably characterize an attribute, should sensory descriptive analysis methods be used. If an instrumental test that accurately characterizes the sensory attributes of a product is available, it makes little sense to use a descriptive sensory panel to quantify the attribute. Physicochemical measurements have been used to attempt to characterize attributes such as color, flavor, tastes, texture, and viscosity. Unfortunately, good correlations are needed but have not been found in the validation of a number of instrumental measurements.

Descriptive analysis. Instruments can be used to a certain extent, and often no further, in describing the sensory characteristics of a product. Descriptive analysis is a widely used technique in the sensory analysis of food materials. Descriptive methods are based on: (1) panelist selection, (2) panelist training, (3) ability of the panelist to develop a descriptive language for the products being evaluated, (4) ability of panelists to provide reliable quantitative judgments similar to an instrument (calibration), and (5) analysis of data (Stone and Sidel, 1993). Panelists who have been trained in descriptive analysis are able to detect, identify, and quantify attributes of a food or product (Hootman, 1992). Human perception and response are critical for descriptive analysis. Panelists are selected on their ability to perceive differences between test products and verbalize perceptions. The descriptive panel is used as an analytical tool in the laboratory and is expected to perform much like one (Muñoz et al., 1992; O'Mahony, 1991). There are several standard techniques available, such as the flavor profile analysis (Cairncross and Sjöström, 1950; Caul, 1957), quantitative descriptive analysis (Stone et al., 1974), spectrum descriptive analysis (Meilgaard et al., 1987), and the texture profile analysis (Brandt et al., 1963; Brody,

1957; Szczesniak et al., 1963). Descriptive analysis testing can be useful in several phases in the product development process.

Panelist selection. The descriptive analysis method requires that a panel be carefully selected, trained, and maintained under the supervision of the sensory practitioner. The panel members must show motivation. The desire to participate is the first and most important prerequisite for a successful panel participant. Panelist availability and demonstrated sensory acuity are likewise needed. In addition, prospective panelists are interviewed and asked about their interests, education, experience, and personality traits. Individuals who are too passive or dominant should be eliminated from the panel (Amerine et al., 1965).

Panel size ranges from 5 to 100 judges (Meilgaard et al., 1991). According to Stone and Sidel (1993), a panel has at least 10 test subjects, but no more than 20. Rutledge and Hudson (1990) and Zook and Wessman (1977) have published methods for selecting panelists.

Screening. Screening is conducted to identify a group of panelists who will undergo training. A series of exercises should be designed, where candidates are shown samples that may be references or products (Einstein, 1991). The purpose of screening is to select candidates with basic qualifications such as normal sensory acuity, interest in sensory evaluation, ability to discriminate and reproduce results, and appropriate panelist behavior, which includes cooperation, motivation, and promptness (ASTM, 1981a).

Training. Training comprises a carefully designed series of exercises that teach, practice, and evaluate the panelists' performance. The purpose of training panel members in sensory analysis is to familiarize an individual with test procedures, improve recognition and identification of sensory attributes in complex food systems, and improve sensitivity and memory so that precise, consistent, and standardized sensory measurements can be reproduced (ASTM, 1981b). One of the main objectives in descriptive analysis training is the development of descriptive language, which is used as a basis for rating food product attributes (Stone and Sidel, 1993). The training session is an important time for the panel leader to measure both individual and total panel performances. Communication within the group is critical to ensure that all attributes are understood and utilized in the same manner (Einstein, 1991).

Training makes a tremendous difference in a panelist's performance. When comparing the performance of consumers and panelists who received 20 h of training, Roberts and Vickers (1994) reported that consumers rated most attributes higher in intensity than did trained panelists. They likewise found that trained panelists found fewer product differences than consumers did. Chambers et al. (2004) compared the performance of descriptive panelists after short-term (4 h), moderate (60 h), and extensive (120 h) training and found that panelists' performance increased with increased training. Panelists reported sample differences in all texture and some flavor attributes after only 4 h of training. More differences were found after 60 h, but panelists were able to ascertain differences in all texture and flavor attributes after extensive training for 120 h. They concluded that extensive training may be required to reduce variation among panelists and increase their abilities to discriminate. Wolters and Allchurch (1994) likewise found that 60 h of training increased the number of attributes that could be discriminated. Trained panelists can discriminate inconspicuous attributes better than an untrained panelist and use a broader range of terminology when describing a product's texture or flavor profile (Cardello et al., 1982; Papadopoulos et al., 1991).

DESCRIPTIVE ANALYSIS TESTING METHODS

There are several testing methods for descriptive analysis. These include the flavor profile method, texture profile method, quantitative descriptive analysis (QDA) method, and spectrum descriptive analysis (Meilgaard et al., 1991; Stone and Sidel, 1993). The flavor profile method, developed by Arthur D. Little, Inc, in Cambridge, Massachusetts (Cairncross and Sjöström 1950), is a method of flavor analysis that makes it possible to indicate degrees of differences between samples based on intensity of individual character notes, degree of blending, and overall amplitude. The flavor profile involves a minimum of four assessors trained over a period of 6 months. The assessors evaluate the food or food product in a quiet, well-lit, odor-free room, usually around a round table to facilitate discussion. Assessors make independent evaluations and rate character note intensities using a 7-point scale from threshold to strong. The texture profile method was originally developed by M.I.T. in conjunction with the Strain Gage Denture Tenderometer (Brody 1957) and later refined by Product Evaluation and Texture Technology groups at General Foods Corp. in Tarrytown, NY (Muñoz et al., 1992), to define textural parameters of foods. The method was developed to focus on aspects overlooked in the flavor profile method (ASTM, 1996). The method is described by Brandt et al. (1963) as the sensory analysis of the texture complex of a food in terms of its mechanical, geometrical, fat, and moisture characteristics, the degree of each present and the order in which they appear from first bite through complete mastication. The objective of the texture profile method is to (1) eliminate problems dealings with subject variability, (2) compare results with known materials, and (3) establish a relationship with instrument measures (Szczesniak et al., 1963). A moderator leads the discussion around a round table to facilitate discussion and evaluation. Assessors are trained on texture definitions, evaluation procedures, and the standard reference scales that have been developed. References are provided as needed and graphical rating scales are used.

Tragon Corp. in Redwood City, California, developed the QDA method (Stone et al., 1974 and 1997), a method where "trained individuals identify and quantify, in order of occurrence, the sensory properties of a product or an ingredient." These data enable development of appropriate product multidimensional models in a quantitative form that is readily understood in both marketing and R&D environments." QDA is led by a descriptive analysis moderator, and requires 10–12 panelists, although in some tests 8–15 panelists may be involved. Training is conducted in a conference style room. Panel members develop terminology, definitions, and evaluation procedures. The training requires 2 weeks of training or approximately 8–10 h. Food products are evaluated in partitioned booths using 15 cm line scales in rating samples. References are provided as needed.

The spectrum descriptive method was designed by Civille, with modifications to the basic spectrum procedure having been made over several years of collaboration with a number of companies (Meilgaard et al., 1991). Spectrum was designed to provide a complete, detailed, and accurate descriptive characterization of a product's sensory attributes. The information received from this method of testing provides both qualitative and quantitative data. A trained moderator and 12–15 trained panelists are required. Training is conducted around a round table where assessors develop the terminology, definitions, and evaluation techniques and agree on references to be used during the evaluations. The trained and calibrated panelists evaluate food in partitioned booths. Evaluation time is approximately 15 min per product. Fifteen-point or 150 mm line scales are used. Descriptive analysis methods that employ a combination of some aspects of both QDA and the spectrum analysis methods are used by a large proportion of sensory practitioners (Einstein, 1991). These are called hybrid methods and are widely used in sensory descriptive analysis (Resurreccion, 1998).

Relating Descriptive Analysis and Physicochemical Measurements to Consumer Acceptance

In food product development, consumer acceptance of the food or food product is measured by affective tests. The characteristics of the product can be quantified, simultaneously, by sensory descriptive analysis ratings or instrumental and physicochemical measurements. The next question is to determine whether a relation exists between consumer acceptance (from affective tests) and what the product is sensed as being (from descriptive analysis), how the product is formulated (in terms of ingredients), and how the product performs (from physical measures). These measurements can be as food product specifications in further development or in the manufacture of a new product that rates high in consumer acceptance.

BIBLIOGRAPHY

1. Amerine, M.A., R.M. Pangborn, and E.B Roessler. 1965. *Principles of Sensory Evaluation of Food*. New York: Academic Press.
2. ASTM. 1981a. "*Selection of Sensory Panel Members in: American Society for Tasting Materials Guidelines for the Selection and Training of Sensory Panel Members*," Pennsylvania: ASTM, pp. 5–17.
3. ASTM. 1981b. "Training of Sensory Panel Members In Guidelines for the Selection and Training of Sensory Panel Members," Pennsylvania: ASTM, pp. 18–29.
4. ASTM. Committee E18. 1979. "Manual on Consumer Sensory Evaluation," Schaefer, E.E. eds., ASTM Special Technical Publication 682, pp. 52.
5. ASTM. Committee E-18. 1996. "Sensory Testing Methods," *ASTM Manual Series: MNL 26*, E. Chambers, IV. and M.B. Wolf, eds., West Conshohocken, PA: American Society for Testing and Materials, pp. 38–53.
6. Bett KL, J.R. Vercellotti, N.V. Lovegren, T.H. Sanders, R.T. Hinsch, and G.K. Rasmussen. 1994. "A Comparison of the Flavor and Compositional Quality of Peanuts from Several Origins," *Food Chemistry*, 51:21–27.
7. Bhattacharya, M., M.A. Hanna, and R.E. Kaufman. 1986. "Textural Properties of Extruded Plant Protein Blends," *Journal of Food Science*, 51:(4) 988–993.
8. Bourne, M.C. 1982. *Food Texture and Viscosity.* New York: Academic Press.

9. Brandt, M.A., E.Z., Skinner, and J.A Coleman. 1963. "Texture Profile Method," *Journal of Food Science*, 28:404–409.
10. Brody, A.L. 1957, Masticatory Properties of Foods by the Strain Gage Denture Tenderometer, Ph.D. Thesis. Cambridge, Massachusetts: Massachusetts Institute of Technology.
11. Bruns, A.J., and M.C Bourne. 1975. "Effects of Sample Dimensions on the Snapping Force of Crisp Foods," *Journal of Texture Studies*, 6:445–458.
12. Cairncross, S.E., and L.B Sjöström. 1950. "Flavor Profiles—a New Approach to Flavor Problems," Food *Technology*, 4:308–311.
13. Cardello, A.V., O Maller, J. G. Kapsalis, R. A. Segars, F. M Sawyer, C. Murphy, and H. R Moskowitz, 1982. "Perception of Texture by Trained and Consumer Panelists," *Journal of Food Science*, 47:1186–1197.
14. Cardello, A.V., H.G., Schutz, C., Snow, and L. Lesher. 2000. "Predictors of Food Acceptance, Consumption and Satisfaction in Specific Eating Situations," *Food Quality and Preference*, 11:201–216.
15. Caul, J.F.1957. "The Profile Method in Flavor Analysis," *Advanced Food Research*, 7:1–6.
16. Chambers, D.H., A.M.A. Allison, and E. Chambers IV. 2004. "Training Effects on Performance of Descriptive Panelists," *Journal of Sensory Studies*, 19(6):486–499.
17. Chambers, E. and M.B. Wolf. 1996. *Sensory Testing Methods*. 2nd ed. West Conshohocken, Pennsylvania: American Society for Testing Materials pp. 58–72.
18. Chen, A.W., A.V.A, Resurreccion, and L.P. Paguio. 1996. "Age Appropriate Hedonic Scales to Measure Food Preferences of Young Children", *Journal of Sensory Studies*, 11:141–163.
19. Christensen, C.M. 1983. "Effects of Color on Aroma, Flavor and Texture Judgments of Foods," *Journal of Food Science*, 48:787–790.
20. Clydesdale, F.M. 1978. "Colorimetry-Methodology and Applications," *Critical Reviews in Food Science and Nutrition*, 10:243–301.
21. Clydesdale, F.M. 1991. "Color Perception and Food Quality," *Journal of Food Quality*, 14:61–74.
22. Clydesdale, F.M. 1993. "Color as a Factor in Food Choice," *Critical Review Food Science Nutrition,* 33(1):83–101.
23. Curley, L.P. and R.C. Hoseney. 1984. "Effects of Corn Sweeteners on Cookie Quality," *Cereal Chemistry*, 61(4):274–278.
24. Duxbury, D.D. 1988. "R&D Directions for the 1990s," *Food Processing* 49(8):19–28.
25. Einstein, M. A. 1991, "Descriptive Techniques and Their Hybridization," *In Sensory Science Theory and Applications in Food,* Lawless, H.T. and Klein, B. P. eds., pp. 317–338.
26. Gacula, Jr. M.C. ed. 1997. *Descriptive Sensory Analysis in Practice*. Trumbull, CT: Foods and Nutrition Press.
27. Galvez, F.C. and A.V.A. Resurreccion. 1990. "Comparison of Three Descriptive Analysis Scaling Methods for the Sensory Evaluation of Noodles," *Journal of Sensory Studies*, 5: 251–263.
28. Gormley, T.R. 1987. " Fracture Testing of Cream Cracker Biscuits," *Journal of Food Engineering*, 6:325–332.
29. Harper S.J. and M.R. McDaniel. 1993. "Carbonated Water Lexicon: Temperature and CO_2 Level Influence on Descriptive Ratings," *Journal of Food Science*, 58(4):893–898.
30. Hashim, I. B., A. V. A. Resurreccion and K. H. McWatters. 1995. "Consumer Acceptance of Irradiated Poultry," *Poultry Science*, 74:1287–1294.

31. Hashim, I.B., A.V.A. Resurreccion and K.H. McWatters. 1996. "Consumer Attitudes Towards Irradiated Poultry," *Food Technology*, 50(3):77–80.
32. Hersleth, M., B. Mevik, T. Naes, and J. Guinard. 2003. "Effect of Contextual Factors on Liking for Wine-Use of Robust Design Methodology," *Food Quality and Preference*, 14:615–622.
33. Hersleth, M., O., Ueland, H., Allain, and T.Naes. 2005. "Consumer Acceptance of Cheese, Influence of Different Testing Conditions," *Food Quality and Preference*, 16:103–110.
34. Hootman, R.C. ed. 1992. "Manual on Descriptive Analysis Testing for Sensory Evaluation," West Conshohocken, Pennsylvania: ASTM Manual Series, MNL 13.
35. IFT/SED. 1981. "Sensory Evaluation Guideline for Testing Food and Beverage Products," *Food Technology*, 35(11):50–59.
36. Jellinek, G. 1964. "Introduction to and Critical Review of Modern Methods of Sensory Analysis (Odor, Taste, Flavor, and Evaluation) With Special Emphasis on Descriptive Sensory Analysis (Flavor Profile Method)," *Journal Nutrition Dietetics*, 1:219–260.
37. Kamen, J.M., D.R., Peryam, D.B., Peryam, and B.J. Kroll. 1969. "Hedonic Differences as a Function of Number of Samples Evaluated," *Journal of Food Science*, 34:475–479.
38. Katz, E.E., and T.P Labuza. 1981. "Effect of Water Activity on the Sensory Crispness and Mechanical Deformation of Snack Food Products," *Journal of Food Science*, 46:403–409.
39. Keane P. 1992. "The Flavor Profile", *In Manual on Descriptive Analysis Testing for Sensory Evaluation,* Hootman RC, ed., Pennsylvania: ASTM, pp. 514.
40. Kimmel, S.A., M. Sigman-Grant, and J.X. Guinard, 1994. "Sensory Testing with Young Children," *Food Technology*, 48(3):92–99.
41. King, S.C., A.J. Weber, H.L Meiselman, and N. Lv, 2004. "The Effect of Meal Situation, social Interaction, Physical Environment and Choice on Food Acceptability," *Food Quality and Preference,* 15:645–653.
42. Kozlowska, K., M. Jeruszka, I. Matuszewska W. Roszkowski, N. Barylko,-Pikielnaand, and A. Brzozowska. 2003. Hedonic Tests in Different Locations as Predictors of Apple Juice Consumption at Home in Elderly and Young Subjects," *Food Quality and Preference,* 14:653–661.
43. Kroll, B.J. 1990. "Evaluating Rating Scales for Sensory Testing with Children," *Food Technology,* 44(11):78–86.
44. Lawless, H.T., and H. Heymann. 1998. "Sensory Evaluation of Food, Principles and Practices," New York, New York: Chapman and Hall, p. 480.
45. Lord, J.B. 2000. "New Product Failure and Success." Developing New Food Products for a Changing Marketplace, Brody, A.L. and J.B. Lord, eds., Lancaster, PA: Technomic Publishing Co., Inc. pp. 55–86.
46. Macfie H.J.H. and D.M.H. Thomson. 1994. "Measurement of Food Preferences," London: Chapman and Hall.
47. Malundo T.M.M., and A.V.A. Resurreccion. 1994. "Peanut Extract and Emulsifier Concentrations Affect Sensory and Physical Properties of Liquid Whitener," *Journal of Food Science*, 59:344–349.
48. Meilgaard, M., G.V., Civille, and B.T Carr. 1987. *Sensory Evaluation Techniques*, Vol. II. Boca Raton, FL: CRC Press. pp. 1–24.
49. Meilgaard, M., G.V., Civille, and B.T. Carr. 1991. *Sensory Evaluation Techniques,* 2nd ed. Boca Raton, FL: CRC Press.
50. Meiselman H.L. 1996. The Contextual Basis for Food Acceptance, Food Choice and Food Intake: The Food, The Situation And The Individual In *Food Choice Acceptance*

and Consumption. H.L Meiselman. and H.J.H. Macfie. eds., London: Chapman and Hall, pp. 239–292.

51. Morris, C.E. 1993. "Why New Products Fail," *Food Engineering*, 65(6):132–136.
52. Moskowitz H.R. 1983. "Product Testing and Sensory Evaluation of Foods," *Journal. Sensory Testing*; Trumbull, Connecticut: Food and Nutrition Press.
53. Moskowitz, H.R. 1974. "Sensory Evaluation by Magnitude Estimation," *Food Technology*, 28 (11):16, 18, 20–21.
54. Moskowitz, H.R. 2000. "R&D-Driven Product Evaluation in the Early Stage of Development,". In *Developing New Food Products for a Changing Marketplace*. A.L. Brody, and J.B. Lord, eds., Lancaster, PA: Technomic Publishing Co., Inc. pp. 277–328.
55. Moskowitz, H.R., A.M. Munoz. And M.C. Gacula, Jr. 2003. *Viewpoints and Controversies in Sensory Science and Consumer Product Testing.* Trumbull, Connecticut: Food and Nutrition Press, p. 212.
56. Muñoz A.M., and G.V.Civille 1992. "The Spectrum Descriptive Analysis Method, In *Manual on Descriptive Analysis Testing for Sensory Evaluation*, RC. Hootman ed., Pennsylvania: ASTM. pp. 22–34.
57. Muñoz, A.M., G.V Civille. and B.T Carr. 1992. *Sensory Evaluation in Quality Control* New York: Van Nostrand Reinhold, pp. 240.
58. O'Mahony, M. 1986. "Sensory Evaluation of Food," New York: Marcel Dekker, p. 487.
59. O'Mahony M., U. Thieme. And L.R. Goldstein. 1988. "The Warm-Up Effect as a Means of Increasing the Discriminability of Sensory Difference Tests," *Journal of Food Science*, 53(6):1848–1850.
60. O'Mahony, M. 1991. "Descriptive Analysis and Concept Alignment," *IFT Basic Symposium Series: Sensory Science Theory and Applications in Foods*, H.T. Lawless and B.P. Klein, eds., New York, Basel, and Hong Kong: Marcel Dekker, pp. 223–267.
61. Papadopoulos, L.S., R.K., Miller, G.R., Acuff, L.M Lucia, C.Vanderzant, and H.R Cross. 1991. "Consumer and Trained Sensory Comparisons of Cooked Beef Top Rounds Treated with Sodium Lactate," *Journal of Food Science*, 56(1141–1146), 1153.
62. Peryam, D.R., and F.J. Pilgrim, 1957. "Hedonic Scale Method of Measuring Food Preference," *Food Technology*, 11(9):9–14.
63. Plemmons L.E., and A.V.A. Resurreccion. 1998. "A Warm-Up Sample Improves Reliability of Responses in Descriptive Analysis," *Journal of Sensory Studies*, 13:359–376.
64. Pound, C., and L. Duizer, 2000. "Improved Consumer Product Development. Part One. Is a Laboratory Necessary to Assess Consumer Opinion?" *British Food Journal*, 102 (11):810–820.
65. Powers, J.J. 1984 *Current Practices and Applications of Descriptive Methods in Sensory Analysis of Foods*, J.R. Piggott ed., London: Applied Science Publishing p. 179.
66. Rainey BA. 1986. "Importance of Reference Standards in Training Panelists," *Journal of Sensory Studies*, 1:149–154.
67. Resurreccion, A.V.A. 1998. *Consumer Sensory Testing for Product Development*, Gaithersburg, Maryland: Aspen Publishers.
68. Resurreccion, A.V.A., and E.K Heaton. 1987. "Sensory and Objective Measures of Quality of Early Harvested and Traditionally Harvested Pecans," *Journal of Food Science*, 52:1038–1040, 1058.

69. Roberts, A.K. and Z.M Vickers. 1994. "A Comparison of Trained and Untrained Judges' Evaluation of Sensory Attribute Intensities and Liking of Cheddar Cheese," *Journal of Sensory Studies*, 9:1–20.
70. Roessler, E.B., R.M. Pangborn, J.L. Sidel, and H. Stone. 1978. "Expanded Statistical Tables for Estimating Significance in Paired-Preference, Paired Difference, Duo-Trio and Triangle Tests," *Journal of Food Science*, 43:940–943.
71. Rutledge KP and JM.Hudson 1990. "Sensory Evaluation: Method for Establishing and Training a Descriptive Flavor Analysis Panel," *Food Technology*, 44(12):78–84.
72. Schutz, H. 1965. "A Food Action Scale for Measuring Food Acceptance," *Journal of Food Science,* 30:365–374.
73. Seymour, S.K., and D.D. Hamann, 1988. "Crispness and Crunchiness of Selected Low Moisture Foods," *Journal of Texture Stud*ies, 19:79–95.
74. Sherman, P., and D.S. Deghaidy. 1978. "Force-Deformation Conditions Associated with the Evaluation of Brittleness and Crispness in Selected Foods," *Journal ofTexture Studies*, 9:437–459.
75. Sokolow, H. 1988. "Qualitative Methods for Language Development," *Applied Sensory Analysis of Foods*, vol. I., H. Moskowitz, ed., Boca Raton, FL: CRC Press, pp. 4–20.
76. Stone, H., J. Sidel, S. Oliver, A.Woolsey, and R.C Singleton. 1974. "Sensory Evaluation by Quantitative Descriptive Analysis," *Food Technology*, 28(11):24–34.
77. Stone H, J. Sidel S. Oliver, A. Woolsey, and R.C. Singleton. 1997. "Sensory Evaluation by Quantitative Descriptive Analysis," MC Gacula, ed., *Sensory Analysis in Practice*, Connecticut: Food and Nutrition Press.
78. Stone, H., and J.L. Sidel, 1985. *Sensory Evaluation Practices*. 2nd ed. San Diego, CA: Academic Press.
79. Stone, H., and J.L. Sidel, 1993. "Sensory Evaluation Practices," San Diego, CA: Academic Press.
80. Szczesniak, A.S., M.A. Brandt, and H.H. Friedman. 1963. "Development of Standard Rating Scales for Mechanical Parameters of Texture and Correlation Between the Objective and the Sensory Methods of Texture Evaluation," *Journal of Food Science*, 28:397–403.
81. Wolters, C. J. and E. M Allchurch. 1994. "Effect of the Training Procedure on the Performance of Descriptive Panels," *Food Quality and Preference*, 5:203–214.
82. Zook K, and C. Wessman. 1977. "The Selection and Use of Judges for Descriptive Panels," *Food Technology*, 11:56–61.

14 The Scale-Up and Commercialization of New Food Products

Traci L. Morgan and Michelle M. Depp

CONTENTS

INTRODUCTION

Taking a new food product from the idea stage all the way through to commercialization is a very painstaking process. Moving successfully from food product development to full scale manufacturing requires proper planning, good communication, and the right amount of scrutiny, to optimize utilization of company resources. The focus of this chapter will be on the scale-up and commercialization aspects of the food product development process.

As indicated by this book, the following steps are involved in the food product development cycle:

- Concept generation and evaluation
- Laboratory formulation
- Consumer sampling/sensory flavor evaluation
- Scale-up
- Pilot testing (could be done in laboratory, production area, or contracted facility)
- Packaging/shelf life studies
- Full scale system development
- Production run
- Test marketing, if warranted
- Commercialization

A new food product is developed for one of several reasons that may include innovation, product variety, convenience, healthier food alternatives, sales, market share, profit, safer products, and so forth. After someone has developed a novel concept, planners determine the steps necessary to manufacture the food product at a benchtop level. Once these methodologies and parameters have been resolved and the process has been proven to be repeatable, it is then introduced to consumers to ensure that there is enough interest in the food product to warrant investing time and resources in pilot and production equipment and facilities. On establishing consumer interest, the pilot plant may then be built and operated to generate information about the system for the purpose of designing larger systems. It is here that the initial details and idiosyncrasies of converting the process from laboratory scale to a pilot facility are resolved. Pilot systems identified as successful with minimal risks to the company are then generated on a much larger scale.

What exactly is scale-up? Scale-up is the collection of changes that occur between the smaller laboratory or kitchen prototype and full scale manufacturing. This does *not* mean to simply fabricate larger pieces of equipment and increase utility/facility capacity.

Factors that must be worked out include

1. Are formulation adjustments required?
2. Where are the ingredient sources?
3. Where are the optimal manufacturing locations?
4. Will this be a batch or continuous process, or a combination of the two?
5. Is it more practical to use single larger units or to have multiple smaller units to meet production needs?
6. Will faster or slower production rates be used on the new system?
7. Does the scale-up increase or decrease cycle times?
8. What impact will variations in ambient conditions have on the new system?
9. Will the process be manual, semiautomatic, or fully automated?
10. Does scale-up impact temperature profiles?

11. Can we incorporate this new development into existing facilities and operations? And if so,
12. What is the appropriate level of staff required to operate the equipment?

These are just a few of a countless questions that must be answered during the design of full scale production areas.

In order to move forward with the technical efforts associated with manufacturing new foods in larger quantities, the scale-up procedure must be well organized and conducted with some overall objectives in mind. For example, the newly developed system must be reliable. This means utilizing the necessary resources during front end design and engineering in order to ensure satisfactory results. The units and facilities constructed must also be able to meet consumer demands based on projected consumer acceptance and sales volumes while keeping the commodity cost competitive. If the resulting new system is not able to adequately manufacture foods while keeping the cost of items competitive with marketing and consumer requirements, then the company will not be able to maximize on the investment dollars used to support the launch of new food products. Project completion timelines must become critical points of consideration that cannot be overlooked. Any delay in getting the system built and operational ultimately results in lost time to the market that drives the company's cost up. Perhaps, the most important goal is to make sure that the overall product that is made meets the agreed on specifications. If the new system is not capable of large-scale production of the same items as the laboratory units within allowable tolerances, then the process needs to either be re-analyzed for reproducibility on a mass level or redefined. The team of researchers, developers, engineers, and operators cannot lose sight of these goals during the entire project.

STARTING POINT

We begin the process of scaling up with final results from the laboratory kitchen, or benchtop stage. It is here that the food product formulation and specifications are finalized and a thorough knowledge of the process is obtained. All steps and parameters required for production are generally denoted in a process flowsheet or process flow diagram (PFD). (Figure 14.1 is an example of a PFD). Product feasibility is also established during which an estimate of investment and operating expenses is completed. This information, coupled with the scale-up factor (SUF), for example, the ratio of commercial plant (or pilot plant) to the small benchtop unit, is used to arrive at a design for the new larger system.

It is a very difficult task for technologists and process engineers to finalize specifications for new food products. The number of variables that exist for a given food is so extensive that only a few of these options can be effectively studied within a reasonable amount of time. This is where knowledge and experience coupled with prototype response and sensory evaluation plays a major role to narrow specifications for product development. From this initial information, the process and necessary equipment is defined.

On completion of final details for the process, the engineer uses these quantitative data to begin designing and/or modifying equipment to meet the need. The

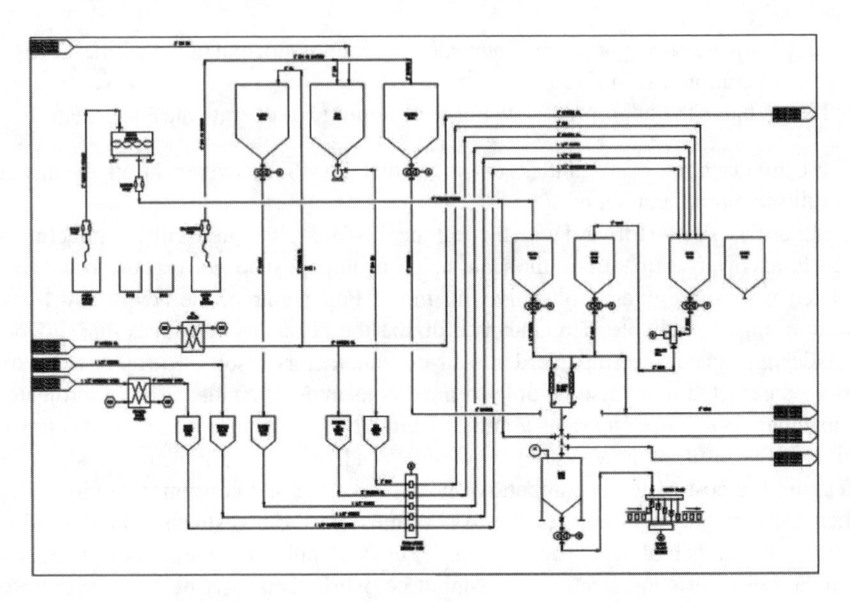

FIGURE 14.1 Example of a process flow diagram.

engineer will work closely with the food scientist/technologist to establish facilities and equipment that are able to result in larger quantities without changing flavor, color, texture, and other characteristics key to the success of the product. Some companies opt to use existing equipment to run batches in much smaller quantities. Many today also build pilot plants to manufacture small quantities of the new food products prior to committing larger amounts of capital to full scale manufacturing. Yet another alternative involves contract processors where the services of an outside company are purchased under agreement of terms to systematically prepare and/or package foods at a separate facility. The objective in each of the aforementioned cases is to commit the inevitable mistakes associated with working through the details of the process to smaller volumes and the benefits associated with an optimally run system to large-scale manufacturing.

DEVELOPING PILOT PLANT FACILITIES

There are many advantages associated with constructing pilot facilities for the manufacture of new food products. The performance data obtained from a properly designed and operated small scale plant is oftentimes more accurate and reliable than theoretical calculations based on laboratory data or general correlations. Safety factors are less significant at reduced levels. Yields as well as efficiencies tend to be higher when exploring the methods for manufacturing new foods. This results in fewer problems to be encountered on a larger scale. The main function of the pilot facility is to provide design criteria for the larger equipment and utilities. Another benefit is that small amounts of food product may be used for consumer trials and test marketing, if used.

When developing production resources, it is good practice to keep emerging technologies in mind as these may greatly impact the streamlining of processes and make operations more efficient. In an effort to accommodate new technologies, major steps can be taken. First, preliminary studies are key to determining the aspects and conditions of food plant design. Close examination should be taken of the products and the raw materials required to make these products, as well as various alternatives available for these two categories. Study of the food products should be as broadly characterized as possible with the overall objective being to establish the technical and commercial aspects of the items manufactured. Next, there must be an exploration of alternatives that exist in food processing. This may be achieved through gathering information from various sources such as institutions and industry leaders, collecting data from process development laboratories, and compiling results from pilot plant studies.

PROJECT DEVELOPMENT AND IMPLEMENTATION: AN OVERVIEW

A project for commercialization of new food products is an organized effort to utilize company resources in order to produce safe, quality food products for consumers. It is an investment in the growth and advancement of the company. The objective of such a capital investment is that the goods or services produced will be of much greater value than the predicted or actual cost associated with the project.

The project life cycle can be divided into five major categories, which are depicted in Figure 14.2.

The project begins with the concept phase during which the organization establishes what exactly is needed and how it may be achieved. The feasibility of completing the project is also determined. This results in a more concentrated and focused application of company resources to complete the work. During these initial

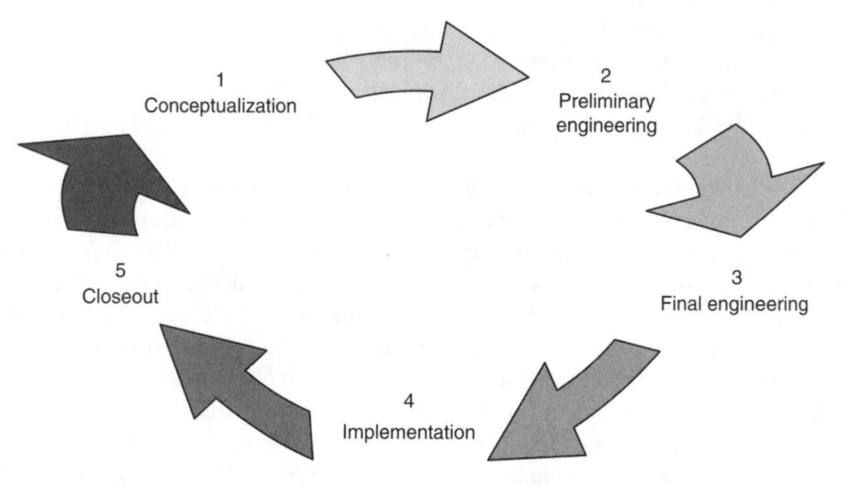

FIGURE 14.2 Project life cycle.

studies, fundamental aspects of the scope of work are established such as similarities/differences between the initial pilot facility, system location, and space layout. Generally, multiple solutions are considered with the most economical and practical plan that meets the company's business objectives being selected. Preliminary engineering is the opportunity for a company to work out the key elements of the project while making a minimal upfront investment.

The basis for initial engineering is outlined by the project definition, which identifies the scope of work. This document states the overall purpose of the business endeavor, that is, which pieces of equipment will be purchased, what facility modifications will occur, and what are the changes in production and personnel flows. An analysis of the flows presented within the document indicates how this new system will impact the existing premises and current manufacturing operations. From this main document, a high-level budget and schedule are established. Equipment and construction Request for Proposals (RFPs) that provide specifications to be met while fabricating equipment and modifying facilities are sent to potential contractors. Subsections of these documents may include the utility requirements, materials of construction, finishes, electricals, flooring, and so forth.

During the final engineering phase, preliminary system analysis results are used as a starting point to arrive at necessary information for the project. Detailed engineering is the bridge between basic engineering and the construction work. Such information as the following is determined:

- Process and instrumentation diagram (P&ID)
- Design of pipe layout and instrumentation
- Instrument specifications (pressure, size, material of construction, etc.)
- Number and location of utility drops
- Required utility parameters
- Airlock dimensions
- Room pressure differentials and classifications
- Operator–system interface
- Safety evaluation
- Programming requirements for controls
- Pump and heat exchanger specifications
- Applicable local and national codes and standards

Figure 14.3 shows a sample P&ID for a manufacturing facility.

Project implementation is where the practical application of algorithms (methodologies) are used to fulfill a desired purpose, which in this case is additional capacity. For the most part, it involves equipment fabrication, construction work, scheduling, and allocation of resources. This is the part of the project life cycle during which working drawings and specifications materialize, contract documents are prepared and awarded, and the construction work is undertaken in order to result in a completed deliverable. Sections of the project can proceed at differing speeds, but the overall project must be completed in order to test the system.

Installation, start-up, commissioning, and validation are the final aspects of project execution. Installation involves the putting into place of processing equipment

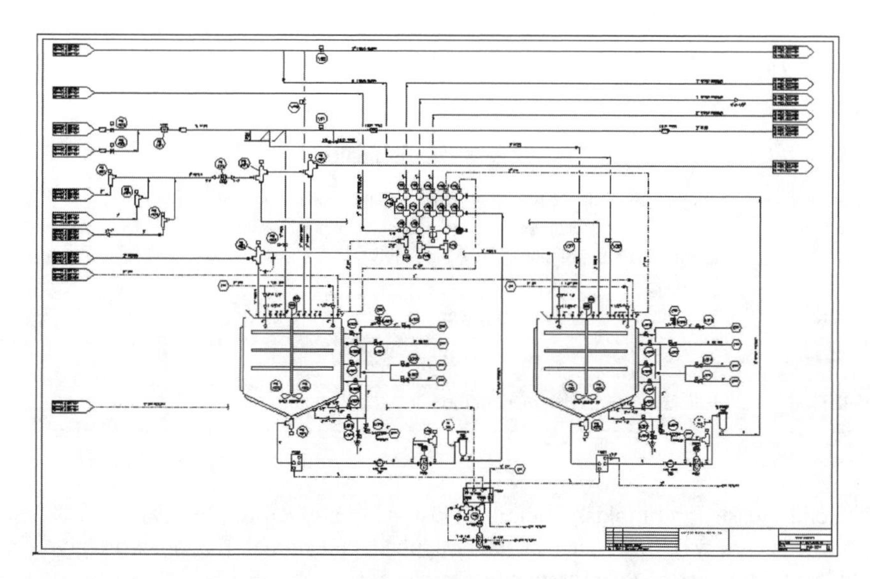

FIGURE 14.3 Process and instrumentation diagram.

by assembly, construction, or set-up. On completion of arranging and connecting all equipment utilities and electrical components, the system is started to ensure proper functioning. Commissioning advances from the stage of static completion to full working order and achievement of the specified operational requirements. Commissioning is defined by the International Society of Pharmaceutical Engineering (ISPE) as "a well planned, documented, and managed engineering approach to the start-up and turnover of facilities, systems, and equipment to the end user that results in a safe and functional environment that meets established design requirements and stakeholder expectations." In short, commissioning establishes total functionality of the system before moving forward with validation. Once the process has been determined to be fully operational, documented testing (validation) must follow in order to verify all aspects of the facility, utility, and equipment that can affect product quality, adhere to approved specifications, operates at all intended levels, and performs as intended by meeting predetermined acceptance criteria.

The completion of all the work for a project marks the end of the life cycle and is referred to as closeout. During this stage, paperwork is finalized and must be accepted and signed off by all responsible parties. Final revision and transmittal of documentation to reflect the "as built" condition are turned over to appropriate departments for retention of files. Any open contracts are concluded, certifications turned over to end users, and a transition of ownership is made to management and operators.

Preliminary Engineering Step 1: Risk Assessment

Some special requirements that must be taken into consideration include microbiological, chemical/biochemical, and physical variables when designing and constructing new food production facilities. Microbiological refers to microorganisms such as

TABLE 14.1
Risk Assessment Table

Type of Risk	Examples
Biochemical	Enzymes to enhance food production, hormone adjustment in animals
Chemical	Spray residues, fertilizers, aflatoxin, benzene hexachloride, cadmium, lead, mercury, methyl alcohol, polychlorinated biphenyls
Physical	
Animal	Excreta, hairs, feathers, insect eggs
Plant	Phytotoxins
Mineral	Foreign metals
Other	Human to food, equipment to food

bacteria, yeasts, and molds. Attention must also be given to physical hazards that occur in a variety of forms such as mineral, plant, and animal. Examples are given in Table 14.1 for each of the aforementioned categories.

Each one of the above categories provides some level of risk, or potential challenge, which may be associated with the development of the new food products. Risk is defined here as the potential negative impact to an asset or some characteristic of value (i.e., final product or system) that may arise from some present or future process. Risk analysis may be done in an effort to identify and negate these risks if done in a comprehensive and thorough manner. The analysis of risks consists of three components: risk assessment, risk management, and risk communication. Risk assessment is a tool used to understand the possible hazards, determine the probability of occurrence, and establish the impact if they do occur. Risk management weighs policy alternatives in the light of risk assessment, and if necessary, selects and implements appropriate control measures. Finally, risk communication is an interactive exchange of pertinent information between risk assessors, risk managers, consumers, and all other parties impacted. All three levels of risk analysis should be developed and maintained throughout the development and use of new facilities in order to minimize product loss and possible hazards to customers.

The Hazard Analysis and Critical Control Point (HACCP) program is a systematic approach to food manufacture in which raw materials and each individual step in a process are considered in detail and evaluated for their potential to contribute to the development of pathogenic microorganisms, and/or chemical or physical hazards. Steps in this analysis may be used as part of the risk assessment process.

Preliminary Engineering Step 2: Processing Requirements

The engineer must work with food technologists and culinologists to gather the data necessary for the design of the food processing system. There may be some estimation along with theoretical modeling and simulations combined with these data to form the overall design criteria for the new equipment and facilities. Optimal processing conditions best suited to producing maximum quality and quantity of product are also determined. The results of the initial findings are then confirmed at a pilot plant or small batch level so as to not commit significant company revenue and resources to erroneous findings and substandard products.

Many stumbling blocks are associated with moving from small scale to mass production in food manufacturing facilities, and so it becomes imperative that the appropriate amount of time and support is dedicated during the concept development phase. The production levels can significantly change from what was originally considered to be quantity sufficient for new food product launch. This makes additional adjustments to the formulation and process a necessity. One then has to consider the following: Do the ingredients behave in a different manner when handled in much larger quantities? Do the flavors, appearance, and texture change at all during the production process? There are sometimes unknowns or insignificant product losses found in product streams when working with smaller quantities. As the level of product increases, these soluble impurities and product losses become vital in that they impact the overall cost of operating the plant. For example, soluble impurities may disappear during the reaction process due to adsorption or reaction with vessel surfaces. Since the surface-to-volume ratio is much lower as the process is increased, the impurities would become more noticeable especially over longer operating times. Another problem is that subtle changes in conditions or parameters can significantly impact operation results. To illustrate this point, let us examine three examples. A company may choose to recycle resources in an effort to improve the economic situation, which could have an impact on product quality. Speeding up production processes may result in difficulties such as reduction in accuracy with regard to variables such as weight control, deposition accuracy, and assembly functions (i.e., cutting, folding, and shaping). Changes in temperature and humidity as well as fluctuations in ambient conditions change not only from one season to the next, but also during the course of one day, which is a major concern for open systems.

If we examine more closely at some important areas of processing, such as mixing, several variables must be taken into account as engineers move toward increasing aspect ratios. One variable that must be considered is dispersion of ingredients inside the vessel. Typically as the volume increases, the surface-to-volume ratio decreases, which may result in dead zones or inadequate mixing inside the unit. Larger volumes also usually mean an increase in mixing times. Depending on the parameters set for the process, this may or may not pose a problem. Whenever mixer sizes are adjusted to compensate for uniform mixing of fluids, this automatically means an increase in power requirements for the system. Much older plants usually encounter problems such as lack of ample electrical capacity within existing facilities, which can pose a serious problem. Materials handling is another issue that must be addressed as operators move from dealing with very small manageable quantities to much larger ones. This proves to be a greater concern for batch processes as opposed to continuous operations that are usually built to allow for expansion. Another point of consideration is incorporation of air. As the quantity of liquid increases, so also does the length of time required for uniform incorporation of gases, or the reverse, avoidance of air incorporation. When significantly changed, all of these factors could greatly impact product quality by altering the physical characteristics of the food.

Another example of how subtle differences in conditions can cause major changes in resulting product involves deviations in temperature, more specifically ambient conditions. Variations in season, time of day, regional location, and altitude are all factors that impact ambient weather conditions. One example of where this parameter would be applicable involves conveyor operations for a new juice beverage. The product leaves the filling area where it is closed under vacuum, travels to the cleaning area where the exterior is rinsed and dried prior to labeling. Heat exchangers are provided as part of the rinse/dry system in order to reduce the bottle temperature to appropriate levels prior to entering the labeler. The load requirements for the heat exchangers will be very different depending on the temperature of the product entering the equipment, although the necessary exit temperature is the same. So, for an identical equipment installation located in Atlanta, Georgia and operating during the summer to yield the same results as the system located in Chicago, Illinois and operating in the winter, allowances in the design must take into account the fluctuations in ambient temperatures.

Ingredient quality variations are also important factors that have great impact on the design of system processes as well as the end product. Numerous factors can affect ingredient quality some of which include age, temperature, moisture, concentration, viscosity, and use of rework. Subtle modifications in these variables could result in loss of nutritional value and sensory characteristics (flavor, color, and mouth feel), and poor overall product results (safety, consistency, reliability, and stability). Not only does raw material variability affect processing, but it also makes the management of manufacturing operation more difficult. Low quality ingredients will result in a substandard quality and reject finished goods and possibly, even items with shortened shelf life. It then becomes necessary to have well-defined specifications and tolerances for ingredients and materials introduced to the system.

Differing ingredients present another set of challenges for processing requirements from the standpoint of inherent characteristics as well as interactions. Anytime there is a change in the composition of ingredients, the microbiological, chemical, and physical interactions between these items must be determined at different volumes. For example, how will the addition of a dry ingredient into a liquid dairy product impact the resulting mixture? Will there be an increased risk of spoilage due to the change in microflora? How will the rheological properties change? How will this impact the mixing requirements? Depending on the percentage of dry ingredients added, there could be a variation in color, texture, and aroma of the resulting solution. Emulsification breakdown is also another possibility. Two items may also become immiscible on trying to incorporate one item into the other leading to solid stratification. Mixing of materials may also result in a thixotropic fluid in which time dependence becomes a very important factor. Then, it becomes critical for all physical properties and reactivity of each ingredient to be known thoroughly during this phase. An inadequate knowledge of these characteristics may result in wasting of some of the company's vested interests. Process development will flow from this information once these facets have been finalized.

PRELIMINARY ENGINEERING STEP 3: PROCESS DEVELOPMENT

Process development begins with the design of steps desired for physical and/or chemical transformation of materials. This can include the design of new facilities, modification of existing areas, or expansion of production capabilities. The layout starts on a conceptual level and concludes with the blueprints for fabrication and construction. The purpose of documents generated during the process development phase is to ensure that all design components fit together and flow properly from introduction through exiting of the system. It also serves as a useful tool for communicating the overall plan to other parties involved, such as quality assurance and regulatory affairs, as well as foundational information for equipment vendors and construction contractors.

It is the process engineer's responsibility to examine all forms of physical food processing involved and break this up into a small number of basic operations that are referred to as unit operations. Each one of these unit operations must have the capability to stand alone with a dependence on coherent physical principles. For example, mass transfer can be considered a unit operation with the underlying physical principle involving molecular and convective transport within physical systems. A practical application of this would be determining the effective diffusion coefficient of dissolved sucrose from beets.

Once the food processes have been divided into basic unit operations, the physical principles supporting the phenomena are established. Quantitative relationships in the form of mathematical equations may already exist or can be built to describe physical principles. These equations are then used to follow the activities of the process and for process modification if necessary. Parameters found within the equations provide points for process control, and the extent to which these factors alter results must be determined by means of mathematical modeling.

Some of the unit operations involved in the food and beverage industry include fluid mechanics (mixing, rheology, viscometry), heat transfer (freezing, drying, evaporation), mass transfer (distillation, extraction, gas absorption, crystallization, membrane processes), mechanical separation (filtration, centrifugation, sedimentation, sieving), and chemical reactions. A sample process mixing cycle is shown in Figure 14.4.

Building on the unit operations, the next step is to construct drawings incorporating all unit operations. The initial drawing generated for process design is the Block Flow Diagram (BFD) that is a simplified schematic composed of rectangles and lines indicating major equipment, material, or energy flows for the system. An example of a BFD is given in Figure 14.5. A more detailed version of the BFD, the PFD or Flowsheet is a figure that includes all equipment and indicates the general flow of production processes along with the relationship between the major equipment required. A more thorough mass balance must be carried out to define all the flows in and out of the system and between individual units. In addition to this, a thermal balance is completed to estimate the heating and cooling loads required in the facility.

Figure 14.6 that shows a PFD for one element of a food production facility. Overall data displayed on the PFD will include operational data (flow rates, temperatures, and pressures), composition of fluids/materials, major equipment involved, bypasses and

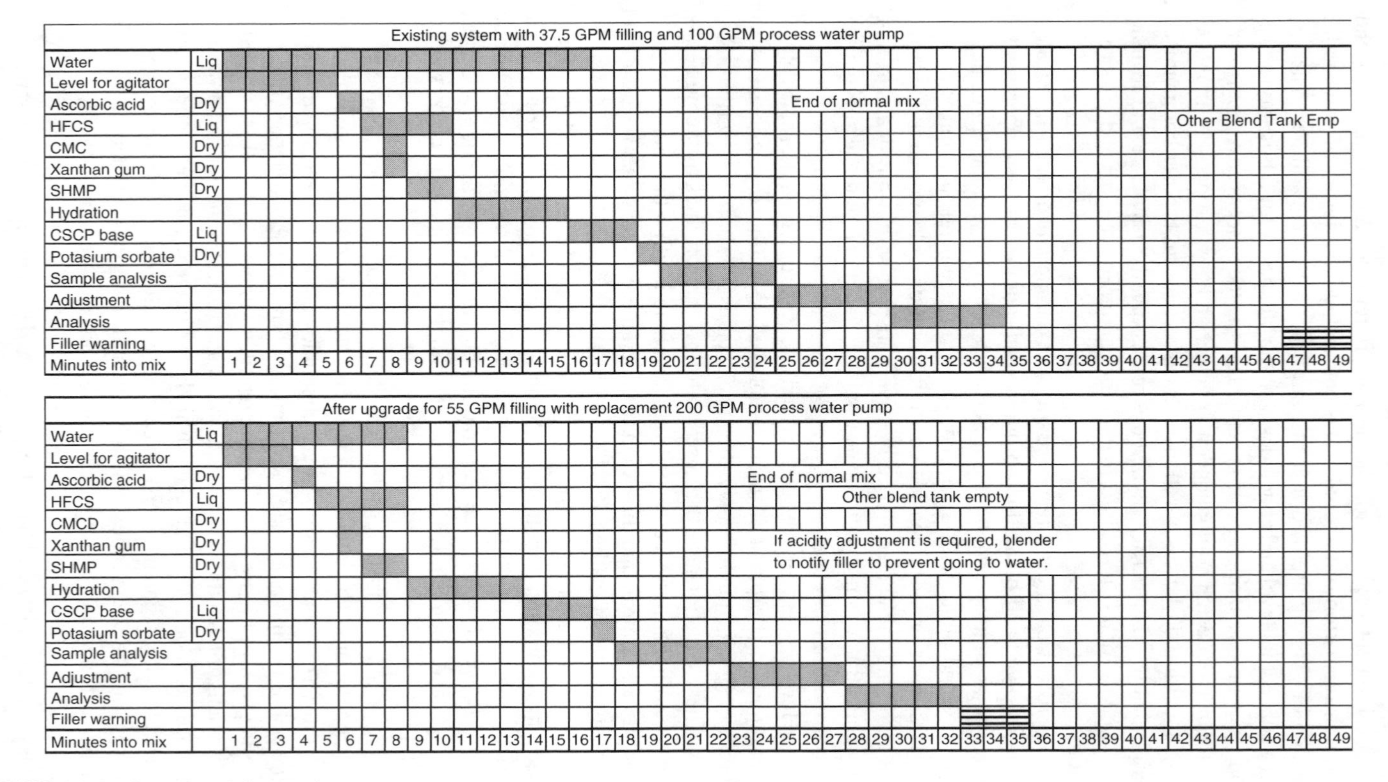

FIGURE 14.4 Sample mixing cycle.

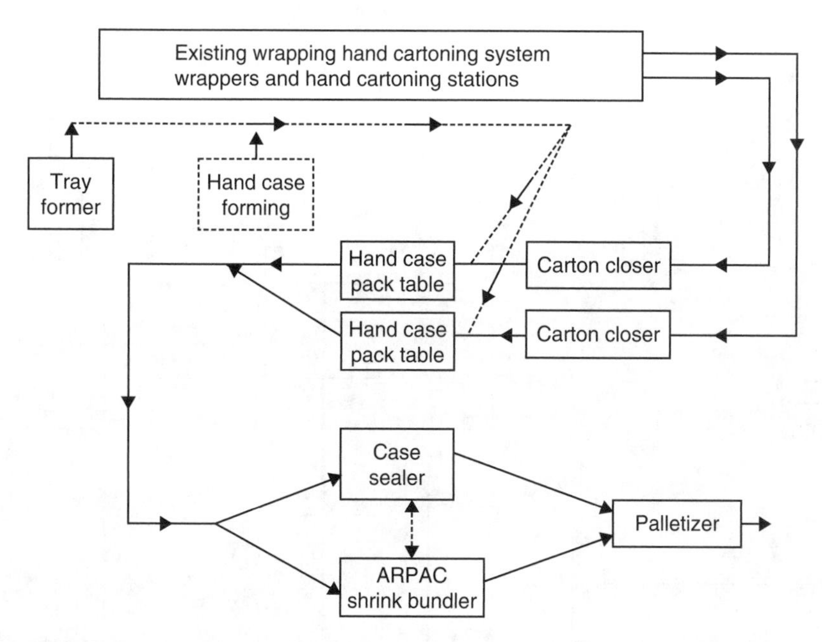

FIGURE 14.5 Block flow diagram.

circulations, and connections with other systems (if applicable). The PFD incorporates all the key elements of the process without covering the specifics of the design. One of the purposes for the drawing is to provide a basis for the detailed design or specifications for equipment and facilities. Another reason for the PFD is to consider alternatives for process flow and to determine feasibility. The objective here is to select the best option with regard to efficiency, quantity, and economics.

Raw material sourcing is an important factor that significantly impacts engineering of the overall system. It can prove to be one of the biggest challenges to food manufacturers. The objective of the producer to provide safe, quality foods that satisfy the expectations of the consumers, and so safety, compliance, reliability, and traceability of raw ingredients becomes mandatory. Another consideration deals with the availability of supplies in which case flexibility of process design would lend itself to a more robust system. Appropriate consideration must be given to sourcing of materials which involves establishing raw ingredient specifications and usage, product specifications, recipe protocols, and batch sizes. Once this information is in place, options such as selection of qualified open market or private sector suppliers become less arduous tasks.

Raw material properties is a second point of evaluation as part of the steps involved with the development of a well-integrated process. Several categories of properties of interest to developers include physical, mechanical, thermal, chemical, and biological, some examples of which are listed in Table 14.2.

The characteristics in Table 14.2 are quantitative variables for raw material quality control. Providing criteria for these ingredient properties to suppliers safeguards the operation and is a driver for superior commodities.

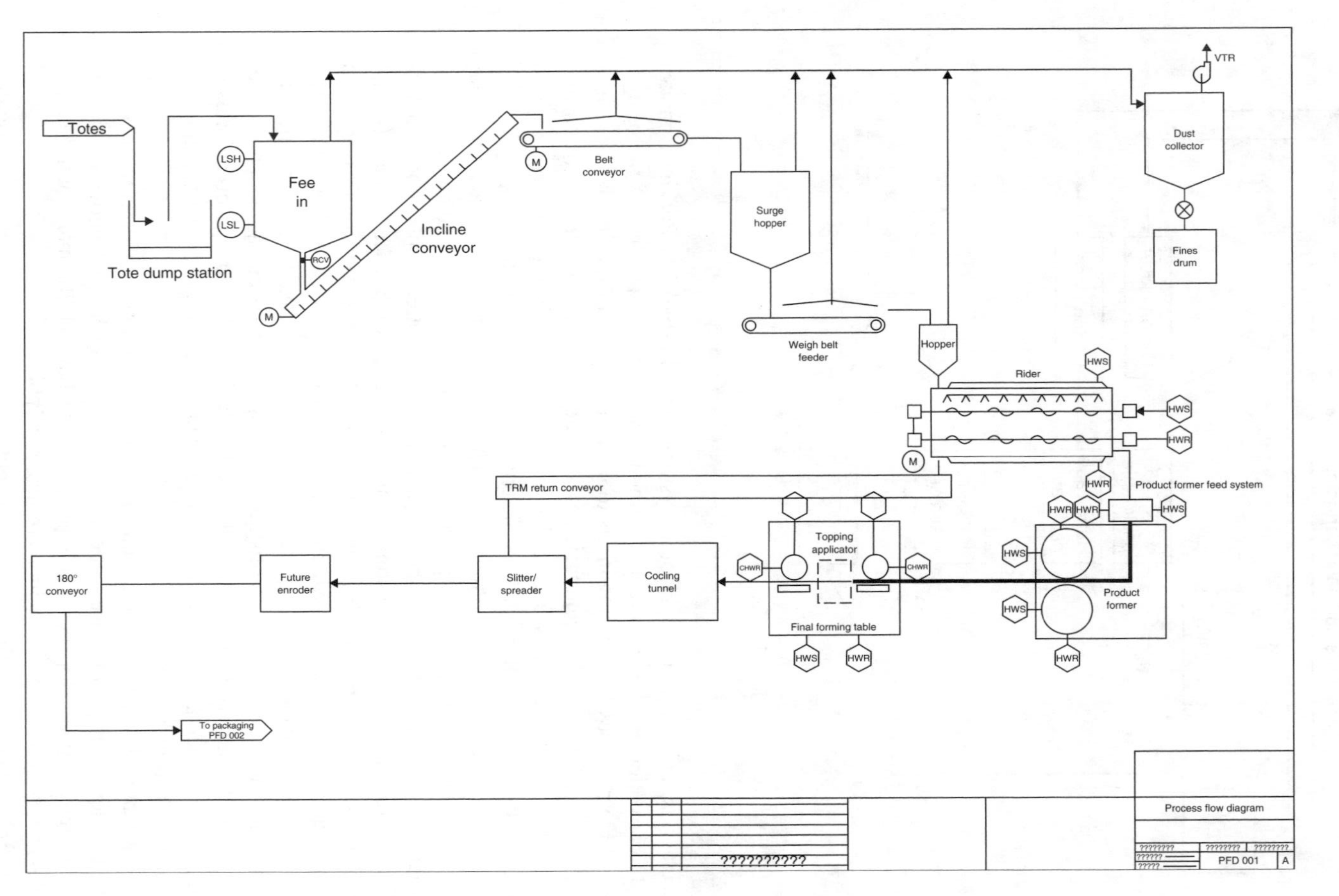

FIGURE 14.6 Process flow diagram.

TABLE 14.2
Raw Material Properties

Raw Material Properties	
Type	Examples
Physical	Size Shape Volume Surface Area
Mechanical	Density Porosity Permeability Rheology
Thermal	Thermal Expansion Specific Heat Flash Point Flammability Thermal Conductivity Thermal Diffusivity
Chemical	pH Reactivity Corrosion Concentration Humidity
Biological	Toxicity

FINAL ENGINEERING

Detailed design and engineering is not an activity that haphazardly occurs, but is the result of cumulative efforts along with strategic planning that makes this an important aspect of building larger manufacturing facilities. The objective of this phase is to develop a compilation of drawings and specifications that completely describe all equipment and facilities involved with the generation of food products. Design activities occur over a period of time and are comprised of step-by-step methodology. The groundwork for this phase of planning is taken from preliminary engineering with the objective being to expound on the information determined.

In preliminary engineering, one of the deliverables was the PFD. This schematic shows the overall steps involved in the production process by indicating the major equipment involved and input/output data from the thermal and mass balances performed on the system. An examination is made at each of the unit operations in order to determine the specifications and sequencing necessary to achieve the required flow rates and product volume. Mathematical modeling is done using the known variables to first conclude if a batch or continuous process is necessary and then what are the appropriate speeds and cycle times. On determining which type of process will accommodate production levels, an effort is made toward finalizing the component specifications. The specification documents usually take the form of piping and instrumentation diagrams, isometric drawings, fabrication drawings, and room layouts. These drawings must provide pertinent information such as materials of constructions, labeling/tagging, equipment dimensions, instrument types and locations, operating parameters, automation requirements, finishes/surface treatments, life of component, utility requirements, piping configurations, electricals, maintenance requirements, and methods of assembly/disassembly.

Equipment Design

What are the steps involved with leading the engineer to the information necessary for the designing of the major equipment? The procedure is summarized in Table 14.3.

1. *Clarify Unit or Component Requirements*: to define and explain clearly the purpose of the piece of equipment and what the demands of this component will be to meet system needs. What is its function? What are the required dimensions? What are the rates of temperature increase or decrease? What are the hold times? What physical, chemical, and microbiological properties of the product must be changed or taken into consideration? Are there special or unusual circumstances? These are just a few examples of numerous questions that must be asked and answered in completing this first step.
2. *Determine the Impact of Environment*: to conclude whether the system surroundings will affect the normal operation of the equipment. If it will, then this must be accounted for in the overall design. Take for example, an open system automated doughnut line. A product such as yeast raised doughnuts

TABLE 14.3
Equipment Engineering Design Steps

1. Clarify Unit or Component Requirements

↓

2. Determine Impact of Environment

↓

3. Conceptualization

↓

4. Identification of Constraints

↓

5. Evaluation of Proposed Solution

↓

6. Communication of Design

is very temperature and humidity dependent. Any time the doughnuts are exposed to the ambient atmosphere along the production line (e.g., cooling before icing application), impacts the resulting product.

3. *Conceptualization*: the act of developing an ideal scheme or plan of action. This is basically starting with a framework for the overall design and building on it utilizing the information gathered from the first two steps.
4. *Identification of Constraints*: recognize and resolve the limitations of the resulting design options. As the engineer progresses through conceptualization to build the unit, not all solutions will be appropriate due to unavailability of resources, lack of time to market, financial impact, or other reasons.
5. *Evaluation of Proposed Solution*: to assess the resultant design. Once a workable solution has been achieved, it must be analyzed for such elements as efficiency, verification that all needs are met, reliability, and optimization.
6. *Communication of the Design*: the descriptions, dimensions, specifications, calculations, and so forth, are placed in a formal document and assigned a document control number. It is then distributed for review to all the appropriate stakeholders.

There are situations when acquisition or design and fabrication of new equipment is not a necessity, but used equipment may be purchased. This option could save considerably on time, financial investment, and the need for a multidiscipline engineering team. In this case, the project team knows exact specifications of the equipment and all components. The only concerns would be equipment modifications, if there are any, along with commissioning and validation.

Instrumentation and Controls

Instrumentation and controls are very important aspects associated with the detailed design of the system. The whole reason for this automation is to improve on the equipment's productivity, reliability and stability. Various components, such as flow control valves, are selected with the objective of enhancing system performance. This must be kept in mind as engineers and Computer Aided Design (CAD) technicians work together in developing the major components list for a piece of equipment.

Looking specifically at the instrumentation, one must begin by assessing the essential features. This can be done by stepping through a functional analysis of the system and examining whether or not automating a process action is worthwhile. Once this is determined, the engineer then has to decide on the level of automation. Oftentimes the level of automation is impacted by the accuracy and output needed from the instrument. After this information is known, the process engineer can then begin to develop specifications for each instrument. Supply catalogs, local vendors, and manufacturers that specialize in the type of components sought are all resources used by the engineer in order to settle on instrument types and availability.

The controls philosophy develops from a general examination of the resulting number of instruments and the amount of automation associated with each one. Controls are used to regulate the influx and outflux of materials though the equipment as well as maintain setpoints. The more automated the instrumentation found on the apparatus, the more likely the system will be a fully automated controls system. If the system is found to have very few instruments, then the engineer weighs the advantages and disadvantages of having a manual versus a semi-automatic versus a fully automated operation.

There are several means for controlling processes all of which can be categorized in one of two ways, local controllers or remote controllers. Operations that are automated have several alternatives for central locations to receive and send input and output from the system. The Supervisory Control and Data Acquisition, or SCADA, is probably the most elaborate method of control. This large centrally located computerized set-up is used to monitor and control all aspects of food manufacturing including supply of utilities, instrumentations, supporting equipment (heat exchangers, pumps, etc.), and electrical power distribution. The Programmable Logic Controller (PLC) makes the control system more localized. A PLC is a microprocessor used for the automation of industrial processes by controlling machinery through a series of programs comprised of ladder logic. Computer programmers and engineers work closely with equipment operators to develop the overall operating principle that includes the desired sequence of operation, matrix, and cycle times.

Systems integration, or the incorporation of controls, equipment, and instrumentation into a single process control line, is one of the most important aspects for applications such as ingredient handling and mixing within food and beverage manufacturing automation. When appropriately selected components are combined with a tightly integrated network, a significant enhancement of overall food processing quality and performance results. Although the use of systems integration is significant, proper design to allow for well functioning interface between all devices can sometimes prove to be a challenging area best suited for controls engineers.

Ingredient Handling and Metering

Achieving and maintaining accurate handling of liquid and solid ingredients requires consideration of several factors: material characteristics, storage requirements, overall operating principle, environment, and control scheme. The vast diversity of materials along with storage and handling practices makes it a necessity to pay close attention to the assessment of these characteristics when selecting the most suitable equipment.

Ingredient handling and metering systems are very diverse in capabilities in that they can be designed to handle both minor quantities for prescaling as well as larger volumes of product for major metering. Figure 14.7 is a diagram of ingredient handling. Use of this type of equipment has numerous advantages. Individual ingredient preparation can be a very slow process and generate substantial amounts of waste. For example, operators making breading by a batch process have to transfer large-volume bags to the mixing station where the ingredients are removed from the package and

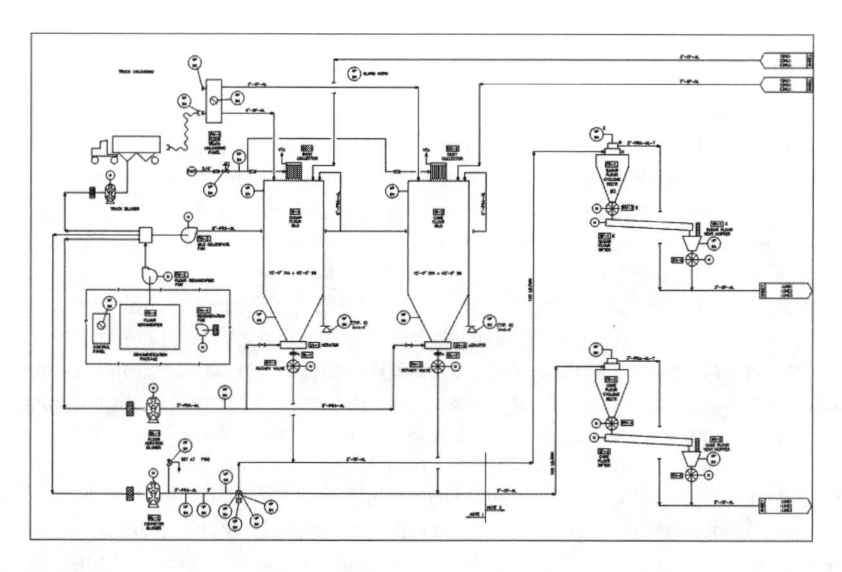

FIGURE 14.7 Ingredient handling diagram.

poured into a mixing vessel. This repetitive process is labor intensive for operators and could possibly result in worker injury due to poor ergonomic situations. Another issue is the generation of waste and its removal from the premises each time a container is emptied. Dispensing of ingredients by operators may lead to product lost from spillage or damaged containers along with microbiological concerns from the package's exterior due to ingredients stored in a warehouse for extended periods.

The use of bulk ingredient dispensing systems has many different advantages: elimination of worker injury (back and shoulder pain, carpal tunnel syndrome) caused by handling raw ingredients, minimization of costs associated with container disposal, higher material handling efficiency, lower labor costs, process expandability, more accurate and repeatable results, dust control (dry handling), easier overall volume tracking, production flexibility, shorter production cycle times, higher product volumes, and additional storage capacity, just to cite a few.

Developing specifications for the appropriate dry or liquid dispensing system can be a difficult and time consuming process. Engineers are tasked with developing equipment specifications based on process parameters, environmental constraints, safety, regulatory requirements, and maintainability. The objective for selecting the appropriate equipment is having total customer satisfaction by reaching the highest level of quality with minimal rejects. Whether newly fabricated, refurbished, or used, the engineer must work very closely with equipment vendors to select reliable systems that meet the need.

The use of controls with bulk ingredient handling and metering systems by food producers makes automated batching more accurate, efficient, and manageable. It makes for a smoother transition as the product continues through further processing downstream to areas such as mixing.

The Right Mix

Ingredient sequencing and mixing can sometimes be a convoluted task for engineers to take on during detailed design due to the fact that the substances of interest may involve all the physical states of matter and every possible combination of these materials. However, much work has been done during the last 20 years on computational fluid dynamics in vessels. Such modeling allows engineers to design or select the appropriate type of mixer and sequence to meet the needs of the process.

It is good practice for the process engineer to walk through a series of questions in an effort to narrow the mixer types to only those appropriate for the application. First, what will the mixer do? Emulsify, provide solid suspension, blend, or disperse? In what type of vessel will the mixing occur? What type of mixing action is required, high or low shear? How will shear stress impact the product? Configuration is another issue. Does the process require a batch or in-line mixer? What is the fabrication material and finish and how will this impact schedule and delivery? What type of drive, seals, and bearings are needed? How will operators clean the component? It is also very important to determine early on the performance level of the mixer along with the maintenance requirements. Interlocks should also be reviewed in to address safety concerns.

Technical Specification Document

All the information gathered for various pieces of equipment and associated components is compiled in a concise Technical Specification Document. This document is the roadmap from the engineer to the company supplying the equipment, and provides a detailed description of the requirements (i.e., materials of construction, dimensions, electrical amperage/phase, etc.) for the proposed system. This same information will be used by the engineering owner's representative to verify acceptable deliverables from the supplier during the Factory Acceptance Test (FAT).

The Technical Specifications Document is very important in that it outlines key information for all stakeholders with regard to process equipment. Examples of the information found in this document include

1. Materials of construction
2. Component specifications—pumps, instruments, heat exchangers, sample ports, sight glass, and so forth.
3. Dimensions, general arrangements, isometric drawings, and elevation drawings
4. Fabrication, installation, and start-up
5. Electricals (amperage, phase, back-up power supply)
6. Control system, number of terminals, and location
7. Tagging system and nomenclature
8. Utility flowrates, temperature, pressure, and line sizes
9. Material finish
10. Vessel pressure rating
11. Clean-In-Place (CIP)/Sanitize-In-Place (SIP)
12. Thermal and special processing requirements

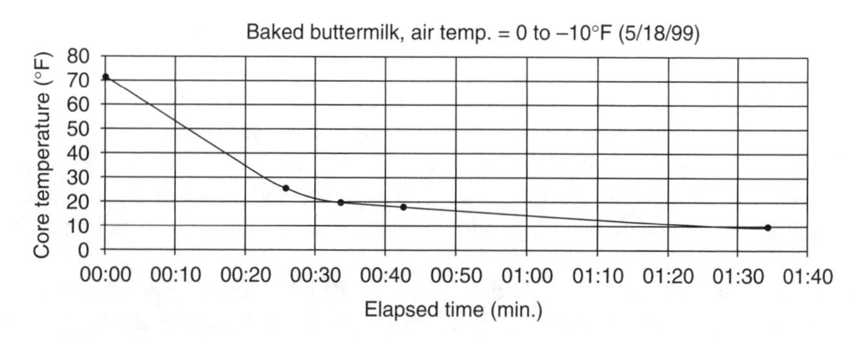

FIGURE 14.8 Sample bakery freezing curves.

Figure 14.8 depicts an example of thermal processing requirements for bakery equipment.

Recycling and Waste Handling

A very important part of designing food processes involves the recycling of by-products and unused ingredients in addition to waste handling. This aspect of design and planning is usually handled by an environmental engineer collaborating with the process engineer to take such factors into account before actual food production begins. The incentives for engineering to consider recycling methods include a reduction in the demand for raw materials, less energy consumption, fewer emissions, and reducing costs associated with waste disposal.

The first step in reduction and elimination of wastes generated on food product lines is to ensure optimally implemented Standard Operating Procedures (SOPs) and Equipment Operating Procedures (EOPs). All personnel must also be properly trained to operate the equipment efficiently. It is the equipment operator's responsibility to ensure that the lines are running smoothly and in the correct mode.

Despite a food manufacturing facility's best efforts to have optimal performance and to make optimal use of ingredients, food processing inevitably generates a certain amount of waste. This may include anything from scrap dough that remains after moving through cutters on an automated doughnut line to fruit and vegetable peels collected while making ready-to-eat fruits and raw vegetables.

Several alternatives exist for scrap materials and residual ingredients. Scrap may be reworked into the process to make more product as long as implementing such steps does not compromise the quality of the resulting food product. This is often done with baked goods such as pastries. The by-products generated on the line may also be used in the manufacturing of a secondary product as is the case with fat renderings from slaughter houses being used to generate biofuels. Scraps may also be sold to other markets in order to be utilized in their processes. The bones and other items found not fit for human consumption, which remain after meat processors have removed selected cuts may be sold to animal feed producers for further processing. Even after all reuse efforts have been exhausted, sometimes there are explanations in which items will ultimately be categorized as waste and therefore must be discarded. This would be the case for livestock condemned at a slaughterhouse because of disease, when safe disposition is dictated.

Handling of solid and liquid wastes must also be resolved by environmental engineers. Wastes encountered may be in the form of process waste water, sludge, oils, and empty raw material containers. The project team comprised of the process engineer, production manager, environmental engineer, and operators work together to explore means, resources, regulations, and associated costs for handling these items.

Waste water reduction and recycling for the food and beverage industry continues to be a major concern as local agencies strictly enforce discharge levels. This is a very important area of concern because water is used in a wide variety of steps throughout the food production process including cleaning, sanitizing, steaming, peeling, cooking, and cooling. For manufacturing facilities having large volumes of waste water, installation of onsite pretreatment systems may be a more economical solution.

Waste water and air emissions generated in food manufacturing facilities differ from that found in other industries in that it typically does not contain toxic chemicals such as those found on the Environmental Protection Agency's (EPA's) Toxic Release Inventory list. The most important considerations are microbiological contamination, Volatile Organic Compounds (VOCs), and odors. Food processing wastewaters are distinguished by high levels of dissolved/suspended solids (fats and oils), nutrients (ammonia), minerals (salts), and Biological Oxygen Demand (BOD). Recovery of these materials may have added value by way of direct reuse, use in energy recovery, or selling to other industries.

No matter what type of recycling and waste handling program is implemented, it requires the cooperative effort of operators, engineering, quality assurance, maintenance, and management. There should also be commitment to seeking out new technologies on recycling and waste reduction initiatives. Less polluting materials, equipment, and procedures are frequently developed, and it remains the responsibility of food producers and their engineers to stay alert for such developments.

Detailed Engineering: Packaging Requirements

Food packaging is the containment of food to protect the product from damage during transport, prolong the shelf life, promote the product to the consumer, and enhance the convenience in use or storage of foods. The package provides information about contents found inside. The use of packaging serves multiple functions for food producers by reducing moisture infiltration process and oxidation, providing protection from biological agents, and safeguarding against mechanical damage. Package labels provide useful information to the buyer such as manufacturer, ingredient list, product size, warnings, and Universal Product Codes (UPC). The engineer must take all this information into account when establishing the packaging requirements for the final product.

The engineer must also work with packaging system operators, quality assurance, packaging development managers, regulatory affairs, and marketing to determine packaging that is most suitable. What are the best choices of materials for packaging? What is the packaging structure and does it involve primary, secondary, and tertiary containers? What methods will be employed to incorporate the final product into enclosures? See Chapter 9. At what rate does this transfer need to occur? Determination or collection of this information allows the engineer to move forward with packaging equipment specifications and line layouts.

Multiple avenues for the engineer are available to aid in the selection of packaging systems that fit the product need. In some cases, standardized equipment currently exists on the market that may be purchased and customized if necessary. The company could also work with reputable packaging equipment manufacturers to customize systems that are unique in design and capabilities. The third option would be for procurement, engineering, and quality assurance to work together in identifying a copacker with the ability to do all final packaging and distribution offsite.

Conveyor systems are the most frequently used form of materials handling equipment found in the food and beverage industry. A wide variety of conveyor types are available on the market depending on the specific needs of the company's distribution system. The roller conveyor, either the electrically powered or free running, consists of rollers mounted horizontally in a frame. Rollers in the electrically powered system are turned automatically with a belt or chain by the control system, while the running type are usually pushed by operators or moved by gravity on an incline. This type of system is most commonly used for the transport of items with a firm base like drums, crates, case and cartons. The belt conveyor can be any reasonable length and the belt is moved by an idler roller located on the underside of the frame. Numerous materials exist for the belt, some of which include wire mesh, plain/rubber-coated canvas, and stainless steel ribbon.

Palletizers are another package handling system frequently seen in the manufacturing of food products. It is especially useful in that it allows the operator to handle very large volumes of product at one time. Final finished goods are delivered to the warehouse, stored, then retrieved for store delivery when necessary.

Plant layout and operating rates for packaging and distribution are very critical pieces of information for the engineer to resolve. The general layout dictates the cost and efficiency of the handling system. Engineers consider surge capacities in the overall design to accommodate the plant operating at full capacity during peak seasons. The ideal scenario is for the finished product to be received at the end of the facility near a loading dock area without being moved back and forth over long distances. The objective is maximum material flow, minimum handling by operators, minimization of bottlenecks, and elimination of unsafe work practices.

Engineering and procurement work closely with local packaging system representatives to finalize packaging equipment specifications, cost, scheduling, and contract terms and conditions. If the equipment will have controls and be integrated into the overall production system, this information must also be outlined in the purchasing agreement. Another important consideration is the need for secondary, or back-up lines in order to allow for preventative maintenance and repair of the primary lines.

Detailed Engineering: Facility Design

The design of facilities for food processing may involve build-out of a new location or renovation of an existing area. The construction of brand new green field facilities has its advantages in that the work may proceed without hindrances to scheduling, ease of incorporating equipment into facility design, and the simplicity of delivering a complete system that meets the needs of the entire team. Scheduling different phases of the work from execution to utility installation may occur at any time without interrupting production processes. This also eliminates the problem of cross contamination that is a possibility when construction zones overlap with operable production areas. It is also much easier to build the utilities exactly at the desired location and the proper elevation to accommodate new equipment as opposed to retrofitting older facilities. Buy in from all project team members, including engineering, quality assurance, manufacturing, health/safety/environmental, and maintenance is much easier when the requirements for all parties is established upfront during the design phase. The other consideration for developing new facilities is the larger investment of funds.

Making modifications to an existing area in a plant is not a simple task, but it requires much coordination between engineering and production. More risk is involved with this scenario. If not planned correctly, this may result in destroyed product as well as delays in bringing new production capacity online. The advantage of making changes to existing production areas is that it generally costs less for the company to implement this option.

It is critical that facility design be completed satisfactorily, so that the intended purpose is achieved in a manner that complies with key objectives and guidelines. General requirements for food plants and production areas are detailed by agencies including the Food & Drug Administration (FDA), American Institute of Baking (AIB), Dairy 3A, Occupational Safety and Health Administration (OSHA), and the United States Department of Agriculture (USDA). Each of these organizations identifies what current good manufacturing practices (GMPs) are for the food industry. Examples of information covered in these documents regarding building and facilities include grounds keeping, storage, drainage system, and materials of construction. The OSHA specifically addresses hazards associated with the workplace so that concerns such as improper ventilation, high decibel noise exposure, inadequate means of egress, and lack of radiation protection may be eliminated.

Federal, state, and local regulations along with requirements to comply with hazard insurance must be observed by construction contractors. Fire codes dictate the number of emergency exits, sprinkler head density, and the number/types of fire extinguishers necessary. Local building codes address such elements as standard door dimensions, elevation and size for steps, and handicap access. Federal guidelines cover lighting and noise levels, ventilation air changes per hour, and separate men's and women's changing rooms.

It is the engineer's responsibility to record the facility general requirements and specifications for the construction company. Documentation should cover at a minimum (1) analysis of types of processing, (2) material flows, (3) equipment arrangement and connections, (4) emergency and safety needs, and (5) room layout drawings. A summary of this information will establish whether or not the current proposed layout or existing room space is logistically feasible. The engineer must also ensure that there is adequate space for storage of materials as well as all process travel (paths of entry and exit).

Food safety and sanitation are very important factors to be considered during the design or renovation of facilities. The construction work should include all rooms and equipment necessary to permit hygienic and safety practices. Food contamination in a plant may result either directly or indirectly from food-contact surfaces, water, or air. To counteract such problems, the engineer must work with contractors to build facilities that allow for personal hygiene and sanitary food handling. This would mean having locker areas for men and women to change outerwear to production uniforms, wash rooms for cleaning equipment, chemical storage closets, properly designed drains and sewer systems, and airlocks/laminar flow hoods/air balancing and filtration for classified areas.

One often-overlooked aspect until all construction work is completed is pest control. It is worthwhile for the design team to take this into consideration earlier during the design phase. One of the best strategies for pest elimination is good housekeeping and sanitation practices. Newly built or refurbished areas should plan strategic locations for waste collection, elimination of secluded shelves and nooks, and replacement of damaged or decaying materials (in renovated areas) in order to minimize the presence of unwanted pests.

Utility loads are calculated to determine needs for accommodating present as well as future equipment needs. Lack of available utilities can sometimes pose a problem for existing plants where utility capacity is almost at maximum level. In such cases, a small production area renovation becomes a costly effort due to the need to generate added capacity at the source. So, it becomes incumbent for the engineer to survey current uses of electricity, chilled water, steam, hot water, compressed air, potable water, thermal fluids, and so forth, and determine the levels necessary to meet the needs of all equipment that will be used in the area. This includes stationary and mobile equipment as well as safety systems. Additional utilities for area cleaning should also be considered.

Given that a good facility layout provides for optimized movement of personnel, flow of materials, and operation of equipment, these points are the motivating factors as the engineer moves forward with layout planning. Examples of the data gathered by the engineer include:

- Equipment size and shape
- SOPs
- Room dimensions (existing production space)
- Existing utility drop locations
- Types and number of electrical outlets
- Material flow diagram (raw material, processing, packaging, and warehousing)
- Personnel flow diagram
- Process adjacency requirements

Room access by the maintenance department will be necessary once the area is in full production mode, and so distance from the shop location may be a valid concern. There will be occasions when the equipment cannot be repaired *in situ* but must be relocated to the maintenance fabrication shop. Situations such as this warrant close examination of travel path to and from the maintenance area. Can maintenance personnel have access to equipment or utilities while in production mode in the event of emergencies? Repair work can be quite messy that conflicts with having a clean room environment while manufacturing food products. Engineers often design new facilities with a walkable mezzanine area, which house a majority of the utility instrumentation (pressure regulators, pumps, pressure gauges, steam traps, thermocouples, etc.) so that maintenance repair work may occur at any time without compromising production processes.

In recent years, there has been an increase in company interest as it relates to providing devices, systems, and physical working conditions that address the needs of the worker with the overall intention of maximizing productivity by decreasing operator discomfort and fatigue. This in part has been driven by a history of work-related injury occurring at manufacturing sites caused by such events as repetitive motion, poor lighting, loud noises, and improperly designed operator work stations. As the design of the overall layout develops and equipment orientation is determined, the engineer must consistently ask the questions "What are the manual labor requirements?" and "Will this minimize work-related injuries?" Proceeding in this manner improves the internal customer satisfaction responses received from operators once the new production area is turned over and in full operation.

Implementation

Implementation involves the engineers overseeing the fulfillment of project plans according to the specifications set forth in the technical documentation and associated drawings, that is, the realization phase of engineering project management. The engineering team is responsible for identification of all resources needed to conduct the project, and so a precise knowledge of key objectives should be in place. Carrying out the projects for scale-up and production of new foods typically includes construction, equipment fabrication, installation, commissioning, and validation.

Construction and fabrication moves forward after a formal notice to proceed with work has been issued, generally in the form of a purchase order to the contractor or vendor. A kick-off meeting is then held to ensure that the work gets off to a good start. The work proceeds according to schedule with periodic reviews held between the manufacturing company's engineering representative and the company providing the deliverables in order to discuss progress. Finalization of all work leads to commissioning followed by the validation phase in which requested specifications and functionality are substantiated.

Validation

The steps involved with validation, including a brief description of each, are listed below:

1. FTA—documented verification of correct functionality and specifications at the vendor's fabrication shop prior to receipt and delivery of equipment onsite.

2. Site Acceptance Test—documented verification that the equipment functions appropriately as assembled, installed, and connected at the manufacturing facility.
3. Installation Qualification—documented verification that all aspects of the installation (equipment or facility) adhere to the manufacturer's recommendations, local codes, and intended design.
4. Operational Qualification—documented verification that the system and any subcomponents (equipment or facility) performs as intended through all specified ranges and conditions without the use of raw materials.
5. Software Qualification—documented verification that the software package/controls system performs as intended without excess ladder logic.
6. Performance Qualification—documented verification that the system and any subcomponents (equipment or facility) performs as intended through all specified ranges and conditions with the use of raw materials to generate product.

During the validation phase, the engineer tests major pieces of equipment and verifies all facility construction work. Several validation options exist for the food and beverage industry depending on the specific needs of the manufacturing company. The most common practice would be testing completed within the existing manufacturer's facility, which has been modified to meet the company's current needs. A second option involves testing of pilot facilities that have been built with the intent of determining the impact of operating at a larger scale. Testing of equipment may also be completed at the equipment supplier's facilities in situations when the manufacturer's facility is incomplete and not ready for receipt and testing of the newly fabricated equipment. Validation of a copacker's facilities and equipment is another alternative for manufacturers desiring to have another company handle certain aspects of the production process.

Methodologies for completing validation testing are usually outlined in the Validation Master Plan (VMP), or validation plan. This document details what equipment will be validated, what tests will be conducted, the resources required for successful completion of all testing, proposed scheduling, and the definition of acceptable criteria based on spcifications from quality assurance. In stepping through such testing as ingredient sequencing, unit operations, and final product packaging, it is important for a HACCP study to be completed as well. Proper application of HACCP during validation establishes at the onset that food safety is being handled effectively. The objective of taking the entire system through the VMP and HACCP studies is ending up with a safe and consistent product that matches the marketing prototype.

It is not uncommon for issues such as unavailable test sites or raw materials to arise during the validation process. Challenges may also develop from environmental differences in the case of open systems. Temperature and humidity are very important environmental factors that could have significant impact on resulting product. Proper planning for environmental control or product adjustment may help to resolve such problems. Yet another point of consideration relates to ingredient physical property changes. Validators should have knowledge of such changes over time and under given conditions so that this will not result in the production of inferior products.

The validation team works in conjunction with engineering to complete all testing as outlined in the VMP. It is the team's responsibility to coordinate and perform the qualification protocol in order to improve the production process. Some points to consider while performing validation are given below:

- Procurement of raw materials should be handled in advance to avoid testing delays.
- Scheduling trial runs especially becomes important when areas are shared between launching new product and maintaining current production levels.
- Avoid duplicate ingredient handling and mixing.
- Avoid duplicate unit operations (mixing, thermal processing).
- Make sure operators receive adequate training before beginning validation work.
- Assure appropriate monitoring of process operations.
- Capture results in the form of recordable data.
- Consistently monitor product quality (flavor, texture, and color).
- Troubleshooting: Make facility, equipment, and recipe adjustments when appropriate.
- Continuous Process Optimization: Always look for ways to improve monitoring/controls, reduce waste, and incorporate rework.

Budget Analysis

Budget analysis is one of the most important parts of project management and engineering, but if done incorrectly, could result in complications for project completion and closeout. The scale-up and commercialization of new food products can be divided into three different budget categories: (1) capital costs, (2) operating costs, and (3) product pricing structure. An example of an overall project budget is shown in Figure 14.9.

Capital cost refers to any expenditure that results in assets, either tangible (raw materials, equipment, and facilities) or services (engineering, validation), available for use in the production of further assets (food products). Operating costs are expenses related to conducting business including charges for labor, utilities, and distribution. Engineering is primarily concerned with capital costs whereas manufacturing is more interested in operating costs. Overlap of these two areas occurs when providing systems for the launch of new food products. Project success depends on minimization of both types of costs while producing a superior product for launch in the marketplace.

Food product pricing structure on the other hand deals with the marketing aspect of new food products. Factors that affect cost of goods sold include ingredient, processing, and packaging costs. Marketing initially establishes pricing based on fair market value and product demand. It is then the responsibility of manufacturing to generate quality product as economically as possible thereby increasing the overall margin for the company.

Project Closeout

Bringing a project to completion is handled through a series of steps referred to as the Closeout Plan. It outlines the tasks and responsible persons for finishing all

Project Budget

Description	Cost
Equipment	
Process equipment total	$2,711,000
Preform container dumper	$17,000
Preform feeding unit	Incl.
Preform orienting rolls	Incl.
Blow molding machine	$1,634,000
Silo	$344,000
Unscrambler	$208,000
Air conveyor system	$508,000
Utlity equipment total	$345,500
Oven exhaust hood	Incl. w/ HVAC
Air compressor - high press.	$253,500
Air compressor - low press.	$28,000
cooling tower	$25,000
Process water chiller	$39,000
Air handling unit	Incl. w/ HVAC
Air handling chiller	Incl. w/ HVAC
Equipment total	$3,056,500
Construction	
Sitework	$3,000
Equipment pads	$1,500
Landscaping	$1,500
Concrete	$105,800
Existing concrete floor demo	$13,500
New reinforced slab	$38,700
Footings for mezzanine	$19,900
Concrete deck for mezzanine	$33,700
Steel	$87,800
Steel for mezzanine	$70,100
Handrail @ mezzanine, ladder	$5,300
Misc. steel	$12,400
Architectural Items	$127,400
Concrete block walls	$21,900
Rebar & grout for block walls	Incl.
Drywall system (Studs, sheetrock & finishing)	$45,100

FIGURE 14.9 Blow mold installation.

deliverables and turnover to the internal client. It is good practice to complete an evaluation of cost, scheduling, project delivery, and the final product in order for any concerns to be addressed. A transition occurs during this phase of the project lifecycle in that responsibility for the food production area transfers from the engineering department to manufacturing and maintenance. The project team led by engineering efforts is disassembled if there is no longer a need for interaction of all key players on another capital investment. Disbursement and filing of project documentation, including validation data and manuals, is also handled during the conclusion of project engineering activities. Lessons learned are then recorded for the purpose of improving future projects and providing even better food products to consumers.

Description				Cost
Finishes (painting)				$25,800
Utility room louver				$600
Overhead doors				$6,300
Main door				$4,500
Hollow metal window frames				$11,900
Glass for windows				$11,300
Mechanical				$157,600
Piping (Included w/ equipment cost				$40,000
HVAC		100,000	6,500	$106,500
Plumbing		3,700	4,400	$8,100
Fire Protection		3,000		$3,000
Electrical				$313,000
Instrumentation and Controls				$25,000
General (note 1)				$11,600
Temp partition				$9,550
Misc. Demo.				$2,050
Construction subtotal				$831,200
Contingency (0%)				$0
Construction total				$831,200
Total capital (Equipment and Construction)				$3,887,700
Project services				
Preliminary engineering			50,000	
Final engineering, PM, CM			198,000	
Total engineering (Includes 10% contingency)				$248,000
Project total				$4,135,700
Operating costs				
Water and sewer	-	-	-	Negl.
Electrical costs	-	-	-	$90,500
Total operating costs	-	-	-	$90,500

FIGURE 14.9 Continued

COMMERCIALIZATION AND PRODUCT SUCCESS

Establishing a multidiscipline team and organizing for launch of new food products are key components to a product's success. The functional groups within a company, which are responsible for some aspect of marketing new products include

- Marketing
- Engineering
- Quality assurance
- New products division
- Packaging
- Research and development

- Manufacturing
- Finance
- Sales
- Legal
- Distribution
- Management

The marketing department historically has had most of the weight involved with ensuring the success of new product launch. In recent years, companies have begun to notice that marketability actually begins much earlier during research and development and proceeds on through engineering execution of projects and manufacturing of foods. Some companies have also begun forming departments that focus primarily on developing new products from concept through product launch.

Preparing for the release of a new product involves many steps, and a summary of each of these steps is provided here:

1. Spend the appropriate amount of time researching and fully developing various concepts for new products.
2. Concept elimination by analytical methods will narrow choices to the option most likely to succeed.
3. The marketing team must then perform preliminary testing of the proposed concept and establish public response to the new food item.
4. A business analysis of the financials must also be done in conjunction with the determination of the public's response. Both aspects have to support moving forward with new product release. It does not benefit the company in any way to send a product to market that is too costly to make.
5. Research and Development or Product Development then performs the task of converting a concept into a tangible food product.
6. Once the manufacturing prototype is formed, it will be reintroduced to the public in order to determine niche and general acceptance.
7. If successfully received by the public, plans for full scale production and sales will proceed.

How do we establish product success? One way is to evaluate the new food product financially as a business endeavor. Sales levels are indicative of new product performance. Another factor that determines development of a successful product is customer satisfaction. Implementation of the new product development plan helps to ensure marketability to the public. It is the sales of novel, that support the financials of the manufacturing company thus allowing for the development and scale-up of additional new food products in an ever changing marketplace.

BIBLIOGRAPHY

American Institute of Baking. 2005. Quality Assurance Manual for Food Processors. Manhattan: Kansas.

Corlett Jr., Donald A. 1998. HACCP User's Manual. Gaithersburg, Maryland: Aspen Publishers, Inc.

Ertas, Atila and Jesse C. Jones. 1993. The Engineering Design Process, New York: John Wiley and Sons, Inc.

Hulbert, Greg. 1998. Design and Construction of Food Processing Operations. The University of Tennessee Agricultural Development Center, 1–2, Knoxville, TN.

Kramer, Franklin. 2000. "New Products Scale-up From Concept to Commercialization." Food Processing August 1, 120.

Lopez-Gomez, Antonio and Gustavo V. Barbosa-Canovas. 2005. Food Plant Design. New York: CRC Press.

Mansfield, Scott. 1993. Engineering Design for Process Facilities. New York: McGraw Hill, Inc.

Mortimore, Sara and Carol Wallace. 1998. HACCP A Practical Approach. Gaithersburg, Maryland: Aspen Publication.

Uhl, Vincent W. and John A. von Essen. 1987. Scale-up of Fluid Mixing Equipment. *Biotechnology Processes*: 155–163.

Valentas, Keneth J., Leon Levine and J. Peter Clark. 1991. Food Processing Operations and Scale-up. New York: Marcel Dekker, Inc.

15 Response Surface Methodology (RSM): An Efficient Approach for Statistical Data Analysis, Modeling, and Process and Product Optimization

Manjeet Chinnan

CONTENTS

INTRODUCTION

Presently, researchers are increasingly depending on statisticians for help both in planning their experiments and in drawing conclusions from the results. Scientific research is a process of guided learning, and the objective of statistical methods is to make that process as efficient as possible. Much of the research in engineering, science, and industry is empirical and makes extensive use of experimentation. Statistical methods can greatly increase the efficiency of these experiments and often strengthen the conclusions obtained. Various statistical methods and software tools are available. Use of computerized techniques is becoming popular. In this chapter major emphasis is given to the use of response surface methodology (RSM) for data analysis, modeling, and optimization, especially in the food product development process. The importance of experimental design, various steps of optimization of product and process, and applications of RSM in product formulation and processing will be explained with the help of illustrations in each category.

DESIGN OF EXPERIMENT

Experiments are performed by investigators in virtually all fields of inquiry, usually to learn something about a particular process or system. When there is considerable variation from observation to observation and it is not feasible to take a large number of observations, the experimenter is forced to refine experimental techniques and/or use an experimental design. The experimental design allows for unbiased estimates of the true treatment differences with a specified degree of precision. An investigator will arrive at a conclusion most quickly and surely if efficient methods of experimental design and sensitive data analysis tools are used. Of these two resources, design is more important.

A designed experiment is a test, or series of tests, in which purposeful changes are made to the input variables of a process or system so that we may observe and identify the reasons for changes in the output response. Factorial designs are extremely useful for this purpose. In such response surface designs, the levels of each factor are independent of the levels of other factors. In mixture experiments, however, the factors are the components or ingredients of a mixture and consequently, their levels are not independent. Simplex designs are used to study the effects of mixture components on the response variable. A three component simplex centroid design is illustrated further in the application of RSM, as reported by Holt and et al. (1992), for optimization of tortilla formulation (*Example 1*). Later in this discussion we shall illustrate the application of fractional factorial design in optimization of peanut butter formulation to attain maximum stability from the report published by Hinds et al. (1994) (*Example 2*). In some mixture problems, when constraints on the individual components arise, designs for constrained mixture spaces are used. One example of such three component constrained mixture design with both lower and upper bound constraints is illustrated in optimization of chocolate-flavored, peanut-soy beverage formulation from the investigation by Deshpande et al. (2004) (*Example 3*). Optimizing a process using RSM is illustrated in a lye-peeling process examined by Floros and Chinnan (1987) (*Example 4*).

Experimental design is a critically important tool in the engineering world for improving the performance of a manufacturing process. It also has extensive application in the development of new processes. The application of experimental design techniques early in food process development can result in improved process yields, reduced variability and closer conformance to target requirements, reduced development time, and reduced overall costs. The use of experimental design can result in products that are easier to manufacture, that have enhanced field performance and reliability, lower product cost, and shorter product design and development time.

RESPONSE SURFACE METHODOLOGY (RSM)

Response surface methodology can be defined as a statistical method that uses quantitative data from appropriate experimental designs to determine and simultaneously solve multivariate equations (Myers et al., 1995). These equations can be graphically represented as response surfaces to

1. Describe how the test (input) variables affect the response
2. Determine the interrelationships among the test variables
3. Describe the combined effect of all test variables on the response

Suppose a food process engineer wishes to find the levels of temperature (x_1) and time (x_2) that maximize the yield (y) of a process. Then the following equation represents process yield (y) as a function of the levels of temperature and time; wheras, ε represents the error observed in the response y:

$$y = f(x_1, x_2) + \varepsilon \tag{15.1}$$

If the expected response is denoted by η then

$$E(y) = f(x_1, x_2) = \eta \tag{15.2}$$

The surface represented by η is called response surface. This is usually represented graphically where η is plotted versus various levels of x_1 and x_2. The response is represented as a solid surface in a three-dimensional space. To help visualize the shape of the response surface we often plot the contours of the response surface. In the contour plot, lines of constant response are drawn in the x_1, x_2 plane. Each contour corresponds to a particular height of the response surface. Such a plot is helpful in studying the levels of x_1 and x_2 that result in changes in the shape or height of the response surface.

RSM is basically a four-step process where, first of all, two or three critical factors (characteristics that can be varied within the system) are identified, for example salt, sugar, or smoke levels. Preliminary experiments help to identify such critical factors. Secondly, the range of factor levels which will determine the samples to be tested are defined, for example, the percentage salt in the product or amount of time the can is in the retort. The specific test samples are then determined by the experimental design and experiments are conducted to test these samples, obtain quantitative data to use in the statistical analysis. These data can be in the form of sensory responses (descriptive or acceptance); physical measurements (e.g., viscosity); chemical analyses (e.g., percent acidity); microbiological assays; and processing information. Lastly, the data from these experiments are analyzed by RSM and then interpreted. Appropriate computer programs are used and conclusions drawn from this analysis should then be confirmed by follow-up experiments with the optimum product. RSM is thus a sequential procedure. The objective is to lead experimenters rapidly and efficiently to the general vicinity of the optimum and to determine optimum operating conditions for the system or to determine a region of the factor space in which operating specifications are satisfied.

OPTIMIZATION

Optimization may be considered as a procedure for developing the best possible product in its class. In spite of the popularity of optimization, there are relatively few procedures for developing an optimal product. They range from the individual specialist's orchestrating optimization on the basis of professional skill and experience to the more structured statistical approaches of RSM developed by Box and Draper (1987) and Box et al. (1978) or multiple regression as described by Schutz (1983) and Stone and Sidel (1983). Response surface methods are techniques employed before, during, and after the regression analysis is performed on the data. Preceding the analysis, the experiment must be designed and after the regression analysis, certain optimization techniques are applied. Thus, the subject of RSM includes application of regression and other techniques in an attempt to gain a better understanding of the characteristics of the response system under study. The focus here will be on RSM and its applications in optimization research.

RSM is a collection of mathematical and statistical techniques that are useful for the modeling and analysis of problems in which a response of interest is influenced by several variables and the objective is to optimize this response. RSM is more efficient than traditional experimental procedures because it decreases the time and cost required to determine the optimum product. Box and Draper (1987) illustrated the advantage of using RSM over classical approach of one-variable-at-a-time strategy using a hypothetical example of finding time (t) and temperature (T) that will produce maximum yield subject to some other constraints.

IMPORTANT STEPS INVOLVED IN OPTIMIZATION

Optimization research can be done by *Planning* (*Steps 1 through 5*), *Testing* (*Steps 6 through 13*), *Validation* (*Steps 14 and 15*), *and Marketing research* as shown in Figure 15.1. The steps 1 through 15 are explained one by one with the help of an illustration of optimization of tortilla containing wheat, cowpea, and peanut flours [*Example 1:* Formulation, evaluation and optimization of tortillas containing wheat, cowpea, and peanut flours using mixture RSM (Holt et al, 1992)]; wherever necessary information from other examples in the literature is used to further explain the concepts and various steps.

1. *Select input variables and their range*

To develop products with optimal quality, various critical properties of the product should be identified. Out of several ingredients and processing steps, the most important variables can be screened and listed as independent variables (also categorized as input variables or controllable factors) such as temperature, time, concentration, pH, ingredient level, water activity. They can be qualitative (e.g., type of flours or enzymes or machine types) or quantitative (e.g., 5% vs. 10% sugar level).

In this example, the performance characteristics of tortillas prepared with different composite blends of flours were of interest. The desirable characteristics of materials selected as the components to be used in composite flour blends were ready

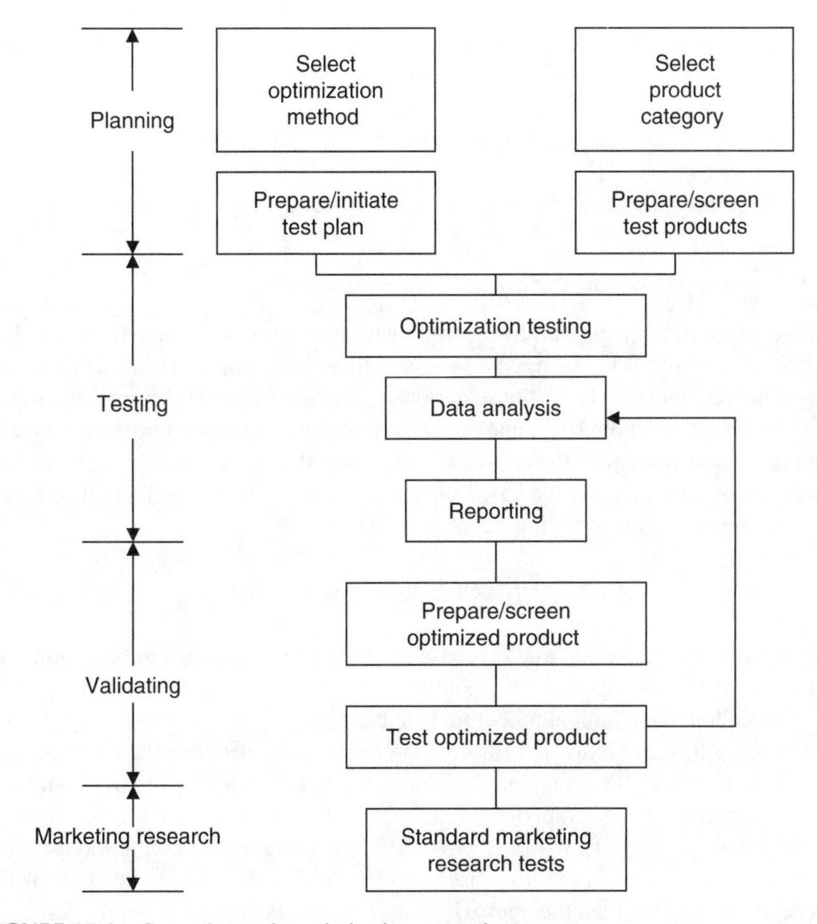

FIGURE 15.1 General steps in optimization research.

availability, cultural acceptability, and increased nutritional quality. In ideal case they should result in the end product that look and taste like traditional tortilla. Based on these requirements and since cowpea and peanut flours have successfully replaced up to 20% wheat flour in baked goods, there were three input variables—wheat, cowpea, and peanut flours-blended together in proportions ranging from 0% to 100%.

2. *Select output variables*

Dependent variables (responses) also known as output variables or measurables are usually properties of the end product. Responses sometimes include sensory scores (sweetness, appearance, flavor, etc.), nutritional value, microbiological stability, shelf life, and instrumental properties (texture, color, viscosity, volume, density, etc.). Since the objective of tortilla formulation was to determine physical, sensory, and compositional characteristics of tortilla prepared from composite flour blends using mixture RSM, the output variables or responses were compositional measurements such as moisture and fat content of tortilla samples, physical measurements such as instrumental color (L, a, b values), line spread (diameter and height), and texture (stickiness and hardness), and sensory attributes including

External appearance (color, lightness, surface appearance, and yellowness)
Flavor (baked wheat, beany, sweet aromatic, doughy)

a. Taste (bitter)
b. Mouthfeel (hardness, fracturability, chewiness, and graininess)

3. *Write the response function (approximating function)*

The response is the measurable quantity whose value is assumed to be affected by changing the levels of the factors and whose values we are most interested in optimizing. The true value of the response corresponding to any particular combination of the factor levels is denoted by η and the objective of the experiment is to maximize or minimize η or to determine the region of best values of η. There exists a mathematical function of x_1, x_2, ..., x_m, the value of which for any given combination of factor levels supplies the corresponding level of η, that is

$$\eta = \varphi(x_1, x_2, \ldots, x_m) \tag{15.3}$$

The function ϕ is called the true response function and is assumed to be continuous function of the x_i. To represent the relation ship $\eta = \phi(x_i)$, a mathematical equation or model called a polynomial model may be used.

In most RSM problems, the form of the relationship between the response and the independent variables is unknown. Thus, critical step in RSM is to find a suitable approximation for the true functional relationship between y and the set of independent variables. Usually, a low-order polynomial in some region of the independent variables is employed. If the response is well modeled by a linear function of the independent variable, then the approximating function is the first-order model

$$y = \beta_0 + \sum_{i=1}^{m} \beta_1 x_1 + \varepsilon = \beta_0 + \beta_1 x_1 + \beta_2 x_2 + \cdots + \beta_m x_m + \varepsilon \tag{15.4}$$

If there is curvature in the system, then a polynomial of higher degree must be used, such as the second-degree model

$$y = \beta_0 + \sum_{i=1}^{m} \beta_i x_1 + \sum_{i=1}^{m} \beta_{ij} x_i^2 + \sum_{i=j}^{m-1} \sum_{j=i=1}^{m} \beta_{ij} x_i + x_j + \varepsilon \tag{15.5}$$

In some situations, approximating polynomials of order greater than two are used. However, the second-order model is widely used in RSM for a variety of reasons, such as

a. The second-order model is very flexible. It can take on a wide variety of functional forms, so it often works well as an approximation to the true response.

b. It is easy to estimate the parameters (the β's) in the second-order model. The method of least squares can be used for this purpose.
c. There is considerable practical experience indicating that second-order models work well in solving real response surface problems.

In this illustration of tortilla formulation, a response surface model, using the Scheffe second degree polynomial was used to fit the treatment means:

$$\eta = \beta_1 + \beta_2 x_2 + \beta_3 x_3 + \beta_{12} x_1 x_2 + \beta_{13} x_1 x_3 + \beta_{23} x_2 x_3 \quad (15.6)$$

where,

η = each sensory and physical characteristic (selected in step 1)

$\beta_1, \beta_2, \beta_3, \beta_{12}, \beta_{13}, \beta_{23}$ = parameter estimates for each linear and cross product term produced from prediction model.

$x_1, x_2, x_3, x_{12}, x_{13}, x_{23}$ = linear terms of wheat, cowpea, and peanut flours and cross product terms of (wheat × cowpea), (wheat × peanut), and (cowpea × peanut).

Almost all RSM problems use one or both of these approximating polynomials (Equations 15.4 or 15.5). The method of least squares is used to estimate the parameters in the approximating polynomials. The response surface analysis is then performed in terms of the fitted surface. If the fitted surface is an adequate approximation of the true response function, then analysis of the fitted surface will be approximately equivalent to analysis of the actual system. Designs for fitting response surfaces are called response surface designs. The model parameters can be estimated most effectively if proper experimental designs are used to collect the data.

4. *Select an appropriate experimental design*

Experimental design methods play important roles in food product development and process troubleshooting to improve the performance. If the experimental design is poorly chosen, so that the resultant data does not contain much information, not much can be extracted, no matter how thorough or sophisticated the analysis. On the other hand, if the experimental design is wisely chosen, a great deal of information in a readily extractable form is usually available, and no elaborate analysis may be necessary. Fitting and analyzing response surface is greatly facilitated by the proper choice of an experimental design. The selection of a response surface design depends upon some important features of an experimental design. The desirable design—

a. Provides a reasonable distribution of data points throughout the region of interest
b. Allows model adequacy, including lack of fit, to be investigated
c. Allows experiments to be performed in blocks
d. Allows designs of higher degree to be built up sequentially
e. Provides an internal estimate of error
f. Does not require a large number of runs
g. Does not require too many levels of the independent variables
h. Ensures simplicity of calculation of the model parameters

TABLE 15.1
A Three Variable Simplex Centroid Design Resulting Into 10 Tortilla Formulations[a] Using the Blends of Wheat, Cowpea, and Peanut Flours

Formula No.	Component Flour (%)[b]		
	Wheat	Cowpea	Peanut
1	100.0	–	–
2	–	100.0	–
3	–	–	100.0
4	50.0	50.0	–
5	50.0	–	50.0
6	–	50.0	50.0
7	33.3	33.3	33.3
8	75.0	12.5	12.5
9	12.5	75.0	12.5
10	12.5	12.5	75.0

[a] Mixture design gave formulations 1, 2, 3, 4, 5, 6, and 7. Additional formulations 8, 9, and 10 were selected as interior points of the design.
[b] Dash represents zero percent of the mixture.

For example, in the optimization of tortilla formulation, a three variable simplex centroid design was used. The number of points in the design ($n = 7$) was obtained from the following equation (since q = number of variable = 3):

$$n = 2^q - 1 \tag{15.7}$$

The seven flour mixtures are shown in Table 15.1. Three additional flour blends were also included to provide extra points within the mixture triangle. Each flour blend was present at levels ranging from 0% to 100% and two replications of the study were conducted.

5. *Develop a table of coded and un-coded levels of factors*

Much of the analytical work in RSM deals with concepts centered on the general linear model (GLM). The polynomial representation of the response surface can be first-degree or second-degree model as explained in step 4. The important step in fitting a model to approximate response surface consists of collecting data and estimating unknown coefficients β_0 and β_i in the model equation. To facilitate the estimation of coefficients in the model equation, the variables in the models are usually re-expressed as coded variables. The most commonly used coding scheme is to define coded variables, x_i, in standardized form as

$$x_{ui} = \frac{X_{ui} - X_i'}{S_i} i = 1, 2 \quad \mathrm{x_{ui}} = \frac{\mathrm{X_{ui}} - \mathrm{X_i'}}{\mathrm{S_i}} \; \mathrm{i} = 1, 2, \ldots, \mathrm{m} \tag{15.8}$$

TABLE 15.2
Process Variables and Their Levels

Independent variables	Symbol Coded (x_i)	Symbol Uncoded	Levels Coded	Levels Uncoded
Concentration, % NaOH	x_1	c	1	12
			0	8
			−1	4
Temperature, °C	x_2	T	1	100
			0	90
			−1	80
Time, min	x_3	t	1	6.5
			0	4.0
			−1	1.5

Where X_i' is the mean of the X_{ui} values ($u = 1, 2, \ldots$, N) and S_i is some scale factor. For example, if each of the k factors is to be set at two levels only (X_{LOW} and X_{HIGH}, say) and the same number of observations is to be collected at each level, then $X_i' = (X_{LOW} + X_{HIGH})/2$ and $S_i = (X_{HIGH} - X_{LOW})/2$. In this case the values of coded variable x_{ui} are $x_{ui} = -1$ when $X_{ui} = X_{LOW}$ and $x_{ui} = +1$ when $X_{ui} = X_{HIGH}$. This coding system produces the familiar ± 1 notation for the factor levels associated with two-level factorial arrangements. Such coding system is widely used in fitting linear regression models, and it results in all the values of x_1 and x_2 falling between -1 and $+1$. One example of assigning coded and un-coded variables is shown in Table 15.2. The first and second order model equations expressed in coded variables can be represented as shown below:

$$Y_u = \beta_0 + \sum_{i=j}^{m} \beta_i x_{ui} + \varepsilon \tag{15.9}$$

$$Y_u = \beta_0 + \sum_{i=j}^{m} \beta_i x_{ui} + \sum_{i=j}^{m} \beta_{ii} x_{ui}^2 + \sum_{i=j}^{m-1} \sum_{j=i+1}^{m} \beta_{ij} xui + \varepsilon_u \tag{15.10}$$

6. *Conduct experiments to collect data*

In any experiment, the results and conclusions that can be drawn depend to a large extent on the manner in which the data were collected. It is crucial to plan the data collection phase of a response surface study carefully. In fact, response surface designs are valuable in this regard. The experiments are usually performed by using the experimental design formulations. Standard experimental procedures and equipments are used to make product samples and their evaluation. Replications are done if necessary.

In *Example 1* of formulating and evaluating tortillas, samples were prepared and cooked using a standardized formula. Flour mixture (as shown in Table 15.1), salt, shortening, and water were blended to develop dough. The dough and cooked tortillas were evaluated for sensory, physical, and compositional analyses (response variables described in step 2). The sensory testing was conducted using descriptive category scaling methods with structured 150 mm line scales having word anchors and references. Eleven experienced panelists were trained to evaluate the quality characteristics of tortilla. Samples were selected randomly and presented to judges using a balanced random order of presentation. Instrumental color values of lightness (L), a, and b were determined using a Gardner XL-845 colorimeter (Pacific Scientific, Bethesda, MD). Color terms of chroma $(a^2 + b^2)^{1/2}$ and hue ($\tan^{-1} b/a$) were also calculated. Textural quality (stickiness and hardness) of dough and tortilla samples was measured with an Instron Universal Testing Machine (Model 1122, Instron Inc., Canton, MA). For compositional analysis triplicate measures of moisture and duplicate measures of fat of cooked tortillas from each formulation were determined using standard AACC(American Association of Cereal Chemists) methods.

7. *Conduct statistical analysis*

The experimenter should be cognizant of techniques for achieving precision and accuracy and how to minimize measurement errors. During statistical analysis sensory impressions are difficult to quantify, but quantification is needed to aid in decision making. The sensory data can be analyzed using appropriate statistical models dictated by the sensory experimental designs. The unique features of sensory data are that they are unstable and skewed; thus special methods of design and analysis are used.

In the illustration using *Example 1*, all analyses were performed using Statistical Analysis System (SAS, 1985). Cluster analysis (VARCLUS) was performed on the sensory data to determine outliers among judges. Analysis of variance (ANOVA) (using the GLMs procedure) and Duncan's multiple range test were used to determine the effects of component flours on the physical characteristics of the tortilla dough. Development of prediction models and model fitting was done. To determine the significant independent component variables, stepwise regression analysis (STEPWISE) and ANOVA using GLM were performed on full models of each sensory and physical quality characteristic as dependent variables and following linear terms, wheat (x_1), cowpea (x_2), and peanut (x_3) flours, as independent variables and the cross product terms (wheat × cowpea), (cowpea × peanut), (wheat × peanut). Multiple regression analysis (REG) was performed on each sensory and physical characteristic using the models containing variables determined to be significant ($\alpha = 0.05$) by stepwise regression analysis. Model significance at the 0.05 level was determined using the F-ratio of means square quantities. Regression analysis was next performed on the means of the sensory and physical characteristics of the fitted models using the no intercept option to determine parameter estimates. To determine the effects of blending the component flours on the quality characteristics of tortillas, response surfaces were generated using PC SAS Graph. Parameter estimates produced from prediction models that were significant ($\alpha = 0.05$) and had R^2 of 0.80 or greater were used. Correlation coefficients (CORR) between sensory, physical, and compositional measures were also determined.

8. *Interpret data*

The interpretation of results require a careful scrutiny of the data. For example, if a new product is being developed, then the best formulation can be obtained after data interpretation such that it is the most acceptable combination of various ingredients. If a process is being modified, then interpretation of data can indicate the best treatment that will result in the maximum process yield. In the study of development of chocolate-flavored, peanut-soy beverage (*Example 3*), various experimental design formulations were subjected to sensory analysis along with the commercial chocolate milk as a control. The sensory data were analyzed and interpreted by comparing various treatments among themselves as well as with the control. Interpretation of data thus gave the best formulation that was comparable to the commercial chocolate milk with respect to sensory and physical characteristics. It was the combination of 44% peanut, 36% soy protein isolate (SPI), and 20% chocolate syrup that had the highest consumer acceptability. As compared to the control, it was rated higher for appearance, color, and sweetness.

9. *Reanalyze data if necessary*

Sometimes it becomes necessary to reanalyze data in order to consider the modifications in the design or the suggested model. In the beverage optimization study (*Example 3*), since the experimental region or the region of interest for optimization purpose was relatively small with lower and upper bound constraints introduction of lower bound pseudo-components (L-pseudo-components) was thought of as an alternative system for model fitting. Hence, using the definition of L-pseudo-components, the components were re-estimated (as per definition of Cornell and Harrison, 1997) and represented as Lpeanut $= X'_1$, Lsoy $= X'_2$, and Lchocolate $= X'_3$. The original lower and upper bound constraints (X_1: 0.306-0.587; X_2: 0.283-0.435; and X_3: 0.130-0.259) were converted into the redefined constraints for the mixture design as Lpeanut or $X'_1 =$ 0-1.0, Lsoy or $X'_2 =$ 0-0.542, and Lchocolate or $X'_3 =$ 0-0.458. The new experimental design points and region bound by these redefined constraints were obtained. The equivalent second degree or Scheffe-type mixture models in the L-pseudo-components were also obtained.

In another case, the optimization of lye-peeling process (*Example 4*), the data was reanalyzed to obtain the best results for model fitting. It was observed that the model chosen for unpeeled skin data did not represent the system appropriately and statistically indicated significant lack of fit. So some mathematical transformations were performed to obtain new model and the data for unpeeled skin were fitted to the new logarithmic model. It was then observed that the transformed logarithmic model provided statistically non-significant lack of fit and explained 97.8% of the variability of the data.

10. *Find parameters (coefficients) of function*

As mentioned in step 3 of selecting approximating response function, the response variable y may be related to *k* regressor variables. The model taken from

Equation 15.4, rewritten below, is called a multiple linear regression model with m regressor variables.

$$y = \beta_0 + \beta_1 x_1 + \beta_2 x_2 + \ldots + \beta_m x_m + \varepsilon \qquad (15.11)$$

The parameters $\beta_j, j = 1, 2, \ldots, m$, are called regression coefficients. The parameter β_j represents the expected change in response y per unit change in x_j when all the remaining independent variables $x_i (i \neq j)$ are held constant. Models that are more complex may often still be analyzed by multiple linear regression techniques. In general, any regression model that is linear in the parameters (the β values) is a linear regression model, regardless of the shape of the response surface that it generates. The method of least squares is typically used to get the regression coefficients in a multiple linear regression model. When parameters of the hypothesized regression response function are replaced by their estimates, the resulting function is said to be a fitted response function.

In tortilla formulation (*Example 1*) Scheffe second degree polynomial Equation 15.6 was used to fit the treatment means. Where, $\beta_1, \beta_2, \beta_3, \beta_{12}, \beta_{13}, \beta_{23}$ were parameter estimates for each linear and cross product term produced from prediction model. Regression analysis was performed on the means of the sensory and physical characteristics of the fitted models using the no intercept option to determine parameter estimates.

Also, in another illustration (*Example 3*), second degree models were fitted in terms of nine sensory attributes for chocolate-flavored, peanut-soy beverage formulations utilizing soy flour as a source of soy protein. Table 15.3 shows the response models with their estimated coefficients for a selected number of attributes.

11. *Develop response surfaces/contours*

The relationship $\eta = \phi(x_1)$ between η and the levels of a single factor may be represented by a straight line or by a curve. However, the relationship

TABLE 15.3
Second-Degree Response Models[1] in Terms of Selected Sensory Attributes and Their Respective Parameter Estimates

Variable	Model Co-efficients	Attribute			
		Flavor	Aftertaste	Mouth Feel	Overall Feel
Peanut (X'_1)	β_1	4.33	4.37	4.87	4.18
Soy flour (X'_2)	β_2	3.06	4.15	2.68	3.70
Chocolate (X'_3)	β_3	5.20	6.16	6.84	4.74
$X'_1 * X'_2$	β_{12}	0.21	1.31	4.13	0.37
$X'_1 * X'_3$	β_{13}	3.34	0.51	0.35	4.37
$X'_2 * X'_3$	β_{23}	7.37	2.19	5.02	5.01

Note: [1] $Y = \beta_1 * X'_1 + \beta_2 * X'_2 + \beta_3 * X'_3 + \beta_{12} * X'_1 * X'_2 + \beta_{13} * X'_1 * X'_3 + \beta_{23} * X'_2 * X'_3 + \varepsilon$

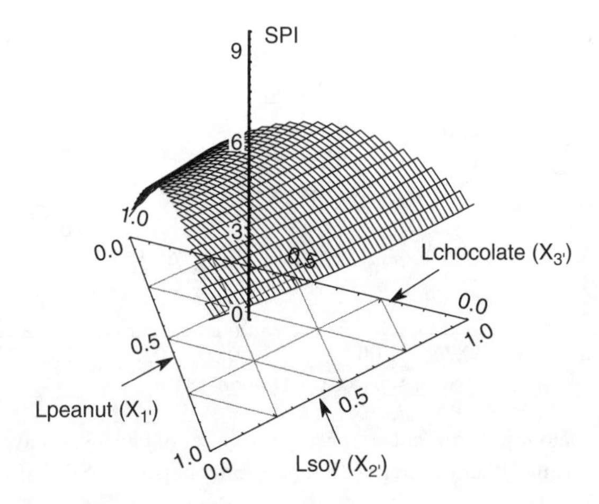

FIGURE 15.2 Three-dimensional response surface for overall acceptability of chocolate-flavored, peanut-soy beverage prepared using soy protein isolate (SPI).

$\eta = \phi(x_1, x_2, \ldots, x_k)$ between η and levels of k factors may be represented by a surface. With k factors, the response surface is of dimensionality k+1, so that the curve can be depicted in two dimensions, whereas the surface can be visualized in three-dimensions. Usually computerized techniques or special software programs are used to generate such response surfaces. For example, to determine the effects of blending the component flours on the quality characteristics of tortillas (*Example 1*), response surfaces were generated using PC SAS Graph. Parameter estimates produced from prediction models that were significant ($\alpha = 0.05$) and had R^2 of 0.80 or greater were used.

Figure 15.2 shows a three dimensional response surface for overall acceptability in case of beverage formulation prepared using SPI as a protein source. The Statistica® software was used to generate these triangular graphs and surfaces. The fitted second degree model representing the response surface is given as Equation 15.12.

Overall acceptability

$$4.561^*X_1' + 4.152^*X_2' + 0.794^*X_3' 1.001^*X_1'^*X_2' + 9.939^*X_1'^*X_3' + 12.581^*X_2'^*X_3' \qquad (15.12)$$

The fitted linear response function is usually represented as

$$Y = b_0 + b_1X_1 + b_2X_2 + \cdots + b_mX_m \qquad (15.13)$$

This can be used to predict responses for desired values of the independent variables. When the fitted response function Y is graphed as a function of independent variables, the resulting graph is called a response surface plot or contour map. A technique used to help in visualizing the shape of a three-dimensional response surface is the plotting of contours of the response surface. Contour maps of responses are an attractive feature

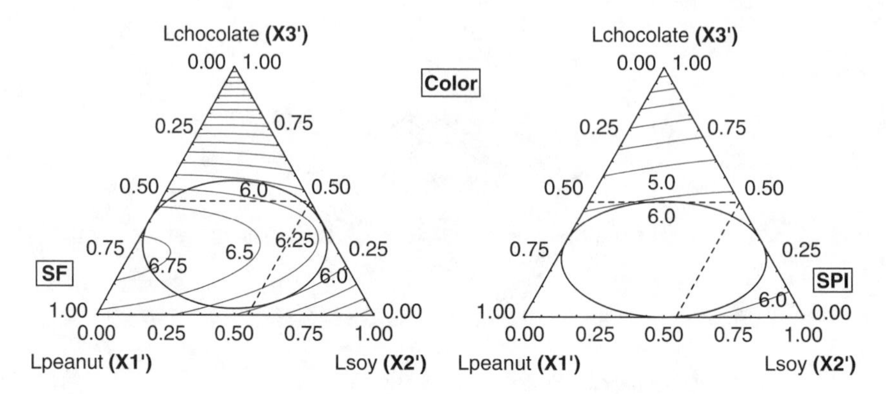

FIGURE 15.3 Contour plots for color (sensory response) of chocolate-flavored, peanut-soy beverage prepared using either soy flour (SF) or soy protein isolate (SPI).

of response surface analysis. By constructing surface plots it is easier to locate and characterize optimum responses. In a contour plot, lines or curves of equal response values are drawn on a graph or plane whose coordinates represent the levels of the factors. Each contour represents a specific value for the height of the surface (i.e., a specific value of Y), above the plane defined for combinations of the levels of the factors. The plotting of different surface height values enables one to focus attention on the levels of the factors at which the changes in the surface height occur. Contour plotting is not limited in depicting surfaces in one and two dimensions. A familiarity with the geometrical representation for two and three factors enables the general situation for $k > 3$ factors to be more readily understood, although they cannot be visualized geometrically.

For example, Figure 15.3 shows contour plots of color (sensory response) in the beverage optimization study (*Example 3*) obtained using Statistica® software. The lines on the triangular graph are contour lines representing respective mean hedonic ratings. The circular regions on the contour plots represent regions of interest or regions of maximum consumer acceptability (where color ratings for the beverage are >5). And the region bound by dashed lines is the experimental region.

Similarly, the contour plots for tortilla optimization study (*Example 1*) can be obtained. Figure 15.4 shows contour plots for beany flavor and sweet aromatic flavor. The shaded regions represent optimum regions for tortillas for those two particular sensory attributes. Such graphs can also be obtained for other attributes—physical characteristics—of interest such as instrumental color.

12. *Identify constraints or limits to determine optimum conditions*

The fitted model may lead to negative or extremely large responses that may not have practical meaning. As a result, the extrapolation of the responses beyond experimental range of the independent variables may not be valid. It is, therefore, important to design experiments properly to allow for exploration of the fitted response surfaces over the desired range of experimental variables. At the preliminary stages of a response surface investigation the experimenter is usually asked to specify the region

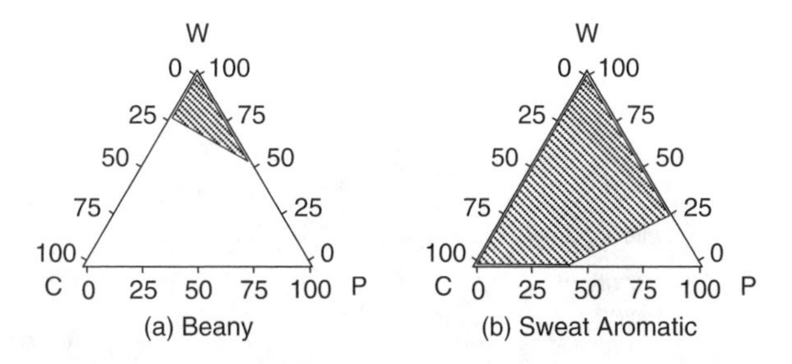

FIGURE 15.4 Contour maps for significant sensory attributes for blends containing wheat (W), cowpea (C), and peanut (P) flours designating the optimum region for (a) Beany flavor, and (b) Sweet aromatic flavor.

of conceivable factor level values that represents the factor combinations of potential interest. For example in Figure 15.3 the parallelogram marked as dashed lines represents the experimental region. It is the region bound by 0–100% peanut, 0–54% soy and 0–46% chocolate syrup (all three components expressed as L-pseudo-components). The optimum should be within this experimental region. Further, during optimization the area within the experimental region is sometimes specified and expressed as the region of interest on contour plots. Such areas of interest marked on contour plots (Figure 15.3) can be selected by specifying the limit of acceptance. For example in case of the beverage study (*Example 3*) response area bound by mean hedonic ratings >5 for all the sensory responses was specified as the region of maximum consumer acceptability since the control ratings were found to be between 6 and 7. So those contour lines representing ratings >5 were marked by circular regions on contour plots called the regions of interest. The optimum was obtained by superimposing these regions of interest. The limit of acceptance can be a standard specified by the experimenter based on the comparison with the control or from a previous experience. Such limits specification will ensure that the optimum conditions lead to the most acceptable/desirable product or process characteristics. Sometimes the constraints are identified based on the desirable characteristics of the best formulation which helps to obtain the optimum formulation. For example in optimization of tortilla (*Example 1*), the 100% wheat formulation produced tortillas with the most desirable quality characteristics.

13. *Find optimum conditions*

The optimum conditions are usually obtained by superimposition of the contour plots. In case of tortilla optimization study, the statistical analysis resulted in significant prediction models for 11 characteristics. Optimum regions for each of these attributes were outlined on surface contour maps. Acceptable regions for formulations producing values within the optimum criteria ranges are shown in Figure 15.5. The optimum was obtained by superimposing all 11 attributes. The hatched area represented the region in which wheat flour could be successfully replaced with cowpea and peanut flours and result in tortillas similar to 100% wheat flour tortillas. Optimum showed that wheat flour could be successfully replaced with up to 24% cowpea flour (C), and as much as 46% peanut flour (P) or combinations of both

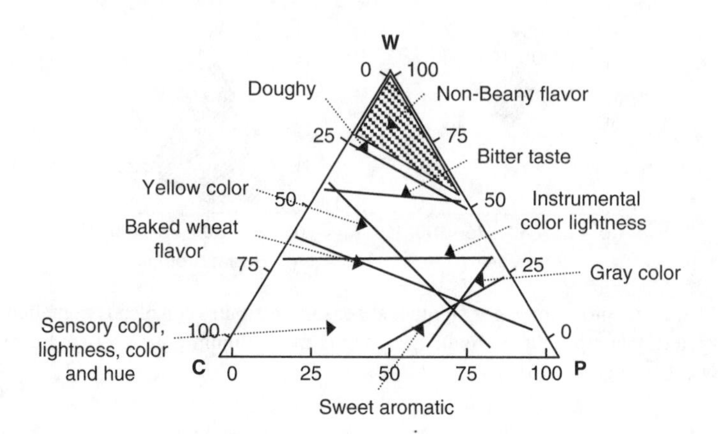

FIGURE 15.5 Optimum region for quality characteristics with a coefficient of determination ($R^2 > 0.8$.), as indicated by the hatched area, obtained by superimposing contour maps of significant attributes for blends containing wheat (W), cowpea (C), and peanut (P) flours.

peanut and cowpea flours. Based on these findings, any blend of cowpea or peanut flours within the ranges of 0–24% or 0–46%, respectively, would produce tortillas with characteristics similar to 100% wheat flour tortillas.

14. *Conduct verification experiments*

The final stage of optimization research is verification stage. The optimum conditions suggested by the response surface method are confirmed by carrying out the verification experiments. For example, in tortilla formulation study (*Example 1*) the third replication of each of the ten mixtures was conducted, and data from the sensory and physical analyses was used to verify prediction models for all attributes with an $R^2 > 0.80$. Experimental errors for the models were determined by comparing observed measures with the predicted quality parameters.

15. *Determine deviation between the predicted optimum responses and experimentally determined response from the verification experiments*

The data from the verification experiments is further analyzed to find the deviation between the predicted response and the observed response values. A comparison of observed measurements and predicted values for attributes with $R^2 > 0.90$ and $R^2 > 0.80$ was done separately for tortilla prepared from wheat, cowpea, and peanut flours (*Example 1*). It was observed that the overall error decreased to 11.5% when all 11 quality characteristics were used for optimization as compared to overall error of 15.2% when optimization was based on only 5 attributes.

SYSTEMATIC APPROACH TO USE RSM TECHNIQUES IN PROCESS AND PRODUCT OPTIMIZATION

Response surface methodology consists of a group of techniques used in the empirical study of relationships between one or more measured responses such as yield, color index, and viscosity, on the one hand, and a number of input variables such as time, temperature, pressure, and concentration, on the other. The techniques have been used to answer questions of a number of different kinds, such as the following:

a. How is a particular response affected by a given set of input variables over some specified region of interest?
b. What settings, if any, of the inputs will give a product simultaneously satisfying desired specifications?
c. What values of the inputs will yield a maximum for a specific response, and what is the response surface like close to this maximum?

We will try to answer these and other such questions with the help of specific examples illustrating application of RSM techniques to product and process optimization.

1. *Product optimization*

Quality improvement is most effective when it occurs early in the product and process development cycle. Biotechnology, pharmaceutical, chemical and process industries are all examples where experimental design methodology has resulted in products that are easier to manufacture, have higher reliability, have enhanced field performance, and meet or exceed customer requirements. RSM is an important branch of experimental design in this regard. For product development, RSM can be used to establish the optimum levels of the primary ingredients in a product. This information helps the product developer to understand ingredient interactions in the product which guide final product formulation and future cost and quality changes. Two special cases of product optimization, one of product improvement category utilizing fractional factorial design, and another of new product development category with mixture design approach, will be illustrated here.

Example 2: Use of unhydrogenated palm oil to improve quality of peanut butter:

Factorial experiments involve the collection of data from a sequence of tests or observations in which levels of one controllable factor, or combinations of the levels of two or more controllable factors, are systematically changed as stipulated in the design of the experiment. A complete factorial experiment includes all possible combinations of the levels of the factors in the experiment. A fractional factorial experiment utilizes only a portion of the possible combinations. Pilot plant studies, screening experiments, and ruggedness testing constitute a few of the many settings in which fractional factorial experiments are commonly used. The careful selection of appropriate combinations of the factor levels to include in the design enables obtaining of estimates of the observed effects and thereby leads to the success of the experiment. The combination of factor levels that are included in an experiment must be selected according to certain well-defined criteria. Failure to do so can result in biased estimates of the effects of the factors, incorrect estimates of the magnitude of the experimental error, and possibly incorrect inferences about the effects of the factors on the responses. Application of RSM in peanut butter quality improvement using fractional factorial design will be illustrated briefly.

TABLE 15.4
Fractional Factorial Design for Peanut Butter Composition and Storage Temperatures

Palm Oil (%)	Peanut Shell Flour (%)	Storage Temperature (°C)
3.0	1.50	25
3.0	0.00	25
2.0	1.50	25
2.0	0.00	25
3.0	0.75	35
3.0	0.75	15
2.0	0.75	35
2.0	0.75	15
2.5	1.50	35
2.5	1.50	15
2.5	0.00	35
2.5	0.00	15
2.5	0.75	25
2.5	0.75	25
2.5	0.75	25

RSM was used by Hinds et al. (1994) to investigate the potential of unhydrogenated palm oil, with and without peanut shell flour, to prevent oil separation in peanut butter. Optimum combinations were determined to produce a stable product. Following input variables were selected in the specified ranges.

Unhydrogenated palm oil:	2.0, 2.5, and 3.0%
Peanut shell flour:	0, 0.75, and 1.5%
Storage temperature:	15, 25, and 35°C

According to Box and Draper (1987) this fractional factorial design necessitated 15 combinations containing palm oil (Table 15.4) Selection of levels for independent variables was based on results from preliminary tests and observations found in the literature.

Preliminary tests were conducted to evaluate the stabilizing effects of palm oil. Peanut butter containing 1.25% Fix-X (a stabilizer from Procter and Gamble in Cincinnati, OH), 93.25% roasted seeds, 4.71% sugar and 0.79% salt was used as a standard. Percent oil separation, texture (apparent viscosity), and color were assessed after 0, 1 and 2 week storage. Triplicate treatments were carried out for texture and color. Six replicates were used for percent oil separation, and mean values were analyzed. Data collected after two week storage was considered representative for reliable information on the stabilizing effects of palm oil. Statistical analysis of data collected was done further. The RSREG procedure of SAS was used to determine the effects of independent variables (palm oil, peanut shell flour and storage temperature) on physical quality characteristics of peanut butter samples after 2 week storage. Percent oil separation, texture, hue angle and L value (lightness) of color were analyzed. Regression analysis was carried out on a second order polynomial equation (see Equation 15.10):

$$Y = b_0 + b_1x_1 + b_2x_2 + b_2x_3 + b_4x_1^2 + b_5x_2^2 + b_6x_3^2 + b_7x_1x_2 + b_8x_1x_3 + b_9x_2x_3 \quad (15.14)$$

Where Y is the response variable and x_1, x_2, and x_3 are independent variables. ANOVA determined the overall effects of the independent variables on response variables of the samples stored for two weeks. Data interpretation indicated that the percent oil separation was affected by all three independent variables, texture was affected by storage temperature only, and peanut shell flour was the main component affecting color. ANOVA indicated that the model developed for texture was adequate and had no significant lack of fit. However, there was significant lack of fit in case of percent oil separation and L value. In an attempt to improve the fit of these models, several expansive transformations were evaluated. The most appropriate equation for calculating percent oil separation giving a statistically nonsignificant lack of fit and explaining 90.2% of the variability was obtained. The establishment of a limit to denote optimum stability in combination stabilized with palm oil was based on several criteria. Based on specifications and observations of peanut butters with commercial stabilizers, a maximum of 0.5% oil separation after 2 week storage at 30–35°C was established as an indicator of stability for the experimental combinations stabilized with palm oil. Also, because commercial peanut butters (which should have remained stable for 1 year at 21–24°C) showed 1% oil separation after 2 week at 35°C, this implied that the palm oil peanut butters should remain stable for ≥1 year at about 21–24°C. Thus, 30–35°C represented the region of interest on contour plots. For color data the limits were arbitrarily to accept stable experimental samples with hue angles and L values that showed ≤2° deviation and ≤5% variation, respectively, from corresponding attributes of the control formulations. The plot of effects at 35°C indicated combinations of palm oil (2.0–2.5%) and peanut shell flour (0–0.8%) should result in optimum oil stability. Comparison of the region of optimum oil stability with color attributes indicated that the combinations of 2.0–2.5% palm oil and 0–0.8% peanut shell flour should produce peanut butter with constraints for acceptable color. Thus, computer-generated contour plots indicated that 2.0–2.5% palm oil should effectively stabilize peanut butter stored at 21–24°C for ≥1 year without affecting color. And incorporation of peanut shell flour ≤0.8% would provide stable combinations that are ≤5% darker than those without peanut shell flour. In this way, an improved peanut butter formulation was obtained by applying stepwise RSM.

Example 3: *Development of a new beverage utilizing peanut flour, soy protein sources, and chocolate syrup*

Many product design and development activities involve formulation problems in which two or more ingredients are mixed together. The product engineer or scientist would like to find appropriate blend of the ingredients so that best overall characteristics of the final product are achieved. In such mixture situations, the response variables depend on the percentages or proportions of the individual ingredients that are present in the product formulation. Unlike factorial experiments, a feature of a mixture experiment is that the independent or controllable variables represent proportionate amounts of the mixture rather than unrestrained amounts where proportions are by volume, by weight, or by mole fraction. The component proportions are nonnegative and if, expressed as fractions of the mixture, they sum to unity. There are special experimental design techniques and model-building methods for mixture problems. Scheffe in 1958 published a theory for mixture experimentation which was further studied and thoroughly explained by Cornell (1990) and Cornell and Harrison (1997). The application of RSM to a mixture problem will be illustrated here referring to the development of chocolate-flavored, peanut-soy beverage (2004).

RSM was used to obtain optimum range of three major components (peanut, soy and chocolate syrup) in a new protein-based beverage formulation (2004). The essential condition of mixture design was

$$X_1 + X_2 + X_3 = 1 \tag{15.15}$$

Where, X_1 represented % peanut concentration, X_2 represented % concentration of soy protein (either soy flour or SPI), and X_3 represented % levels of chocolate syrup used in the mixture formulations. The ranges of all three components used for experimental design were X_1 =30.6–58.7%, X_2 =28.3–43.5%), and X_3 =13.0–25.9%. A three-component constrained mixture design resulted into nine design points. Five additional points were selected within the experimental region. All these 14 experimental formulations were prepared using defatted soy flour and SPI as a source of soy protein in the beverage. Thus, a total of 28 experimental formulations were obtained which were subjected to sensory analysis. The response was determined in terms of nine different sensory attributes which were overall acceptability, appearance, color, aroma, consistency, flavor, sweetness, mouthfeel, and aftertaste. The sensory data obtained after evaluation of 28 different samples along with commercial chocolate milk by 41 untrained consumers was analyzed further. The ANOVA and Duncan's Multiple Range Test were used to compare different attributes at the 95% significance level ($\alpha = 0.05$). A second-degree or quadratic Scheffe-type mixture model given below in L-pseudo-components was fitted to raw data on nine sensory attributes.

$$\begin{aligned} Y = {} & \beta_1 * X_1' + \beta_2 * X_2' + \beta_3 * X_3' + \beta_{12} * X_1' * X_2' + \beta_{13} * X_1' * X_3' \\ & + \beta_{23} * X_2' * X_3' + \varepsilon \end{aligned} \tag{15.16}$$

Two data sets, one for soy flour and another for SPI, were analyzed separately. The fitted models for all the attributes were used to generate three-dimensional response surfaces as well as contour plots using software. The region of interest was determined from the acceptability of commercial chocolate milk. Since the mean hedonic ratings of commercial chocolate milk for all nine attributes was in the range of 6 (like slightly) and 7 (like moderately), the limit of maximum consumer acceptability of experimental formulations was set at 5.0 (neither like nor dislike). Superimposition of contour plot regions of interest (within which each attribute received hedonic ratings ≥5.0) resulted in optimum regions. It was observed that the optimum regions for soy flour (SF) were combinations of 34.1–45.5% peanut, 31.2–42.9% soy, and 22.4–24.1% chocolate syrup; whereas, in case of SPI formulations 35.8–47.6% peanut, 31.2–43.5% soy, and 18.3–23.6% chocolate syrup, resulted into optimum region (Figure 15.6). In this way, the optimum range of three major components that will

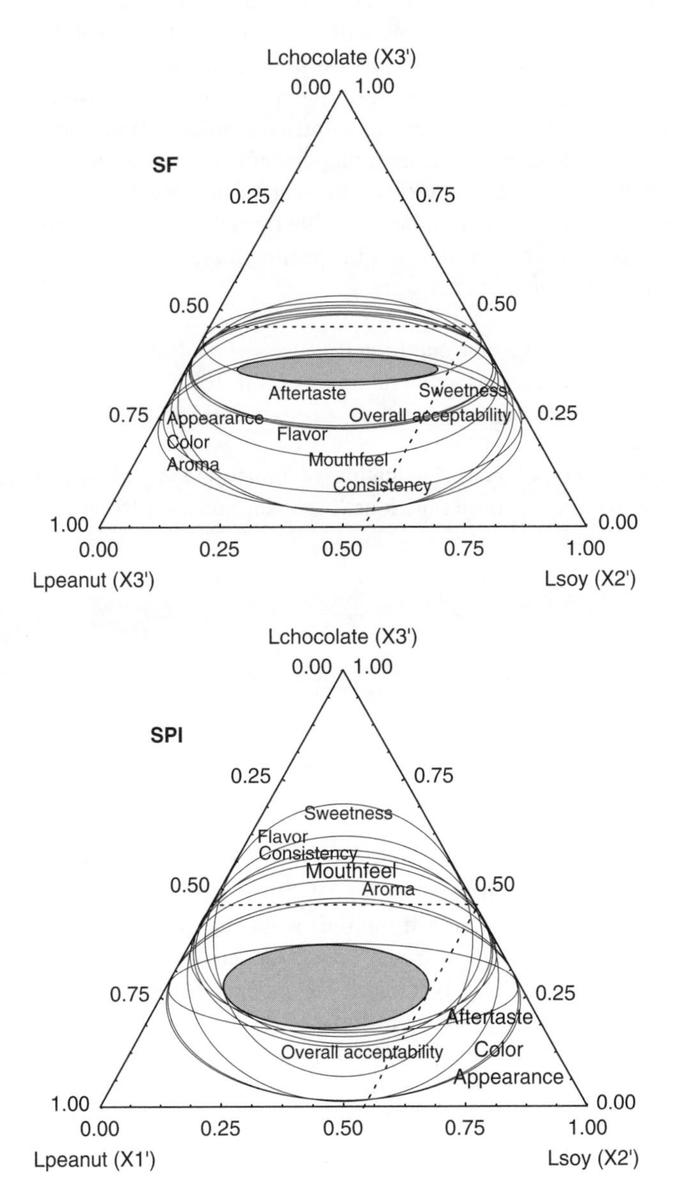

FIGURE 15.6 Optimum regions (shaded areas) obtained by superimposing contour plots for all nine attributes for soy flour (SF) and soy protein isolate (SPI) formulations.

result in a beverage having most acceptable characteristics was obtained by applying RSM for new product development.

2. *Process optimization*

When many factors and interactions affect desired responses, RSM is an effective tool for optimizing process. To optimize lye-peeling process, Floros and Chinnan (1987) used RSM. Optimization was performed to result maximum removal of the skin and minimum loss of the fruit. We will discuss the application of RSM for process optimization taking this lye-peeling process as an example.

Example 4: The objectives of the lye-peeling optimization study were to understand the relationships between the factors affecting a lye-process (time, concentration, and temperature) and the responses determining the effectiveness of the process (peeling loss, unpeeled skin and process yield) (Floros and Chinnan, 1987). A set of optimum processing conditions was determined suitable for removing practically all skin from the pimiento pepper while minimizing the peeling loss.

The ranges of input variables were as

Lye-concentration (c):	4–12% NaOH
Process temperature (T):	80–100°C
Time (t):	1.5–6.5 min

A second order polynomial equation of the following form was assumed to approximate the true response functions (based on Equation 15.10):

$$\eta_K = \beta_{K0} + \sum_{i=j}^{3} \beta_{ki} x_i + \sum_{i=j}^{3} \beta_{kii} x_i^2 + \sum_{i=j}^{3} \sum_{j=i+1}^{3} \beta_{kij} x_i x_j \tag{15.17}$$

where,

$$\eta_1 = \textit{peeling loss (PL)}, \eta_2 = \textit{unpeeled skin surface are (US)},$$

$$\eta_3 = \textit{product yield (Y)};$$

$\beta_{k0}, \beta_{ki}, \beta_{kij}$ were constant coefficients and x_i were the coded independent variables.

The Box and Behnken (Box et al., 1978) experimental design as shown in Table 15.5 was selected because it minimizes the sum of the two main sources of error, variance or sampling error and bias error, and is specific for RSM studies which employ second order polynomials. The independent variables, the coded variables (x_i) and their levels are shown in Table 15.2. The pimiento peppers were treated with the lye solution and were processed further by a standardized pilot-plant processing method. The data on peeling loss (%), unpeeled skin (cm^2/pepper) and yield (%) were obtained for the designed experimental runs. This data were then used to fit second order polynomial equation using RSREG procedure for SAS. The regression coefficients were (β_{ki}) obtained as shown in Table 15.6.

The ANOVA for the three response variables indicated that the models developed for peeling loss and product yield were adequate. However, in case of unpeeled skin

TABLE 15.5
Box and Behnken Design for Three Variables-Three Levels Response Surface Analysis

x_1	x_2	x_3	# of Runs
±1	±1	0	Combinations = 4
±1	0	±1	Combinations = 4
0	±1	±1	Combinations = 4
0	0	0	1 × 3 (replications) = 3
Total runs = 15			

TABLE 15.6
Regression Coefficients for Lye-Peeling Process Optimization

Coefficient	Peeling Loss (PL), $k = 1$	Unpeeled Skin (US) $k = 2$	Yield (Y)$k = 3$
β_{k0}	39.85	0.09	43.60
β_{k1}	11.75	11.63	−8.12
β_{k2}	5.29	−4.06	−3.86
β_{k3}	17.29	−13.03	−12.91
β_{k11}	−7.50	9.16	5.59
β_{k22}	1.64	−6.11	−0.80
β_{k33}	−8.72	11.48	7.81
β_{k12}	−2.36	3.06	0.61
β_{k13}	3.67	20.07	−1.95
β_{k23}	−1.46	4.99	1.83

the highly significant lack of fit suggested that the chosen model did not represent the system appropriately. Thus, sometimes if necessary the data needs to be reanalyzed. Several mathematical transformations were performed on the experimental data to give the logarithmic model for unpeeled skin yielding the best results:

$$A = \text{In}(\text{US}) = -2.55 - 2.39x_1 - 2.11x_2 - 3.74x_3 + 0.70x_1^2 + 0.12x_2^2 + 0.25x_3^2 + 0.69x_1x_2 - 0.80x_1x_3 - 0.90x_2x_3 \quad (15.18)$$

Further statistical analysis was performed to reveal that the three process variables had a significant overall effect on the three responses. Computer generated three-dimensional response surfaces were obtained using predictive models of PL (peeling loss), Y (product yield), and US (unpeeled skin surface). The predictive models of unpeeled skin and peeling loss were used for further examination of the system behavior and localization of the optimum conditions. In attaining optimum region, the ranges of limit of acceptance were decided based on total removal of the skin (US $\leq$ 1 cm^2/pepper) with lowest possible peeling losses (PL = minimum).

The upper limit of 1 cm^2/pepper of unpeeled skin was set so that the product would be graded as "US. Grade A" according to the US. Standards for Grades of Canned Pimientos. Computer generated response surfaces, canonical analysis and contour plot interpretation revealed that relatively high lye concentration (12% NaOH) combined with short processing time (1.6–2 min) at a moderate temperature of around 90°C should yield an optimum process with practically all the skin removed and peeling loss as low as 20%. The adequacy of the model equations was tested in the pilot plant. The comparison of predicted and observed/experimental values of PL, Y, and US indicated that the values of PL and Y were not significantly different at 5% significance level; whereas, the values of US were statistically different. The reason explained was peppers used in the verification models were harvested late in the season giving slightly different results than predicted. Overall, the optimum conditions were therefore, applicable only when pimiento peppers of the "Truhart" cultivar were processed and general trend of relatively high concentration and short time combined with moderate temperature was advised for any other kind of peppers. In this way, the optimization of lye-peeling process was done by following a systematic 15 step optimization approach of applying RSM as described earlier.

SUMMARY

Response surface methodology is a useful statistical technique for investigation of complex processes. It consists of a group of mathematical and statistical procedures that can be used to study relationships between one or more responses (dependent variables) and a number of factors (independent variables). Most of the RSM applications come from areas such as chemical or engineering processes, industrial research and biological investigations, with emphasis on optimizing a process or system. The main advantage of RSM is the reduced number of experimental runs needed to provide sufficient information for statistically acceptable results. It is faster and less expensive method of performing scientific research compared to classical one-variable-at-a-time or full-factorial method.

A brief outline of how to apply RSM techniques for optimization research was given by explaining fifteen important steps using illustrations at each step. An example of formulation, evaluation and optimization of tortilla from wheat, cowpea, and peanut flours (Holt et al., 1992) (*Example 1*) was used for this purpose. The importance of design of experiment and RSM techniques was also demonstrated through product and process optimization. Implementation of RSM for process optimization was illustrated using pimiento pepper lye-peeling process (Floros and Chinnan, 1987) (*Example 2*). Product optimization technique was explained using fractional factorial design for improving peanut butter stability (Hinds et al., 1994) (*Example 3*) and three-component constrained mixture design in development of a chocolate-flavored, peanut-soy beverage (Deshpande et al., 2004) (*Example 4*).

Thus, RSM is a very useful and critical tool in developing new processes, optimizing their performance, and improving design and/or formulation of new products. The objectives of quality improvement, including reduction of variability and improved product and process performance can often be accomplished directly using RSM.

BIBLIOGRAPHY

1. Box G.E.P. and Draper N. R., *Empirical model building and response surfaces*, Wiley, New York, 1987.
2. Box G.E.P., Hunter W.G., and Hunter J.S., *Statistics for experimenters: an introduction to design, data analysis, and model building*, John Wiley & Sons, Inc., New York, 1978.
3. Cornell J.A., *Experiments with mixtures: Design, models, and analysis with mixture data*, 2nd ed., John Wiley and Sons, Inc., New York, 1990.
4. Cornell J.A. and Harrison J.M., Models and designs for experiments with mixtures: part II-exploring a subregion of the simplex and the inclusion of other factors in mixture experiments. *Technical Bulletin*, 899, Florida Agricultural Experiment Station, Institute of Food and Agricultural Sciences, University of Florida, Gainesville, FL., 1997.
5. Deshpande R.P., Chinnan M.S., and McWatters K.H., Optimization of a chocolate-flavored, peanut-soy beverage using response surface methodology (RSM) as applied to consumer acceptability data, In *Development of chocolate flavored peanut-soy beverage*, Master's Thesis, University of Georgia, Athens, Chap. 5, pp 149–192, 2004.
6. Floros J.D. and Chinnan M.S., Optimization of pimiento pepper lye-peeling process using response surface methodology, *Trans. Amer. Society Agricultural Eng.*, 30(2), 560–565, 1987.
7. Hare L.B., Designs for mixture experiments involving process variables, *Technometrics*, 21(2), 159–173, 1979.
8. Hinds M.J., Chinnan M.S., and Beuchat L.R., Unhydrogenated palm oil as a stabilizer for peanut butter, *J. of Food Sci.*, 59(4), 816–820 & 832, 1994.
9. Holt S.D., Resurreccion A.V.A., McWatters K.H., Formulation, evaluation and optimization of tortillas containing wheat, cowpea and peanut flours using mixture response surface methodology, *J. Food Sci.*, 57(1), 121–127, 1992.
10. Myers R.H. and Montgomery D.C., *Response surface methodology: process and product optimization using designed experiments*, John Wiley & Sons, Inc., New York, 1995.
11. Schutz H.G., Multiple regression approach to optimization. *Food Tech.*, 37(11), 46–48 & 62, 1983.
12. Sidel J.L. and Stone H., An introduction to optimization research, *Food Tech.*, 37(11), 36–38, 1983.

16 Shelf Life of Packaged Foods: Its Measurement and Estimation

Gordon L. Robertson

CONTENTS

INTRODUCTION

The deterioration of packaged foods (and this includes virtually all foods because today very few foods are sold without some form of packaging) depends largely on transfers that may occur between the internal environment inside the package, and the external environment that is exposed to the hazards of storage and distribution. For example, there may be transfer of water vapor from a humid atmosphere into a dried product, or transfer of an undesirable odor from the external atmosphere into a high-fat product. In addition to the ability of packaging materials to protect and preserve foods by minimizing or preventing these transfers, packaging materials must also protect the product from mechanical damage, and prevent or minimize misuse by consumers (including tampering).

Although certain types of deterioration will occur even if there is no transfer of mass (or heat, since some packaging materials can act as efficient insulators against fluctuations in ambient temperatures) between the package and its environment, it is possible in many instances to prolong the shelf life of the food through the use of packaging.

It is important that food packaging not be considered in isolation from food processing and preservation, or indeed from food marketing and distribution: all interact in a complex way, and concentrating on only one aspect to the detriment of the others is a surefire recipe for commercial failure.

The development of an analytical approach to food packaging is strongly recommended, and to achieve this successfully, a good understanding of food safety and quality is required. The more important of these is without question food safety, which is the freedom from harmful chemical or microbial contaminants at the time of consumption. Packaging is directly related to food safety in two ways.

Firstly, if the packaging material does not provide a suitable barrier around the food, microorganisms can contaminate the food and make it unsafe. However, microbial contamination can also arise if the packaging material permits the transfer of, for example, moisture or oxygen from the atmosphere into the package. In this situation, microorganisms present in the food but presenting no risk because of the initial absence of moisture or oxygen may subsequently be able to grow and present a risk to the consumer.

Secondly, the migration of potentially toxic compounds from some packaging materials to the food is a possibility in certain situations and gives rise to food safety concerns. In addition, migration of other components from packaging materials, while not harmful to human health, may adversely affect the quality of the product.

The major quality attributes of foods are texture, flavor, color, appearance, and nutritive value and these attributes can all undergo undesirable changes during processing and storage. With the exception of nutritive value, the changes that can occur in these attributes are readily apparent to the consumer, either prior to or during consumption. Packaging can affect the rate and magnitude of many of the quality changes. For example, development of oxidative rancidity can often be minimized if the package is an effective oxygen barrier and oxygen has been initially removed from the package interior; flavor compounds can be absorbed by some types of packaging material; the particle size of many food powders can increase (i.e., clump) if the package is a poor moisture barrier. This chapter addresses the issue of shelf life, which is very clearly related to food quality.

Knowledge of the kinds of deteriorative reactions that influence food quality is the first step in developing food packaging that will minimize undesirable changes in quality and maximize the development and maintenance of desirable properties. Once the nature of the reactions is understood, knowledge of the factors that control the rates of these reactions is necessary in order to fully control the changes occurring in foods during storage, that is, while packaged.

DEFINITIONS

The term "food quality" has a variety of meanings to professionals in the food industry, but the ultimate arbiters of food quality must be consumers. This notion is embodied in the definition of food quality as "the combination of attributes or characteristics of a product that have significance in determining the degree of acceptability of the product to a user." Another definition of food quality is "the acceptance of the perceived characteristics of a product by consumers who are regular users of the product category or those who comprise the market segment." The phrase "perceived characteristics" includes the perception of the food's safety, convenience, cost, value, and so forth, and not just its sensory attributes.

The quality of most foods and beverages decreases with storage or holding time. It therefore follows that there will be a finite length of time before the product becomes unacceptable. This time from production to unacceptability is referred to as shelf life. Although the shelf lives of foods vary, they are generally determined routinely for each particular product by the food manufacturer or processor. Quality loss during storage may be regarded as a form of processing at relatively low temperatures that goes on for rather a long time. It is therefore not surprising that many of the concepts developed in connection with food processing find application in shelf life studies. Such studies are an essential part of food product development, with the food processor attempting to provide the longest practicable shelf life consistent with costs and the pattern of handling and use by distributors, retailers, and consumers.

Inadequate shelf life will often lead to consumer dissatisfaction and complaints. At best, such dissatisfaction will eventually affect the acceptance and sales of brand name products, while at worst it can lead to malnutrition or even illness. Therefore, food processors give considerable attention to determining the shelf lives of their products.

Despite its importance, there is no simple, generally accepted definition of shelf life. The Institute of Food Technologists in the United States has defined shelf life as "the period between the manufacture and the retail purchase of a food product, during which time the product is in a state of satisfactory quality in terms of nutritional value, taste, texture, and appearance." This definition overlooks the fact that the consumer may store the product at home for some time before consuming it yet will still want the product to be of acceptable quality.

The Institute of Food Science and Technology in the United Kingdom has defined shelf life as "the period of time during which the food product will remain safe; be certain to retain desired sensory, chemical, physical, microbiological, and functional characteristics; and comply with any label declaration of nutritional data when stored under the recommended conditions."

Another definition is that "shelf life is the duration of that period between the packing of a product and the end of consumer quality as determined by the percentage of consumers who are displeased by the product." This definition accounts for the variation in consumer perception of quality (i.e., not all consumers will find a product unacceptable at the same time) and has an economic element in that since it is not possible to please all consumers all of the time, a baseline of consumer dissatisfaction must be established. In the branch of statistics known as survival analysis, consumer dissatisfaction can be related to the survival function, defined as "the probability of a consumer accepting a product beyond a certain storage time." Models permitting the application of survival analysis to the sensory shelf life of foods have been published.

Simply put, shelf life is the time during which all of the primary characteristics of the food remain acceptable for consumption. Thus, the shelf life refers to the time for which a food can remain on both the retailer's and consumer's shelf before it becomes unacceptable.

Given the variety of definitions, it is not surprising that there is no uniform or universally accepted open dating system for packaged foods. In some countries, mandatory open dating of all perishable and sometimes semiperishable foods is required, while in other countries such requirements are voluntary. Arguments can be advanced both for and against the open dating of foods. However, there is an increasing quantity of open dated food on sale throughout the world, and this trend is likely to continue.

FACTORS CONTROLLING SHELF LIFE

The shelf life of a food is controlled by three factors:

1. Product characteristics including formulation and processing parameters (intrinsic factors);
2. Properties of the package; and
3. Environment to which the product is exposed during distribution and storage (extrinsic factors).

Intrinsic factors include pH, water activity, enzymes, microorganisms, and concentration of reactive compounds. Many of these factors can be controlled by selection of raw materials and ingredients, as well as the choice of processing parameters.

Extrinsic factors include temperature, relative humidity, light, total pressure and partial pressure of different gases, and mechanical stresses including consumer handling. Many of these factors can affect the rates of deteriorative reactions that occur during the shelf life of a product.

The properties of the package can have a significant effect on many of the extrinsic factors and thus indirectly on the rates of the deteriorative reactions. Thus, the shelf life of a food can be altered by changing its composition and formulation, processing parameters, packaging system, or environment to which it is exposed.

PRODUCT CHARACTERISTICS

PERISHABILITY

Based on the nature of the changes that can occur during storage, foods may be divided into three categories—perishable, semiperishable and nonperishable, or shelf stable, which translate into very short shelf life products, short-to-medium shelf life products, and medium-to-long shelf life products.

Perishable foods are those that must be held at chill or freezer temperatures (i.e., 0 to 7°C or −12 to −18°C, respectively) if they are to be kept for more than short periods. Examples of such foods include milk; fresh meat, poultry, and fish; minimally processed foods; and many fresh fruits and vegetables.

Semiperishable foods are those that contain natural inhibitors (e.g., some cheeses, root vegetables, and eggs), or those that have received some type of mild preservation treatment (e.g., pasteurization of milk, smoking of hams, and pickling of vegetables), which produces greater tolerance to environmental conditions and abuse during distribution and handling.

Shelf stable foods are considered "nonperishable" at room temperatures. Many unprocessed foods fall into this category, and are unaffected by microorganisms because of their low moisture content (e.g., cereal grains and nuts, and some confectionery products). Processed food products can be shelf stable if they are preserved by heat sterilization (e.g., canned foods), contain preservatives (e.g., soft drinks), are formulated as dry mixes (e.g., cake mixes) or processed to reduce their water content (e.g., raisins or crackers). However, shelf stable foods only retain this status if the integrity of the package that contains them remains intact. Even then, their shelf life is finite due to deteriorative chemical reactions that proceed at room temperature independently of the nature of the package, and the permeation through the package of gases, odors, and water vapor.

BULK DENSITY

For packages of similar shape, equal weights of products of different bulk densities will have different free space volumes, and as a consequence, package areas and package behavior will differ. This has important implications when changes are made in package size for the same product, or process alterations are made, resulting in changes to the product bulk density.

The bulk density of food powders can be affected by processing and packaging. Some food powders (e.g., milk and coffee) are instantized by treating individual particles so that they form free-flowing agglomerates or aggregates in which there are relatively few points of contact.

The free space volume has an important influence on the rate of oxidation of foods, because if a food is packaged in air, a large free space volume is undesirable since it constitutes a large oxygen reservoir. Conversely, if the product is packaged in an inert gas, a large free space volume acts as a large "sink" to minimize the effects of oxygen transferred through the package. It follows that a large package surface area and a low food bulk density result in greater oxygen transmission.

CONCENTRATION EFFECTS

The progress of a deteriorative reaction in a packaged food can be monitored by following the changes in concentrations of some key components. In many foods, however, the concentration varies from one point to another, even at zero time. Because most of these compounds have little opportunity to move, the concentration differences increase as the reactions proceed out from isolated initial foci.

Furthermore, several different deteriorative reactions may proceed simultaneously, and different stages may have different dependence on concentration and temperature. Such a situation is frequently the case for chain reactions and microbial growth that have both a lag and a log phase with very different rate constants.

Thus for many foods, it may be difficult to obtain kinetic data useful for predictive purposes. Sensory panels to determine the acceptability of the food is the recommended procedure.

PACKAGE PROPERTIES

Foods can be classified according to the degree of protection required, as shown in Table 16.1. The advantage of this sort of analysis is that attention can be focussed on the key requirements of the package such as maximum moisture gain or oxygen uptake. This then enables calculations to be made to determine whether or not a particular package structure would provide the necessary barrier required to give the desired product shelf life. Metal cans and glass containers can be regarded as essentially impermeable, while paper-based packaging materials can be regarded as permeable. Plastics-based packaging materials provide varying degrees of protection, depending largely on the nature of the polymers and their package structures.

The expression for the steady state permeation of a gas or vapor through a thermoplastic material can be written as:

$$\frac{\delta w}{\delta t} = \frac{P}{\delta t} \frac{A}{X} (p_1 - p_2) \tag{16.1}$$

where P/X is the permeance (the permeability constant P divided by the thickness of the film X), A is the surface area of the package, p_1 and p_2 the partial pressures of water vapor outside and inside the package, and $\delta w/\delta t$ is the rate of gas or vapor transport across the film, the latter term corresponding to Q/t in the integrated form of the expression.

WATER VAPOR TRANSFER

The prediction of moisture transfer either to or from a packaged food requires analysis of Equation 16.1 given certain boundary conditions. The simplest analysis requires

TABLE 16.1
Degree of Protection Required by Various Foods and Beverages (Assuming 1-year Shelf Life at 25°C)

Food/Beverage	Maximum Amount of O_2 Gain (ppm)	Other Gas Protection Needed	Maximum Water Gain or Loss	Requires High Oil Resistance	Requires Good Barrier to Volatile Organics
Canned milk and flesh foods	1–5	No	3% loss	Yes	No
Baby foods	1–5	No	3% loss	Yes	Yes
Beers and wine	1–5	<20% CO_2 (or SO_2) loss	3% loss	No	Yes
Instant coffee	1–5	No	2% gain	Yes	Yes
Canned soups, vegetables, and sauces	1–5	No	3% loss	No	No
Canned fruits	5–15	No	3% loss	No	Yes
Nuts, snacks	5–15	No	5% gain	Yes	No
Dried foods	5–15	No	1% gain	No	No
Fruit juices and drinks	10–40	No	3% loss	No	Yes
Carbonated soft drinks	10–40	<20% CO_2 loss	3% loss	No	Yes
Oils and shortenings	50–200	No	10% gain	Yes	No
Salad dressings	50–200	No	10% gain	Yes	Yes
Jams, jellies, syrups, pickles, olives, and vinegars	50–200	No	10% gain	Yes	No
Liquors	50–200	No	3% loss	No	Yes
Condiments	50–200	No	1% gain	No	Yes
Peanut butter	50–200	No	10% gain	Yes	No

the assumptions that P/X is constant, that the external environment is at constant temperature and humidity, and that p_2, the vapor pressure of the water in the food, follows some simple function of the moisture content.

External conditions will not remain constant during storage, distribution, and retailing of a packaged food. Therefore, P/X will not be constant. However, using Water Vapor Transmission Rates (WVTRs) determined at 38°C (100°F)/90% Relative Humidity (RH) gives a "worst-case" analysis, but if the food is being sold in markets in temperate climates, use of WVTR's determined at 25°C (77°F)/75% RH would be more appropriate. WVTRs can be converted to permeances by dividing by Δp.

A further assumption is that the moisture gradient inside the package is negligible, that is, the package should be the major resistance to water vapor transport. This is the case whenever P/X is less than about 10 gm^{-2} day^{-1} $(cm\ Hg)^{-1}$, which is the case for most films but not paperboard under high humidity conditions.

The critical point about Equation 16.1 is that the internal vapor pressure is not constant but varies with the moisture content of the food at any time. Thus the rate of gain or loss of moisture is not constant but falls as Δp gets smaller. Therefore some function of p_2, the internal vapor pressure, as a function of the moisture content, must be inserted into the equation to be able to make proper predictions. If a constant rate is assumed, the product will be overprotected.

In low- and intermediate-moisture foods, the internal vapor pressure is determined solely by the moisture sorption isotherm of the food. Several functions can be applied to describe a sorption isotherm, although the preferred one is the GAB (from Guggenheim-Anderson-de Boer) model. If a linear model is used, the result can be integrated directly, but if the GAB model is used, it must be numerically evaluated.

In the simplest case, the isotherm is treated as a linear function:

$$m = b_{aw} + c \tag{16.2}$$

where m = moisture content in g H_2O per g solids; aw = water activity; b = slope of curve, and c = constant.

The moisture content can be substituted for water gain using the relationship:

$$m = \frac{W_{\text{(weight of water transported)}}}{W_{\text{s(weight of dry solids enclosed)}}} \tag{16.3}$$

$$\therefore W = mW_s \tag{16.4}$$

and:

$$\delta W = \delta m\ W_s \tag{16.5}$$

By substitution,

$$\frac{\delta W}{\delta t} = \frac{\delta m W_s}{\delta t} = \frac{P}{X} \bullet \quad A \bullet \quad \left[\frac{p_0 m_e}{b} - \frac{p_0 m}{b}\right] \tag{16.6}$$

which on rearranging gives:

$$\frac{\delta m}{m_e - m} = \frac{P}{X} \bullet \frac{A}{W_s} \bullet \frac{P_0}{b} \bullet \delta t \tag{16.7}$$

and on integrating:

$$\text{In} \frac{m_e - m_i}{m_e - m} = \frac{P}{X} \frac{A}{W_s} \frac{P_0}{b} t \tag{16.8}$$

where m_e = equilibrium moisture content of the food if exposed to external package RH; m_i = initial moisture content of the food; m = moisture content of the food at time t; p_o = vapor pressure of pure water at the storage temperature (NOT the actual vapor pressure outside the package).

A plot of the log of the unaccomplished moisture change (the term on the left-hand side [LHS] of Equation 16.8) vs. time is a straight line with a slope equivalent to the bracketed term on the right-hand side of the equation.

The end of product shelf life is reached when $m = m_c$, the critical moisture content, at which time $t = \theta_s$, the shelf life. Thus, Equation 16.8 can be rewritten as:

$$\text{In} \frac{m_e - m_i}{m_e - m_c} = \frac{P}{X} \frac{A}{W_s} \frac{P_0}{b} \theta_s \tag{16.9}$$

The relationship between the initial, critical, and equilibrium moisture contents for a snack bar is illustrated in Figure 16.1.

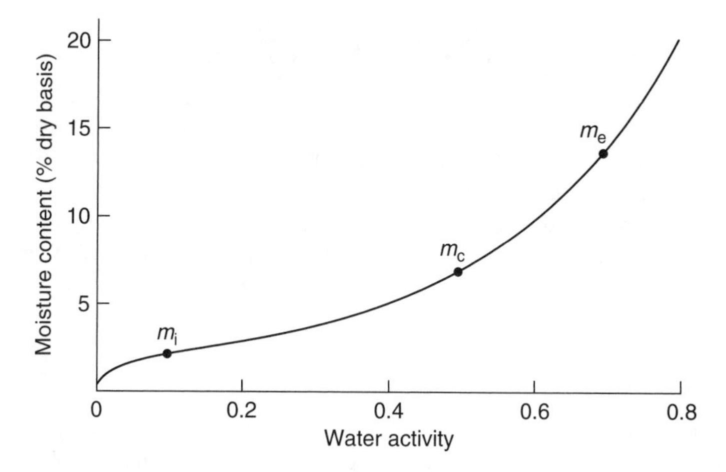

FIGURE 16.1 Typical moisture sorption isotherm for a snack bar, where mi = initial moisture content; mc = critical moisture content of product; me = equilibrium moisture content.

To simplify matters, the packaging parameters can be combined into one constant as:

$$\Omega = \frac{P}{X}\frac{A}{W_s} \tag{16.10}$$

Using Equation 16.10, one can calculate a minimum Ω, given a critical moisture content and maximum desired shelf life. Then from Equation 16.9 for a given package size and weight of product, the permeance can be calculated and a packaging film(s) to satisfy this condition selected.

Equation 16.8 and the corresponding one for moisture loss:

$$\text{In}\frac{m_{\text{i}} - m_{\text{e}}}{m - m_{\text{e}}} = \frac{P}{X}\frac{A}{W_s}\frac{p_0}{b}t \tag{16.11}$$

have been extensively tested for foods and found to give excellent predictions of actual weight gain or loss. These equations are also useful when calculating the effect of changes in the external conditions (e.g., temperature and humidity), the surface area-to-volume ratio of the package, and variations in the initial moisture content of the product.

Given specific external conditions and a critical a_{w} for moisture gain, the shelf life is:

$$\theta = \Phi\frac{W_{\text{S}}}{A} = \Phi'\frac{\Phi''}{A}\frac{V}{r} \tag{16.12}$$

where Φ, Φ', and Φ'' are constants proportional to

$$\left[\text{In}\frac{m_{\text{e}} - m_{\text{i}}}{m_{\text{e}} - m}\right] + \left[\frac{P}{X}\frac{p_0}{b}\right] \tag{16.13}$$

where W_{s} = weight of food solids = $\rho \times V$, ρ = density of food, V = volume of food, r = characteristic package thickness, and θ_{s} = time to end of shelf life.

Because the $V:A$ ratio decreases as package size gets smaller by a factor equivalent to the characteristic thickness of the package, the shelf life using the same film will decrease directly by this thickness. Thus to ensure adequate shelf life for a food in varying sizes of packages, shelf life tests should be based on the smallest package.

Gas and Odor Transfer

The gas of major importance in packaged foods is oxygen since it plays a crucial role in many reactions that affect the shelf life of foods, for example, microbial growth, color changes in fresh and cured meats, oxidation of lipids and consequent rancidity, and senescence of fruits and vegetables.

The transfer of gases and odors through packaging materials can be analyzed in an analogous manner to that described for moisture vapor transfer, provided that values are known for the permeance of the packaging material to the appropriate gas, and the partial pressure of the gas inside and outside the package.

Packaging can control two variables with respect to oxygen, and these can have different effects on the rates of oxidation reactions in foods:

1. Total amount of oxygen present. This influences the extent of the reaction, and in impermeable packages (e.g., hermetically sealed metal and glass containers), where the total amount of oxygen available to react with the food is finite, the extent of the reaction cannot exceed the amount corresponding to the complete exhaustion of the oxygen present inside the package at the time of sealing. This may or may not be sufficient to result in an unacceptable product quality after a certain period of time dependent on the rate of the oxidation reaction. Such a rate is, of course, temperature dependent. With permeable packages (e.g., plastic packages), where ingress of oxygen occurs during distribution, two factors are important: sufficient oxygen may be present inside the package to cause product unacceptability when it has all reacted with the food, or there may be sufficient transfer of oxygen through the package over time to result in product unacceptability through oxidation.
2. Concentration of oxygen in the food. In many cases, relationships between the oxygen partial pressure in the space surrounding the food and the rates of oxidation reactions can be established. If the food itself is very resistant to diffusion of oxygen, then it will probably be very difficult to establish a relationship between the oxygen partial pressure in the space surrounding the food and the concentration of oxygen in the food.

The principal difference between predominantly water vapor-sensitive and oxygen-sensitive foods is that the latter are generally more sensitive by 2–4 orders of magnitude. Thus, the amount of oxygen present in the air-filled headspace of oxygen-sensitive foods must not be neglected when predicting their shelf life. This amount is actually 32 times higher per unit volume of air than per unit volume of oxygen-saturated water. A further complicating factor with oxygen-sensitive foods is that a concentration gradient occurs in them much more frequently than in moisture-sensitive foods.

Prediction of the shelf life of food products that deteriorate by two or more mechanisms simultaneously (e.g., oxidation due to ingress of oxygen and loss of crispness due to ingress of moisture) is more complex. Some general approaches that can be applied have been proposed the amount of data necessary to develop the equations required for predictive purposes are such that the food industry is generally unable to afford to utilize such an approach to shelf life testing. Reliance is being placed on the use of accelerated shelf life testing (ASLT) procedures as a more cost-effective and simpler method for the determination of product shelf life.

DISTRIBUTION ENVIRONMENT

CLIMATIC

The deterioration in product quality of packaged foods is often closely related to the transfer of mass and heat through the package. Packaged foods may lose or gain moisture; they will also reflect the temperature of their environment, because very few food packages are good thermal insulators. Thus, the distribution environment has an important influence on the rate of deterioration of packaged foods.

MASS TRANSFER

With mass transfer, the exchange of vapors and gases with the surrounding atmosphere is of primary concern. Water vapor and oxygen are generally of most importance, although the exchange of volatile aromas from or to the product from the surroundings can be important. Transmission of nitrogen and carbon dioxide may have to be taken into account in some packages, for example, in modified atmosphere packaging.

Generally, the difference in partial pressure of the vapor or gas across the package barrier will control the rate and extent of permeation, although transfer can also occur due to the presence of pinholes in the material, channels in seals and closures, or cracks that result from flexing of the package material during filling and subsequent handling. In contrast to the common gases, the partial pressure of water vapor in the atmosphere varies continuously, although the variation is generally much less in controlled climate stores. Thus, mass transfer depends on the partial pressure difference across the package barrier, and on the nature of the barrier itself.

HEAT TRANSFER

One of the major determinants of product shelf life is the temperature to which the product is exposed during the time from production to consumption. Without exception, food products are exposed to fluctuating temperature environments, and it is important, if an accurate estimation of shelf life is to be made, that the nature and extent of these temperature fluctuations are known. There is little point in carefully controlling the processing conditions inside the factory and then releasing the product into the distribution and retail system without knowledge of the conditions that it will experience in that system. The storage climates inside buildings such as warehouses and supermarkets are only broadly related to the external climate.

If the major deteriorative reaction causing end of shelf life is known, expressions can be derived to predict the extent of deterioration as a function of available time–temperature storage conditions. Fundamental to such an analysis is that the particular food under consideration follows the "laws" of "additivity" and "commutativity." Additivity implies that the total extent of the degradation reaction in the food produced by a succession of exposures at various temperatures is the simple sum of the separate amounts of degradation, regardless of the number or spacing of each time-temperature combination. Commutativity means that the total extent of the degradation reaction in the food is independent of the order of presentation of the various time-temperature experiences.

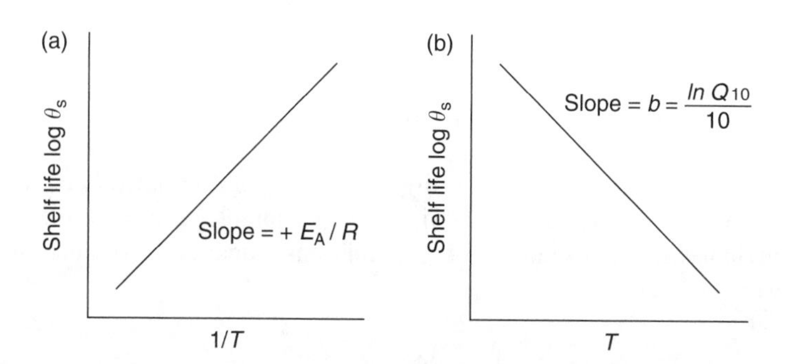

FIGURE 16.2 Arrhenius plot of log shelf life (θ_S) versus reciprocal of the absolute temperature (K) showing a slope of EA/R, and (b) linear plot of log shelf life versus temperature (°C) showing a slope of b.

Shelf Life Plots

One useful approach to quantifying the effect of temperature on food quality is to construct shelf life plots. Several models are in use to represent the relationship between the rate of a reaction (or the reciprocal of rate that can be time for a specified loss in quality or shelf life) and temperature. The two most common models are the Arrhenius and linear, and these are shown in Figure 16.2.

The equations for these two plots are:

$$\theta_s = \theta_0 \exp \frac{EA}{R}\left[\frac{1}{Ts}\frac{1}{T_0}\right] \qquad (16.14)$$

and:

$$\theta_s = \theta_s e{-}^{b(T_s - T_0)} \qquad (16.15)$$

where θ_s = shelf life at temperature T_s and θ_0 = shelf life at temperature T_O.

If only a small temperature range is used (less than ±20°C or ±40°F), there is little error in using the linear plot rather than the Arrhenius plot.

Most deteriorative reactions in foods can be classified as either zero- or first-order reaction. The way in which these two reaction orders can be used to predict the extent of deterioration as a function of temperature will be outlined.

Zero-Order Reaction Prediction

The change in a quality factor A when all extrinsic factors are held constant is expressed in Equation 16.16:

$$A_e = A_0 - k_z\theta_s \qquad (16.16)$$

and:

$$A_0 - A_e = k_z\theta_s \tag{16.17}$$

where A_e = value of A at end of shelf life, A_0 = value of A initially, k_z = zero-order rate constant (time^{-1}), and θ_s = shelf life in days, months, years, and so forth.

For variable time-temperature storage conditions, Equation 16.16 can be modified as follows:

$$A_e = A_0 - \sum(k_i\theta_i) \tag{16.18}$$

where $\Sigma k_i\theta_i$ = the sum of the product of the rate constant k_i at each temperature T_i times the time interval θ_i at the average temperature T_i for the given time period $\Delta\theta$.

To apply this method, the time-temperature history is broken up into suitable time periods and the average temperature in that time period determined. The rate constant for that period is then calculated from the shelf life plot using a zero-order reaction. The rate constant is multiplied by the time interval θ_i, and the sum of the increments of $k_i\theta_i$ gives the total amount lost at any time.

Alternatively, instead of calculating actual rate constants, the time for the product to become unacceptable (i.e., for A to become A_e) can be measured, and Equation 16.18 modified to give:

$$f_c = \text{fraction of shelf life consumed}$$

$$= \text{change in A divided by total possible change in } A$$

$$= \frac{A_0 - A}{A_0 - A} \tag{16.19}$$

$$= \frac{\sum(k_i\theta_i)}{\sum(k_i\theta_s)} \tag{16.20}$$

$$= \sum\left[\frac{\theta_i}{\theta_s}\right]T_i \tag{16.21}$$

The temperature history is divided into suitable time periods and the average temperature T_i at each time period evaluated. The time held at that temperature θ_i is then divided by the shelf life θ_s for that particular temperature, and the fractional values summed up to give the fraction of shelf life consumed.

The shelf life can also be expressed in terms of the fraction of shelf life remaining, f_r:

$$f_r = 1 - f_c. \tag{16.22}$$

Thus, for any temperature T_s,

$$f_r\theta_s = (1 - f_c)\theta_s = \text{shelf life left at temperature } T_s \tag{16.23}$$

In other words, the shelf life left at any temperature is the fraction of shelf life remaining times the shelf life at that temperature.

The above method is referred to as the TTT or time-temperature-tolerance approach. To use this method, the period of time (designated as the high-quality life or HQL) for 70–80% of a trained sensory panel to correctly identify the control samples from samples stored at various other temperatures using the triangle or duo-trio test was determined. The change in quality at this stage has been designated the "just noticeable difference" (JND). The HQL has no real commercial significance and is quite different from the "practical storage life" (PSL) that is of interest to food processors and consumers. The ratio between PSL and HQL is often referred to as the "acceptability factor" and can range from 2:1 up to 6:1.

Generally, the HQL varies exponentially with temperature. When overall quality rather than just one single quality factor is measured, however, a semilogarithmic plot results in curved rather than straight lines.

TTT relationships are not strict mathematical functions but empirical data subject to large variability, particularly because of variations in product, processing methods, and packaging (the PPP factors). Therefore, any shelf life prediction made will be specific for a particular product processed, packaged, and stored under specific conditions. Predictions cannot be made with any precision on the quality or quality change in a food from knowledge of its time-temperature history and TTT literature data only. Therefore, in determining the shelf life of foods, the PPP factors must be taken into account in addition to the TTT relationships.

First-Order Reaction Prediction

The expression for a first-order reaction for the case in which all extrinsic factors are held constant is:

$$A_e = A_0 \exp(-k\theta s) \tag{16.24}$$

or:

$$\text{In } A_e = \text{In } A_0 - k\theta_s \tag{16.25}$$

From this an expression can be developed to predict the amount of shelf life used up as a function of variable temperature storage for a first-order reaction in the form:

$$A = A_0 \exp\left(-\sum k_i\theta_i\right) \tag{16.26}$$

where A = the amount of some quality factor remaining at the end of the time-temperature distribution, and $\Sigma k_i\theta_i$ has the same meaning as in Equation 16.18.

If the shelf life is based simply on some time to reach unacceptability, Equation 16.26 can be modified to give an analogous expression to that derived for the TTT method. Note that because of the exponential loss of quality, A_e will never be zero. Thus,

$$\text{In}\,\frac{A}{A_0} = -\sum k_i\theta_i \tag{16.27}$$

and:

$$k_i = \frac{\text{In}\,^{A_e}}{A_0} \tag{16.28}$$

where $\ln A/A_o$ = fraction of shelf life consumed at time θ and $\ln A_e/A_o$ = fraction of shelf life consumed at time θs.

The fraction of shelf life remaining, f_r is:

$$f_r = 1\frac{\text{In}\,^{A_0}/A}{\text{In}\,^{A_0}/A_e} = 1 - \Sigma\left[\frac{\theta_i}{\theta_s}\right]T_i \tag{16.29}$$

SEQUENTIAL FLUCTUATION TEMPERATURES

Although the above analysis can be applied to any random time-temperature storage regime, in practice, many products are exposed to a sequential regular fluctuating temperature profile, especially if held in trucks, rail cars and uninsulated warehouses. This is because of the daily day-night pattern resulting from exposure to solar radiation. Many of these patterns can be assumed to follow either a square or sine wave form.

Equations have been developed for both zero-order and first-order reactions that enable calculation of the extent of a degradative reaction for a food subjected to either square wave or sine wave temperature functions. It can be shown that the extent of reaction after a period of time will be the same as it would have been if the food had been held at a certain steady "effective" temperature for the same length of time. This effective temperature will be higher than the arithmetic mean temperature. Comparisons for losses in a theoretical temperature distribution showed that for less than 50% degradation, the losses were about the same for zero- and first-order reaction at any time, and thus determination of the reaction order is not critical. However, the temperature sensitivity (Q_{10}) of the reaction is very important in making predictions.

SIMULTANEOUS MASS AND HEAT TRANSFER

In the majority of distribution environments, many packaged foods undergo changes in both moisture content and temperature during storage as a result of variable temperature and humidity conditions in the environment. This has the effect of complicating the calculations for prediction of shelf life of packaged foods.It is unlikely that a package would be totally impermeable to water vapor, and therefore a_W would change with time. This complicates the calculation of quality loss, since the rate is now dependent on both temperature and a_W.

A further complication is that data on the humidity distribution of environments where foods are stored are scarce and not as easily predicted as the external temperature distribution. Therefore, prediction of the actual shelf life loss of packaged foods will only be approximate. More complete data about the humidity distribution of food storage environments is required, so that shelf life predictions can be further refined.

SHELF LIFE ESTIMATION

There are at least three situations when shelf life estimation might be required:

1. To determine the shelf life of existing products;
2. To study the effect of specific factors and combinations of factors such as storage temperature, package materials, processing parameters, or food additives on product shelf life; and
3. To determine the shelf life of prototype or newly developed products.

Several established approaches are available for estimating the shelf life of foods:

1. *Literature study*: the shelf life of an analogous product is obtained from the published literature or in-house company files. Examples can be found in recent books on the shelf life of foods;
2. *Turnover time*: the average length of time that a product spends on the retail shelf is found by monitoring sales from retail outlets, and from this the required shelf life is estimated. This does not give the "true" shelf life of the product but rather the "required" shelf life, it being implicitly assumed that the product is still acceptable for some time after the average period on the retail shelf;
3. *End point study*: random samples of the product are purchased from retail outlets and then tested in the laboratory to determine their quality; from this a reasonable estimation of shelf life can be obtained, since the product has been exposed to actual environmental stresses encountered during warehousing and retailing;
4. *Accelerated shelf life testing*: laboratory studies are undertaken during which environmental conditions are accelerated by a known factor, so that the product deteriorates at a faster than normal rate. This method requires that the effect of environmental conditions on product shelf life can be quantified.

Regardless of the method chosen or the reasons for its choice, sensory evaluation of the product is likely to be used either alone or in combination with instrumental or chemical analyses to determine the quality of the product. Because human judgment is the ultimate arbiter of food acceptability, it is essential that the results obtained from any instrumental or chemical analysis correlate closely with the sensory judgments for which they are to substitute.

Correlation of values of individual chemical parameters with sensory data is often not straightforward because overall organoleptic quality is a composite of a number of changing factors. The relative contribution of each factor to the overall quality may vary at different levels of quality or at different storage conditions. Other problems with sensory evaluation include the high cost of using large testing panels and the ethics of asking panelists to taste spoiled or potentially hazardous samples.

Three experimental designs are commonly used for the purpose of shelf life estimation: the paired comparison test; the duo-trio test, and the triangle test. Further details about these tests can be found in standard texts on sensory evaluation. Descriptive methods are used to measure quantitative and/or qualitative characteristics of products and require specially trained panelists. Affective methods are used to evaluate preference, acceptance, and/or opinions of products and do not require trained panelists (see Chapter 13).

The selection of a particular sensory evaluation procedure for evaluating products undergoing shelf life testing is dependent on the purpose of the test. Acceptability assessments by untrained panelists are essential to an open dating program, while discrimination testing with expert panels might be used to determine the effect of a new package material on product stability. However, an expert panel is not necessarily representative of consumers, much less different consumer segments. Even if that assumption can be made, a cut-off level of acceptability has to be decided. The time at which a large (but predetermined) percentage of panelists (e.g., 50%) judge the food to be at or beyond that level is the end of shelf life.

In shelf life testing, there can be one or more criteria that constitute sample failure. One criterion is an increase or decrease by a specified amount in the mean panel score. Another criterion is microbial deterioration of the sample to an extent that renders it unsuitable or unsafe for human consumption. Finally, changes in odor, color, texture, flavor, and so forth, that render the sample unacceptable to either the panel or the consumer, are criteria for product failure. Thus, sample failure can be defined as the condition when the product exhibits either physical, chemical, microbiological, or sensory characteristics that are unacceptable to the consumer, and the time required for the product to exhibit such conditions is the shelf life of the product.

However, a fundamental requirement in the analysis of data is knowledge of the statistical distribution of the observations, so that the mean time to failure and its standard deviation can be accurately estimated, and the probability of future failures predicted. The length of shelf life for food products is usually obtained from simple averages of time to failure on the assumption that the failure distribution is symmetrical. If the distribution is skewed, estimates of the mean time to failure and its standard deviation will be biased. Furthermore, when the experiment is terminated before all the samples have failed, the mean time to failure based on simple averages will be biased because of the inclusion of unfailed data.

In order to improve the methodology for estimating shelf life, knowledge of the statistical distribution of shelf life failures is required, together with an appropriate model for data analysis. Five statistical models—normal, lognormal, exponential, Weibull and extreme-value distributions—have been fitted to failure data using the method of hazard plotting that provides information about the adequacy of fit of the observed data to the proposed model, the mean or median time to failure, and the probability of future failures. The Weibull distribution was suggested as the most appropriate shelf life model, and it has been used to predict end of shelf life for foods.

One challenge with shelf life testing is to develop experimental designs that minimize the number of samples required (thus minimizing the cost of the testing) while still providing reliable and statistically valid answers.

ACCELERATED SHELF LIFE TESTING (ASLT)

Basic Principles

The basic assumption underlying ASLT is that the principles of chemical kinetics can be applied to quantify the effects which extrinsic factors such as temperature, humidity, gas atmosphere, and light have on the rate of deteriorative reactions. By subjecting the food to controlled environments in which one or more of the extrinsic factors is maintained at a higher than normal level, the rates of deterioration will be speeded up or accelerated, resulting in a shorter than normal time to product failure. Because the effects of extrinsic factors on deterioration can be quantified, the magnitude of the acceleration can be calculated and the true shelf life of the product under normal conditions calculated. Thus, a shelf life test that would normally take a year can be completed in about a month if the storage temperature is raised by 20°C (40°F).

The need for ASLT of food products is simple: since many foods have shelf lives of at least 1 year, evaluating the effect on shelf life of a change in the product, the process, or the packaging would require shelf life trials lasting at least as long as the required shelf life of the product. Companies cannot afford to wait for such long periods before knowing whether or not the new product/process/packaging will give an adequate shelf life, because other decisions have lead times of months and/or years. The use of ASLT in the food industry is not as widespread as it might be, due in part to the lack of basic data on the effect of extrinsic factors on the rates of deteriorative reactions, in part to ignorance of the methodology required, and in part to a skepticism of the advantages to be gained from using ASLT procedures.

Quality loss for most foods follows either a zero-order or first-order reaction. Figure 16.3 showed the logarithm of shelf life versus temperature and the inverse of absolute temperature. If only a small range of temperature is considered, the former shelf life plot generally fits the data for food products.

For a given extent of deterioration and reaction order, the rate constant is inversely proportional to the time to reach some degree of quality loss. Thus by taking the ratio of the shelf life between any two temperatures 10°C (18°F) apart, the Q_{10} of the reaction can be found. This can be expressed by Equation 16.26 assuming a linear

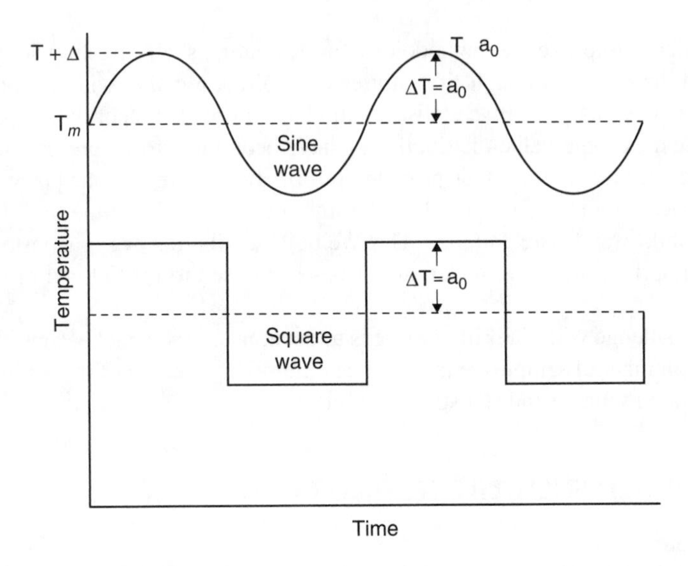

FIGURE 16.3 Square and sine wave temperature fluctuations of packaged foods where a_0 is the amplitude.

shelf life plot:

$$Q_{10} = \frac{k_{T+10}}{k_T} = \frac{\theta_{s_T}}{\theta_{s_{T+10}}} \quad (16.30)$$

where:

θ_{s_T} = shelf life at temperature T°C and

$\theta_{s_{T+10}}$ = shelf life at temperature $(T + 10)$°C.

The effect of Q_{10} on shelf life is shown in Table 16.2, which illustrates the importance of accurate estimates of Q_{10} when making shelf life estimations. For example, if a product has a shelf life of 2 weeks at 50°C and a Q_{10} of 2, it has a shelf life of 16 weeks at 20°C. However, if Q_{10} were 2.5 rather than 2, the shelf life at 20°C

TABLE 16.2
Effect of Q_{10} on Shelf Lives

	Shelf Life (Weeks)			
Temperature (°C)	$Q_{10} = 2$	$Q_{10} = 2.5$	$Q_{10} = 3$	$Q_{10} = 5$
50	2*	2*	2*	2*
40	4	5	6	10
30	8	12.5	18	50
20	16	31.3	54	4.8 years

* Arbitrarily set at 2 weeks at 50°C. Shelf lives at lower temperatures are calculated on this.

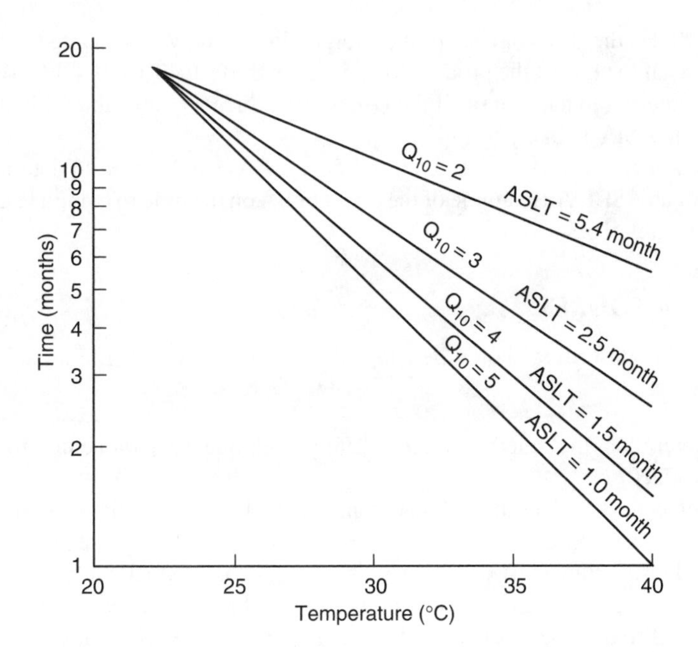

FIGURE 16.4 Hypothetical shelf life plot for various Q_{10}'s passing through a shelf life of 18 months at 23°C. Accelerated shelf life times (ASLT) are those required at 40°C for various Q_{10}'s.

would be almost twice as long (31 weeks). Thus, a small error in Q_{10} can lead to huge differences in the estimated shelf life of the product. Typical Q_{10} values for foods are 1.1–4 for canned products, 1.5–10 for dehydrated products, and 3–40 for frozen products.

A further use for Q_{10} values is illustrated in Figure 16.4 that depicts a shelf life plot for a product that has at least 18 months shelf life at 23°C (73°F). To determine what the shelf life of the product is likely to be at 40°C (104°F), lines are drawn from the point corresponding to 18 months at 23°C (73°F) to intersect a vertical line drawn

TABLE 16.3
Effect of EA of the Key Deteriorative Reaction on the Time to Complete an ASLT Test for a Low-Moisture Food Product with a Targeted Shelf Life of 2 Years at Ambient Storage

E_A (kJ mol^{-1})	Testing Time at 40°C (Days)	Testing Time at 45°C (Days)
45	224	171
85	78	47
125	28	13

at 40°C (73°F); the slope of each of the straight lines so drawn is dictated by the Q_{10} value. Thus, if the Q_{10} of the product were 5, its shelf life at 40°C (104°F) would be 1 month, increasing to 5.4 months if the Q_{10} were 2. Such a plot is helpful in deciding how long an ASLT is likely to run.

Instead of Q_{10} values, the Activation Energy (EA) can be used to determine the duration of an ASLT. An example of the effect of EA on the time to complete an ASLT is given in Table 16.3.

ASLT PROCEDURES

The following procedure should be adopted in designing a shelf life test for a food product:

1. Determine the microbiological safety and quality parameters for the product.
2. Select the key deteriorative reaction(s) that will cause quality loss and thus consumer unacceptability in the product, and decide what tests (sensory and/or instrumental) should be performed on the product during the trial.
3. Select the package to be used; often a range of packaging materials will be tested so that the most cost-effective material can be selected.
4. Select the extrinsic factors that are to be accelerated. Typical storage temperatures used for ASLT procedures are shown in Table 16.4 and it is usually necessary to select at least two.
5. Using a plot similar to that shown in Figure 16.4, determine how long the product must be held at each test temperature. If no Q_{10} values are known, then an open-ended ASLT will have to be conducted using a minimum of three test temperatures.
6. Determine the frequency of the tests. A good rule of thumb is that the time interval between tests at any temperature below the highest temperature should be no longer than:

$$f_z = f_1 Q_{10}^{\Delta T/10} \tag{16.31}$$

TABLE 16.4
Experimental Design for Accelerated Shelf Life Tests for Various Food Categories

Product	Test Temperature (°C)	Control (°C)
Frozen	−7, −11, −15	< −40
Chilled	5, 10, 15, 20	0
Dry and Intermediate Moisture Food (IMF)	25, 30, 35, 40, 45	−18
Canned	25, 30, 35, 40	4

where f_1 = the time between tests (e.g., days, weeks) at the highest test temperature T_1, f_2 = the time between tests at any lower temperature T_2, and ΔT = the difference in degrees Celsius between T_1 and T_2.

Thus, if a product is held at 40°C (104°F) and tested once a month, then at 30°C (86°F) with a Q_{10} of 3, the product should be tested at least every

$$f_2 = 1 \times 3^{(10/10)}$$
$$= 3 \text{ months.}$$

More frequent testing is desirable, especially if the $Q_{10}$is not accurately known, because at least six data points are needed to minimize statistical errors, otherwise the confidence in θs is significantly diminished.

7. Calculate the number of samples that must be stored at each test condition, including those samples that will be held as controls;
8. Begin the ASLTs, plotting the data as it comes to hand so that, if necessary, the frequency of sampling can be increased or decreased as appropriate.
9. From each test storage condition, estimate k or θs and construct appropriate shelf life plots from which to estimate the potential shelf life of the product under normal storage conditions. Provided that the shelf life plots indicate that the product shelf life is at least as long as that desired by the company, then the product has a chance of performing satisfactorily in the marketplace.

Problems in the Use of ASLT Conditions

The potential problems and theoretical errors that can arise in the use of ASLT conditions have been described as follows:

1. Errors in analytical or sensory evaluation. Generally, any analytical measure should be done with a variability of less than ±10% to minimize prediction errors.
2. As temperature rises, phase changes may occur (e.g., solid fat becomes liquid) that can accelerate certain reactions, with the result that at the lower temperature the actual shelf life will be longer than estimated.

3. Carbohydrates in the amorphous state may crystallize out at higher temperatures, with the result that the estimated shelf life is shorter than the actual shelf life at ambient conditions.
4. Freezing "control" samples can result in reactants being concentrated in the unfrozen liquid, creating a higher rate at the reduced temperature and thus confounding estimates.
5. If two reactions with different Q_{10} values cause quality loss in a food, the reaction with the higher Q_{10} may predominate at higher temperatures while at normal storage temperatures the reaction with the lower Q_{10} may predominate, thus confounding the estimation.
6. The a_w of dry foods can increase with temperature, causing an increase in reaction rate for products of low aw in sealed packages. This results in overestimation of true shelf life at the lower temperature.
7. The solubility of gases (especially oxygen in fat or water) decreases by almost 25% for each 10°C (18°F) rise in temperature. Thus, an oxidative reaction such as loss of ascorbic or linoleic acid can decrease in rate if oxygen availability is the limiting factor. Therefore at the higher temperature, the rate will be less than theoretical that in turn will result in an underprediction of true shelf life at the normal storage temperature.
8. If the product is not placed in a totally impermeable pouch, storage in high-temperature/low-humidity cabinets will generally enhance moisture loss, and this should decrease the rate of quality loss compared to no moisture change. This will result in a shorter estimated shelf life at the lower temperature.
9. If high enough temperatures are used, proteins may become denatured, resulting in both increases or decreases in the reaction of certain amino acid side chains, leading to either under- or overprediction of true shelf life.

Therefore, the use of ASLT to estimate actual shelf life can be limited except in the case of very simple chemical reactions. Consequently, food packaging technologists should always confirm the ASLT results for a particular food product by conducting shelf life tests under actual environmental conditions. Once a relationship between ASLT and actual shelf life has been established for a particular product, then ASLT can be used for that product when process or package variables are to be evaluated.

PREDICTING MICROBIAL SHELF LIFE

Microbial spoilage of food is an economically significant problem for food manufacturers, retailers, and consumers. Depending on the product, process, and storage conditions, the microbiological shelf life can be determined by either the growth of spoilage or pathogenic microorganisms. In the case of spoilage microorganisms, the traditional method for determining microbiological shelf life involved storing the product at different temperatures and determining spoilage by sensory evaluation or microbial count. Where the microbiological shelf life is determined by the growth of pathogenic microorganisms, challenge testing of the product with the organism of concern followed by storage at different temperatures and microbial analysis at intervals has been the traditional approach. For processes such as heat treatments where the elimination of particular microorganisms is required (e.g., canning), the use of inoculated packs is common.

Over recent years, the development and commercialization of predictive models has become relatively widespread. The use of such models can reduce the need for shelf life trials, challenge tests, product reformulations, and process modifications, thus saving both time and money. Although there are both mechanistic and empirical predictive models, it is the latter that predominate. The ultimate test for predictive models is whether they can be used to predict reliable outcomes in real situations. For a detailed discussion the reader is referred to the standard text in this area.

Predictive models have been used to determine the likely shelf life of perishable foods such as meat, fish, and milk and recent publications in the area have been reviewed, together with a discussion of the limitations of such models. Despite their increasing sophistication and widespread availability, models should not be relied on completely but rather be used as a tool to assist decision making. Models do not completely negate the need for microbial testing, nor do they replace the judgment of a trained and experienced food microbiologist.

BIBLIOGRAPHY

Blackburn, C.deW., Modelling shelf-life, In *The Stability and Shelf-Life of Food*, Kilcast, D. and Subramanian, P., Eds., CRC Press, Boca Raton, Florida, 2000, chap. 3.

Cardelli, C. and Labuza, T.P., Application of Weibull hazard analysis to the determination of the shelf life of roasted coffee, *Lebensm. Wiss. u Technol.*, 34, 273, 2001.

Dens, E.J. and Van Impe, J.F., Modelling applied to foods: predictive microbiology for solid food systems, In *Food Preservation Techniques*, Zeuthen, P. and Bøgh-Sørensen, L., Eds., CRC Press, Boca Raton, Florida 2004, chap. 21.

Eskin, N.A.M. and Robinson, D.S., Eds., *Food Shelf Life Stability: Chemical, Biochemical and Microbiological Changes*, CRC Press, Boca Raton, Florida, 2001.

Hough, G., Langohr, K., Gómez, G., and Curia, A., Survival analysis applied to sensory shelf life of foods, *J. Food Sci.*, 68, 359, 2003.

Kilcast, D. and Subramaniam, P., Eds., *The Stability and Shelf-Life of Food*, CRC Press, Boca Raton, Florida, 2000.

Labuza, T.P. and Kamman, J.F., Reaction kinetics and accelerated tests simulation as a function of temperature, in Computer-Aided Techniques In *Food Technology*, Saguy, I., Ed., Marcel Dekker, New York, 1983, chap. 4.

Labuza, T.P. and Schmidl, M.K., Accelerated shelf life testing of foods, *Food Technol.*, 39(9), 57, 1985.

Labuza, T.P. and Schmidl, M.K., Use of sensory data in the shelf life testing of foods: principles and graphical methods for evaluation, *Cereal Foods World*, 33, 193, 1988.

Labuza, T.P. and Szybist, L.M., Playing the open dating game, *Food Technol.*, 53(7), 70, 1999.

Man, C.M.D., Shelf-life testing, In *Understanding and Measuring the Shelf-Life of Food*, Steele, R., Ed., CRC Press, Boca Raton, Florida, 2004, chap. 15.

Man, C.M.D. and Jones, A.A., Eds., *Shelf-Life Evaluation of Foods*, 2nd ed., Aspen Publishers, Gaithersburg, Maryland, 2000.

McMeekin, T.A., Olley, J.N., Ross, T., and Ratkowsky, D.A., *Predictive Microbiology: Theory and Application*, John Wiley, New York, 1993.

Meilgaard, M.C., Civille, G.V., and Carr, B.T., *Sensory Evaluation Techniques*, 3rd ed., CRC Press, Boca Raton, Florida, 1999.

Mizrahi, S. Accelerated shelf-life tests, In *Understanding and Measuring the Shelf-Life of Food*, Steele, R., Ed., CRC Press, Boca Raton, Florida, 2004, schap. 14.

Robertson, G.L., *Food Packaging: Principles and Practice*, 2nd ed., Marcel Dekker, New York, 2005.

Salame, M., The use of low permeation thermoplastics in food and beverage packaging, In *Permeability of Plastic Films and Coatings*, Hopfenberg, H.B., Ed., Plenum Press, New York, 1974, 275.

Singh, T.K. and Cadwallader, K.R., The shelf life of foods: an overview, In *Freshness and Shelf Life of Foods, ACS Symposium Series #836*, Cadwallader, K.R. and Weenen, H., Eds., American Chemical Society, Washington, DC, 2003, chap. 1.

Taoukis, P.S. and Giannakourou, M.C., Temperature and food stability: analysis and control, In *Understanding and Measuring the Shelf-Life of Food*, Steele, R., Ed., CRC Press, Boca Raton, Florida, 2004, chap. 3.

Taoukis, P.S., Labuza, T.P., and Saguy, I.S., Kinetics of food deterioration and shelf-life prediction, In *Handbook of Food Engineering Practice*, Valenta, K.J., Rotstein, E., and Singh, R.P., Eds., CRC Press, Boca Raton, Florida, 1997, chap. 9.

Van Arsdel, W.B., Estimating quality change from a known temperature history, In *Quality and Stability of Frozen Foods*, Van Arsdel, W.B., Copley, M.J., and Olson, R.L., Eds., Wiley, New York, 1969, chap. 10.

17 Toward the Development of an Integrated Packaging Design Methodology: Quality Function Deployment—An Introduction and Example

Stephen A. Raper

CONTENTS

Although Quality Function Deployment (QFD) is more typically used for product development, the methodology is equally as versatile for use as a packaging design/development tool. QFD is essentially a method of "mapping" the elements, events, and activities necessary throughout the development process to achieve customer satisfaction—a techniques-oriented approach using surveys, reviews, analyses, and robust design all centered on the theme of translating the "voice of the customer" into items that can be measured, assessed, and improved. It is a planning tool for translating customer needs and expectations into appropriate organizational requirements; that is, a system for translating consumer/customer requirements into company requirements at each stage from research and product development to engineering and manufacturing to marketing/sales and distribution. A common theme among each definition of QFD is the customer and his/her requirements and satisfaction. Though the method offers many benefits, there are also challenges in implementing it. QFD can be applied to packaging design and development and serve as an effective communications tool among diverse internal and external customers.

INTRODUCTION

The design and development of food products is an area well-researched and, for the most part, well understood with a multitude of tools and techniques to assist with the product development process. However, tools and techniques for food packaging design and development are relatively few and far in between. Moreover, those which do exist are sometimes directed toward a specific area of packaging. Most individuals involved in packaging are fully aware of its diverse nature, and also recognize it as among the most interdisciplinary of disciplines. The package and packaging system must fulfill many different functions driven by sometimes conflicting demands. A comprehensive design process for packaging must address the total spectrum of these demands and involve both internal and external customers. Furthermore, this design process should include a communications mechanism for the voices of these customers, and should also be able to prioritize the diverse needs of the customers involved.

As noted above, there are few widely used models for packaging design and development. Those that exist usually relate to general areas of mechanics, distribution, cost, and environmental impact (Topi, 1997). For instance, methodologies such as the 4-Step and 6-Step methods (Bresk, 1992) provide good insight and a logical approach for protective packaging design. Software driven approaches such as those espoused by CAPE Systems, Inc. (CAPE), and TOPS Engineering Corporation (TOPS) also help in the design of all levels of packaging. More recently an approach proposed by Sun (Sun, 1991) provides a framework which integrates packaging design with product design, manufacturing systems design, and distribution systems design. However, none of these methods or approaches provides a simple, efficient way to accommodate the voice of the customers and prioritize the various needs expressed.

One method traditionally used in the area of product design, shows strong potential for use as an integrated packaging design methodology. This method, is referred to as Quality Function. QFD was used because of its ability to incorporate internal and external customer needs into the packaging design process. The remainder of this chapter will provide a brief overview of QFD, present one representative application, and provide conclusions with regard to its merit as a packaging design methodology.

QUALITY FUNCTION DEPLOYMENT

Pioneered and developed in Japan in the early 1970s at the Mitsubishi Kobe shipyard, QFDs major thrust is to make sure that the voice of the customer is heard throughout the organization. This is accomplished by providing a framework for communications with emphasis on the functions responsible for designing, manufacturing (packaging included) and marketing the product or service.

QFD has been successfully used for a wide range of applications in service, manufacturing, and government organizations. Some of the companies include Ford, General Motors, Boeing, Hewlett Packard, Westinghouse, and 3M to name just a few (Bahill, 1993; Jacobs, 1995).

QFD has been defined in several ways. Three such definitions are as follows:

1. "A method of mapping the elements, events and activities necessary throughout the development process to achieve customer satisfaction. A techniques oriented approach using surveys, reviews, analyses, and robust design all centered on the theme of translating the "Voice of the Customer" into items that can be measured, assessed, and improved." (Hauser, 1988)
2. QFD is defined at the Ford Motor Company as "a planning tool for translating customer needs and expectations into appropriate company requirements." (Ford Motor Company, 1989)
3. The American Supplier Institute defines QFD as "A system for translating consumer/customer requirements into company requirements at each stage from research and product development to engineering and manufacturing to marketing/sales and distribution." (American Supplier Institute, 1991)

A common theme among each definition is the customer; this includes the voice of the customer, customer requirements, and customer satisfaction. A basic description of QFD is presented here.

Although QFD takes different forms, all use some type of a matrix representation to portray the information being used and the results of the decisions made. This matrix representation is often called the house of quality. The name derives from the basic shape of the matrix shown in Figure 17.1. The names of the various matrix components are also shown in Figure 17.1 and indicate the type of information contained. A more detailed representation of the house of quality is shown in Figure 17.2. The essence of this representation is its focus on the needs of the customer and its ability to provide a simple and useful tool for planning, communication, and coordination during the design process (Hauser, 1988).

The development of the house of quality begins with determining the customer needs. This is reflected in Block 1 and is often termed customer requirements or customer attributes, or simply "Wants". The requirements are stated in the customer's terms, and may be bundled into groups of requirements. The customer requirements may be determined from focus groups, in-depth interviews, or similar techniques. It should also be noted that customers include a wide range of interested parties such as functional areas within an organization, end-users, regulatory agencies, and distributors. Block 1A allows the requirements to be prioritized. The weights used to establish priorities may be derived from customer input, or from direct experience of those involved in developing the matrix.

Block 2 of the house helps to determine whether or not satisfying perceived customer needs will yield a competitive advantage. If a company wants to match

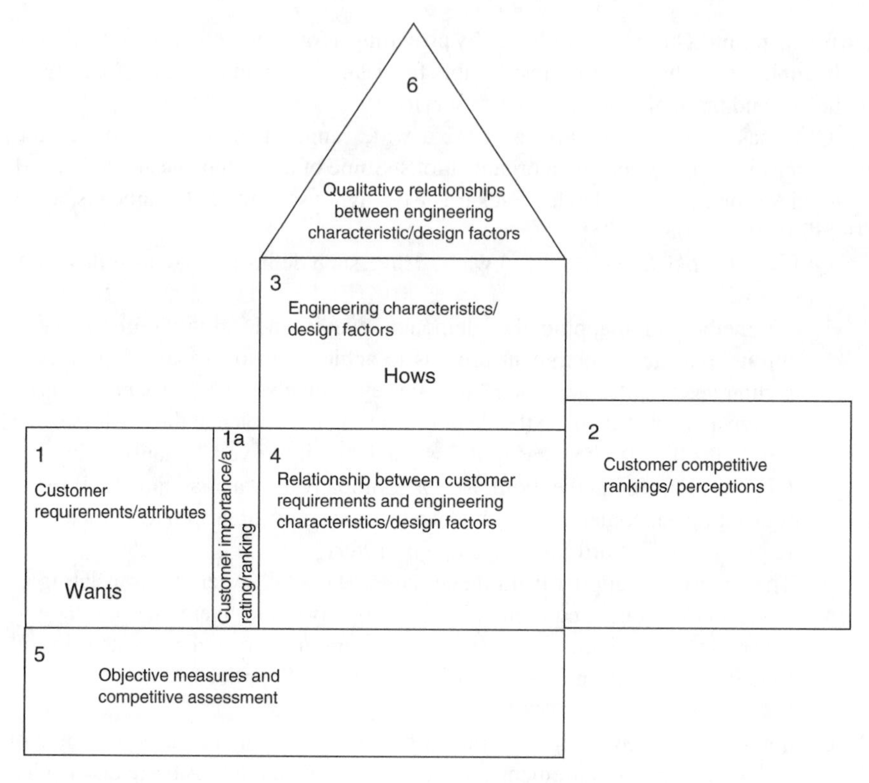

FIGURE 17.1 House of quality.

or exceed the competition, they must first determine their relative standing among competitors. The relative standing can be determined by a number of ways, including customer evaluations and benchmarking. A two-fold by-product of this exercise is the identification of your organization's areas of strength and also areas of weakness as compared to the competition.

Block 3, engineering characteristics/design factors describes "How" the engineer can meet the customer's needs. Thus, the voice of the customer is translated into engineering terminology by highlighting engineering characteristics, which affect the customer attributes. Engineering characteristics should describe the product in measurable terms, and should directly affect customer perceptions (Hauser, 1988). It should also be noted that a single engineering characteristic may affect more than one customer attribute.

The next step in building the house of quality is shown in Block 4. This is a relationship matrix which indicates the degree to which each engineering characteristic impacts each customer attribute. These relationships are usually stated symbolically in some manner, as shown in Figure 17.1; however the symbols may be assigned a numerical value to indicate relative importance. For example, referring to Figure 17.2, the three customer "Wants" for a package have been determined to be a package that is easy to open, yet resealable, and also one with appealing graphics. The engineering

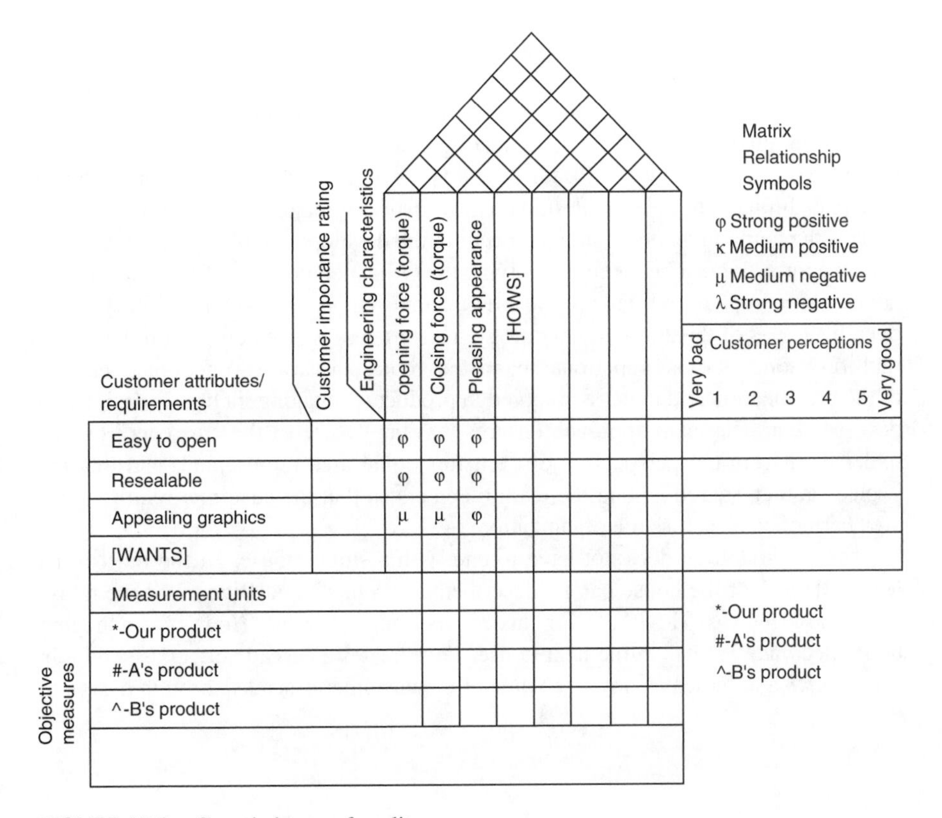

FIGURE 17.2 Generic house of quality.

requirements, or "Hows" may be translated as opening and closing force, and pleasing appearance. The symbols shown within the matrix indicate the relationship between the "Wants" and "Hows". For instance, appealing graphics and opening/closing force show a strong negative relationship (as usually expected), and a strong positive relationship for pleasing experience. When the matrix is completed, objective measures are included in Block 5 of the house. These measures help to establish target values for the engineering characteristics, and also compare the organization to the competition.

The top portion of the house of quality is its "roof" and is shown as Block 6. The roof is also often referred to as the correlation matrix (Ford Motor Company, 1989). The correlation matrix serves to identify the qualitative correlations between the various engineering characteristics and is accomplished by use of the relationships symbols. These correlations may be either positive or negative and may range from weak to strong. Too many positive interactions may indicate redundancy in critical product requirements or technical characteristics. Negative interactions point to the need to consider engineering trade-offs to address customer requirements. These trade-offs can be considered based upon company priorities, competitive strategies, and so on. In particular, this segment of the house of quality enables the team involved in the process to note how one engineering change may affect other characteristics.

The completed house provides a number of benefits such as measurable target values (Block 5), competitive assessment and competitive position (Blocks 2 and 5), relationships between customer requirements and engineering attributes (Block 4), and the relationship between engineering characteristics (Block 6). If numerical emphasis is used, the house can readily identify the most important customer attributes to pursue through engineering design characteristics. A major benefit from the use of QFD is the communication that must occur in the development of the house. Clearly a team approach must be taken in applying the QFD methodology. This point should not be under emphasized. That is, the use of teams in this process can be critical to the overall success of the entire methodology. This process is oriented toward the use of multifunctional or cross-functional teams that represent various functional elements of the organization and that are involved in product (packaging included) design and development. The team approach ensures that the "voice of the customer" can be understood from the perspective of each functional area represented, and that the "voice" is then shared among the team members so that the customer requirements are "harmonized" across the organization.

The methodology does not have to end with a single house. At the most basic level, Block 1 of the house can be described as "Wants", and Block 3 of the house can be described as "Hows". Using this convention, important "Hows" from the first house become "Wants" in the next house. This creates a cascade effect as shown in Figure 17.3 and indicates how the voice of the customer cascades through design to

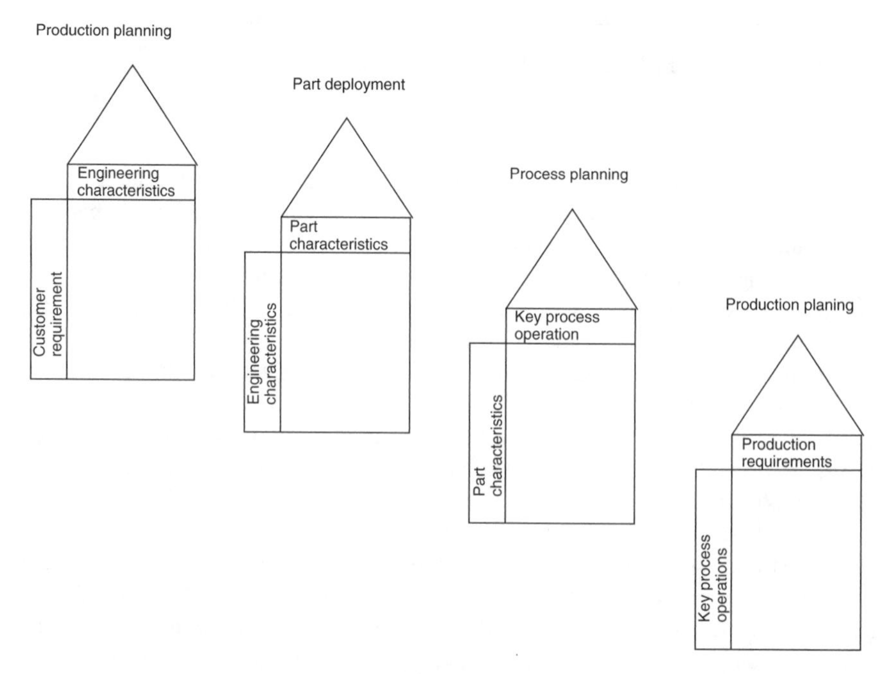

FIGURE 17.3 Cascade effect (Modified from, Hauser, J. R., and Clausing, D., *Harvard Business Review*, 1988.)

manufacturing and the ultimate end-user. The deployment may occur as shown, but the cascade flow may occur in general terms as follows: customer requirements, design requirements, engineering design, product characteristics, manufacturing characteristics, and quality characteristics. This logic, or deployment process, could easily include separate houses for the deployment of packaging processes, packaging machinery systems, packaging logistics systems, and packaging environmental systems. Moreover, where product and package synergy exists, such as the case for consumer foods and consumer products, the QFD methodology allows for the full integration of packaging systems in the process. In other words, the QFD methodology could be used for product design and for package design, then in an integrative fashion for the deployment of the product-package system. The illustrative example presented in this chapter illustrates the case of package design using only the initial house of quality. Examples actually using a package as an illustration of a product can be found in some of the references.

ENHANCED QUALITY FUNCTION DEPLOYMENT

QFD can be used as the central mechanism for integrated product/packaging development with the addition of other product design concepts and philosophies. Some researchers and practitioners refer to this as enhanced quality function deployment (EQFD). This approach consists of five parts as follows:

1. Concept selection
2. Deployment through the levels
3. Contextual analysis
4. Structured specifications
5. Static/Dynamic status evaluation (Clausings).

With further study and perhaps refinement, EQFD may well serve as an integrated packaging or product-packaging design model, thus overcoming the deficiency initially noted in this paper. The five parts of EQFD are briefly described below.

Concept selection is another matrix based technique developed by Stuart Pugh in the 1980s. In this process/method, a multifunctional team that is usually made up of the same team that develops the basic QFD house of quality, convenes to generate ideas or concepts used to initially address customer wants. Further deployment of concepts or solutions may occur as QFD is cascaded through the levels. A list of criteria is developed by the team to be rated against various concepts. A datum concept is selected to compare against all others. Each alternative is then evaluated against the criteria list, usually as worse than (-), better than (+), or the same as (S). An initial tally of the alternative concepts is made, which often provides an initial view of the concepts versus the criteria. The team then selects the best of the concepts and looks for potential opportunities to create hybrid solutions. These may then be put in the matrix and evaluated again against the datum and other concepts. In this way, bad points or potential bad points of alternatives may be eliminated or minimized. More advanced analysis of the concepts may occur in the form of finite element analysis, failure analysis, design of experiments, simulations, etc. The success of the Pugh selection concept, much like basic QFD, depends on the input and participation of the team members and their ability to communicate their ideas and understand the ideas of others leading to common understanding, consensus, and commitment (Biren, 1996; Ulrich, 1995). The Pugh Concept Selection technique offers tremendous potential for the evaluation of various packaging concepts, and perhaps development of new or hybrid concepts.

Contextual analysis refers to conducting an analysis to determine better the context of your product or package in a new market. Structural specifications is a detailed process for developing specifications in a life-cycle concept, and the results may be integrated into the house of quality. Structured specifications also are used in the overall guide of the design of the product or package. Static/dynamic status refers to evaluating designs and concepts in terms of the competitive market, and the technology streams present in the market place and those under development by the organization. In other words, with regard to design, what is the correct balance between current (static) and new (dynamic) technologies.

PACKAGE DESIGN USING QFD

Case Application: To serve as an example, a packaging design problem for a major manufacturer of remanufactured diesel engine components (obviously, not a food product, but used here to illustrate the technique) was addressed with the basic QFD methodology. One plant within the firm was receiving complaints from marketing concerning a single part package used for some fuel systems products. A marketing director within the firm had attempted to develop a survey based on the complaints received in order to determine the extent of the problem and develop solutions. This effort did not solve the problem. QFD was then applied in order to address the intra-functional communication problems as well as the packaging design problem. Internal and external customers were involved and included fleet users, dealers, distributors, warehousing and logistics personnel, and plant personnel from the production, operations, purchasing, and marketing areas. Dealers, distributors, and warehousing personnel were chosen from several geographic locations to insure that any data gathered was of national scope, and not limited to any particular region.

Customer requirements ("Wants") were determined through personal interviews. Individuals interviewed were asked to elaborate on several major issues and sub-issues within each category as shown in Table 17.1. Consolidated customer "Wants" were developed for the five major categories of packaging design issues related to unitization, communication, protection, shipping, and operations as shown in Table 17.2. The customers were also asked to rank the "Wants" on a 5-point scale, where a score of five was high and a score of one was low. A customer analysis was conducted which compared the package currently used to three competitors. Two competitors were external to the firm, and one was a multiple pack used internally by the firm.

A list of technical requirements or "Hows" necessary to satisfy the "Wants" was developed. As a matter of note, this was the most difficult part of the model application. Common mistakes included mixing of general and specific solutions and restating the

TABLE 17.1
Package Design Issues

I. Unitization	Number used per order
	Ease of handling
	Ease of palletizing
	Determination of number remaining in package
	Optimum number per unit: 1, 2, 6, 8, 12
	Other unitization problems
II. Communication	Brand identification consistent/clear
	Part identification consistent/clear
	Product information
	Warranty information
	Core return instructions
	Bar code
III. Protection	Damage Oil/soil Core return Method of closure
IV. Shipping	Problems Distribution chain
V. Manufacturing	Ease of assembly Ease of handling Method of closure Cost Vendor information
VI. Other	Miscellaneous complaints Environmental concerns

desired "Wants." The customer requirements, their ranking of importance, competitive assessment information, and technical requirements were then used to construct a house of quality. This effort resulted in the matrix shown in Figure 17.4 which indicates the correlation between the respective "Wants" and "Hows". Each of those who participated in determining "Wants" and "Hows" was given a copy of the initial matrix and allowed to fill in the relationships from their viewpoint. Correlation relationship matrix values were assigned as strong, medium, or weak. In this case, strong was equal to 9 points, medium was equal to 5 points, and weak was equal to 1 point. A weighted scoring system was developed to help prioritize or weigh the importance of the "Hows." The scores in Figure 17.4 are the sum of the product of the "Wants" rank and the matrix value. The value of the "Hows" score associated with the person surveyed was determined from their "Wants" ranking and matrix values. The final scores were determined by calculating the percentages of the total and the average of rank/matrix products of all surveyed. Based on those scores, the requirements, in order of importance, were rigidity, package breakability, simplicity of design, internal stability, information location, dirt resistance, compression load support, and standard dimensions.

Target values for the "Hows" were developed based on the survey comments and through interactions with engineering personnel. In some cases, a target value was replaced by an upper or lower threshold. Target values are also included in Figure 17.4. A competitive analysis of the "Hows" was also conducted and is shown here. Generally, the external competitor's packages were judged to satisfy these demands more completely than the current package, and the internal competitive package.

TABLE 17.2
Consolidated Customer Wants

U (Unitization)	Sell mostly 6X
	Easy to count
	Easy to handle full
	Standard box
	Easy to palletize
	Fits in wire basket
	Sell singles
	Looks good alone
	Minimal pack parts
	Stackability
P (Protection)	Individual protection
	Oil/soil protection
	Strong box
	Tight closure
	Box protection
	Box core returnable
	Product fits in box
	No axial movement
	No rolling movement
C (Communication)	Clear information
	Label sticks to box
	Easy to read
	Untorn label
	Read when stacked
	Distinct product package
	Dirt doesn't show
	Full box indicator
O (Operations)	Easy to assemble
	Easy to handle
	Filling speed
	Low inventory cost
S (Shipping)	Individual shipping

Case Summary. The knowledge gained from this particular QFD application was used to ultimately develop a package that met customer requirements ("Wants"). Although the package design was not finalized for this case, the QFD process did identify critical information. The determination of the customer "Wants", technical "Hows", weighted scores, and competitor analyses all provided direction to alternative package development. The customer wanted a package that had clear information, was easy to read from a distance when stacked, protected from the environment, and allowed for easy counting. To meet these requirements the QFD analysis indicated the

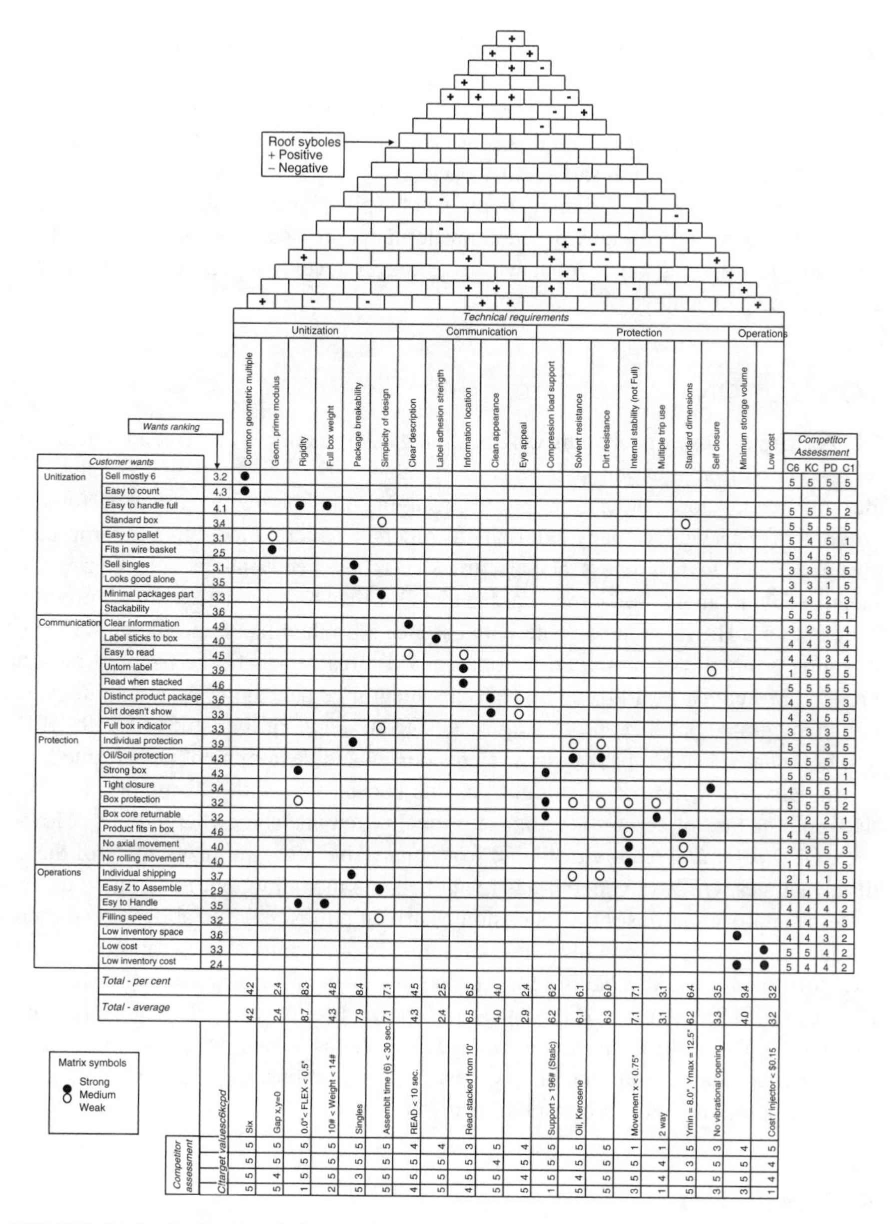

FIGURE 17.4 Packaging design application—house of quality.

package should be rigid, simple, and strong while allowing breakability (multipack) and clear labeling.

In this application, the direct survey process used to determine the voice of the customer worked well although extensive customer contact was required. This experience indicated that a positive relationship must exist between those working on the project and those whose opinions are used to solve the problem. The most difficult step in the process involved the determination of the technical "Hows." As mentioned previously, general requirements were often mixed with characteristics of specific alternatives. However, despite these difficulties, preliminary responses to design alternatives derived from this process indicated that the customers voice was successfully translated into satisfactory packages.

The QFD Process is surprisingly popular at the plant location where it was applied. It was implemented in all new product development projects. Furthermore, the plant personnel felt that customer input was essential in developing successful products, and QFD helped achieve desired results.

CONCLUSIONS

QFD can be used as a packaging design strategy. The example described here, along with successful application at Procter & Gamble (1992), supports that conclusion. Because the development of appropriate packaging involves many internal customers within a firm, as well as many external customers, a general model that incorporates customer needs in the process, and also translates those needs into technical terms is a positive development. Some might argue that QFD can only be successful for product development. However, a package is in essence a product itself and the application of sound principles of product development will greatly benefit the package design process. Moreover, EQFD, can serve as an integrative packaging design strategy, if not an integrated product and packaging design strategy. Further study of both QFD and EQFD as related to packaging systems design and development is warranted.

Initially, the QFD process might be time consuming, with information hard to find, and the task of creating the house of quality may be somewhat tedious . However, QFD software, readily available from many sources, minimizes some of these disadvantages. QFD, in general, has many benefits including, but not limited to, its systems approach philosophy, its customer driven philosophy, its ability to shorten product/package development cycles, its team/communications driven approach, and its ability to serve as a mechanism for translating knowledge to new teams (Ford Motor Company, 1989). Clausing shares a similar view as follows: "QFD is a useful tool for bonding the multifunctional team. It also serves as a training tool because each team member learns from the other's views and limitations. QFD becomes an asset to the corporation … it is a historical record if you will."

BIBLIOGRAPHY

1. Anonymous. 1991. Quality Function Deployment: Excerpts from the Implementation Manual for Three-Day QFD Workshop, The Third Symposium on Quality Function Deployment, American Supplier Institute, pp. 19–40.
2. Bahill, A. T., and W. L. Chapman. 1993. A Tutorial On Quality Function Deployment, *Engineering Management Journal*, Vol. 5, No. 3, Sept., pp. 24–35.
3. Biren, Prasad. 1996. *Concurrent Engineering Fundamentals: Integrated Product and Process Organization*, Vol. 1, Prentice Hall PTR.
4. Bresk, F. C. 1992. Using A Transport Test Lab To Design Intelligent Packaging For Distribution, *TEST Engineering & Management.* Oct/Nov., pp. 10–17.

5. CAPE: Computer Assisted Packaging Evaluation, CAPE Systems Inc., Plano, Texas.
6. Clausings, Don P. *Enhanced Quality Function Deployment*: *Video Series*, MIT Center for Advanced Engineering Study.
7. Ford Motor Company. 1989. *Quality Function Deployment Awareness Seminar, Reference Guide.*
8. Hauser, J. R., and D. Clausing. 1988. "The House of Quality," *Harvard Business Review*, May–June, pp. 63–73.
9. Jacobs, D. A., S. R. Luke, and B. M. Reed. 1995. Using Quality Function Deployment as a Framework for Process Measurement, *Engineering Management Journal*, Vol. 7, No. 2, June, pp. 5–9.
10. Procter & Gamble, Inc., Guest Lecturer, Associate Director of Packaging, Cincinnati, Ohio, from lecture and notes on package design, The University of Missouri-Rolla, March 20, 1992.
11. Raper, S. A., and M. R. Sun. 1994. Understanding the Role of Packaging in Manufacturing and Manufacturing Systems, In *Handbook of Design, Manufacturing and Automation*, Richard C. Dorf and Andrew Kusiak, eds. Wiley-Interscience Publications, pp. 331–342.
12. Sun, M. R. 1991. Integrating Product and Packaging design for Manufacturing and Distribution: A Survey and Cases, Ph.D. Dissertation, The University of Missouri-Rolla.
13. Topi, M. A. 1997. Using Quality Function Deployment to Establish Package Design Requirements, M.S. Thesis, The University of Missouri-Rolla.
14. TOPS: Total Optimization Packaging Software, TOPS Engineering Corporation, Plano, Texas.
15. Ulrich, Karl T., and Steven D. Eppinger. 1995. *Product Design and Development.* McGraw-Hill, Inc.

18 Shaping a Brand through Package Design

Christopher K. Bailey and Geralyn Christ O'Neill

CONTENTS

In this chapter, we will discuss ways that you can make your brand stand out by developing not only a functional package that will be noticed on the shelf, but one that works in conjunction with the rest of the components in your marketing mix to create an emotional connection with your customer. We will outline the process involved in designing a package, and discuss the elements of package design, which are the most effective in helping you communicate your selling message and capture the minds and hearts of the consumer.

INTRODUCTION

FIRST IMPRESSIONS = INITIAL JUDGMENTS

We all know that first impressions are critical. Whether you are meeting someone, dining in a restaurant, or visiting a new place, your first impression will always influence the way you think about that person, place or experience. It takes just a split second for you to form an opinion, based solely on what you see. Good or bad, you make a judgment without any long-term assessment of the facts.

The Same Holds True for Your Brand's Package

Now think about a consumer walking into a grocery store. Perhaps she has an idea of what she is looking for, but maybe not. In just a split second, she will be bombarded with shapes, sizes, colors, pictures,and illustrations—all converging together at one time on thousands of packages. All competing for her attention and hoping they have one thing that will make her take a closer look.

When customers enter a store and know exactly which product, brand and size they want, there is no question in their head. They just grab it and go.

But if they are not sure—or they want to compare brands—or their kids are running through the store—or their minds are on a million different things—then they need a little help.

That Is Where the Package Comes In

Packaging plays a very important role in the purchase decision. It is the execution of your brand at the retail level, and, as such, the primary vehicle consumers use to make a judgment about you and your brand.

Packaging is your silent sales person. Advertising may have drawn the consumer into the store, but your package is what will seal the deal. If it does not meet their expectations, they will continue searching until they find something else that does.

And That Is Not a Hard Thing to Do

An average shopping trip takes 30 minutes. In that half hour, consumers are confronted with more than 30,000 products.

That means more than 300 items per minute or 1 product in 2/10 of a second.

You have only 2/10 of a second to get your consumer's attention.

Packaging—From Brown Box to Brand

Packaging has evolved from a simple box or bag to a powerful vehicle that conveys the brand's message. The package still maintains a functional use—it houses the product and helps deliver it from manufacturer to consumer—but it serves a much different purpose when it becomes a part of the consumer's lifestyle. The structural form, aesthetic look and functional aspects of the package work together and create an emotional connection with the consumer. By filling an unanswered need, making a task easier or offering convenience in a harried day, the package gives consumers a compelling reason to gravitate toward it and choose to purchase it.

Packaging continues to influence consumers well after the initial purchase. It is a more pervasive and permanent part of the brand communication platform than any other marketing tool. Advertising flashes by on a television screen or billboard. In-store promotions draw fleeting attention. But the package is what the consumer interacts with every day. Therefore, it must not only achieve the brand's goals in terms of sales and profitability, but also effectively communicate the brand's promise and connect emotionally with the consumer.

Packaging does not stand alone in doing this. It is only one tool in the marketing mix that includes advertising, public relations, promotions, and every other way your brand touches the consumer. These are the brand touchpoints, and to be most powerful, they need to work together to send one continuous message to your consumer as demonstrated in Figure 18.1.

The packaging is one component of the marketing mix that contributes to the consumer's experience with your brand.

Figure 18.2 shows how deeply packaging can connect with consumers when it is part of a cohesive marketing plan.

When a change is made to only the packaging, perhaps an enhancement to the design, but that same look and feel is not carried out through the rest of the brand touchpoints, the new packaging has less of an impact on the consumer's overall impression of the brand. However, the more integrated the brand touchpoints become, the greater impact they can have on the consumer, creating a holistic experience and, thus, working harder to create strong brand loyalty. The brand essence of P.I.N.K.vodka was

FIGURE 18.1 Elements of marketing integrate into a simple continuous message.

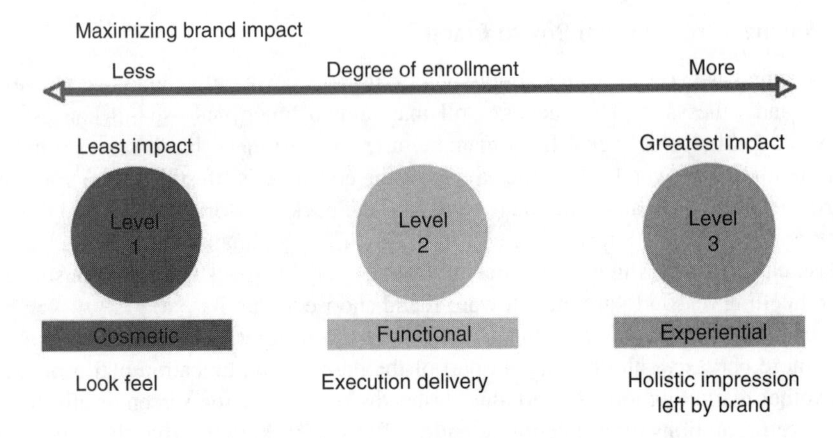

FIGURE 18.2 The impact of packaging on a consumer in a cohesive marketing plan.

FIGURE 18.3 Integration of packaging into the marketing mix.

created by the bottle graphics as illustrated in Figure 18.3. To create a unified brand presence, this look and feel was carried out to other elements of the marketing mix, including environmental and magazine advertising, and the brand's website.

Engaging the *Six* Senses

Successful packaging should engage a consumer and all their senses—sight, smell, taste, and touch. When packaging is effective, it also makes "noise" on the shelf—figuratively speaking to consumers and attracting their attention.

Packaging can also engage a consumer's *sixth sense*: consumer intuition. Consumer intuition is the immediate expectation that your packaging evokes in consumers—that quick and ready insight consumers have about a product before using it. It is also the holistic element that helps create not only a physical, but an emotional connection with them.

When your packaging touches all six of these senses, it becomes a truly powerful weapon in your marketing mix. Why? Because from the time your product arrives on-shelf until it is purchased and used in the consumer's home, your package reinforces your brand promise and generates consumer awareness and loyalty—24 h a day, 7 days a week.

Packaging Is not One-Size-Fits-All

In today's marketplace, manufacturers cannot take a one-size-fits-all approach to packaging. Changing consumer lifestyles and the addition of shopping channels including warehouse clubs and the Internet are just two reasons why brand managers need to evaluate their packaging to ensure that they are meeting the needs of their customers.

The same shopper can have many different reasons, or occasions, for purchasing the same product. One day, she fills the various needs of all her family members by shopping at her local grocery store. Another time, she is on-the-go, and stops for a quick bite at a convenience store. And a third occasion, she stocks up for a party at a warehouse club. These situations represent different times she may encounter your product in different locations to fill different needs. While these environments and occasions are all different, your product remains the constant in this equation. Therefore, it needs to be easy for the consumer to find your product. The exact size or configuration of your packaging may vary to suit the different retail outlets (e.g., single-serve vs. bulk) but the overall look and feel of the brand must remain the same to ensure that the customer has a consistent and positive experience with your product.

Pepcid Complete is available in single bottles in grocery and drug stores as seen in Figure 18.4. For the club store market, it is shelved in a coordinating shipper.

ASSEMBLING THE BRAND AND CREATIVE TEAM

Designing and executing a package design project requires a team effort, with both the client (Brand Team) and design firm (Creative Team) working together. Optimally, both sides are engaged in the process from the beginning of the project, from concept development through finalization of the package.

The Brand Team is led by the brand manager, along with assistant brand managers, in-house graphics personnel, and packaging engineers. The Creative Team typically includes an account manager, brand strategist, designers and production specialists.

FIGURE 18.4 Two different display systems for the same product for two different distribution environments.

Choosing a Design Firm

Choosing the right design firm to work with your Brand Team is an important step in the package design process. The brand manager should decide what is needed from the Creative Team before starting the search. Do you want a consultant who can be a strategic partner with you and help you plan and grow your brand? Or do you simply want a tactician who can execute your own vision? Whichever you choose, it is important to understand, through a written proposal, what services they will offer, along with costs and timing.

Roles and Responsibilities

A successful relationship between the Brand Team and Creative Team is based on collaboration and responsiveness. The brand manager needs to provide as much information initially, not just about the specific project, but about the brand's objectives, the short- and long-term plans and how this project fits into the overall objectives of the brand. In essence, the brand manager is the brand steward and is responsible for the ultimate success of the design project, ensuring it stays on time and on budget.

The Creative Team is charged with developing the packaging graphics. They creatively interpret the brand's essence into a successful package that maximizes the brand's message and works in conjunction with the rest of the touchpoints.

As the design project evolves, all members of both the Brand and Creative Teams needs to have a clear understanding of the Four Cs: client, category, competition, and consumer. At minimum, there should be a general knowledge of consumer preferences, a review of competitive products, and a consideration of current and future trends that may impact the brand. Some of this information will be presented in the Creative Brief developed by the Brand Team, and more will be uncovered during the research phase (covered in section Gathering and Analyzing Information).

THE PACKAGE DESIGN PROCESS

Once the team is in place, but before undertaking the design process, it is important to understand the specific objectives of the project. Several questions must be answered. This information is best communicated to the design team through a Creative Brief.

The Creative Brief

The Creative Brief is a blueprint to help everyone on both the Brand and Creative Teams understand the goals and objectives of the project. The following steps will help you create an effective Creative Brief.

1. *Set Clear Objectives for the Project.* This puts you and the designers on the same page and acts as both a roadmap and a contract for expectations. It also helps ensure that design concepts remain on track.
2. *Background Information.* Before beginning, you need to provide all necessary background information to help the designer understand the context

within which he or she is working, as well as the history of the product and brand, including business considerations. You can answer the "Why?" of the project—the circumstances that led you to recognize the need to begin the package design project. Are you new to a category trying to compete with established leaders? Are you the leader but losing ground to new entries and need to enhance your brand by developing a new product?

3. *Research Recommendations/Rationale*. Review any current research, and assess the need to conduct new research. (For a discussion of various research methods, see the section of *Consumer Research*, below in this chapter, and also Chapters 6 and 13 of this book.)
4. *Description of the Product's Image and the Target Market*. Explain the image that you want the package to convey, such as premium quality, a good value, or therapeutic benefits. Be specific in identifying your target audience, including age group, economic group, education, household size, gender, income levels and other demographics and psychographics.
5. *Package Strategy*. Outline how the package should achieve the desired image and how it should relate to, or differ from, the competition.
6. *Design Elements*. Clearly communicate the brand equity and what your customers think of when they encounter your brand. Communicate which elements of the brand must be retained (logo, color, symbol) and which you would like to explore.
7. *Specific Development Criteria*. Specify guidelines for designing your package, including size, special printing methods, number of colors to use and manufacturing constraints. Ensure that the design can be produced. Pretty does not matter if it can't be manufactured.
8. *Timing*. Develop a schedule based on approved marketing plans, key launches, production and inventory requirements.

It is just as important to make sure you have both a short-term and a long-term strategy for product packaging. To stay relevant in today's competitive marketplace, you need to refresh your packaging every 2 years at minimum. Otherwise, you will get lost in a sea of change. new developments in retail environments and consumer trends, as well as the actions of your competitors.

DESIGNING THE PACKAGE

The following time-honored approach will help ensure the package design process is successful.

Phase 1: Gathering and Analyzing Information

As discussed earlier, an understanding of the Four C's—the client, the category in which they market their product, the competition and the consumer or marketplace—are essential to develop a success package. While the Brand Team may have provided some of this information in the Creative Brief, the Creative Team will seek to learn all they can about these elements during this phase. A thorough examination of these elements and a complete understanding of the strengths, weaknesses, challenges, and opportunities of the brand can influence the design and start the process off on the right foot.

A variety of research methods, including primary and secondary techniques, can be employed during this phase to discover the following:

1. *Client*. What makes this brand unique? Here you seek to learn the brand's unique selling proposition (USP), that one thing that makes your brand stand out and gives consumers a reason to choose your product over another. In addition to reviewing current packaging, advertising and other marketing materials, interviews with key management and employees will help uncover internal perceptions of the brand as well as future plans and expectations for the brand.
2. *Competition*. What is everyone else doing? It is important to visit the retail outlets to get an understanding of the environment in which the product will be sold. A trip to the store will reveal what competing products look like, how they are shelved and which brands seem to have a prominent display. You can answer the questions: Who are the leading players in the category, and do they have any new entries? Are they using unique structures? Have they updated their look recently? What message are they sending to the consumer? It may be helpful to purchase a full array of products and recreate the complete shelf set for designers to reference while designing. New packages can also be placed in the shelf set and photographed to give the client an idea of its appearance in the store.

An illustration like Figure 18.5 was used by the Creative Team to show the Kozy Shack Brand Team how a new design would appear on the shelf.

3. *Category*. A thorough grasp of what trends are influencing the category in which the brand competes is as important as understanding the competition. In the food category, there is a constant stream of research and media attention surrounding low-carb, low-fat, no trans fat, and a host of other topics. Just as prevalent is the topic of childhood obesity and a general concern for the health of all consumers. These are important issues that you must remain aware of when developing new food products and packaging. Not only will they influence product development, these trends and values can greatly affect how you position your product and the messages you want your packaging to send.
4. *Consumer*. You want your packaging to create a lasting impression and connection with the consumer. At this early stage, therefore, feedback from consumers should be used to help you determine their perceptions about your brand and the competition. Review existing consumer research and conduct more, if needed. When introducing a new product, you can gauge up front, potential interest in the concept. For the redesign of an existing

FIGURE 18.5 Retail display showing relative positioning of various brands and products in a category.

brand, you can determine what equity lies in your brand, including logo and icons.

PHASE 2: SYNTHESIZING INFORMATION AND ESTABLISHING THE BRAND

The information gathered in the first phase will become the basis for the formulation of a brand strategy for the product. The Creative Team begins to determine exactly what the brand stands for, and distinguishes the brand's personality—what makes it different from other brands and the emotional connection customers will make when they interact with the brand. Whether you are building a new brand or reinvigorating an existing *one, it is important* to clearly understand—and articulate—its brand essence.

The *brand essence* reflects the promise of the brand—what you want people to believe about your brand. To determine a brand's essence, you must first understand and articulate its positioning, values and character. You cannot create an essence until you have a firm handle on those key elements.

Every brand needs a brand essence to help unify its image. Also an excellent mechanism for brands that have lost a clear positioning or differentiation in the market, a new or reinvigorated brand essence is a great way to get everyone back on track.

A brand essence synthesizes your positioning, values, and character and communicates them via words, colors, images, textures, and sounds. To communicate

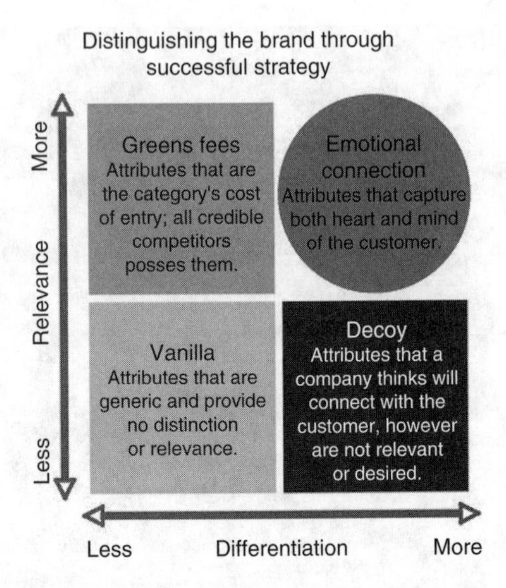

FIGURE 18.6 Distinguishing the brand through successful strategy.

an essence to internal stakeholders, you can create and distribute books, collages, and/or multimedia pieces. Although customers probably will not come in contact with these physical items, they should sense your essence in every interaction they have with your brand—whether in an environmental setting, on a website, or through your advertising, packaging, and other brand communications.

Creating an essence puts a proverbial "stake in the ground" as to what the brand stands for. As a result, a brand essence becomes the blueprint or road map that enrolls everyone in a brand vision. A brand essence can help educate and align internal employees and external partners with the brand—enabling them to consistently communicate the brand message.

A classic example is the Starbucks brand. Starbucks is about more than a cup of coffee. This brand encompasses the multisensory experience of entering one of their café-atmosphere stores. They established a clear brand essence early on and have embraced a core brand essence as they evolve to address changes in the marketplace that is reflected in everything they do, from developing new seasonal beverages to altering advertising campaigns to appeal to new audiences.

Figure 18.6 shows how implementing a successful strategy based on a solid brand essence can help you make that emotional connection with the consumer. In this example, the more you can differentiate your brand, the stronger the connection will become.

Greens Fees: There are certain attributes that all products must possess. A juice drink targeted to children, for example, must be tasty and healthy simply to be considered by the main purchaser, usually Mom. But these qualities alone do not make your brand stand out.

Vanilla: That same juice drink could claim to be "fortified with vitamins and minerals." Again, a nice benefit, but it doesn't distinguish the brand from the dozens of other brands on-shelf.

Decoy: So maybe you add some bright colors and a few fun games to the package, thinking this will attract moms to purchase it for their kids. It may, at first, but this alone is not enough of a distinction to generate brand loyalty, especially if a competitor is using the same technique.

Emotional Connection: Digging deep through consumer research and testing, you realize that your target market really wants a wholesome drink that will energize kids without extra calories, yet is considered "cool" to drink among peers. Because the competition is not filling this need, you can design a package that maximizes these attributes through graphics, color and typography, and supplement it with ad campaigns, public relations events and promotions that send the same successful message with the same brand essence. By tapping into the qualities that are most relevant to your consumer, you will differentiate your brand and generate a connection that leads to brand loyalty.

PHASE 3: CREATING THE PACKAGE

Once the brand's values, positioning and essence have been adopted, the Creative Team actually designs the package. They use a variety of tools such as illustration, photography, typography and colors to create a graphic look and feel that are consistent with the brand's image, but can differentiate it from the competition. Typically, a variety of options are presented, all of which meet the objectives as set forth in the Creative Brief. The design phase is usually the lengthiest portion of the package design process, and could involve several rounds of creative designs.

These four design options shown in Figure 18.7 for Kozy Shack Rice Pudding were presented to the Brand Teams.

When it comes time to review the options, the Brand Team should try to leave personal preference out of the decision process and make decisions based on the objectives of the project. They should offer open, honest, and critical feedback, which enables the designers to refine the designs.

The Brand Team may choose to send a few design options to consumer testing, to see which, if any, of the designs resonate with consumers. Designs are usually presented to several focus groups in a representative number of markets (see section of Consumer Research and in Chapters 6 and 13). Along with the Brand Team's comments, feedback from this research can help the Creative Team further refine the designs.

FIGURE 18.7 Four design options presented to the Brand Team.

Phase 4: Implementing the Design

During this next phase, commonly called the Production phase, the design is applied to all packaging components of the design project. Packaging components can include a primary structure such as a tube or bottle, and an outer, or secondary structure, such as a carton or box. Other components include shipping containers and display cases. The chosen design must be used throughout all of these pieces to achieve a consistent look and feel for the brand.

Mechanical art is developed electronically and then sent to printers for production of the final package. If the printer's package specifications have not been supplied to the production staff at the beginning of the project, they must first verify these before mechanical production can begin.

During this phase, quality assurance checks are performed before the package is printed. Proofreaders, production personnel, the Creative Team, and the Brand Team will conduct a thorough review of all mechanicals to be sure that mistakes, from typos to missing artwork, are caught and corrected.

To ensure that the final printed piece matches the one that has been approved, the production staff will attend press checks. During these checks, the printer will show the production specialist a printed sample. The specialist checks the sample against the final mechanical and completes one final proof before authorizing the full print run.

Even though production work takes place primarily toward the end of the project, it is beneficial to include production personnel in discussions before design work actually begins. By including them early on, they can meet with the client's packaging engineers and preferred suppliers to determine any printing parameters before the designers start working. As stated earlier, a package is not successful if it cannot be produced.

ENSURING BRAND CONSISTENCY

As you can see from the package development process, developing a brand takes work and commitment. Nothing can undermine these efforts more than an inconsistent presentation of the brand logo, colors, and other assets. To ensure consistency across all communication touchpoints, including advertising, promotions, and signage, the Brand Team should consider developing a Brand Graphic Standards Manual. This guide is used by everyone who comes in contact with the brand and includes instructions for accurately using the logo, PMS colors, and other brand assets. It is a step-by-step guide that lists how the logo can appear in print, what colors can be used with the brand, type fonts and sizes, and other key components.

The example in Figure 18.8 shows the correct usage of the logo in printed material.

DEVELOPING A BRAND NAME

The name of a product can add greatly to, or take away from, the success of the brand. The name should be unique enough to stand out in the consumer's mind, but make

Identity-color specifications

Color and size specifications for primary brand mark

Specs for a full-color logo using Pantone® colors:
Background Diamond: PMS 288 C
Double Diamond Icon:
Left Edge: PMS 3005 C
Center Diamond: PMS 2945 C
Right Edge: PMS 370 C

Specs for a CMYK logo should match the specified Pantone® colors according to printer settings and substrate being used.

Specs for one-color applications using the primary Pantone® brand color:
Background Diamond: 100% PMS 288 C
Double Diamond Icon:
Left Edge: 55% Screen of PMS 288 C
Center Diamond: 75% Screen of PMS 288 C
Right Edge: 45% Screen of PMS 288 C

Specs for black & white applications:
Background Diamond: 100% Black
Double Diamond Icon:
Left Edge: 55% Screen of Black
Center Diamond: 75% Screen of Black
Right Edge: 45% Screen of Black

The minimum application size of the Sam's Club® logo is .75"
In this application, the ® should be 1/4 the size of the lowercase "s" in the word Sam's to ensure readability.

Distance between the brand identity and any other type or symbol is the "x" height of the lowercase "m" in Sam's.

If you have any questions regarding the use of the Sam's Club® identity that are not addressed within this document, please call The Bailey Group at 610.940.9030

FIGURE 18.8 Example of instructions for correct usage of a logo in all situations.

sense for the product. Depending on the client's goals and project objectives, the name can be functional or fanciful, easy-to-understand or made-up. No matter what name chosen, however, several checks should be performed to ensure its availability and appropriateness.

Pronunciation Study: To check for ease of pronunciation, the name is put in front of consumers who are asked to read it as they think it should be pronounced.

Consumer testing: can offer an initial guide to the consumer's reaction to and perceptions of the name.

Global Language Check: If the brand will have a global presence, a language check can help avoid embarrassing or inappropriate translations.

Linguistic Review: Professional linguists can interpret the name and determine how it meets the project's communication objectives and provide any common meanings associated with the name.

Trademark searches: Preliminary trademark searches can be conducted online to eliminate any names that conflict with products already in the market and is a first step before conducting a full trademark review.

THE ELEMENTS OF EFFECTIVE PACKAGE DESIGN

A successful package design should achieve seven key objectives:

1. *Know your customer*. Packaging is more effective when it is targeted specifically to your customer. Understand the demographic profile, specifically gender, age, and household income level. How and why will they use your product in their lives? What needs does this product fulfill (emotional, nutritional, personal)?
2. *Take a big picture perspective*. When conducting research for your project, explore possibilities outside your product segment. If, for example, the product is geared toward kids, examine toys, books, electronics, and games that appeal to children. In other words, extend your thinking and allow nontraditional ideas to influence your strategy.
3. *Accurately communicate the product's selling message*. Besides meeting certain guidelines established by the Nutritional Labeling and Education Act [The Nutrition Labeling and Education Act of 1990 (NLEA) provides the Food and Drug Administration (FDA) with specific authority to require nutrition labeling of most foods regulated by the Agency; and to require that all nutrient content claims (i.e., "high fiber," "low fat," etc.) and health claims be consistent with agency regulations (http://www.fda.gov/ora/inspect_ref/igs/nleatxt.html)], and a package must display enough information through its copy and graphics to quickly and effectively tell the brand's story.
4. *Compete effectively on the shelf*. The package needs to give consumers a compelling reason to purchase the product over the one sitting next to it. One way to achieve this is to leverage aesthetics and function. Your package needs to look good and appeal to consumers, but must also be functional. The prettiest package in the world may win design contests, but will not promote repeat purchases if it cannot be opened. Identify innovative packaging structures that fill an unmet need.
5. *Be "ownable" in architecture, logotype, symbol, graphic elements, structure and color*. Packaging should be unique in terms of its look so that,

when customers see it, they immediately recognize the product as different and special. It should not look like every other package on the shelf. If everyone else is using blue, think about choosing red. If the competition is housed in cartons, investigate a different packaging format.

6. *Have a strong-line look, yet clearly differentiate products within your product line.* If there are several products in the brand line (different flavors or sizes, for example), all the products must look similar, representing themselves as part of the same family. Colors, names, or graphics can be used to tell each product apart.
7. *A consistent brand message is key to establishing awareness and recognition.* Everything in your marketing mix should work together to send a consistent message. When consumers interact with your package, it should be easy for them to identify it as a part of your brand because it looks and feels the same as your advertising, merchandising, promotions and website.

CONSUMER RESEARCH

Throughout the design process, it is important to keep in touch with consumers and get their perspective and thoughts about your package. Consumer research offers a pulse on what the end user thinks.

There are two types of consumer research, quantitative and qualitative.

1. *Quantitative research* leads to data that are presented as numbers. For example, attitudinal data are gathered using scales where the scale points are numbers such as 1–5. These data can be analyzed statistically. The most common forms of quantitative research include surveys and tracking (Chapter 13).

(a). *Surveys* are typically conducted via the Web, phone or through the mail. Most consumer products-goods companies use the Internet for their research, although the decision to conduct an interview via the phone or online will depend mostly on the audience and the subject matter. Mail surveys are used least frequently because of mailing costs and a low response rate.

(b). *Tracking* can be used to monitor the purchase behavior of consumers. Online tracking of website visits, and using a scanner at the point-of-purchase are two examples. Tracking can also be used to determine prepurchase behavior. Eye tracking studies are often used to gauge response to package designs before a design is chosen for finalization. Eye tracking uses sophisticated equipment to track what respondents actually see, rather than what they say.

2. *Qualitative Research* measures consumers' opinions and sentiments. Qualitative data are not presented as numbers. As such, the results of qualitative research cannot be analyzed using traditional statistical methods. Qualitative research includes focus groups, interviews, and store intercepts.

(a) A *focus group* brings together a small number of people, usually 8–12, and asks their opinions about a product, service, concept, advertisement, idea, or packaging. A trained moderator leads the group and encourages active participation from all members.

(b) *One-on-one interviews* are conducted via phone or in person. Respondents are asked the same questions and the interviews are designed to take less than 30 min each.

During *store intercepts*, customers are approached while they are shopping. The most common forms of intercept interview are the *mall intercept* (conducted on the floor of a mall) and the *store intercept* (conducted while the respondent is shopping in a retail store), which could be conducted while the respondent is shopping, when she enters the store, or as she is leaving.

PACKAGING TRENDS

Packaging will continue to be affected by trends in the consumer landscape. A few of the most prevalent trends include highly defined targets, convenience, meal replacement and sustainability.

Highly defined targets: The most successful brands will take time to get to know their consumers so well that they can separate the market into highly defined target groups known as segments. Then, these brands can develop products and design packaging, which fill the needs of these different segments such as women, baby boomers, and ethnic groups—even before the consumers knows they need it. Consumers will choose a brand because it appears tailored just for them. This strategy, known as segmentation, is one way an established brand can grow their business and branch out into other markets.

Welch's squeezable jams as seen in Figure 18.9 were developed to appeal to the changing needs of an older female audience.

For decades, Welch's has sold their products as what they call "all family"—products that appeal to everyone. When they wanted to grow their brand with new

FIGURE 18.9 Bread/cracker spread moved from glass jars to inverted wide mouth plastic jars with disperse closures and full panel printed shrink film labels.

offerings, they chose to develop a more premium preserve to target a more mature woman whose children are raised and who finally has some time for herself. Welch's envisioned target consumers purchasing the preserves as something special—a treat for themselves. As such, this product required unique, high-end structure and graphics, which the consumer would be very honored and proud to have on a special table.

To achieve the desired affect, the package structure was changed to a smaller size—16 ounces instead of Welch's customary 22- and 32-ounce sizes for jams and jellies. The package itself is made of clear plastic that mimics the cut glass jars commonly used for homemade and high-end gourmet preserves. Shrink film labels spotlight the preserve flavors—strawberry, orange marmalade, red raspberry, and Concord grape. They are designed to look like old-fashioned, hand-written paper labels, which highlights the special, premium aspects of this line.

Convenience: Consumers are always looking for ways to make life easier. Whether it is quick meal preparation at home, or on-the-go eating, convenience through single-serve and prepackaged goods will continue to influence what consumers purchase.

In the Welch's example, consumer research had indicated that this older target found it difficult to get the last bit of jelly out of a traditional jar, and disliked the waste of leaving it in there. So, the structure incorporated the convenience of a flip-top format. A wide opening and unique valve system allowed the preserves to come out easily—and without making a mess.

Meal Replacement: For busy consumers who do not have the time—or the desire—to cook at home, meal replacement options offer the chance to create balance by enjoying a full meal without the hassle of preparation. Rotisserie chicken and prepacked fresh cooked meals, for example, can be purchased in grocery stores and wholesale clubs, and then reheated at home.

Sustainability: In the past, manufacturers often took a "bigger is better" approach to their packaging. The product was housed in an internal package and then placed in a larger external structure. More and more manufacturers are thinking about environmental consequences when designing their packaging. With the introduction of so many products in the marketplace, retailers, too, will be demanding a minimalist approach to packaging so they can continue to offer consumers a wide variety on-shelf.

A GLOBAL VIEW OF PACKAGING

Around the world, brands recognize the benefits that strong packaging can play in creating a strong brand.

AUSTRALIA

Tempus Two—a wine that is sold in an Italian bottle with a pewter label—is a highly successful brand in Australia. The reasons for its success have been clearly articulated by the brand's creator, Lisa McGuigan. Her philosophy is simple yet powerful: do things differently, make an impact, go beyond limitations set by the industry, and ignore the doubters.

Tempus Two illustrates that packaging is the pioneer of brand communication. Both powerful and persuasive, packaging is the "skin" of the brand—it's what people buy. Great design combined with a great product is a potent marketing mix.

FRANCE

Actimel is fermented drinking milk. It contains a probiotic with two traditional yogurt ferments in addition to living bacillus. Launched in 1997, Actimel is now consumed by 6 million people in France. It is also one of Danone's leading products.

The brand has succeeded by providing an answer to new consumption trends—namely, the modernity and protection of health. Only through this kind of innovative marketing strategy can brands succeed in saturated markets.

GERMANY

One brand that has achieved great success in recent years is Beck's Gold, part of the German Inbev Group. For years, their normal pils in the green bottle was known as a bitter and strong beer. But when they introduced their mild version—called Beck's Gold—they broke every rule. They developed a beer that targets women and nonbeer drinkers. And they ramped up production by around 400,000 hectoliters, which is as much as a medium-sized German brewery produces. Both turned out to be very good bets. In the years that followed, every brewery—from the biggest to the smallest—came out with mild beers.

19 Public Policy Issues

Eric F. Greenberg

CONTENTS

Laws and regulations cover far more than labeling on package surfaces. Various governmental agencies have authority to ensure food safety, an increasingly important facet of delivered food product; the prospect of terror-inspired contamination; intentional or unintentional adulteration; misbranding; misrepresentation; and good manufacturing practices (GMPs) in food processing operations. Hazard analysis critical control points (HACCP) programs are now mandatory in meat plants to try to control food safety and are being considered for the entire food chain. Nutritional labeling is increasingly a key contributor and element of food package labeling.

INTRODUCTION

A study of the legal and regulatory requirements, or more broadly, the public policy issues applicable to food products, can be viewed as an inquiry into (i) requirements relating to food as such and (ii) requirements relating to food packaging. The universe of food packaging-related requirements is broader, but, as explained below, less well defined, than the universe of substantive food laws and regulations. Nevertheless, because a packaged consumer food product contends with more than just food laws and regulations, some understanding of that broader realm is crucial.

First, therefore, below is an overview of the key elements of U.S. food law and regulation. Second, a discussion of the wider realm of packaging law of which food law is a component, and an overview of its other elements.

FOOD LAWS AND REGULATIONS

FDA/USDA

When it comes to summarizing food legal requirements, the primary focus will be the requirements of the United States Food and Drug Administration (FDA) for most food products, and those of the United States Department of Agriculture (USDA) for meat and poultry products. Other agencies, such the U.S. Environmental Protection Agency (EPA), which regulates food-related pesticides, also have a hand in food regulation. Other examples are the U.S. Customs and Border Patrol service, which works closely with FDA to inspect imported foods, and the Treasury Department's Bureau of Alcohol, Tobacco, Firearms and Explosives which regulates alcoholic beverages including beer, wine and distilled beverages. In addition, many states have counterpart agencies to these federal bodies, which often cover similar ground.

Some observers consider this FDA/USDA division of labor to be an odd historical accident, and suggest that it be remedied by creating a single food regulatory body. The somewhat anomalous circumstance of having two important agencies regulating the nation's food supply, operating under different statutes and in different ways, can be mitigated when the agencies work closely together and coordinate their requirements as much as possible. Recent areas in which these two agencies have worked closely include nutritional labeling and food safety.

Conceptually, it is best to think of those marketing food in the United States as operating in a heavily regulated realm. Through the prohibitions against adulteration and misbranding, federal law and regulations impose strict sanitation standards and complex labeling specifications. There is no federal requirement that food makers apply for and receive preapproval of foods, as there is for makers of new drugs and medical devices. Still, there are many other forms of regulatory control. For some categories of food, such as seafood and low-acid canned foods, complex quality control programs for manufacturing are mandatory.

These systems are the wave of the future, or, for some industry segments, the main story of the present. The modern regulatory approach to controlling food safety is a framework referred to as Hazard Analysis and Critical Control Points (HACCP).

(see Chapter 8). A HACCP system requires manufacturers to identify the likely hazards, such as contamination from outside sources, or deterioration owing to failure to maintain refrigeration. These "critical control points" (CCPs) are points in the process where a failure could lead to a hazard, most notably, unsafe levels of dangerous microorganisms or other contaminants. Manufacturers establish critical limits, and then, if a manufactured product falls outside the limits, the maker knows that something needs correction. HACCP systems require extensive record keeping, so the history of manufacturing processes can be easily reviewed and monitored.

Adulteration

Food is one of the primary product categories under the jurisdiction of the FDA. It is defined as including (i) articles used for food or drink by man or other animals, (ii) chewing gum, and (iii) components of any of these.

The conceptual framework regulating foods is probably best summarized as a series of prohibitions against adulteration or misbranding. Adulteration can mean a number of different things, generally summarized as something being wrong with the product. For example, adulterated food may contain a poisonous or deleterious substance that renders it injurious to health; or may be filthy or otherwise unsanitary; or does not meet the standards it purports to meet.

Once a food product is in production, compliance with current Good Manufacturing Practices (GMPs) is a central concern. GMPs are a body of requirements, set forth in regulations, that are designed to assure that food is manufactured in a consistently sanitary manner. Building such steps into the manufacturing process is considered a preferable technique to merely testing some end products for compliance with contamination or sanitation standards. The GMP concept builds preventive measures into production of every unit, instead of relying on periodic testing at the end of the line.

It is useful to think of food as a combination of food components, that is, ingredients traditionally consumed as food or drink, together with food additives. "Food additive" is a very important term of art. The legal concept of "food additives" is quite broad in scope. Food additives must be used in compliance with a fairly complex web of requirements, or else the food in which they are found will be considered adulterated, and subject to enforcement action.

Food additives are defined as substances added to food, or, substances that, when used as intended, are reasonably expected to become components of the foods, and which are Generally Recognized As Safe (GRAS) by qualified scientists. Color additives, which are intended to impart color to food, are not included in the definition of food additive, and are regulated separately. Food additives must be the subject of an existing FDA regulation or effective Food Contact Notification permitting their use and used consistently with it, or be otherwise exempted by virtue of a prior sanction.

The "reasonably expected" clause encompasses, at least theoretically, virtually all components of food packaging, since components of packaging that contact food have the potential of migrating into the food. Because the components of packaging are not directly added to food, they are referred to as "indirect food additives." Food package manufacturers and users must be alert to the fact that components of their packaging are candidates for treatment as food additives, and will be considered food additives, unless one of the relevant exceptions applies, such as an existing approval, general recognition of safety, or a prior sanction issued by the government.

When a new use of an unapproved food additive is contemplated, submission to FDA of a food additive petition (FAP) may be required before its use is permitted. An FAP must contain a variety of types of information about the product and its use, including its chemical identity and composition; its physical, chemical and biological properties; specifications for use, with identification and limits on reaction by products; the amount proposed for use, and directions, recommendations and suggestions for use, and labeling specimens; data establishing that the additive will have the intended effect; practical methods to determine the amount of the food additive in food; reports of investigations of the safety of the use; and proposed tolerance (limit) for the additive, if needed, to assure safety; among other details. FDA provides scientific guidance documents to assist applicants with petition preparation, for example regarding the relevant chemistry considerations. The crux of the FAP process is always on the *safety* of the proposed use of a material.

FDA reviews the petition, and, if the agency approves it, a regulation is added to the Code of Federal Regulations describing the new approved use. Other manufacturers can also use the material as described in the regulation. FDA has 90 days to review the FAP, plus another 90 days if needed. Frequently (i.e., almost always), FDA fails to meet this deadline.

In late 1997, Congress created a new, alternative food additive approval procedure for many indirect food additives, now referred to as "food contact substances," such as packaging components. The FDA Modernization Act of 1997 created this new "notice" system, in which marketers submit notice of their intended new use of a food contact substance, called a Food Contact Notification. The submission details information similar to that in a FAP. The major difference under the notice system is that the proposed use is permitted unless, within 120 days, FDA gives notice to the petitioner of questions it may have about the safety of the proposed use. If the 120 days pass without action by FDA, the use of the substance is automatically approved. This new notice system adds considerable predictability to the timing of the petition review process by FDA, which is a boon for new food packaging research and development. The new law also made implementation of the notice system dependent on certain FDA funding conditions in future years; these conditions are expected to be met.

In recent years, FDA has added another ground for exemption, its Threshold of Regulation rule, which is a cutoff line for very small exposures to certain substances.

The new rule simplifies the approval process for food-contact articles such as packaging that migrate in very small amounts. Under the Threshold of Regulation procedure, packagers submit a petition to the agency and receive its approval, but the information that must be submitted can be much simpler, and the agency review time could be much shorter, more a matter of months than years.

One important feature of the rule is that substances being considered for exemption may not have been the subject of any toxicological testing at all. Their safety comes in the low level of their migration, which is so low that the agency is comfortable with it despite the paucity of data.

Another important point is that once a substance has been exempted, other manufacturers may use the substance in accordance with the exemption. "Other manufacturers may use exempted substances in a food-contact article as long as the conditions of use (e.g., use levels, temperature, type of food contacted, etc.) are those for which the exemption was issued," states FDA. In this sense, the exemptions are the same as food additive approvals.

FDA makes available a list of the exemptions it grants at its Dockets Management Branch, and updated lists are available from FDA's Office of Premarket Approval, HFS-200, 200 C-Street, SW, Washington, DC 20204. FDA does not publish detailed descriptions of each exemption in the *Federal Register,* though detailed information can be obtained, minus confidential business information, through a request under the Freedom of Information Act.

Misbranding

Misbranding can also mean a variety of items, best summarized as something being wrong with the product's labeling. A product is misbranded, for example, if its labeling is false or misleading in some detail; or misrepresents the quantity of its contents; or fails to reveal material information; or contains some other defect.

The prohibitions against misbranding state that a food whose label that fails in any detail to comply with FDA's labeling requirements is misbranded, and subject to enforcement action. Certain information is required to appear on most food labels: a statement of identity of the product; its net quantity of contents; the name of the manufacturer, packer, or distributor; a list of ingredients, in descending order of predominance; and nutrition labeling. Other requirements apply if the label makes specified claims about the attributes of the food. Under a new law effective from the beginning of 2006, the presence of one or more specified allergens in the food product, including its packaging, must be revealed on the label if it is not already clear from the list of ingredients. The covered allergens are milk, eggs, peanuts, tree nuts, fish, shellfish, soy, and wheat (see Appendix).

Virtually anything incorrect on a label makes the product misbranded. False labeling is prohibited, but so is merely misleading labeling. Failure to present information on a label that is required to be there renders the product misbranded. Misstated or omitted ingredients; over-stated product attributes; false claims of 100% purity; or false claims of geographic origin ("Imported cheese"), are just some of the types of label problems that would render a food misbranded.

Among the most significant developments in this area was the Nutrition Labeling and Education Act (NLEA) of 1990, and the regulations it spawned. These are responsible most notably for the appearance of the familiar Nutrition Facts box on the labels of most packaged foods. Other highlights of the NLEA and its regulations included standardized definitions for commonly used terms like "lite," "high," "low," and others; newly standardized food servings sizes, and new rules for the making of claims about the nutrient content of food, and so-called "health claims" making a connection between a food and a disease-related condition.

The NLEA was inspired by the recognition by both Congress and FDA of the increasingly strong evidence of the connection between diet and health. This made it more important than ever that consumers be provided with accurate and complete nutrition information on their food labels so they could use them to control their intake of certain nutrients or coordinate their overall dietary patterns. Adjustments continue to be made to the requirements on the nutrition information on the food label. Separate statements of "trans fat" content and origins of fat and juices were added.

The outlawing of misbranding can be seen as achieving a variety of public policies. It prevents fraud and deception of the public. This is really a type of theft by deception, so its prevention is obviously desirable. Standardization of food labeling also reduces confusion about the differences between different products, and makes side-by-side comparisons of foods easier. In addition, now that the connection between diet and health is better understood, the current, more informative food labels are designed to allow consumers to make more informed food choices with specific health goals in mind, such as reducing their intake of fat and trans fats, sodium, or calories, or increasing their intake of vitamins, minerals, or dietary fiber.

FDA Enforcement Powers

Food and Drug Administration's (FDAs) primary powers for enforcing its adulteration and misbranding prohibitions, including its labeling requirements, are its powers of seizure, injunction, and criminal prosecution. It initiates its enforcement mechanisms most commonly through a plant inspection. FDA has the right to inspect food plants, without advance warning, and examine only (i) pertinent equipment; (ii) finished and unfinished materials; (iii) containers and (iv) labeling in the plant. FDA has broader inspection powers with respect to inspections of plants where seafood is packed, or where infant formula or low acid canned foods are manufactured, because their HACCP programs are subject to inspection. Pursuant to a new post-9/11 law, FDA can also see documentation including records identifying suppliers and consignees, if FDA suspects the food may be a hazard to health.

It is a violation of the law to refuse to permit an FDA inspection. Congress determined that this relatively free reign for FDA was necessary to allow it to effectively protect the public against adulterated or misbranded foods.

If an FDA inspector finds what he or she thinks are violations, FDA commonly follows up with a Warning Letter giving the company notice that violations were found. If more serious violations are discovered, product seizure, or injunction against manufacture and distribution, or criminal prosecution, seeking fines against the company and fines and imprisonment against involved individuals, can be sought.

Those seeking to understand the legal significance of FDA's requirements should not fail to recognize that the penalties built into the Federal Food, Drug and Cosmetic Act include court actions for criminal penalties of fines and imprisonment. More importantly, these penalties can be imposed on companies or even individuals regardless of whether the violations were committed intentionally.

This is directly opposite the traditional rule studied by law students that crimes are a combination of some overt action, taken with the requisite intent. In FDA law (and nowadays in other regulatory contexts as well), it is possible to be prosecuted, and, if convicted, to be fined and imprisoned, simply because the product you are responsible for is in violation of the Federal Food, Drug and Cosmetic Act—regardless of how it happened, regardless of what your intention was, regardless of how hard you tried to get it into compliance.

Food packagers and others subject to FDA jurisdiction must rely on the good sense and reasonableness of prosecutors and judges to avoid having to face these most serious consequences for trivial violations. Under the strict language of the statute, however, the most serious consequences are possible; it is only FDA's prosecutorial discretion that keeps it from using its most serious weapon of criminal prosecution in matters of trivial violations.

The underpinning for this state of affairs is the Congressional recognition of the importance of a safe and wholesome food supply to the American public. Given the lack of control that the average consumer has over the safety of products that are packaged and distributed over long distances, Congress determined that it would place the highest possible duty of vigilance on those responsible for food in interstate commerce. If an individual shares some of the responsibility for a violative product's appearance in interstate commerce, they can be held criminally liable for the violation. Only if the person with responsibility can prove that it was objectively impossible for him or her to have done anything to prevent the violation, will there be a sufficient legal defense to the charges.

How USDA Regulates Meat and Poultry

USDA regulates meat and poultry products. The Department's Food Safety and Inspection Service (FSIS) carries out the legally required, mandatory inspection of meat and poultry production. Labeling and packaging of meat and poultry are also governed by USDA. More recent developments include promulgation of "safe handling" labeling requirements for retail packages of raw meat and poultry (1993), and an overhaul of the food safety standards to require implementation of Hazard Analysis and Critical Control Points ("HACCP") systems. Under these systems, hazards that could affect products are identified and evaluated, control systems to avoid those hazards are put into place, and then compliance with the control systems is carefully documented and monitored.

USDA borrows from FDA requirements with respect to the approval status of food additives, including components of packaging for meat and poultry.

USDA had long been known for its pre-approval of meat and poultry labels. Effective July 1, 1996, USDA changed its long-standing practice of pre-approving labels for many meat and poultry products. In fiscal 1991, USDA says its Food Labeling Division processed 167,500 labels. Individual on-site USDA inspectors, called Inspectors In Charge (IICs), who visit or work full-time at plants, also have the authority to approve certain types of labeling. IICs approved another 43,000 labels in fiscal 1991.

The USDA always had a class of labels that were passed without pre-approval, called generic labels. USDA has now made most of the system generic. Instead of submitting labels to a USDA headquarters office for review, or letting the individual IIC approve labels, no prior submission is needed. The individual IIC at each registered official establishment will have the responsibility to make decisions on label compliance, along with product and process compliance.

Under the old system, IICs could approve labels for single ingredient products for which no claims were made. Labels for products that contain a single ingredient, such as beef steak or chicken legs, no longer require approval unless the manufacturer makes specific claims about quality, nutrient content, and so forth.

Even the system for labels that must obtain prior approval received an overhaul. Manufacturers would only need to send USDA one set of sketches of printer's proof quality or equivalent, instead of first sending sketches and later sending final labels. Also under the new system, if a manufacturer has several locations that make the same product, it may submit only one label instead of one label for each location.

USDA says it will try to minimize variation in interpretation through guidance to inspectors, "a notice to field personnel that will clearly describe how to respond to and report label deficiencies." Some in industry worry that product will be unnecessarily detained over label issues by ill-informed inspectors.

Food Safety: A New Emphasis

Whereas traditionally, concern about food safety primarily centered on preventing contamination, ever since 9/11 the prospect of terrorist contamination of the food supply has been a new concern. FDA issued detailed guidance documents to help all players in the food supply think though how they might enhance security.

Moreover, pursuant to a 2002 law, FDA was given several new powers to help it monitor and maintain the safety of food made domestically and imported. Food makers in the U.S. and abroad are now required to register with FDA, and importers are required to provide prior notice of imports. Further, FDA can inspect documents including those identifying suppliers and consignees, or can place food under temporary detention, if the agency suspects the food is a hazard to health.

Traditional concerns with contamination have not been forgotten, however. In an era when many consumers are fearful of minuscule amounts of what are, to them, mysterious and unidentified chemical residues such as pesticides in their foods, government has instead focused much of its attention, with a good deal of justification, on more traditional food-borne hazards. Food processing requirements and controls aimed at reducing or eliminating hazards like *Salmonella, Listeria*, botulism and *Escherichia coli* 0.157:*H*7 have grown in recent years. Both USDA and FDA have expanded the applicability of the HACCP principle, requiring it for more and more categories of food. In 1997, seafood processors were, for the first time, required to impose HACCP systems.

Packaging Law Generally

Summarizing Packaging Law

The other major category of laws applicable to food are those applicable to food packaging as such.

There are two primary obstacles facing those who seek to study the legal requirements applicable to packaging. First, the body of federal, state and local laws and regulations applicable to packaging has not historically been referred to as "packaging law." Only recently, perhaps as an outgrowth of the increasing self-identity of the packaging industries, has demand grown for a systematic, orderly study of the legal and regulatory pressures on packaging generally.

The second obstacle is that, as a practical matter, many packages and products face private standards just as onerous, complex and important as any government-derived legal requirement. Sophisticated operations of a variety of types are controlled by private standards, or demanded by customers or industry pressures relating to quality or other parameters.

In short, not only is the law applicable to packaging, and food packaging in particular, difficult to summarize because it has not been summarized very often in the past, it is also not the whole story, since private standards and requirements also affect virtually all modern operations.

Summarizing packaging law is, then, a pioneering activity. It is logical to begin with a list of the subject matter areas within packaging law. Not included are areas of law that affect packaging and its businesses, but not specifically because they are packaging; so worker safety regulations, corporate law (the rules about stock issuance, corporate organization and so on), and even product liability to some degree are certainly substantive areas of law, but are not "packaging law" as such. These substantive areas are part of packaging law:

1. Regulatory law, including requirements of the FDA, USDA, Federal Trade Commission, Consumer Product Safety Commission, and others, concerning primarily labeling, but also food additives, and other topics
2. Environmental law, encompassing specific requirements of the EPA and states regarding solid waste handling and disposal, laws controlling hazardous wastes, laws limiting heavy metals, and other topics
3. Intellectual property law, the umbrella term referring to patents, trademarks, copyrights, the increasingly important doctrine of trade dress, and unfair competition.

Intuition has probably suggested that this list of subject areas is incomplete as an attempted expression of the universe of packaging law. It does not account for the extraordinary breadth and depth of the packaging industries, nor of the legal requirements that fall upon them. Some of the other factors that contribute variations to specific products or classes of products are:

1. The material used to make the package
2. The contents of the package.

And the list still is incomplete, for an accurate list of the elements that contribute to packaging law must include recognition of forces such as the following:

1. Politics, as the vagaries of modern politics is as responsible as anything else for why specific requirements and exceptions read as they do;
2. Scientific advances, which are the mothers of new packaging law invention; and

3. Federalism, the philosophy that calls for state law and policy to control unless there is a good reason for federal, nationwide action; unique, or at least varied, state legal requirements that do not match federal requirements or those of other states, are a crucial component of the packaging law landscape.

Toward a Practical Approach

The question of how to think about food law and regulation in a total quality environment can be summarized in one word: early. Consistent with the total quality philosophy of minimizing defects and maximizing consistency and predictability, consideration of legal burdens should be among the earliest activities in the development of a food product. Questions such as these should be made part of the product development strategy for a new product almost before the inventor's shouts of "Eureka!" fade away:

1. Is any aspect of the new product able to be patented? Is the packaging of a unique design such that it might be able to be patented?
2. Are any of the ingredients in the food product potential "food additives" under the federal Food, Drug, and Cosmetic Act? If so, do any approvals for its use already exist? Will they be used in accordance with existing approvals for those additives? Alternatively, are the ingredients GRAS; are they candidates for treatment under FDAs Threshold of Regulation; or are they subject to a prior sanction permitting their use?
3. Which FDA or USDA requirements will control the manufacturing processes employed to manufacture this product? Are they subject to HACCP?
4. Are any of the components controlled by other, more general regulatory requirements, such as control on handling of hazardous materials, worker-protection right-to-know laws or OSHA requirements?
5. Can the product be labeled as desired? Can the marketers' intended label claims for the virtues of the product pass muster under government requirements? If not, should changes be made to the label, or the product?

When questions like those listed above are handled as afterthoughts, companies too often find themselves quite far along in the product development process before first discovering a significant hitch in their plans. Production and marketing schedules have to be pushed back to accommodate the time it takes to examine questions of these types, and, in some cases, to obtain necessary government approvals or make necessary changes in product, labeling, or packaging to achieve compliance.

BIBLIOGRAPHY

Regarding FDA:

Federal Food, Drug And Cosmetic Act, 21 USC §301 et seq.

Code Of Federal Regulations, Title 21 Food and Drugs.

Regarding USDA:

Federal Meat Inspection Act, 21 USC §601, et seq.

Poultry Products Inspection Act, 21 USC §451, et seq.

Regarding PACKAGING:

Greenberg, Eric F., (1996) *Guide to Packaging Law, A Primer for Packaging Professionals*, Institute of Packaging Professionals, Herndon, VA.

Regarding FOOD PACKAGE LABELING:

Storlie, Jean, *Food Label Design: A Regulatory Resource Kit*, 1996, Institute of Packaging Professionals, Herndon, VA.

Olsson, Frank, and P. C. Weeda, *U. S. Food Labeling Guide*, 1998, The Food Institute, Fair Lawn, NJ.

Shapiro, Ralph, *Nutrition Labeling Handbook*, 1996, New York, Marcel Dekker.

APPENDIX: FOOD ALLERGENS

INTRODUCTION

For a small and unfortunate group of consumers eating certain foods that the rest of us enjoy everyday can have deadly consequences. About 11 million Americans are allergic to one or more types of food leading to as many as 200 deaths each year from allergic reactions to food. Awareness of food allergies has increased in recent years and government agencies charged with overseeing the safety of our food supply have focused more attention on the problem.

CHEMICAL BASIS OF ALLERGIC REACTIONS

Food allergies are primarily related to the body's immune system that normally serves to protect us from harmful "invaders" such as viruses and bacteria that are responsible for diseases. In the case of an allergy, the body's immune system mistakes a component of the food, most often a protein, as a threat to the body and "attacks" it. The physical symptoms of an allergic reaction can vary widely from a mild rash, all the way to death in the most severe cases. The more common allergic reactions to foods are shown in Table 19.1.

The severity of allergic symptoms varies from one individual to another, so it appears that there can be degrees of sensitivity, and for most individuals their symptoms are proportional to the amount of the allergen to which they are exposed. Most people who are allergic to a particular food must actually eat that food to cause a reaction, however highly sensitive individuals can have a reaction by simply coming in contact with an allergen.

What Foods Cause Allergies?

Over the years, a number of foods from both plant and animal sources have been identified that can cause allergies owing to some naturally occurring component in them. Table 19.2 lists the more common foods that have been associated with allergic reactions, but others that have been identified include green peas, sunflower seeds, rice, cottonseed, tomatoes, and sesame seeds.

Most often a protein in the food is responsible for the allergic reaction, but there are other classes of ingredients that have been shown to cause reactions as well. Compounds such as sulfites, lactose, monosodium glutamate (MSG), and aspartame (an artificial sweetener), and others, have been shown to have negative effects but the mechanism by which this occurs in sensitive individuals is different than the immune related response that allergens induce, and they are treated separately from a regulatory standpoint.

Food Allergy Labeling

As the result of the growing number of food allergy cases being reported, and the frustration of affected individuals trying to determine the allergy content of foods available, the government passed a new law in 2004 entitled The Food Allergen Labeling and Consumer Protection Act (FALCPA) (Title II of Public Law 108–282). The reason for this law was that it was often difficult to tell if a food had allergens in it when a manufacturer could legally list an allergen under an unfamiliar or unrelated name. For instance, a manufacturer could legally list "casein" as an ingredient instead of calling it a milk protein. Another example of this problem is that in some areas of the world peanuts are called arachis nuts and may be labeled this way. Being unaware of this fact can be deadly for someone with a peanut allergy. Another "loophole" in

TABLE 19.1
Symptoms Commonly Associated with Food Allergies

Skin related	Hives, eczema, dermatitis, rash
Digestive system	Nausea, vomiting, diarrhea, abdominal cramps
Respiratory	Asthma, wheezing, rhinitis, bronchospasm
Other	Anaphylactic shock, hypertension, swelling of the tongue/larynx

Schirer, 1999.

TABLE 19.2
Common Foods Associated with Allergic Reactions

Infants and children	Milk, eggs, peanuts, soybeans, wheat, tree nuts
Adults	Wheat, eggs, tree nuts, fish, clams, oysters, scallops, shrimp, crab, lobster, peanuts, soybeans

Schirer, 1999.

TABLE 19.3
The Major Food Allergens Covered by the 2004 FALCPA Regulations (Pub. L. 108-282)

Milk
Eggs
Fish (by common name)
Crustacean Shellfish (crab, lobster, shrimp)
Tree Nuts (almonds, pecans, walnuts, etc.)
Wheat
Peanuts
Soybeans

the old regulations was that certain types of ingredients could be grouped together in an ingredient statement under generic terms such as "spices" and "natural flavors." Sometimes there was an allergen "hiding" in these generic terms that were perfectly legal. The new law was designed to correct these two issues so that all allergens present were identified in "plain English," and so that there could be no "hidden allergens" on a food label. The FALCPA law was signed in 2004 and gave manufacturers until January 1, 2006 to bring their labels into compliance.

As part of the FALCPA legislation, the FDA identified the eight "major food allergen" foods and food groups that must be clearly labeled on a food package. These eight allergen categories (shown below in Table 19.3) are thought to be responsible for over 90% of the known reactions, including the most serious ones.

It is important to note that the regulations pertain not just to the specific items listed in Table 19.3, but also to any food ingredient that is made from, or contains the protein from them. Therefore, a common ingredient such as lecithin, which is derived from processed soybeans, would be included, as would whey that is a component of milk. The one category of processed products that was excluded was highly processed oils derived from nuts, peanuts, or soybeans because they do not contain the protein fractions from the original source that causes the allergic reaction.

The new law gives manufacturers two labeling options with which to identify the potential allergens in their products. These are referred to as the "Contains" option and the "Parentheses" option. With the "Contains" option, the manufacturer can place a separate statement immediately after or adjacent to the normal ingredient statement that lists the major allergens present. For instance, if the ingredient statement of a product included "sodium caseinate," "egg yolks," and "natural peanut flavor," then it would have to be followed by a statement in the same typeface and font size as the ingredient statement that said "Contains: milk, egg, peanuts." Using the "Parentheses" option, the word "milk" would be inserted into the ingredient statement in parentheses immediately following the ingredients whey and sodium caseinate, and the word "egg" in parentheses directly after the ingredient egg yolks. Table 19.4 shows how the ingredient label for the same product would look using each of these options.

Table 19.4 illustrates one difference between the two options. Specifically, with the "Contains" option all allergens present are listed even if it is the same exact word(s) appears in the ingredient statement (e.g., "pine nuts"), but with the "Parentheses"

TABLE 19.4
The Same Ingredient Statement with the "Contains" and "Parentheses" Labeling Options Illustrated

"Contains" Option	"Parentheses" Option
Ingredients: semolina, rice flour, rolled oats, pine nuts, tomato juice whey, sodium caseinate, tuna gelatin, natural peanut flavoring Contains: wheat, milk, pine nuts, tuna and peanuts.	Ingredients: semolina (wheat), rice flour, rolled oats, pine nuts, tomato juice, whey (milk), sodium caseinate (milk), tuna gelatin, natural peanut flavoring

FDA, 2006.

option, if the word that would be placed in the parentheses is exactly the same as the word it would follow, then it is not necessary to include it as a parenthetical item. For instance, it is not necessary to say "pine nuts (pine nuts)," nor would it be necessary to say "tuna gelatin (tuna)" or "natural peanut flavoring (peanuts).

Unintentional Allergens and Labeling

Another issue that arises relative to allergen labeling is that of "unintentional allergens," meaning those that might "accidentally" end up in the product but were not included as an ingredient. This situation might occur when a manufacturer produces two similar products using the same machinery—one with an allergen, and one without. It is possible that residue from the product containing the allergen may still be present on the production line when the product without it is made. This can even happen when the products are made on different production lines in the same plant and "dust" from one line is carried to another as might happen with flours made from wheat or soy. In these cases the government has allowed labeling that states that a product "May Contain" a known allergen. For instance the statement "May Contain: peanuts." In other cases manufactures might include a statement that says "This product was manufactured in a facility that also uses...," where any known allergens present in the plant would then be listed. While these statements are still allowed, regulatory agencies seem to be emphasizing that effective cleaning and the segregation of allergen containing and nonallergen containing products would be viewed more favorably than just the use of these statements.

Differences Between FDA and USDA

In the current food regulatory system in the United States, the USDA is responsible for the production of all meat, poultry, and dairy products. The FDA has regulatory authority over all other types of food products. The Food Allergen Labeling and Consumer Protection Act of 2004 described in this article is an FDA law and is only mandatory for those products produced under FDA jurisdiction, however, the USDA issued an official statement endorsing the practice of plainly identifying all known allergens in a product, but as of this date it is only a "voluntary" requirement for products produced under USDA authority (USDA, 2004). The USDA has indicated that they are in the process of developing new rules to provide the same "plain English" labeling of allergens, so from a practical standpoint, it would seem prudent for all USDA manufacturers to follow the same guidelines as those mandated for FDA products until the new USDA rules are enacted.

BIBLIOGRAPHY

FDA, 2006. Questions and Answers Regarding Food Allergens, including the Food Allergen Labeling and Consumer Protection Act of 2004 (ed. 3)—Final Guidance. April 6, 2006. http://www.cfsan.fda.gov/~dms/alrguid3.html

Sicherer, S.H., 1999. Manifestations of Food Allergy: Evaluation and Management. American Family Physician. January 15, 1999. Vol. 59, No. 2, pp. 415–429. http://www.aafp.org/afp/990115ap/415.html

USDA, 2004. Labeling and Consumer Protection; Allergens—Voluntary Labeling Statements. January 14, 2004. http://www.fsis.usda.gov/OPPDE/larc/Ingredients/Allergens.htm

www.fda.gov

www.usda.gov

http://peaches.nal.usda.gov/foodborne/fbindex/Food_Allergy.asp

20 Launching the New Product

John B. Lord

CONTENTS

INTRODUCTION

At this stage, development work on the product and package should have yielded a product/package combination that delivers the benefits promised by our concept and has met requirements for safety, integrity, quality, and shelf life. At the same time, work on the introductory marketing program has proceeded, and individual elements of that program, including advertising copy and media, package and label graphics, and price, have been evaluated. We are now ready to put the final marketing plan together, conduct final market testing, if that step is included in our development process, and, if all proceeds according to plan, launch our new product. In this chapter, we will cover three major topics: (i) the introductory marketing program; (ii) market testing; and (iii) the launch, including retail sell-in, execution of the launch program, and monitoring results during and after launch.

THE INTRODUCTORY MARKETING PROGRAM

The two most important decisions that any food manufacturer makes about a marketing program are those related to target market and positioning. Logically, all elements of the marketing program—product design, package design, pricing, distribution and merchandising, advertising and promotion—follow from the specification of our target market and positioning strategy.

These two decisions should be made very early in the development process. The target audience should be clearly specified during the opportunity analysis and ideation stages, and the positioning strategy should evolve and solidify during concept testing and subsequent product testing.

THE TARGET MARKET

The basic premise underlying both the opportunity analysis and idea generation stages of new product development is that we have identified a gap in the market. This gap is a combination of product benefits that a specified group of consumers demand but that is not currently being provided by existing products or competition. To the extent that we have followed this logic, the task of identifying our target audience is one of fine-tuning and elaboration. That is, we have a basic notion of our target customers, for example, mothers of school-aged children, and the task is one of expanding that definition to include appropriate demographic and lifestyle variables, and then overlaying media habits. Target marketing is only actionable when we have the ability to selectively and efficiently reach the target audience with marketing communications.

Target market information is available through our own sales data if the new product category is one in which we have prior experience. Results of early stage testing should yield some insight into segmentation patterns to the extent we have asked the right questions, particularly classification questions, and to the extent our sample is adequately diverse and our sample size large enough to find significant patterns in the data.

Several sources of syndicated marketing research can help us in this phase of new product development. Household panel data from a variety of sources, most notably A.C. Nielsen and IRI, will provide an overview of buying and consumption patterns along with demographic and geographic correlates, such as size of household, number and age of children, income level, region of the country, and so on. Data provided by Mediamark Research Inc. (MRI) yield even richer information. These data show, by-product category, purchasing indices according to demographic and geographic groupings, as well as the media habits, for both broadcast and print vehicles, of these customer segments. Consumer targeting analyses from MRI allow marketers to specify and profile user segments, as well as identify variables that distinguish heavy, medium, and light users.

A population subgroup with a purchasing index of over 100 is a group that purchases more than the average for the total population of a given commodity. For example, a Mediamark report on Red Bull (Mediamark Research, Inc., Spring 2004) shows purchase indices for two different age cohorts, adults 18–34, and adults 35–54. The data in Table 20.1 below show that the younger group is 99% more likely than consumers as a whole to drink Red Bull (having a purchase index of 199 on a base of 100), while the 35–54 group is 19% less likely (with a purchase index of 81.) Table 20.2 shows how the numbers are derived. The conclusion is obvious: consumption of Red Bull skews heavily toward younger consumers, and the marketers of Red Bull are well-served to focus their marketing and promotional efforts on those younger consumers.

In the example provided by MRI above, there is a strong age skew to the consumption of Red Bull, as anyone familiar with the product would expect. Those in the 18–34 year-old cohort are twice as likely (purchasing index of 199) than the population as a whole to drink Red Bull, while those in the 35- to 54-year-old group are about four-fifths as likely to drink Red Bull as the population as a whole (index of 81). These indices clearly pinpoint the demographic groups with the greatest potential in a given category. Once the primary and secondary targets have been identified, we can use other data from MRI to specify the media and vehicles that will most efficiently reach target consumers.

TABLE 20.1
MRI Report on Consumption of Red Bull

	All	A 18-34	A 35-54
All			
Unwgtd	25639	6654	10787
(000)	211845	66588	84359
Horz %	100.00	31.43	39.82
Vert %	100.00	100.00	100.00
Index	100	100	100
Energy drinks: red bull: all users: drank in Last 6 months			
Unweighted	1149	638	424
(000)	10126	6341	3260
Horizontal %	100.00	62.62	32.19
Vertical %	4.78	9.52	3.86
Index	100	**199**	**81**

TABLE 20.2
Derivation of Data of Table 20.1

	How the Numbers are Derived
Unweighted = 424	The number of MRI respondents who meet the qualifications specified (in this case, A 35-54 who drank Red Bull in the last 6 months).
(000) = 3,260	After applying each respondent's weight, the "(000)" value is the number of thousands of adults in the 48 contiguous United States represented by the MRI respondents who met the qualifications specified. Expressed in terms of individuals, this means 3,260,000 people.
Horizontal % = 32.19	The percent calculated by dividing the "(000)" value in the cell by the "(000)" value in the base column = 3260/10126 = 32.19%.
Vertical % = 3.86	The percent calculated by dividing the "(000)" value in the cell by the "(000)" value in the base row = 3260/84359 = 3.86%.
Index = 199	The percent calculated by dividing either the horz % in the cell by the horz % in the base row (62.62/31.43) or by dividing the vert % in the cell by the vert % in the base column (9.52/4.78). Either calculation generates the same result, because, when the horz % numbers and vert % numbers are expressed in terms of "(000)", the relationship is identical.

Source: Mediamark Research Inc., Spring 2004. Used with permission.

POSITIONING

Positioning refers to the way consumers perceive our brand relative to competitive brands. Consumer behavior theory tells us that consumers develop beliefs about the characteristics and potential benefits of products through a combination of marketing communications (advertising, promotions, packaging, price, and display), word-of-mouth, and actual usage experience. We compare and contrast different products on the basis of these beliefs, which are filtered by our attitudes and experiences. Product characteristics and benefits define the mental framework we use to compare and evaluate brands. Thus, we can affect the way in which consumers fit a new product into that mental framework by associating specific characteristics and benefits with it, primarily through advertising but also through all aspects of product design and marketing strategy. Note that the verb in the preceding sentence is "affect." Many phenomena impact the meaning a product ultimately has for the consumer, most of them uncontrollable from the point of view of the marketer. The major implication here is that we must find the right combination of benefits, and communicate them clearly, unambiguously and consistently to our target audience via all aspects of our product and marketing program.

The positioning statement succinctly notes the combination of sensory, rational and emotional benefits offered by the brand. Sensory benefits refer to taste, appearance, texture, aroma, and so on. Rational benefits refer to what the product does for the consumer, such as being a "source of Vitamin C" or "can be prepared in 5 min." Emotional benefits refer to psychological benefits, such as those provided by eating "comfort foods" or foods that are good for our children's health, or the way in which we think others see us—image—because we consume a certain brand. Every positioning statement must cite one or more of these kinds of benefits. Yoplait Nouriche, a drinkable yogurt, was positioned as a positive solution to women's guilt about not eating healthily, following research by General Mills that showed a significant number of women realized they were skipping meals and choosing unhealthy fast food alternatives because their busy lifestyles did not lend themselves to giving the time to plan, prepare, and consume healthy meals. This positioning provided for all three benefits: sensory, taste, and texture; rational, nutritional content; and emotional, in terms of doing some good for yourself.

Patrick (1997) noted that value-added positioning requires four ingredients. *Simplicity* means making our positioning understandable to the customer. *Specificity* means relating the brand to the specific needs of the target customer with specific benefits that will address these needs. *Durability* means developing a positioning strategy with staying power, meaning that we are addressing underlying and long-term consumer trends, not fads. *Advertisability* means that the strategy must lend itself to a number of different executions that can be communicated effectively via different media. The positioning statement provides the platform upon which all communication about the brand via advertising, promotion, publicity, price, and distribution must be based.

The Product Name

Great product names come from a variety of sources, and require a great deal of research. A name should be simple, memorable, easy to pronounce, and indicative of product benefits. Strategic Name Development (SND), a brand name consultancy, identifies five key characteristics of names. The first is *memorability*: can the target market remember the new product name after seeing it just once? Second is *latent association*: what negative or positive associations does the name connote? The third factor is *pronounceability*: if the target market consumers cannot pronounce the brand name, how will they ask for it? *Emotional bonding* is the fourth factor: does the name emotionally connect with the target market? If the new product sells imagery, is the name aspirational? Finally, *fit to concept* is important; the name should best position the product, that is, associate the appropriate benefits.

Brand names are legally protected, so we must do due diligence to ensure that the name we are planning to use is neither owned by someone else nor will it create some type of trademark infringement. Names must be tested on both an absolute basis and in a competitive setting. The name can make a big difference. A national online study conducted in 2004 showed that only 14% of consumers tested associated Swoops, a Hershey chocolate candy launched in 2004, with a candy. Almost 50% of the respondents, unaided, associated Campbell's Kitchen Classics with kitchen appliances and utensils while only 13% thought it was a food product (Strategic Name Development, 2004). The issue is straightforward: the less the association of name with product, the greater the demand on advertising and promotion to help establish that relationship. SND's data show that companies need to spend time with target customers when doing brand name research. Hershey's Nutrageous candy bar was originally named "Acclaim" but was changed after testing because while the candy tested very positively, consumers—kids—found no relevance in the name; in fact, many consumers associated the term "Acclaim" with a sedan, hardly the type of image Hershey wanted for their new candy bar aimed at kids and teens. Nutrageous is a play on "outrageous," at the time a commonly used kids' term to describe something they think is really different and "cool."

THE ADVERTISING PROGRAM

Consumer advertising creates consumer awareness, a relationship that is central to the development of consumer launch programs. Historical data that shows the linkage between the amount of advertising and awareness created. The amount or extent of advertising is measured by Gross Rating Points (GRPs), basically reach times frequency, or Target Rating Points (TRPs), reach of target consumers times frequency. Awareness is measured by the percentage of the population who become aware of the existence of the new product. The launch of Milk 'n Cereal Bars by General Mills was accompanied by a mix of prime time, early morning, and cable TV, which generated over 1000 TRPs over a 7-month period. Advertising also builds understanding of the brand, its key benefits and points of differentiation, which ultimately can lead to positive affect and perhaps purchase. Major advertising decisions are shown in Table 20.3.

Setting Objectives

New product advertising must capture the customer's interest up front and build involvement with the ad and the product. Our objectives must be stated clearly and quantitatively, such as, to "create awareness among 60% of our target customers within the first six weeks of the program." Marketers have developed some principles of new product advertising, based on experience. Successful new product advertising copy must position the brand clearly in a specific product category, communicate

TABLE 20.3
Major Advertising Decisions

- Objectives, typically stated in terms of awareness and trial levels
- What we will say about the brand, that is, our creative or copy strategy
- How much we plan to spend, that is, our budget
- Which media and vehicles we will use to reach our target customers, that is, our media plan
- How expenditures will be allocated to different geographic areas
- How the messages will be scheduled over the introductory campaign
- How to measure the effectiveness of our ads.

product benefits, demonstrate how the product is different from competitors and how it will benefit consumers, and provide some type of evidence that the product will deliver the benefits promised. One leading consumer products advertiser maintains that new product advertising must have three specific qualities: *relevance*, *originality*, *and impact*. Advertisers face significant challenges today as viewership of network TV has fallen, new forms of advertising have emerged, and costs of media have skyrocketed. According to industry statistics, advertising recall fell from 18% in 1965 to 4% in 1990, and things only have become worse since then. The average American is exposed to 500,000 advertising messages per year including 100,000 for new products. The challenge clearly is to create memorable advertising that is quickly recognized, and which serves to educate consumers about a product's important differences.

Creative—Advertising Copy

An interesting example of a new copy approach was reported in the *Wall Street Journal* (Parker-Pope, 1998). While marketers of toilet tissue have traditionally used softness, communicated by images of puffy clouds or squeezable packages, or economy, communicated by comparing number of sheets per roll, Kimberly-Clark employed a novel approach. The copy strategies for a new toilet tissue involved communicating an important product difference, specifically, that a new "rippled texture" was "designed to leave you clean and fresh." Results of advertising focus groups indicated that consumers responded negatively to words such as hygiene and cleansing, but not to the final benefit of being clean and fresh. This copy strategy is unique and even risky, but may be necessary in a competitively crowded and mature product category such as bathroom tissue.

Advertising Budgets

Setting advertising budgets is, at best, an inexact science. The most effective approach is to test different expenditure levels in a controlled experiment. The next best approach is to use historical comparative data that show statistically reliable relationships between spending levels and awareness and trial levels. Certain principles guide the determination of introductory marketing expenditures:

- The first year budget should permit a heavy introductory schedule ("heavy up") followed by ongoing expenditures equal to or greater than the second year budget.
- For a new product, share of voice (i.e., the brands' advertising expenditures divided by advertising expenditures for all brands in the category) should be approximately two times the market share objective. For example, if we are targeting a 10% share of market, we will need to spend to achieve about a 20% share of voice during the introductory period.

The short life cycle for most new food and allied products necessitates that trial be created through heavy advertising expenditures in the first few months. The most important period in the life of a brand is during and just after introduction, when sales build to an initial peak. The higher that peak share, the higher the probability the new brand will be a success.

Information Resources (*New Product Trends*, 2004) provide data on new product launches that clearly indicate the importance of advertising and building awareness in the overall success of new consumer packaged goods. IRI studies have found that as much as 35%of a new brand's year-one sales may be attributed to advertising, versus 5–10% for most established brands. Spending on marketing will never be more efficient than during the introductory period; the equity created and the buyer base established at introduction are irreplaceable. Of the top 10 new food brands as reported by IRI (2004) based on 2003 data, 8 spent between $13 million and $40 million on first year advertising, an average in the 10–20 million dollar range. Moreover, advertising support should extend "through year-two to ensure enough repeat purchases to firmly establish (the new) brand in the consumers' 'competitive set.'" Since the product is only young once, the cautionary tale is, "do not under-spend." The equity created and buyer base established at introduction cannot be replaced. Marketing spending will never be more efficient than during the trial period. Finally, short-term advertising return on investment (ROI) (dollar profit returned per dollar spent in the same year) is highest for new product launches. IRI has found that as much as 35% of a new brand's year-one sales may be attributable to advertising, as compared to 5–10% for most established brands.

We allocate consumer advertising expenditures in three ways, first, across media and vehicles; second, across regions; and third, across time. Media planning is more complex today because of the proliferation of media choices and the fragmentation of audiences. The principle is simple. We must place our messages where our target audience will be exposed to them. But the execution is difficult because there are so many choices. Media advertising is still the best way to build awareness quickly, and to build brand loyalty and equity through consistent messages focusing on product benefits. But gross audience numbers are no longer relevant. The key is to use targeted communications to reach targeted audiences. These will necessarily include electronic (Internet) advertising as well as more traditional broadcast, print, outdoor, transit, and specialty media.

The most logical decision rule for geographical allocation of advertising dollars is straightforward: allocate according to market potential. Firms often resort to category development index (CDI) and brand development index (BDI) in allocating all types of marketing effort. A CDI is an index number representing the sales of a product category in a specific geographic market relative to the average in all markets. As with all index numbers, the average is equal to 100. A BDI is analogous to a CDI except that it is used to measure the sales of a brand in a category rather than a category. The rule-of-thumb is to allocate more advertising dollars in areas where there is greater penetration of the category, ergo greater sales potential.

With respect to the timing of promotional expenditures, several points must be kept in mind. First, there is a threshold level of advertising below which it will lack the intensity to break through the communications clutter all consumers face. Second, once advertising has had an impact in terms of creating awareness and comprehension, decay or forgetting will take place in the absence of reminder advertising. Third, rarely is it financially feasible to maintain high levels of advertising expenditure over an extended period of time. Fourth, advertising messages must be properly scheduled with respect to when product is available in the store, when promotions are to be run, seasonal and/or special event sales patterns, and so on. Just as with expenditures, there is no magic or unique formula to guide our timing decisions; experience is the best teacher.

Reaching Consumers in the Twenty-First Century

Television networks used to deliver most of America to advertisers as audiences, but 30 years later, networks' audience share is down to 40% (Schneider, 2004). A study conducted in 2004 by Jupiter Research found that among 1803 U.S. adults who consume household information online, approximately 44% prefer the Internet as their leading information channel, compared to just 20% who prefer magazines and television (McGann, 2005). Stanton (2004) wrote a provocative piece in Food Processing Magazine entitled, "The End of TV Advertising," in which he argued that new technologies, such as TiVo and personal video recorders "make the 15-sec commercial a dinosaur." Not only have television audiences shrunk and become more fragmented, now technology way beyond fast forwarding through a VHS tape makes it substantially easier to avoid commercials. Roper reported (Schneider, 2004) that the 21 million "influentials" that shape national opinions and trends are far less likely than the population as a whole to watch commercial television. The 18–34 year-olds, which most marketers want to reach, watch the least TV. And a study by Forrester Research and Intelliseek in 2003 found that 76% of those responding disagreed with the statement companies tell the truth in ads, meaning that less than one-quarter found credibility in commercial advertising. What is a marketer to do?

A good example of a large CPG company forsaking traditional media to reach a specific demographic is the Pepsi Cola campaign to reintroduce a reformulated version of Pepsi One in early 2005. Pepsi is noted for elaborate and expensive television advertising campaigns featuring celebrities such as Britney Spears, Michael Jackson, and many others. However, in an attempt to reach males and younger consumers who normally shun diet drinks, Pepsi skipped the typical TV campaign and instead employed a website dedicated to Pepsi One (oneify.com), oversized billboards, trading cards in magazines, print ads in magazines such as Giant, Stuff and Sync, and promotional events planned to resemble art shows, and featured oddball characters that appeal to the younger consumer (Elliott, 2005).

The Internet

First, the Internet has become a significant source of new product information, both commercial and non-commercial. More and more companies have created destination websites that someone will visit to get information on, for example, household cleaning problems (e.g., P&G) or healthy eating (e.g., Kellogg), and these websites are designed to provide information on new products that fit the consumer's interest. The Internet is also an effective outlet for distributing targeted coupons as well as allowing consumers to sign up on site for samples they want, and then receive them in the mail. By making Crest Whitestrips available for trial initially on the Internet, P&G used the Internet very effectively to build initial demand as well as to identify four core groups who were very interested in having whiter teeth, teenage girls, brides-to-be, young Hispanics and gay men. Frito-Lay uses an interactive Web site to reach teenage consumers of its snack chips, including the announcement of new product launches. Using a campaign called "inNw?" which is shorthand for "If not now, when?" Frito-Lay targets the "millennials," that is, the first generation to grow up in the twenty-first century, with messages about living life in the present and taking advantage of every opportunity possible, an attitude important to millennials and how they view popular brands, including Doritos. The campaign uses innovative approaches, combining text messaging, an interactive web venue featuring a full motion "window to the world" where visitors can grab coded icons and drop them into an instant messaging interface to unlock video and audio clips and games (Williamson, 2005.) Frito-Lay augments the Web site with TV spots and outdoor advertising, all directed at the target audience.

A powerful new medium has emerged on-line, known as "commercial generated media" (Schneider, 2004), which consists of on-line discussion groups, chat rooms, and blogs. There are numerous Internet sites on which "bloggers" post messages which describe their opinions about and reactions to, among other things, new products. One of the characteristics of the information shared by bloggers is that it tends to be very honest, and therefore very credible. Marketers need to tap into this important new medium to (1) be aware of what consumers are saying about your brand through tracking consumer impressions of new products, and responding if the talk becomes negative, and (2) reach out to selected bloggers, give them an early opportunity to use and evaluate your product, and start a conversation about your product which will spread like a virus, hence the term "viral marketing."

Buzz Marketing

A relatively new term in the marketing lexicon is "buzz marketing." The idea behind buzz marketing is that in many instances marketers can no longer reach their target consumers through tried and true traditional media, so those marketers need to try something else. Also, as people increasingly perceive word-of-mouth to be a credible source of information, it is essential for new product marketers to put their new products and launch messages in front of the influentials, "the opinion leaders, trendsetters, innovators and first-users, who are most effective at generating person-to-person buzz" (Schneider, 2004).

Buzz marketing embodies the following principles: (1) be where your consumers tend to congregate; (2) use any way possible to bring attention to your product and show its relevance to the target consumer, and (3) make sure that individuals who are considered influencers within the target population are conspicuous consumers of your product. Buzz marketing involves recruiting influentials to try products, than sending them out into the marketplace to talk up the product with people they encounter. Companies, especially those targeting that elusive 18–34 year-old demographic, that want to "build a buzz" about their new product use a variety of event sponsorships, sampling and giveaways, wacky stunts, and similar techniques to make consumers aware of new items. Pfizer used creative sampling and "built a buzz" for Listerine PocketPaks by sponsoring parties around the Emmy Awards and Golden Globes, and distributing PocketPaks to modeling agencies, the New York Mets, and to NBA players. The idea is that people will want to use products that "cool people" use (Stagnito's *New Products Magazine*, 2004).

Other Nontraditional Communication Vehicles

Guerilla marketing is defined as "Unconventional marketing intended to get maximum results from minimal resources." A term coined by Jay Conrad Levinson, guerilla marketing is more about matching wits than matching budgets. Guerilla marketing can be as different from traditional marketing as guerilla warfare is from traditional warfare. Rather than marching their marketing dollars forth like infantry divisions, guerilla marketers snipe away with their marketing resources for maximum impact. (http://www.marketingterms.com/dictionary/guerilla_marketing/).

Glaceau, a small company that markets Vitamin Water (acquired by Coco Cola in 2007), represents a classic guerilla marketer, relying on sampling and from a van in high traffic locations combined with promotions such as "spin the bottle" which gives target consumers a chance to win products and prizes, deejay events with local radio stations, in-store tastings, and unique in-store displays, including irreverent signage such as "OJ Found Guilty" (of being high in sugar).

When Stouffer's (Nestlé) launched Corner Bistro, a line of frozen dinners designed to compete with casual dining establishments, their introductory marketing program included traditional national television and print advertising, a national free standing insert (FSI), Catalina targeted couponing, in-pack coupons, and in-store tactics such as floor ads, secondary freezer display units, and point-of-purchase (POP) signage to drive awareness and trial. But in order to reach their target consumers, Stouffer's also employed a combination of nontraditional vehicles including event marketing featuring taste-testing, sampling and product giveaways at events such as "the Taste of Chicago"; commuter sampling at transit terminals; menu door hangers placed on front doors in Del Webb communities with a high concentration of 55+ consumers; advertising on elevators in office towers to reach the younger end of their market as these consumers head home; and use of the Internet to build awareness and distribute coupons.

Publicity and Public Relations

Schneider (2004) stresses the importance of a strong public relations program to get the message about your new product out to both the trade and consumers. Publicity—well placed stories in newspapers such as the Wall Street Journal and USA Today, that have significant reach and significant credibility, not only helps to generate some excitement about your brand, but allows the marketer to tell a compelling story about the product's innovative features and benefits that will reach people that traditional media advertising cannot reach, plus potentially do it more effectively. Companies which can reach out to writers with news about a new product that will change our lives in some dramatic way or achieves a significant improvement over previous products in the category can take advantage of public relations much more readily that companies launching a me-too item with no real story to tell.

Trade Advertising

Although a relatively small part of any product launch budget, manufacturers often prepare the trade for new product launches by placing announcement type of advertisements in trade periodicals. These ads tend follow a common approach, and include two key types of information: (1) description of the product and why it will appeal to consumers, and (2) the marketing support that will be provided to the trade. An ad by Mars, Incorporated for the launch of reformulated and repackaged Whiskas dry cat food, used the headline, "Our best dry cat food ever." The ad included bullet points about the product ("double-basted for a meatier flavor"), the packaging ("boldly redesigned packaging") and details about an aggressive advertising campaign (over 1200 TRPs from January to June), direct mail sampling and a national FSI, and in-store support including trial-size shippers and a National Catalina program (Supermarket News, 2005.). The role of trade advertising is to create some awareness among retail merchandise and category managers so that when food companies make new item presentations, they are not completely "cold calls."

Packaging and Labeling

Packaging strategy and design plus labeling have been covered in depth elsewhere in this book (see Chapters 9 and 18). As part of our introductory marketing program, packaging plays a vital role in accomplishing several objectives. The package must provide visual impact at point of purchase, clearly identify the brand, inform the consumer about the product and the package contents, provide adequate protection for the product, and provide functionality for the consumer in terms of opening and reclosing, dispensing, preparation, storage, and so on. The label must also provide ingredients, nutritional information as required by law (the Nutrition Labeling and Education Act), and storage, preparation and serving instructions.

Consumer Promotions

Consumer promotions, such as coupons, cents-off deals, money back and refund offers, premiums, and so on, are commonly used and in many cases necessary to build awareness and especially trial to targeted levels. The most common type of consumer promotion is couponing. Coupons can be distributed through FSIs (in newspapers); in newspaper (run-of-press) and magazine advertising; via direct mail; on packages, either "on-pack" or "cross-ruff" (cross-ruff coupons are coupons for one product which are contained on the package of another product); in-store, or via the Internet. In store coupons can be distributed via coupon kiosk, as a tear sheet on the shelf, or using Catalina Marketing coupons, which are targeted coupons delivered at checkout. Many companies post coupons on their websites. Coupon redemption is typically very low, around 3% for run-of-press and FSI coupons, while targeted coupons can have redemption rates approximately twice as high. Nevertheless, FSI coupons remain an industry standard to support new product launches.

Since consumers love to save money, money-saving offers such as cents-off, rebates and money back offers, and BOGOs (buy one, get one for free or at a reduced price) can be used to help incent trial and early repeat purchase. Manufacturers may also use premiums, giving away something of perceived value for free or having an item redeemed at a very low price, and events such as sweepstakes, games and contests, that give the consumer a chance to win something of value while highlighting and drawing attention to the new item.

Sampling

Another key consumer promotion is sampling—it is important to get the consumer to taste the product. Samples of packaged goods can be distributed with newspapers delivered to homes, via direct mail, in or on packages of other products the manufacturer distributes, or in-store. Sampling can be accomplished through in-store demos, which are extremely popular with the trade. Combining an in-store demo with a coupon can lead to significant product sales during the promotional period. Sampling can also be very effective as part of event marketing, one of the non-traditional communications vehicles.

Trade Promotion

Trade promotion is a generic term used by the consumer products industry to refer to funds budgeted and spent by manufacturers to support the retail launch of new products and to reduce the list cost of existing products. These allowances both add to the retailer's bottom line as well as "pay down" the cost of offering the new item at a special low price during the introductory or promotional period. Consumer packaged goods companies have increasingly relied and trade spending and in-store activity in the hopes of gaining consumer attention, trial and repeat purchase, largely because expensive traditional media advertising is less effective and efficient in reaching targeted consumers, and there is so much new product activity that leads to a lot of "noise" at retail. While the percentage of marketing dollars spent on trade promotions rose dramatically over the last three decades of the twentieth century, that growth has apparently slowed. Cannondale Associates (2005) reported that consumer packaged goods companies spent 48% of their total marketing budget on trade promotions in 2004, compared to 49% in 2002 and 51% in 2001. However, Cannondale attributes these shifts to changes in accounting practices rather than actual decreases (Retail Wire, 2005).

New Item Funds

During the past couple of decades, it has become standard industry practice among most traditional supermarket operators to demand slotting fees (also called "new item funds") from vendors of new products. While most supermarkets charge slotting fees, other operators, notably Wal-Mart, and Whole Foods, for example, do not. Slotting fees are payments that allow the retail operator to recover the various costs of taking on the new product, particularly documentation, warehousing, and shelf maintenance costs. But a significant portion of slotting money is designated for the bottom line, and slotting dollars represent an important component of food retailer profits. The amount of a slotting fee varies greatly according to the category, the size and policies of the retailer, the ability of the vendor to pay (slotting fees are often waived for small local suppliers), the manufacturer's track record with previous new product launches, whether the product is carried by competitive retailers in the same market, and the amount of space and retail support the vendor is requesting. Typical slotting fees in the United States may range from as little as several hundred dollars for a single store or a small regional chain to $40,000 (even higher in New York City) for a large chain, per stock keeping unit (SKU or individual item), per chain. In some instances, manufacturers provide free cases of new products to help retailers gauge consumer demand. In any event, slotting money represents a significant element in overall retail profitability.

Customer-Specific Trade Programs

Manufacturers are interested in driving profitable sales volume at the retail level and want to ensure that their promotional dollars are doing just that. In addition, most large vendors have developed customer-specific marketing and sales programs. So it is the norm today for a company, such as General Mills, to allocate a specific amount of promotional dollars to an individual retail customer for a period, usually a year, and the purpose of these promotional dollars is to give the vendor salesperson assigned to that customer ability to utilize these funds to achieve volume and margin targets by category and by customer. A common allocation technique is for the manufacturer to designate a specific per case promotional allowance which is multiplied by a forecasted quantity of total case sales over a designated period, usually a year. This computation yields a specific amount of promotional dollars in an accrual fund for each trade customer. The salesperson negotiates with his/her retail counterpart, the category or merchandise manager, regarding how these funds will be used to support promotional programs for existing items as well as the launch of new items during the year. Promotional dollars are divided into two pots: new item funding or slotting dollars and dollars used for specific promotions, which typically are set up on a regular periodic basis, such as once per quarter.

These dollars may be paid in the form of different types of allowances (off-invoice, billback, display, advertising) or free goods. The term "performance funds" is now used to describe the current manner by which new products are promoted to the retail trade. Performance implies that retailers are using the funds for more than just the bottom line, but are reducing price, including the new item in displays, and/or doing other things to push the product out the door. Companies such as Nestlé, Kraft Foods and Hershey Foods all stipulate minimum performance requirements that retailers must meet to receive funds that in the past mainly were paid by vendors with "few strings attached." Nestlé, for example, dictates that they will pay full slotting fees only if the trade customer takes on all of SKUs in a multi-SKU launch.

The new item promotional program will often include innovative in-store merchandising, such as the use of shipper-display units to create secondary display space, shelf-talkers that stand out on the shelf and draw the consumer's attention, plus other in-store vehicles. These POP incentives are funded by the vendor's promotional dollars. They should be included along with the rest of the promotional program in the new item presentation.

Other Elements

Other elements of trade promotion include such things as in-store signage, POP and point-of-sale (POS) displays, sales kits, sell-in brochures and sell sheets. The common purposes of these elements are to help educate the retailer and to help announce to consumers that the new product exists. What manufacturers often call "collateral material," sell-in brochures and sell sheets, should be designed to be consistent with package and display graphics, and should contain compelling highlights of what the product will do for the trade.

New Accounting Standards Affect Promotion Planning and Execution

In January 2002, a new FASB (Federal Accounting Standards Board) accounting standard dealing with how promotional dollars appear on accounting statements took effect. Essentially, the change means that promotional dollars will appear on profit and loss statements as reductions of net income instead of SG&A or some other type of expense. All sales incentives voluntarily offered by manufacturers without charge, including bill-backs from invoice allowances, buy-downs, free-product deals, rebate and coupon costs, and price reductions, must now be subtracted from sales totals, as must slotting fees, co-op advertising, and price reimbursements. The impact on income reported by CPG manufacturers can be significant. For example, Kraft Foods was forced to reduce its previously reported 2001 sales total of $33.9 billion by $4.6 billion, according to the company's annual report (Hanover, 2002). Bishop Consulting predicts that these FASB regulations may lead manufacturers to move to different types of promotional activity, for example, scan-downs, coupons, and comarketing (Willard Bishop Consulting, 2004).

Price

There are a few tried and true pricing principles. One is that marketers must consider the price responsiveness of consumers in setting price. Several aspects of price elasticity should be evaluated including:

- Base price elasticity, the impact of a given item's everyday shelf price on its own sales and profits
- Promoted price elasticity, the impact of a given item's promoted price on its own sales
- Price elasticity variation, price elasticity reported by outlet, region, retailer, or season
- Cross price elasticity, the impact of item A's price on item B's sales
- Price thresholds, the price point at which elasticity, the consumer response to price, increases significantly
- Price elasticity is dependent to a great extent on availability of substitutes. So, the idea then is to launch a product that is distinctive and meaningfully different than those of competitors, and you will have much more freedom to set price. Consumers will pay more for higher quality, greater convenience and other important benefits, but the perceived differences among brands must be significant.

The price for a new product does not have to be the lowest on the market but it must be set with reference to perceived product quality and benefits so that consumers perceive higher value than they could obtain with the purchase of a competitive item. Therefore, competitive prices must be considered in this context as well. If consumers do not perceive meaningful differences among competing brands, then price becomes one of the key factors in determining consumer choice. However, pricing a product too low will decrease its perceived value for many consumers. Pricing a product too low at introduction makes it almost impossible to raise price later because consumers will have established a reference price for the item. Launching the product with a higher initial price point and promotions that lower the effective price for trial and early repeat purchase is much more effective than launching with a "lowball" price and then attempting to recover margin with a higher price later. Research reported by Marn et al. (2003) showed that once prices hit the market it is difficult, even impossible to raise prices, and up to 80–90% of all poorly chosen prices are too low.

The pricing structure for the new item must allow for adequate trade margins, without which the item will never reach the retail shelf. For each product category, the retailer has a profit objective that, along with sales velocity and cost factors related to stocking and displaying the product, drives the margins required for that category. Ultimately, the pricing structure must provide both adequate sales velocity and adequate margins for the manufacturer so that objectives for profitability and return on investment can be met.

Psychological factors in responding to price must be considered. For instance, in certain product categories, there is a standard price level. In many instances, especially in cases where consumers have few objective cues with which to evaluate quality, consumers use price level as an indicator of product quality. And there is still a tendency on the part of marketers to price at $2.99 instead of $3.00 because of the belief that consumers will perceive the $3.00 price point to be significantly higher than the $2.99 price point, a practice known as psychological pricing.

Financial Evaluation of New Product Projects

The bottom line for new product development is the "bottom line." Every new product project involves expending dollars for the purpose of making money. In publicly held companies, there is extreme pressure to meet short and intermediate term revenue and profit targets, as well as a need for projects to meet corporate targets for return on investment. While companies normally want line extensions to achieve bottom line profitability within the first year of introduction, really new products may take longer. For instance, Oscar Mayer Lunchables did not start making money until year three on the market. Pepcid AC, a prescription pharmaceutical product converted (after FDA approval) to over-the-counter and launched through a partnership between Merck and Johnson & Johnson, took over 3 years on the market to turn a profit. This is because the initial marketing expenditures to build awareness and trial, by plan, were extremely high. Companies with deep pockets and many successful brands at different stages of the life cycle can sustain new products through periods of heavy losses if there is the promise of long-term viability and profitability.

Except in the case of very straightforward line extensions, most new product development projects require investment capital, primarily for new or refitted plant and equipment, as well as other resources. As capital projects, product development projects must undergo financial evaluation for the primary purpose of determining whether it makes sense to commit resources to the project initially and at decision points throughout the process. All one-time expenditures to support a new product launch are classified as investment, even if these expenditures, such as package design, are expensed for accounting purposes. Three major criteria are evaluated: projected demand, profitability and return on investment.

Estimating demand is an exercise in sales forecasting. We have discussed projecting sales for new food products in Chapter 2. Profitability is always a primary consideration. Profits equal revenues less expenses. Estimates of revenue are derived from the sales forecast. Expenses include manufacturing costs (raw materials, supplies, labor, utilities, depreciation on plant and equipment, etc.), packaging costs, distribution costs (transportation, warehousing, inventory, order processing), marketing expenditures (measured media advertising, promotions, publicity, customer service) and general and administrative expense. According to Boike and Staley (1996), several key financial measures must be estimated and validated against corporate standards and funds availability at the onset of the new product development project. These measures are shown in Table 20.4.

Measuring ROI involves determining the amount of net returns, for which we use cash flows after taxes (CFAT) plus their timing and duration. Scarce capital resources must be put to use in projects that justify their deployment. In other words, new product development projects must be attractive projects. Attractiveness is evaluated by several methods. We will discuss four commonly used techniques, including the

TABLE 20.4
Financial Measures in NPD

- Development cost (e.g., hours, capital)
- Prototype and pilot plant costs
- Manufacturing costs: tooling and scaleup, in addition to ongoing manufacturing cost
- Related costs (e.g., advertising, packaging, promotion)
- Pricing
- Anticipated sales (e.g., units, revenue)
- Payback measures (e.g., ROI, profit contribution, anticipated margin)

average rate of return method, the payback method, the internal rate of return method and the present value method.

The average rate of return method is very simple to employ. We begin by calculating the ratio of average annual earnings after taxes to average investment in the project. For example, if average CFAT are $4 million and average annual investment is $48 million, the average rate of return is 8.3%. Note that this method does not consider the timing of either the investment or the profits. The projected return rate can be compared to some standard such as an investment hurdle rate to determine if we should fund the project.

The payback method uses as an investment decision criterion the amount of time (number of years) we project that it will take to recover the initial investment in the project. The calculation employs as a numerator the initial capital investment, and as a denominator the annual cash flows. This computation yields the number of years to recovery of the initial investment. The computed payback period can be compared to a target to determine if the capital expenditures should be made. Like the average rate of return method, the payback method in its simplest form does not consider timing of the cash flows.

The internal rate of return method involves the discounted cash flow for the project. The internal rate of return is the interest rate that equates the initial investment and the discounted cash flow of the project according to the following formula:

$$A_0 = A_1/(1+r) + A_2/(1+r)^2 + \ldots + A_n/1 = r)n$$

where r is the rate of interest, A_0 represents the initial investment occurring at time period 0 (at the beginning of the project) and A_1 through A_n represent cash flows accruing in periods 1 through n. This "r" must be greater than or equal to a hurdle rate set by management. The hurdle rate should be set in accordance with the riskiness of the project, so that riskier projects must surpass higher hurdle rates to be funded.

The present value method uses a criterion of net present value (NPV) for a project. NPV is computed by summing cash flows discounted by the rate of return required for a project by the company. The present value method uses a decision rule that if the NPV of the project is greater than zero, the project should be approved. In other words, if the present value of cash generated by the project is greater that the present value of cash expended, we should go ahead with the project. The NPV criterion can be used to prioritize multiple projects, because those projects with higher NPVs will be ranked ahead of those with lower NPVs.

Another type of analysis useful in financial evaluation of new product projects is break-even analysis. Break-even provides a very simple calculation to determine how much of a new product we need to sell to recover fixed costs of the project. The break-even formula is:

$$\text{B/E point} = \text{fixed costs/contribution margin (SP-VC)}$$

where SP is selling price and VC represents variable costs. We can expand the analysis to include a desired profit, by adding desired profit to the fixed costs to be recovered. We can also do sensitivity analysis for different price levels to determine the impact of a price change on the breakeven point.

The "financials" are detailed proforma profit and loss statements for the new product for year one, year two, year three. They represent the financial plan for the project. The financial plan is important for the following reasons (Cooper, 1993):

1. 1. It serves as a budget for the new product—an itemized accounting of how much will be spent, and where.
2. 2. It is the critical input for the final go/no go decisions as the project moves closer to launch.
3. 3. It provides benchmarks critical to the control phase of the launch plan, making sure that the launch is on course.

We generally include both manufacturing and distribution costs in calculations of cost of goods sold or cost of sales which is a key entry in the income statement, which is used to communicate sales, expense and profit data. A simplified income statement can be formatted as follows:

	Period 1	Period 2	Period n
Income			
Net sales			
Other			
Less: Cost of goods Sold			
Equals: gross profit			
Less: general and administrative expense			
Less: marketing expense			
Equals: net profit before taxes			
Less: taxes			
Equals: net earnings			

Net sales are total sales revenue minus returns and allowances. General and administrative expenses typically include expenses not directly related to production output and include salaries, rent, insurance, utilities, supplies, and so on. Net profit—earnings before taxes—measures performance of a firm's operations. The operating profit margin, which is calculated by dividing operating profit by net sales, shows the relationship between operating profit and net sales. In order to establish cash flow after taxes (CFAT) for investment analyses, both the interest expense and the depreciation allowance on new investment must be considered in the analysis.

For most new food products, sales build slowly as distribution is achieved, dollars are granted to the trade via promotions, and consumers are incented to try the new product. Usually a brand will hit its peak share 6–12 months after introduction in a new area, and subsequently level off or even decline. Much of the initial sales volume for a new product is promotion-driven. Even bad new products can achieve high initial sales, but only new products that deliver significant perceived benefits and substantial customer value will succeed after the initial flurry of sales activity.

Market Testing

Market testing is all about beating the odds in new products. All food processing firms are looking to identify high potential opportunities, reduce the risk of costly new product failures, develop the strongest marketing plan possible, and execute effective launches. Market testing is designed to aid firms in achieving these objectives.

We pointed out earlier that the new product development process consists of a series of screens and evaluations—of the idea, the concept, versions of the prototype and finished product, package structure and graphics, plant and/or equipment (if new), and the marketing program. Having completed development, we are now ready for launch. Only one more preliminary step may be completed—market testing. Market testing refers to evaluation of the entire launch program through taking the finished product and marketing program to the final consumer, and testing consumer reaction to the new product. The firm must first decide whether to launch without a formal market test. For simple line extensions, such as new flavors of an existing product line, market tests may be unnecessary provided that early stage testing yielded necessary information on consumer reaction to the new item(s) as well as data to be used in forecasting sales and performing financial analysis. As pressure to get to market quickly has increased, companies have looked for ways to shorten the development cycle, and traditional test marketing is used much less frequently than it was in times past. Especially with new products that do not require significant investment in production technology, plant and equipment, many firms opt to "roll and fix." This means that the firm will launch initially in a limited geographical area, and roll the product into new markets while gathering consumer and trade feedback, making necessary adaptations to our program, and building up production capacity as demand increases.

As the new product takes on more dimensions of newness, such as a new brand name, new category, new technology, new plant and equipment, both capital investment and risk increase, and the need to perform some type of market testing increases as well. Let us examine market testing in some detail.

Data Requirements

Market testing uses research methods designed to provide several important types of data. Sales volume and market share projections provide important financial

TABLE 20.5
Test Market Data Sources

- Point-of-sale transactions data, captured by retail scanners
- Store audits of displays and in-store promotions
- Household panel data showing purchase behavior by household
- Salesperson call reports
- Factory shipments and warehouse withdrawal data
- Tear sheets from retail feature advertising
- Results of surveys which measure awareness, attitudes, and response to advertising

information plus direction for the sourcing, production, and logistical operations regarding expected demand. The test market allows evaluation of vendor, logistical, and manufacturing performance. We test consumer reaction along several dimensions, including awareness, attitudes, trial, repeat purchase, satisfaction with the product, brand-switching behavior, and response to promotions such as coupons. We test trade reaction to our launch program in terms of adoption, distribution penetration, number of SKUs and facings, shelf position, extent of display and point of purchase activity, and feature advertising. By evaluating trade response, we also evaluate sales force or broker performance in executing the launch strategy. By examining consumer purchase dynamics and category sales figures, we can determine source of volume for the new brand. That is, to what extent does the new brand bring in new buyers to the category, raise usage among consumers in the category, take sales from competitive brands in the category, and take sales from our existing brands?

Sources of Data

Just as there are several important types of data to be gained from a market test, there are multiple sources of these data. These are listed in Table 20.5.

Market test methodology will determine which and how many of these data sources are employed.

Market Test Alternatives

Having made the decision to market test the launch program, the next step is to decide testing methodology. There are three major classes of market testing, with several variations of each: (i) simulated test markets, (ii) controlled testing, and (iii) traditional sell-in test marketing. These are not mutually exclusive categories of testing, and there are numerous instances of companies employing more than one type of market testing methodology, such as conducting a "pretest market" early in the process, and a controlled test market or sell-in test market prior to launch.

Simulated Test Markets (STMs)

Simulated test marketing provides volume forecasts, assessment of consumer reaction to the concept and product, and diagnostics of different elements of the overall program without some of the problems that can are typically associated with sell-in test marketing. The primary advantage of a simulated test market is preservation of confidentiality for the product development firm in that we do not expose the product to our competition. Relative to traditional sell-in tests, STMs are less expensive, can be accomplished more quickly, and allow "what if" analysis in terms of testing different levels of price, advertising weight, and so on without compromising the launch. Several competing models were created during the 1970s, including ASSESSOR, BASES, DESIGNOR, LITMUS and Simulator ESP. According to Sorenson & Associates (1999), over the years, many aspects of the various models have converged.

However, STM's are limited by their very nature. First, the purchase situation into which we place respondents is "simulated" or contrived. While we have control over the variables, any laboratory experiment, by definition, removes the evaluation from the real world. In this case, the real world includes competition and the trade, and simulated test markets evaluate consumer response in a competitive vacuum. Further, we do so without assessing trade response to the program; instead the test establishes assumptions about trade penetration and distribution coverage. In addition, the assumptions about the effects of advertising and promotions on awareness may be unrealistic and therefore unreliable. Finally, the commitment to the launch, as measured by the level of spending, may not be sustainable. That is the bad news. The good news is that STMs represent possibly the best-validated tool of marketing research. Because several STM providers can claim highly accurate and projectable results, food and allied companies commonly use simulated test markets to generate sales volume estimates, especially when confidentiality is critical and some investment will be required.

According to Sorenson & Associates (1999), main reasons for the popularity of STMs are: (i) there is an absolute need for sales projections as a basis for financial evaluation of the project, production and distribution planning, and preparation of the marketing plan; (ii) an STM provides an organized way to think about all of the components of sales, and how sales will be affected by changes in any of these components; and (iii) the STM provides a good deal of data about "why" in terms of how category usage, consumer attitudes, attribute data, and other factors explain the sales dynamics. The most popular STM tool is BASES, a subsidiary of ACNielsen.

Sell-In Tests

Whether or not the project incorporates a simulated test market, the firm may decide to conduct an in-market test. An in-market test provides the opportunity to assess consumer and competitive reaction to our launch, but on a limited scale to avoid the investment, expense, and risk of a larger scale introduction. In-market tests come in two varieties: traditional sell-in tests and controlled tests. Traditional sell-in tests, which are conducted under normal market conditions and use the company's regular sales force and advertising and promotional programs, allow a much broader evaluation of the launch program than controlled tests. However, traditional sell-in tests have been falling out of favor during the 1990s, because they have several potential disadvantages, which are shown below in Table 20.7.

TABLE 20.6
Benefits of Test Marketing

- Reducing risk exposure, both in terms of avoiding potential damage to the company's image with the trade and consumers if the product does not succeed, and in terms of limiting investment to small-scale production
- Achieving success that can be used as a tool or showcase to gain trade acceptance when we launch the new item
- Gaining experience with and fine-tune sourcing, manufacturing, logistical and sales operations via a pilot test
- Assessing consumer and trade reaction to specific elements of the marketing program for diagnostic purposes

Test Market Methodology

If the nature of the project dictates traditional sell-in test marketing, we must make several decisions about the design of the test. These include choice of test cities, length of the test, and specification of the data to be collected and the sources of that data.

Test cities should possess certain characteristics. First, they should be representative of the broader geographic distribution target in terms of (i) population demographics, (ii) category development and brand development, as measured by CDI and BDI, and (iii) structure of the food retailing market. Second, the cities should be reasonably isolated in terms of media spill-in and spill-out. Media spill-in refers to a situation in which households in the test city have ready access to broadcast media from other cities. Spill-out refers to households in other cities having access to broadcast media from the test city. Limiting both spill-in and spill-out is necessary to accurately gauge the relationship between advertising weight and consumer response.

Choosing the length of the test always represents a series of trade-offs. For instance, the longer we are in test, the more reliable the data become because we "average out" seasonal sales trends as well as any short-term factors such as stock-outs. Longer tests also provide the opportunity to assess more repeat purchase cycles and give the firm a more in-depth read of both consumer and trade reaction. However, longer tests cost more, represent higher opportunity cost, and results in greater loss of lead-time.

Table 20.6, based on the work of Olson (1996), shows the benefits of test marketing.

However, test markets have limitations and can cause problems. The limitations of test markets are shown in Table 20.7.

Sell-in test markets are most often utilized today in situations in which there is significant risk because the new product incorporates new technology and requires large capital investment in plant, equipment and processing technology. With significant technical and/or production barriers to entry, firms do not necessarily sacrifice the first-mover advantage by conducting a lengthy test market.

TABLE 20.7
Limitations of Test Markets

- Cost a lot of money, both in terms of the costs of the test, which include production and distribution of the product, marketing costs and research costs, and in terms of opportunity cost of limiting distribution and sales to test areas instead of a larger market area
- Expose new products and accompanying marketing programs to competition, a lack of confidentiality that may cause the firm to lose its first mover position by giving competitors an opportunity to develop their own versions of the item and get to full-scale launch before the original firm can
- Provide an opportunity for competition to interfere through such actions such as lowering price, engaging in heavy promotional activity and even buying up large quantities of the test item, all of which lead to unreliable test data
- Sometimes competitors sabotage results by tampering with the product on the shelf, so that it is less noticeable by consumers
- May lead to unrealistic levels of sales force effort which cannot be sustained in a larger scale launch, creating unreliable data

Despite the potential downside, some manufacturers use traditional tests, especially when launching a really new item about which there is little track record. Dannon's Yogurt Shake Drinkable Lowfat Yogurt was tested in about 15% of the country, including Chicago, Detroit, Baltimore and Washington, D.C. prior to a decision as to whether to launch nationally. Frito-Lay extensively tested olestra-based fat free salted snacks before rolling out. And, in a different situation, ConAgra tested Banquet Bakin' Easy, a seasoned dry baking mix designed to compete with Kraft's Shake 'n Bake, in Phoenix, Minneapolis, the Carolinas and Georgia followed by rollout to the Midwest.

STMs Versus Sell-In Tests

Packaged goods firms use different decision rules when deciding upon market tests. If STM results are positive, investment requirements are low to moderate, and there is a substantial risk of being beaten to market by competitors learning about the product in test market, we may skip formal test marketing. On the other hand, in very high-risk situations or when a simulated test market uncovered potential problems, both the STM and the test market may be conducted. In situations where it is important to do in-market testing, but we want to conserve both time and expense, an alternative to sell-in testing is controlled testing.

Controlled Testing

Controlled testing is a term that incorporates different types of market tests that share the same basic methodology. The key difference between controlled testing and sell-in tests is that controlled tests employ outside companies hired by the manufacturer to handle the product's distribution, merchandising and, in some cases, consumer advertising and promotions. A research supplier administers the entire test, allowing for control of all marketing variables, such as delivery of specific levels of media weight or advertising copy alternatives, and in-store execution. IRI's Controlled Store Test "accurately quantifies the effectiveness of in-store marketing variables with far less cost and risk than a national rollout." A field experiment matching test and control stores yields higher data accuracy than typical in-market tests.

BehaviorScan

BehaviorScan is a product of Information Resources Incorporated. BehaviorScan employs live, in-market testing methodology; however, IRI controls the retail environment including product distribution and placement as well as control over television advertising. The industry refers to this type of testing as "single source testing." BehaviorScan is available in five markets: Pittsfield, MA; Eau Claire, WI; Cedar Rapids, IA; Grand Junction, CO; and Midland, TX. Using its own production studio, IRI provides "targetable" television advertising in these markets to individual households by cutting over regular advertising on broadcast and cable channels. Different advertising copy and/or media plans can be delivered to two or three groups of households within the same market. In each market, IRI has recruited a demographically balanced and representative household panel. Members of the panel use a shopper's card when they purchase food and allied products in supermarkets, drug stores and mass merchandise outlets, and they scan all other purchases with a handheld scanner at home, enabling IRI to track household level purchasing, including individual household trial and repeat. IRI has created partnerships with local retailers to permit both broad outlet coverage and almost immediate distribution. New products receive 100% distribution in one week, eliminating the typical 8–12 week distribution build, allowing much faster reading of in-market results. IRI can also target newspaper advertising, both run of press and FSIs, and direct mail to the test and control panelists.

A Behavior Scan new product test provides the ability to forecast year one national volume potential within plus or minus 10% in 90% of cases (IRI data); measure cannibalization to allow calculation of incremental profit; analyze the new item's contribution to category build; evaluate the effectiveness of each element of the marketing mix in driving trial and repeat purchase; measure the impact of alternative marketing plans on results; understand consumer response to the full line and each individual SKU; and review trial and repeat rates by demographic segment:

- Forecast national year-one volume potential within plus or minus 10% in 90% of cases.
- Measure cannibalization of existing products for incremental profit calculations.
- Analyze the new product's contribution to the category to build a persuasive selling story for the trade.
- Evaluate the effectiveness of each marketing mix element in driving trial and repeat purchases.
- Review trial and repeat sales by demographic segment to identify targeted growth opportunities.

Other advantages of using BehaviorScan to test new products include

- Contained costs—lower media costs, less test product, no sales force involvement, no slotting allowances or failure fees
- Confidentiality of test results
- Ability to improve the marketing plan through use of simulation modeling techniques and manipulating marketing variables including media weight, copy, shelf location, and so on.
- More efficient use of new product and marketing funds, because BehaviorScan results give marketers information which guides better launch decisions. Validation: with no controlled market test, IRI data indicates that 72% of new products failed; of products launched after a BehaviorScan test, 84% were successful. For all BehaviorScan tests, 55% of the tested items were not rolled out, 38% were successful rollouts and 7% were failures.

The use of BehaviorScan also involves some methodological limitations, which include limited choice of markets, the fact that the test does not provide for the assessment of either competitive or retailer response, and research costs, which can range as high as $500,000 for a test.

Sell-In to the Trade

The retail environment presents significant challenges to the new product marketer. Retailers face increasing competitive pressures from alternative formats as well as from other supermarkets. Operators are driven to maximize their return on assets, at least at the headquarters level, and so are increasingly concerned about the high rate of new product failure. At the store level, the major emphasis is on increasing product turns and minimizing shrink. But the bottom line is the bottom line: new items must add profitability to the category. The new item must have both sales velocity and adequate margin, and therefore the manufacturer must provide significant support for the launch, which falls into two categories, consumer support and trade support. Consumer support includes advertising, promotional activities, and public relations. Trade support includes all promotional incentives designed to (i) cover the wholesale/retail costs of accepting the new item and getting it on the shelf, (ii) pay down the cost of price promotions for consumers that are designed to increase sales volume for the new item, and (iii) add to retailer profitability.

Many retailers charge not only placement/slotting fees to get initial distribution, but also failure fees if the product is de-listed. For chain operators, the net effect of these fees is to help defray setup costs, help the distribution function and to increase profits. Shelf space is growing in value because the number of new products and the number of SKUs stocked by supermarkets have grown faster than store size and the amount of shelf space available. Increasing pressure on shelf space means that shelf space is available to manufacturers only at a premium. At the same time, there is a demand for special package sizes and variety packs, particularly by club stores, which causes increasing item proliferation and higher costs for both the processor and the trade channel.

Hill (1997) identified three consistent trends prevalent in the food business: (i) general agreement that new products are the lifeblood of the business; (ii) increasing focus on managing manufacturer product lines for efficiency and cutting back on new item proliferation; and (iii) increasingly stringent trade hurdle rates by retailers for new items. Suppliers emphasize category value as the key metric in determining the makeup of product lines and the decisions to launch new products. Retailers are working with suppliers in category management partnerships to develop approaches to efficiently managing product mix, shelf space and location, promotion planning, and pricing. The key is to eliminate unproductive SKUs and emphasize product launches that build categories and create customer value.

New product hurdle rates are defined in terms of sales and profit contribution, sales per square foot and return on inventory investment. Manufacturers attempting to launch new items must incorporate these hurdle rates into the process. Hill suggests that to be successful, manufacturers must carry out account-specific analysis to understand the existing category hurdle rates of retail customers and prepare a pro-forma analysis for key customers to project the profit and productivity of each SKU, including projections for the new items. Manufacturers must understand the specific impact new items will have on the total category; in order to do this, manufacturers must understand consumer purchase dynamics in the category, including the interactions among brands and brand-switching behavior. They must be empathetic to the food retailers' perspective of efficient product assortment and efficient product introduction. Manufacturers should be able to answer for the retail customer questions that the retailer must ask such as: should we carry the new product? If so, where do we shelve it and how many facings do we give? What products lose facings? What will be the effect on the category, in terms of both sales and profits, if we cut in the new item? Carrying out this type of analysis successfully requires a commitment to true strategic partnering between manufacturer and retailer.

New Product Selection Criteria

Retail buyers and category managers specify a large number and wide range of criteria for evaluating new products, but the decision process should, if carried out objectively, boil down to basics. These include: projected sales volume of the new item, effect of the new item on category sales and profits, and the amount of money to be made on the sale of each unit and from stocking and advertising the item, and merchandising requirements. Individual decision criteria used by buyers are related to one or more of these basic factors. Unfortunately, both retailer and manufacturer executives will tell you that the one-on-one dynamic between the manufacturer sales rep and the retail buyer/merchandise manager can lead to some very subjective decisions on the part of the retailer.

New products can be thought of in four general categories:

Exclusive or signature products—This item is usually the most looked for but the hardest to find. This type of item brings the customer back to the same retailer due to the exclusivity. For a manufacturer it limits ability to increase volume.

Unique items—This is the second most valuable item to the retailer. There is less cannibalization and a greater chance of increased total sales.

TABLE 20.8
Factors Affecting Sales Volume for a New Product

- What consumer need is being addressed?
- Whether the new product builds on an emerging and significant consumption trend
- For a chain, in which particular stores in our trading area will this product fit best (for instance, supermarkets in college towns need to stock more snacks, health foods and vegetarian items while stores in areas with a heavy concentration of certain ethnicities must stock more ethnic foods)
- Performance of the category
- Marketing research information, such as the results of concept tests, product tests and market tests that suggest strong consumer purchase intent
- Manufacturer's track record with recent new product launches
- Projected source of new product volume and impact on the category, both volume and profitability
- Uniqueness of the item relative to items currently stocked; are there similar items already available?
- The amount of consumer value delivered by the item
- The strength of the manufacturer's advertising program to create awareness
- The strength of the manufacturer's consumer promotion program—coupons, deals, sampling, demos, and so forth—to induce trial
- Whether the manufacturer will provide shippers, display units, which make it easy for the store to put a display on the floor
- Whether the manufacturer will sponsor some type of local event

Line extensions—These are more likely to cannibalize other varieties in the same line. More must be done by the representatives to show growth in the category. This may be done by deleting other items in the line or prove the value of the new items.

Duplicate items—Certainly the hardest sale. Proof must be shown how your item is better or different, and what will happen to total category sales.

Projected sales volume is a function of the several different variables and factors. These are shown in Table 20.8.

Retail margins and profitability are both dependent on the following:

- Gross margin for the item, both at regular cost and during deal periods
- Introductory trade incentives: off-invoice allowances, bill-back allowances, display allowances, advertising allowances, and so on
- Placement/Slotting or performance-based monies available
- Pallet or shipper programs offered to build in-store displays
- Will the manufacturer pay the ad fee and provide the means for the retailer to feature the item in their circular as a "hot deal?"
- Special terms available, for instance if normal terms are 2/10 net 30, will the manufacturer offer 2/40 net 41 (for example) during the introductory period?

Merchandise managers and buyers want several questions answered regarding the merchandising plan for a new item. These include:

- How does the item fit into the retailer's variety strategy?
- Does the item attract the customer the retailer is trying to reach?

- Does the quality meet the expectations of the customer the retailer is going after?
- Is there proper product liability insurance behind the company?
- Where (in which category; where in the store) should the product be merchandised?
- What distribution channel (warehouse to store, direct store delivery, distributor) will be used?
- If a multi-item launch, how many SKUs?
- What type of shelf placement is preferred?
- Will an existing item or items be deleted to make room on the shelf?
- How many facings needed?

The New Product Presentation

New item presentations by manufacturers to the trade do not follow a standard format. Every retailer, every retail buyer or merchandise manager and every category is somewhat unique. There are, however, some general guidelines for planning a sales presentation.

The first issue involves which information the sales representative will provide to the buyer. The vendor sales team must convince the retail merchandise manager that the new product takes advantage of a significant market opportunity, filling a void in the market. This is the rationale or "reason for being" for the new item. The sales team must first specify clearly which consumer trend(s) provide the basis for the opportunity. For example, recently published USDA dietary guidelines calling for an increase in the number of servings of fresh fruits and vegetables create a real opportunity for food processors to create new fresh fruit and vegetable combinations and new packaging so that consumers can consume more of these items in more places and on more occasions. The sales team must also provide category data that indicate future growth in the category and how the item will contribute to that growth, or at least how the item will create greater margins for the retailer. These data are typically sourced from IRI and ACNielsen scan and panel data. Strong consumer interest in the new item must be demonstrated at this juncture.

Many manufacturers proclaim (or at least would like to proclaim), "we don't pay slotting." The fact remains that most food retailers demand some type of financial incentive to take on a new product. The new item allowance that is collected is often determined by the category. For example, bottled water demands significantly higher new item fees than, for instance, canned vegetables.

Several companies, such as Con Agra and Kellogg's, have packaged their new item allowances by making them introductory offers. They will offer the new items with a standard "off invoice" deal and then will add on an introductory deal that can be quite substantial. The retailer can lower the price point to drive trial, let the funds drop directly to the bottom line, or find a happy medium. For example, a new item with an everyday cost to the trade of $3.00 may have an introductory offer of $1.00 (per unit). The retailer can decide to price the item at the regular ad price of $3.99, which creates a 25% gross margin, and bank the extra $1 per unit. Alternatively, the retailer can take the $1 offer as an offset to item cost, creating an effective cost of

$2 per unit, and then feature the item at "2 for $5" and really move cases out the door. The "happy medium" might be to price at "2 for $6," which is still a substantial introductory saving for consumers. While the penny profit for the retailer will be only literally a penny, the introductory deal will move enough cases to drive the $1 introductory allowance to a really substantial number.

Firms may provide a fixed percentage per case for performance funds for the retailer to put towards advertising, demos, and so on. For example, Helene Curtis set aside a significant amount of money to promote an extension of its Degree line of antiperspirants to retailers. The company was flexible with its guidelines for using the money to allow customization for the retailers' needs. A budgeted amount was set for each retailer, followed by a joint decision on how the investment would be used. Typically, funds were used for distribution allowances, promotion allowances, and feature allowances. Smaller companies may not be able to duplicate large company performance funds. However, if the retail believes its customers will buy the product, most retailers will "sacrifice" smaller dollars to take on the product. Another possible use for the new item funding is to guarantee the profitability of the new item. This is a unique and is used relatively infrequently. If a manufacturer is not willing to pay slotting fees but has the confidence that the new item will add profit to the category, the manufacturer and retailer would agree on a profit amount and if the new item does not perform to that expectation, the manufacturer would pay the difference in a lump sum payment.

Acceptance of a new item requires the shelf to be re-set, "cutting in" the new item and eliminating other items to create space. The key to this step is "fact-based selling" by which the vendor presents the retail customer with specific category data that shows sales velocity and gross profit for the items in the category. Variables to the shelf set or "plan-o-gram" include which items to be stocked, how many facings for each, and the shelf position (e.g., eye level, center) for each SKU. Computerized shelf management programs such as Apollo or Spaceman analyze data on items and sales to create a shelf set designed to achieve key retail objectives. Some retailers delegate the responsibility of managing the shelf to a manufacturer category captain, while other retailers handle shelf management for themselves.

The recommended content of new item sales presentations follows:

- Discuss the category in terms of size, growth, and how the new item will help the retailer to increase product movement in the category.
- Review the manufacturer's position in the category, overall category consumption and any relevant information about sub-categories, such as chocolate versus nonchocolate in the candy segment.
- Provide the consumer rationale for new item in terms of basic demographic, lifestyle, and consumption trends, and the type of opportunity that exists for the new item.
- Discuss points of difference between your product and competitive products.
- Identify who is doing the shopping in this category, and specify their needs, wants, and unsolved problems.

- Discuss the track record of new product launches by the manufacturer, and the proven winners from that company.
- Provide results of marketing research, such as results of concept tests, packaging and product tests that justify the contention that the product will move off of the shelf.
- Use "fact-based selling" providing a thorough review of category data, especially that specific to the retail customer. Companies practicing category management must share information, helping the retailer and manufacturer to work as partners to grow the category. Provide a plan-o-gram with a recommended shelf set, and project the impact on category movement and margin for the retail customer.
- Present the consumer program, which includes advertising and promotions. Provide the specifics of the advertising program: budget, weight, media mix and schedule. Explain consumer promotion tactics including FSI coupons and other promotions, including the schedule. Discuss any retail-specific programs including in-store demo program.
- Present the trade program including pricing, terms and margin structure plus all monies and incentives available: discounts, allowances, free goods, and slotting fees or performance funds. When presenting introductory deals, there is a need to be specific, for example: "We are offering an off-invoice allowance of $3.40 per case during the introductory buy-in period, along with a bill-back of $1.50 per case, providing a gross margin of 26% on the intro at $2.49 suggested retail." This plan should be for a minimum of 1 year.
- Discuss possible advertising and display allowances and programs.
- Include an overview, typically in the form of a calendar with months in the columns and consumer and trade events along the rows, of all of the key dates in the launch program: start ship, advertising flights, FSI's, and so forth.
- Discuss product specifics—packaging, case pack and size, shippers, display modules available.
- Provide a product fact sheet that includes information about the package, including weight, dimensions, and the UPC code; the case, including pack and size, case weight and dimensions, cube configuration, and truckload quantity in cases; and the pallet, including weight, pallet pattern—number of layers, cases per layer, cases per pallet, and pallet dimensions—length, width, height, cube, plus truckload quantity in pallets. Also include order number, shelf life if appropriate, availability, and pricing.
- Discuss merchandising recommendations for the customer—distribution, display, plan-o-gram, and a recommendation for removal of slow moving items in the category. This should include a suggested initial quantity for stores, based on anticipated movement. Look for cross-merchandising promotional opportunities in the store, for example, the manufacturer's brand merchandised with a complementary private label item.
- Most importantly, do not forget to include contact information.

In order to attract consumer attention, build consumer awareness and incent consumers to try the new item, vendors sometimes want secondary display space in the store. Secondary display space can be at the end of the aisle (an "endcap"), in the aisle, or perhaps in the front of the store. Manufacturers want to merchandise new items in areas of high traffic and high visibility. One way to achieve secondary display is to provide the retail customer with shipper display units, self-contained units that are placed in the aisle. Incremental display space costs money, typically in the form of promotional incentives paid by the vendor to the retailer.

One of the most important visuals that a manufacturer salesperson can provide his or her trade customer during the sell-in presentation is a promotional calendar, which portrays all of the consumer and trade marketing activity that the manufacturer plans for the introductory period. Trade calendars do not follow a standard format though the most common is a Gantt Chart set up using Excel. The columns represent time periods, usually weeks during the introductory period. Normally the trade calendar supporting the launch shows the first year's advertising and promotional activity. The rows represent the specific elements, initiatives and events that make up the launch, including consumer advertising flights, FSI drops, and all other consumer, trade, and account-specific activity.

The sales function of most major CPG companies is now organized along customer lines, and sales teams are dedicated to major wholesale and retail accounts. This means that all promotional activity, including new item presentations, are customer-specific, with the limitations imposed by the Robinson-Patman Law. For example, when Nestlé launches a new brand, such as Corner Bistro Panini Sandwiches, the marketing department will have created a sell-in presentation in general terms for the use of all the sales teams, which each customer sales team will then customize to the specific characteristics and demands of a trade customer. The Nestlé team calling on Giant of Carlisle will make a new item presentation that incorporates category data specific to Giant. Each sales team has been allocated an annual promotional budget for each customer on the basis of projected sales volume, and the promotional dollars that support the new item launch are drawn from this budget, which is managed by the customer sales team.

Finally, food manufacturers sell products through various classes of trade, including supermarkets, club stores, mass merchandisers, drug stores and convenience stores. Manufacturers must be aware of the need to tailor pack sizes and merchandising recommendations based on the class of trade. It is important that manufacturers present all new items to each class of trade and allow that class to make the decision of whether or not they would like to carry it. Nothing frustrates a retailer more than when they see an item at their competition that was never presented to them, especially at Wal-Mart or Target. In addition, supermarkets follow different pricing and merchandising strategies, with "high-low" operators acting differently with regard to passing along promotional dollars in the form of lower consumer prices than "everyday low price" or EDLP operators. The use of performance funds allows manufacturers to tailor their promotions to the specific needs of the customer. The risk lies in retailers "knowing" that someone else is getting a better deal.

The Launch

The launch involves executing all of the activities involved in manufacturing, distributing and selling the new product to both the trade and the final consumer. Each functional area—sourcing, production, distribution, sales force activity, and marketing—must effectively and efficiently perform its designated activities; just as importantly, these activities must be properly timed and coordinated. The manufacturer must make sure that product is in the stores in the intended time frame and that advertising and promotional events are scheduled accordingly. This means that retail sell-in and distribution of the product will have to be scheduled with adequate lead-time to ensure product availability in the store. As much as 40 weeks may be needed to accomplish all of the tasks needed to move a new product from the end of production and ultimately onto store shelves and then to begin advertising. Having scheduled when product must be in the store, the start ship date can be determined, then production runs and sourcing of ingredients and supplies must be scheduled. Considerations such as any seasonal patterns of product sales, the geographic extent of the launch (crash vs. rollout), and possible supply and production problems must be taken into account. We also must plan for fine-tuning and follow-up work with the trade after the initial sales presentations are completed.

An article in Stagnito's *New Product News* (2003) gives an example from IRI of a novel strategy to get a new product on the shelf. George Weston Bakeries, Inc. merchandised Boboli pizza crust on its own free-standing rack in stories, because the product did not fit into an existing category, either fresh bread or pasta sauce. Given its own space in the store, the product caught on with consumers and became a winner.

Tracking the Launch

The *Progressive Grocer Report* (Mathews, 1997) cited a study conducted by Efficient Marketing Services (EMS) of a sample of new items launched in 1995. Results of the study showed that for the 300 new items examined, success was most negatively affected by four factors: slow distribution, incomplete distribution, lost distribution and poor merchandising decisions. All of these problems are correctable, but only if the manufacturer has sufficient and timely knowledge of the situation and can take corrective action within a narrow window of time. Considering that a new product launch has about 12 weeks to take hold, the window is extremely tight. An industry study conducted in the mid 1990s by Deloitte & Touche measured average lost sales per SKU per store due to slow, incomplete, and lost distribution at almost $90. With an average launch of 4.3 SKUs, the loss averages almost $400 per store per launch. All of the above point to the critical importance of the sell-in and monitoring stages of new product launch.

A quick read of the new product's performance, using objective scan data from a provider such as Nielsen or IRI, is necessary soon after launch. The trade-off between early information and accurate information is important, as early buyers, especially those buying only because of novelty considerations or because of large promotional incentives, may not be fully representative of buyers who adopt the product for regular use. The tracking program must be designed to provide clear and actionable indications about both areas of success and specific problems.

Syndicated Tool for Managing the Launch

IRI offers a product called IntroCast (Pre/Post) Launch Forecasting, an "interactive tool that leverages concept scores and category-specific historical trial and repeat benchmarks to break down year one goals into period-by-period goals" which allows for early warning if the introduction is not on track. Within 12–24 weeks after launch, IntroCast provides forecasts of year one and year two sales, under both current and potential marketing scenarios. This product allows companies not only to assess where the launch stands early in the launch relative to first year sales, but to simulate the impact on sales of alternative marketing approaches such as increasing advertising and/or promotional activity or modifying price points. This product also gives marketers the ability to assess within the first 12–24 weeks whether the product is likely to achieve its first year sales objective, the potential impact of increasing or reducing advertising and/or promotional support, and how much production capacity will be needed to support a national rollout in year one and in year two.

Both the types and sources of data we use to track the launch are similar to those we use in test markets. We need to track consumer data such as sales volume, share and share trends, reasons for purchase, likes and dislikes, awareness, trial and repeat purchase levels, consumer attitudes, and source of volume, including the effect of the new item on the category and brand shifting patterns. It is absolutely critical to be able to quantify a new item's increment to the existing franchise and category sales, for production planning and financial purposes, and to gain additional shelf space.

We need to track wholesale and retail data such as wholesale inventory levels and sell-through, retail distribution penetration, shelf location, number of facings and display activity. We should disaggregate statistics by market area and customer segments to determine any patterns that may exist. We should attempt to determine the portion of our initial sales that is accompanied by a coupon redemption or other promotional incentive. And if we are executing a multiple SKU launch, we need to gather this data for each individual item.

Gaining and Holding Distribution

IRI has determined through analysis of hundreds of new product launches that the first key to success is gaining and holding distribution. It is absolutely essential that accurate reports of ACV build be available early and often. According to a Willard Bishop Consulting test (2003), only 28% of targeted stores had cut-in a new item 2 weeks after authorization, and after 13 weeks, 26% of SKUs still had not made it to the shelf. The Bishop Consulting report identified two key metrics to be tracked. Assessing *retail readiness* includes asking if the product is on the shelf, whether it will be reordered when it sells out (proper shelf tag), and if consumers will be able to find it (right department.) *Distribution quality* relates to in-store positioning of the product in terms of both number of facings and shelf position, and whether retail stores are supporting the launch in terms of promotional pricing and proper display.

Initial consumer sales will reflect primarily early consumer trial of the new item, which are normally inflated by consumer promotion events such as FSI coupons and samples. As a result, sales velocity inevitably peaks in the first couple of months and then falls off as trial flattens. Understanding the month-by-month sales growth and decline curves is essential to projecting sales accurately (Olson, 1996). It is also critical to ascertain which sales are trial sales and which sales are due to repeat purchase. Heavy trial with low repurchase means that sales may decline rapidly after we have tapped trial households. Sources of data include scan data, such as InfoScan from IRI; household panel data; and results of consumer surveys.

You Have the Data...What Now?

Finally, the firm needs to have some type of contingency and postlaunch plans in place for responding to the results of our launch. Accurately diagnosing the launch data is a critical step. High trial and poor repeat might indicate a strong concept and strong trial-inducing promotions but a weak product or package. Lower than expected sales may be attributable not to poor product quality but instead to poor distribution penetration. Successful companies anticipate problems that may occur, gather necessary data and analyze the source of problems that do occur, and have remedial programs in place to respond. Successful companies also have what might be termed "year two" strategies as a follow-up to a successful launch. These might involve introduction of new flavors, rollout into new geographic territories, or taking a product from retail to foodservice distribution.

Getting to Market for Smaller Companies

Bob Messenger, in his *Morning Cup* (July 12, 2004) said:

"Hardly a week goes by when I don't talk to someone from a small food company who says it's pretty much impossible to get their products on shelves in the mainstream supermarket chains. Well, I know it doesn't seem right, but I tell these folks with dreams of competing fair and square that, in the real world, they're better off putting their effort and resources into developing alternative retail channels and just forget the supermarket chains. Whether the small player wants to hear it or not, so-called unfair practices such as slotting fees actually work to the benefit of the industry's Top Guns. They pay, they play, and they end up owning space that you, the small company, cannot afford. So what I'm saying is, take your opportunities where they are, seek out other distribution channels, and build up equity that way..."

Not all retailers demand slotting or new item fees. As previously mentioned, true EDLP operators such as Wal-Mart and stores such as Whole Foods do not charge slotting. And companies like Acme-Albertsons and other supermarket chains which normally want slotting money sometimes allow smaller, local companies access to store shelves without slotting dollars, giving these small players some access to the market. Alternatively, smaller manufacturers can use websites, smaller specialty stores, specialty catalogs and even home shopping channels to reach consumers, sometimes establishing a market base that can be used as evidence of consumer demand to entice supermarkets and other large retail formats to take on the product.

SUMMARY

This chapter covers the final stage of the new product process. We discussed the introductory marketing program, including positioning, advertising, and promotional strategy. We discussed options for market testing, which is the final type of dress rehearsal prior to launch. The sell-in process was covered, including specific criteria buyers use to evaluate new product programs and the elements of a productive sales presentation. Finally, the need to properly time, coordinate and track the launch was discussed. According to IRI (2003), a successful launch is one where all elements work. These elements are: a unique idea that fills a consumer need with a meaningful and sustainable benefit which appeals to a "large enough" market, which fulfills its promise with no drawbacks, is superior to competition, and makes both a rational and emotional connection that leads to long-term repeat purchasing, provides good value for the money, is on time to market with consumer trends, with advertising that persuasively explains the concept and enough marketing dollars to break through the clutter in year one and continue in year two, is easy for consumers to adopt, creates volume incremental to the category, and is accepted by the trade. Sounds simple.

BIBLIOGRAPHY

1. Ailloni-Charas, D. 1997. In-Store Sampling Accelerates the Establishment of New Products, *New Product News*, August 11, p. 15.
2. Anonymous. 2002 Trade Promotion Spending and Merchandising Study, Evanston, IL: Cannondale Associates, Inc.
3. Anonymous 2003. An Expert's Rules For New Product Launch Success, *Stagnito's New Products Magazine*, 3(8):26–27.
4. Anonymous 2004. "Best New Products of 2004," *Stagnito's New Products Magazine*, 4(12):36–42.
5. Anonymous 2004. *NameWire*. Minneapolis, MN: Strategic Name Development, Inc.
6. Anonymous *New Product Trends 2004*. Information Resources, Inc. Presentation at Saint Joseph's University, October 2004; also on www.infores.com.
7. Anonymous 2002. "Slotting Allowances in the Supermarket Industry," *FMI Backgrounder*. Washington DC: FMI.
8. Anonymous 2005. *Sunday Circulars Becomes Passe as Cairo.com Launches Services to Help Shoppers Find the Best Prices on Frequently-Purchased Grocery Items,* http://home.businesswire.com/2005, accessed January 19, 2005.
9. Anonymous 1999. *The Sorenson In Store Sales Forecast*, www.sorenson-associates.com/volume_estimate.htm, accessed January 10, 2005.
10. Anonymous 2002. *Trade Promotion Practices Study*. www.acnielsen.com/pubs/ci/2003/q2/features/tpp.htm, accessed March 15, 2005.
11. Anonymous *WBC Executive Summary*, www.bishop–consulting.com, accessed January 11, 2005.

12. Anonymous 2005. www.foodinstitute.com/private/pdf_frameset.cfm?pdf, accessed January 5, 2005.
13. Bishop, W. 2004. Key Developments Impacting Large North American Retailers. Presentation at MEI's Trade Practices Meeting, New Orleans, LA, March 15, 2004.
14. Boike, D.G. and J.L. Staley. 1996. "Developing a Strategy and Plan for a New Product," In *The PDMA Handbook of New Product Deveopment*, Rosenau, M.D., A. Griffin, G.A. Castellion, and N.F. Anschuetz, eds. New York: John Wiley & Sons, pp. 139–152.
15. Cooper, R.G. 1996. "New Products: What Separates the Winners from the Losers," in *The PDMA Handbook of new Product Development*, Rosenau, M.D., A. Griffin, G.A. Castellion, and N.F. Anschuetz, eds. New York: John Wiley & Sons, pp. -18.
16. Cooper, R.G. 1993. *Winning at New Products*, 2nd ed. Boston: Addison-Wesley.
17. Elliott, S. 2005. Pesi One Goes on a Television-Free, Celebrity-Free Commercial Diet. www.NewYorkTimes.com, March 16, 2005, accessed March 20, 2005.
18. Fusaro, D. 2004. What Grocers Want. *Food Processing*, 65(6):8.
19. Hennessey, J. 2005. Trade Dollars Not Very Loyal. www.retailwire.com, accessed March 21, 2005.
20. Hill, J. 1997. The New Product Hurdle That May Cost You the Race, *New Product News*, 33(8):13.
21. Hoban, T.J. 1998. Improving the Success of New Product Development, *Food Technology*, 52(1):46–49.
22. Kirkpatrick, D. 2005. "Want Truth in Advertising? Try a Blog," *Fortune*, 151(5):44.
23. Hanover, D. 2002. *Full Disclosure*, http://promomagazine.com, accessed January 6, 2004.
24. Marn, M., X. Roegner and X. Zawanda. 2003. Pricing New Products, *McKinsey Quarterly*, Issue 3:40–49.
25. Mathews, R. 1997. Efficient New Product Introduction, July 1997 Supplement to *Progressive Grocer.*
26. McGann, R. 2005. *Study: Magazine Risk Losing Core CPG Advertising to Web*, www.clickz.com, accessed January 5, 2005.
27. Olson, D.W. 1996. Postlaunch Evaluation for Consumer Goods, In *The PDMA Handbook of New Product Development*, Rosenau, M.D., A. Griffin, G.A. Castellion, and N.F. Anschuetz, eds. New York: John Wiley & Sons, pp. 395–410.
28. Ottum, B.D. 1996. Launching a New Consumer Product, In *The PDMA Handbook of New Product Development*, Rosenau, M.D., A. Griffin, G.A. Castellion, and N.F. Anschuetz, eds. New York: John Wiley & Sons, pp. 381–392.
29. Parker-Pope, T. 1998. The Tricky Business of Rolling Out a New Toilet Paper, *Wall Street Journal*, January 12, pp. B1, B8.
30. Patrick, J. 1997. *How to Develop Successful New Products*. Lincolnwood, IL: NTC Business Books.
31. Schneider, J. and J. Yocum. 2004. *New Product Launch: 10 Proven Strategies*, Deerfield, IL: Stagnito Communication, Inc.
32. Stanton, J. 2005. The End of TV Advertising, *Food Processing*, June 2004, 65(6):26.
33. Wells, M. 2005. Have It Your Way, www.forbes.com, accessed February 2, 2005.
34. Wells, M. 2004. Kid Nabbing, www.forbes.com, accessed January 11, 2005.
35. Williamson, R. 2005. Frito-Lay Spells Crunch Time for Teens, www.adweek.com, March 1, 2005, accessed March 3, 2005.

36. ICI 2006 http://www.infores.com/public/us/analytics/marketingmix/controlledstoretest.htm
37. IRI 2004 http://www.infores.com/public/us/analyitics/productportfolio/bscannewprodtest.htm
38. IRI 2004c http://www.infores.com/public/us/analytics/productportfolio/introcastlaunch.htm

Index

C

D

E

P

Q

T

U

V

W

Z